AF167658

Communications
in Computer and Information Science 2238

Rationale

The CCIS series is devoted to the publication of proceedings of computer science conferences. Its aim is to efficiently disseminate original research results in informatics in printed and electronic form. While the focus is on publication of peer-reviewed full papers presenting mature work, inclusion of reviewed short papers reporting on work in progress is welcome, too. Besides globally relevant meetings with internationally representative program committees guaranteeing a strict peer-reviewing and paper selection process, conferences run by societies or of high regional or national relevance are also considered for publication.

Topics

The topical scope of CCIS spans the entire spectrum of informatics ranging from foundational topics in the theory of computing to information and communications science and technology and a broad variety of interdisciplinary application fields.

Information for Volume Editors and Authors

Publication in CCIS is free of charge. No royalties are paid, however, we offer registered conference participants temporary free access to the online version of the conference proceedings on SpringerLink (http://link.springer.com) by means of an http referrer from the conference website and/or a number of complimentary printed copies, as specified in the official acceptance email of the event.

CCIS proceedings can be published in time for distribution at conferences or as postproceedings, and delivered in the form of printed books and/or electronically as USBs and/or e-content licenses for accessing proceedings at SpringerLink. Furthermore, CCIS proceedings are included in the CCIS electronic book series hosted in the SpringerLink digital library at http://link.springer.com/bookseries/7899. Conferences publishing in CCIS are allowed to use Online Conference Service (OCS) for managing the whole proceedings lifecycle (from submission and reviewing to preparing for publication) free of charge.

Publication process

The language of publication is exclusively English. Authors publishing in CCIS have to sign the Springer CCIS copyright transfer form, however, they are free to use their material published in CCIS for substantially changed, more elaborate subsequent publications elsewhere. For the preparation of the camera-ready papers/files, authors have to strictly adhere to the Springer CCIS Authors' Instructions and are strongly encouraged to use the CCIS LaTeX style files or templates.

Abstracting/Indexing

CCIS is abstracted/indexed in DBLP, Google Scholar, EI-Compendex, Mathematical Reviews, SCImago, Scopus. CCIS volumes are also submitted for the inclusion in ISI Proceedings.

How to start

To start the evaluation of your proposal for inclusion in the CCIS series, please send an e-mail to ccis@springer.com.

Meenu Khurana · Abhishek Thakur ·
Praveen Kantha · Chin-Shiuh Shieh ·
Rajesh K. Shukla
Editors

Machine Learning Algorithms

First International Conference, ICMLA 2024
Himachal Pradesh, India, February 23–24, 2024
Proceedings

 Springer

Editors
Meenu Khurana ⓘ
Chitkara University
Himachal Pradesh, India

Abhishek Thakur ⓘ
Chitkara University
Himachal Pradesh, India

Praveen Kantha ⓘ
Chitkara University
Himachal Pradesh, India

Chin-Shiuh Shieh ⓘ
National Kaohsiung University of Science
and Technology
Kaohsiung, Taiwan

Rajesh K. Shukla ⓘ
Oriental Institute of Science and Technology
Bhopal
Bhopal, Madhya Pradesh, India

ISSN 1865-0929   ISSN 1865-0937 (electronic)
Communications in Computer and Information Science
ISBN 978-3-031-75860-7   ISBN 978-3-031-75861-4 (eBook)
https://doi.org/10.1007/978-3-031-75861-4

Preface

The Springer Scopus-indexed first International Conference on Machine Learning Algorithms (ICMLA 2024) organized by Chitkara University, Himachal Pradesh India, on February 23–24, 2024, provided a great platform for researchers to engage in prolific discussions and exchanges in the field of machine learning, contributing significantly to the success of the conference. Participants from four countries and ten Indian states brought a plethora of experience and varied perspectives to solve real-world problems through machine learning. The conference received an overwhelming response, with 400 paper submissions out of which 40 were selected through a rigorous review process. All the submissions underwent a strenuous peer-review process (double blind) by subject matter experts. The papers were reviewed based on the novelty, technical content, originality and clarity. The entire process, including the submission, review and acceptance, was done electronically. All these efforts undertaken by the organizing and technical committees led to an exciting, rich and a high-quality technical conference program, that featured high-impact presentations for all attendees to understand, explore and expand their expertise in the latest developments in machine learning. The selected papers were categorized in three main conference areas: image processing, Internet of Things (IoT), and core machine learning (ML).

The conference featured six plenary lectures on "Continual Learning in the Open World"; "Machine Learning and Deep Learning Application"; "IoT Governance Research Directions"; "ChatGPT and Transformer Algorithms"; "Machine Learning for Healthcare" and "AI and Ethics,", from eminent experts, which were highly appreciated by the participants.

Overall, the conference met its' aim to connect and bring together researchers on one platform. The plenary lectures, along with the progress and special reports, bridged the gap between different fields of machine learning, enabling non-experts in specific areas to gain insights into new domains.

We extend our special thanks to the Chief Patron, Ashok K. Chitkara, Chancellor, Chitkara University, Himachal Pradesh; Madhu Chitkara, Pro-Chancellor, Chitkara University, Himachal Pradesh; and Patron, Rajnish Sharma, Vice Chancellor, Chitkara University, Himachal Pradesh, for their support and guidance that helped in making the conference memorable.

We would also like to thank the speakers, authors and participants of this conference, who prioritized the conference despite all hardships. Finally, we express our gratitude to all the committee members, program chairs, session chairs, reviewers, and volunteers for their tireless efforts in meeting deadlines and ensuring the smooth conduct of the conference.

July 2024

Meenu Khurana
Abhishek Thakur
Praveen Kantha
Chin-Shiuh Shieh
Rajesh K. Shukla

Organization

Chief Patrons

Ashok K. Chitkara   Chitkara University, Himachal Pradesh, India
Madhu Chitkara   Chitkara University, Himachal Pradesh, India

Patron

Rajnish Sharma   Chitkara University, Himachal Pradesh, India

General Chairs

Meenu Khurana   Chitkara University, Himachal Pradesh, India
Abhishek   Chitkara University, Himachal Pradesh, India

Advisory Committee Chairs

Sumeet Dua   Louisiana Tech University, USA
Bing Liu   University of Illinois, US

Technical Advisory Chairs

Minho Jo   Korea University, South Korea
Son. N. Tran   Deakin University, Australia

Program Chair

Meenu Khurana   Chitkara University, Himachal Pradesh, India
Abhishek   Chitkara University, Himachal Pradesh, India
Chin-Shiuh Shieh   National Kaohsiung University of Science and Technology, Taiwan
Rajesh K. Shukla   Oriental Institute of Science and Technology, India
Praveen Kantha   Chitkara University, Himachal Pradesh, India

Advisory Committee

Aditya Nigam	Indian Institute of Technology Mandi, India
Amit Doegar	NITTTR Chandigarh, India
Anil Kr. Gourishetty	Indian Institute of Technology Roorkee, India
Ankur Choudhary	Dharampal Satyapal Limited (DSL), India
Chinmay Chakraborty	Birla Institute of Technology Mesra, India
Deden Witarsyah	Telkom University, Indonesia
Marijn Janssen	Delft University of Technology, Netherlands
Neeraj Kumar	Thapar Institute of Engg. and Technology, India
Neeru Jindal	Thapar Institute of Engg. and Technology, India
P.S. Rana	Thapar Institute of Engg. and Tech, India
Partha Pratim Roy	Indian Institute of Technology Roorkee, India
Rohit Salgotra	AGH University of Kraków, Poland
Ruchika Gupta	ABES Business School, India
Shahrina Md Nordin	Universiti Teknologi PETRONAS, Malaysia
Shashi Shekhar Jha	Indian Institute of Technology Ropar, India
Sunil Kumar	AURO University, India
Vishakha Sood	Indian Institute of Technology Ropar, India
Vivek Singh	Oxford Brookes University, UK

Technical Advisory Committee

Aditya Nigam	Indian Institute of Technology Mandi, India
Amitanshu Pattnaik	DRDO, Ministry of Defence, India
Dac-Nhuong Le	Haiphong University, Vietnam
Fitri Yakub	University Teknologi Malaysia, Malaysia
G. Jaya Lakshmi	Indira Gandhi Centre for Atomic Research, India
M. S. V. K. V. Prasad	Swarnandhra College of Engg. and Technology, India
Max P.	Louisiana Tech University, USA
Robbie L. Watson	Louisiana Tech University, USA
Md Tounsi Prince	Sultan University, Saudi Arabia
Partha Pratim Roy	Indian Institute of Technology Roorkee, India
Raman Singh	University of the West of Scotland, Scotland
Shashi Shekhar Jha	Indian Institute of Technology, Ropar, India
Gaurav Singh	Stanford University School of Engineering, USA
Nadeem Sarwar	Bahria University, Pakistan

Program Committee

A. Meenakshi	Kamaraj College of Engg. and Technology, India
Abhishek	Chitkara University, Himachal Pradesh, India
Ambuj Kumar Misra	Mahatma Gandhi Kashi Vidyapith, India
Amit Doegar	NITTTR Chandigarh, India
Amitanshu Pattnaik	DRDO New Delhi, India
Anil Kumar	Indian Institute of Technology Roorkee, India
Arun Lal Srivastav	Chitkara University, Himachal Pradesh, India
Arvind Mahindru	DAV University, India
Ashok Kr. Thakur	Chitkara University, Himachal Pradesh, India
Ashutosh Kr. Dubey	Chitkara University, Himachal Pradesh, India
Bhavya Deep	Delhi University, India
Bikromadittya Mondal	B. P. Poddar Institute of Management and Technology, India
Chinmay Chakraborty	Birla Institute of Technology Mesra, India
Deepa Negi	Chitkara University, Himachal Pradesh, India
Deepak Kr. Rout	IIIT Bhubaneswar, India
Dileep Kr. Sharma	Chitkara University, Himachal Pradesh, India
Divya Sharma	Institute of Engineering and Technology, India
Gaurav Garg	Chitkara University, Himachal Pradesh, India
Hakam Singh	Chitkara University, Himachal Pradesh, India
Harish Ch. Anandaram	Amrita School of Artificial Intelligence, India
Harpreet Kaur	Chitkara University, Himachal Pradesh, India
Honey Sharma	Gulzar Group of Institutions, India
Javed Akhtar Khan	Gyan Ganga Institute of Technology and Sciences, India
Kapilya Gangadharan	Saveetha Institute of Medical and Technical Sciences, India
Karan Bajaj	Chitkara University, Himachal Pradesh, India
Latika Kakkar	Chitkara University, Himachal Pradesh, India
M. S. V. K. V. Prasad	Swarnandhra College of Engg. and Technology, India
Madhu Nakirekanti	Sreyas Institute of Engg. and Technology, India
Meenakshi A Kamaraj	Kamaraj College of Engg. & Technology, India
Minaxi Dassi	Chitkara University, Himachal Pradesh, India
Murali Kalipindi	Vijaya Institute of Technology for Women, India
Nagesh Kumar	Chitkara University, Himachal Pradesh, India
Narayan A. Joshi	Dharmsinh Desai University, India
Navneet Kaur	Chitkara University, Himachal Pradesh, India
Navneet Verma	Institute of Engineering and Technology, India
Neeraj	Chitkara University, Himachal Pradesh, India

Pankaj Pratap Singh	Central Institute of Technology kokrajhar, India
Pooja Kamat	Symbiosis Institute of Technology Pune, India
Pradeepta Kr. Sarangi	Chitkara University, Himachal Pradesh, India
Praveen Kantha	Chitkara University, Himachal Pradesh, India
Ramamani Tripathy	Chitkara University, Himachal Pradesh, India
Raman Singh	University of the West of Scotland, UK
Rani Kumari	Chitkara University, Himachal Pradesh, India
R. A. Vishwakarma	Institute of Technology Pune, India
Ravi Prakash Jaiswal	Mahatma Gandhi Kashi Vidyapith, India
S Ajitha	Acharya's Bangalore Business School, India
Sachin Minocha	Amity University Noida, India
Sandhya Sharma	Chitkara University, Himachal Pradesh, India
Sangeeta Singh	Vardhaman College of Engineering, India
Saurabh Tewari	Gyan Ganga Institute of Technology and Sciences, India
Seeta Devi	Birla Institute of Technology Mesra, India
Shahbaz Afzal	Chitkara University, Himachal Pradesh, India
Sheerin Zadoo	Central University of Jammu, India
Shilpa Choudhary	Neil Gogte Institute of Technology, India
Shivendu Prashar	Chitkara University, Himachal Pradesh, India
S. Kaur Samagh	Shaheed Bhagat Singh State University, India
Sunil Kumar	AURO University, India
T Vijay Muni	K L University, India
T. K. Senthil Kumar	Anna University, India
Triveni Lal Pal	Pranveer Singh Institute of Technology, India
Uma Maheswari	Gurusamy Kamaraj College of Engg. & Technology, India
Vijay Kumar Sinha	Chitkara University, Himachal Pradesh, India
A M Abirami	Thiagarajar College of Engineering, India
Abhinav Jain	ITER India Institute for Plasma Research, India
Fitri Yakub	University Teknologi Malaysia, Malaysia
Gubba Naveen	Joginpally B R Engineering College, India
Neelam Joshi	Sinhgad Institute of Technology, India
Ronak R Pansara	Pace University, USA
Sachin Jaiswal	Sheat Group of Institution, India
Sachin Kumar	Ashoka Institute of Technology and Management, India
Vishakha Sood	Indian Institute of Technology, India
Vivek Singh	Oxford Brookes University, UK

External Reviewers

Aakanksha Lakhanpal	Thapar Institute of Engg. and Technology, India
Abhishek Bhatt	Symbiosis Skills and Professional University, India
Aishwarya Kadu	D.M Institute of Higher Education and Research, India
Ajay Kumar	Gopal Narayan Singh University, India
Ajay Kushwaha	Rungta College of Engineering and Technology, India
Ajay Thakare	Sipna College of Engg. and Technology, India
Ajitha Savarimuthu	M S Ramaiah University of Applied Science, India
Amit Gudadhe	D.M Institute of Higher Education and Research (DU), India
Amit Singh	DRDO, India
Amlan Raychaudhuri	B.P. Poddar Institute of Management & Technology, India
Antriksh Goswami	National Institute of Technology, India
Anupama Phakatkar	SCTR's PICT, India
Archana Shirke	FCRIT, India
Arun Kumar	Ramco Institute of Technology, India
Arun Kumar Jhapate	SIRT, India
Arvind Mahindru	DAV University, India
Ashok Sahoo	Graphic Era Hill University, India
Ashwini Pradhan	Galgotias University, India
Awanit Kumar	Sangam University,India
Babita Verma	Bhilai Institute of Technology, India
Baibaswata Mohapatra	Greater Noida Institute of Technology, India
Bhushan Deore	RAIT, India
Bikrambir Dhillon	Thapar Institute of Engg. and Technology, India
Bikromadittya Mondal	B.P. Poddar Institute of Management & Technology, India
Chandan Behera	VIT Bhopal University, India
Chinmay Chakraborty	BIT Ranchi, India
Deepak Jayaswal	ST. Francis Institute of Technology, India
Deepak Khantwal	University of Johannesburg, India
Deepak Rout	International Institute of Information Technology, India
Dhanashree Wadnere	Sandip University, India
Dhruva Ghai	Oriental University, India
H. Anandaram	Amrita Vishwa Vidyapeetham, India

Mandadapu Prasad	Swarnandhra College of Engg. & Technology, India
Pravin Pokle	Priyadarshini J.L. College of Engg., India
Reji R	Carmel College of Engineering, India
Vikas Gupta	Vidyavardhini College of Engg. in Technology, India
Chandresh Chhatlani	JRN Rajasthan Vidyapeeth, India
Deepika Bhatia	VIPS-TC, School of Engineering & Technology, India
Deepti Khanna	Jagan Institute of Management Studies, India
Dhawaleswar Rao	Centurion Univ. of Technology and Management, India
Jameel Qurashi	Poornima College of Engineering, India
Karuna Sharma	Arya College of Engineering & IT, India
Mohit Tomar	Sagar Institute of Research & Technology, India
Neetu Rani	Chandigarh University, India
Ramjeet Yadav	Dr. Rammanohar Lohia Avadh University, India
Ranjan Bala Jain	V.E. Society's Institute of Technology, India
Ravindra Jogekar	S B Jain Institute of Technology, Mgmt. & Research, India
Samarjit Roy	SCIT, EIU, Becamex IDC Corp, India
Shruti Aggarwal	Chandigarh University, India
Suhaib Ahmed	Model Institute of Engg. and Technology, India
Sunil Wanjari	St. Vincent Pallotti College of Engg. and Technology, India
Triveni Pal	Chandigarh University, India
Varun Saxena	Govt. Women Engineering College, India
Basant Sah	Koneru Lakshmaiah Educational Foundation, India
Madhu Nakirekanti	Sreyas Institute of Engg. and Technology, India
Nilesh Ingale	G.H. R. Institute of Engg. & Business Mgmt., India
Sujatha E.	Saveetha Engineering College (Autonomous), India
Gajanan Nagare	Vidyalankar Institute of Technology, India
Gaurav Bharti	IIIT Bhopal, India
Gaurav Garg	Chandigarh Engineering College, India
Gaurav Srivastav	Datta Meghe Institute of Medical Sciences, India
Gursimran Bakshi	Chandigarh University, India
Honey Sharma	Gulzar Group of Institutions, India
Jacob Davis	Loyola College, India
Jagriti Saini	Eternal RESTEM, India
Jahangeer Sidiq	Vellore Institute of Technology, India

Jai Bhaskar	University College of Engg. and Technology, India
Javed Khan	Gyan Ganga College of Technology, India
Jitendra Samriya	Graphic Era Deemed University, India
K Singh	Babu Banarasi Das Institute of Technology, India
K. Murali	Vijaya Women's Engineering College, India
Kanwarpreet Kaur	Chandigarh University, India
Kapil Aggarwal	K L Deemed to be University, India
Kapil Gupta	SVPCET, India
Kapilya Gangadharan	SIMATS, India
Ketan Sarvakar	Ganpat University, India
Kiran Kaware	Sandip University, India
Leelkanth Dewangan	DMIHER, Sawangi (Meghe), India
Mahesh Kumar	NIET, India
Mahesh Singh	Aditya Engineering College, India
Mahima Pandey	Galgotia College of Engg. & Technology, India
Manoj Chaudhari	Priyadarshini Bhagwati College of Engg., India
Manoj Mahto	Vignan Institute of Technology and Science, India
Manoj Kumar Patra	GITAM (Deemed to be University), India
Mayurkumar Patil	MIT-AOE, India
Megha Kolhekar	Fr. C. Rodrigues Institute of Technology, India
Monika	NIFT, India
Munish Kumar	Maharaja Ranjit Singh Punjab Technology Univ., India
Narayan Joshi	Dharmsinh Desai University, India
Naveen G	Joginpally B R Engineering College, India
Navneet Verma	PIET, India
Neelam Joshi	Sinhgad Institute of Technology, India
Neeraj Sharma	Jagran Lakecity University, India
Neetu Mittal	Amity Institute of Information Technology, India
Nilamadhab Mishra	VIT Bhopal, India
Nileshsingh V. Thakur	Yeshwantrao Chavan College of Engineering, India
Nishant Tripathi	Lovely Professional University, India
Nupur Chugh	Bharati Vidyapeeth College of Engineering, India
Palash Gourshettiwar	DMIMS, India
Pankaj Bhambri	GNDEC, India
PKP Gaitry Chopra	D.M Institute of Higher Education and Research, (DU), India
Pooja Pandya	Noble University, India
Prachi Sharma	SIRT, India
Pramod Bokde	Priyadarshini Bhagwati College of Engg., India

Pranshu Saxena	ABES Engineering College, India
Prasana Lakshmi Balaji	King Khalid University, Saudi Arabia
Prateek Singhal	Christ University, India
Prateek Verma	D.M Institute of Higher Education and Research, (DU), India
Praveen Gupta	KIET Group of Institutions, India
Praveen Kumar	D.M Institute of Higher Education and Research, (DU), India
Praveen Kr. Ramagiri	Nalla Malla Reddy Engineering College, India
Pulok Mohanta	IIT Kharagpur, India
PV Chandrika	Welingkar I.M.D Research, India
Radha Pavanasam	Mepco Schlenk Engineering College, India
Ragini Verma	IIT Roorkee, India
Rahul Bhandari	Chandigarh University, India
Rajanesh Kaushal	Chandigarh University, India
Rajeev Ranjan	Chandigarh University, India
Rajendra Kachhava	Sangam University, India
Rajinder Kaur	Punjabi University, India
Rajkumar Patra	Netaji Subhash Engineering College, India
Rakesh Ranjan	University of Petroleum and Energy Studies, India
Ram Prajapati	Chandigarh University, India
R. M. Dwarkadas J.	Sanghvi College of Engineering, India
Ravindra Rathod	Walchand College of Engineering, India
Revathy B	National Institute of Technology, India
Richa Vyas	Sushila Devi Bansal College of Technology, India
Rishabh Sharma	Amity School of Engineering and Technology, India
Rohit Jain	Lakshmi Narain College of Technology, India
Saiteja Kagitha	National Institute of Technology, India
Sakshi Raturi	Guru Nanak Dev Engineering College, India
Saloni	Guru Nanak Dev Engineering College, India
Sandeep Kharb	Amity University Haryana, India
Sanjay Kumar	Shoolini University, India
Santhosh G	PES College of Engineering, India
Saurabh Sahu	GITAM, India
Saurabh Verma	Chandigarh University, India
Seema Sharma	Thapar Institute of Engg and Technology, India
Shaik Abdul Hafeez	VFSTR Deemed to be University, India
Shaik Masthan Basha	Era Hill University, India
S. Singh Bhadoria	Institute of Technology and Management, India
Shailesh Sharma	IFTM University, India
Shaik Saleem Basha	Chaitanya Bharathi Institute of Technology, India

Shashank Pareek	JECRC University, India
Sheela Ganesh	KGiSL Institute of Technology, India
Shekhar Shukla	Babasaheb Bhimrao Ambedkar University, India
Shivam Shrivastava	KIET Group of Institutions, India
Shreya	Symbiosis Institute of Technology, India
Shrikant Yerpude	T.G. Patil College of Engg. and Technology, India
Shubham Balhara	Geeta University, India
Shriram Nene	ABMSP's APCOER, India
Shubham Panchal	T John Institute of Technology, India
Shweta Agarwal	IIMT University, India
Shrikrishna Lale	Sardar Patel Institute of Technology, India
Shyama Shyam	KIET Group of Institutions, India
Siddharth Sharma	KIET Group of Institutions, India
Sohail Ali	Institute of Management Technology, India
Sourav Kumar	IIFT, India
Sreelekshmi N	SCMS, India
S.Rao Samudrala	Aditya Institute of Technology and Mgmt., India
Subodh Mahadik	Dr. D.Y. Patil Institute of Engg. Mgmt. & Research, India
Suchismita Dash	Centurion University of Technology and Mgmt., India
Sudipta Sahana	Birla Institute of Technology Ranchi, India
Sukanta Tripathy	Veer Surendra Sai University of Tech, India
Suman Chatterjee	RCC Institute of Information Technology, India
Sumit Agrawal	Malviya National Institute of Technology, India
Suneet Sood	Chandigarh Engineering College, India
Surender Kumar	Maharaja Agrasen University, India
Surinder Singh	NIT Kurukshetra, India
Supratim Banerjee	RCC Institute of Information Technology, India
Tirath Kumar	Central University of South Bihar, India
Upender Yadav	IMS Engineering College, India
Upendra Kumar	Jagran Lakecity University, India
Vaibhav Mishra	BBDITM, India
Vaibhav Saxena	JSS Academy of Technical Education, India
Vaishali Chavan	FCRIT, India
Vikas Kumar	Kalinga University, India
Vikash Meena	Indian Institute of Technology Bombay, India
Vikram Kumar	Indian Institute of Technology Kharagpur, India
Vimal Kumar	KIET Group of Institutions, India
Vinay Kumar Pandey	Motilal Nehru National Institute of Technology, India
Vineet Kallurkar	Don Bosco Institute of Technology, India

Vinit Dinkar Dhikale	Pimpri Chinchwad College of Engineering, India
Vishal Pareek	J.K. Lakshmipat University, India
Vishal Pathak	Chandigarh University, India
Vishwajeet Verma	Institute of Technology and Management, India
Vivek Chaturvedi	University Institute of Technology RGPV, India
Vivek Kumar	Birla Institute of Technology Ranchi, India
Yadu Nandan Dwivedi	Nirma University, India
Y. Singh Thakur	Birla Institute of Technology Ranchi, India
Yogesh Prasad Katiyar	RKGIT, India
Yogesh Sharma	Maharishi Markandeshwar University, India

Contents

Image Processing

Deep Learning

Machine Learning

Performance Evaluation of Hybrid Machine Learning Models for Prediction of Coronary Disorder in Smart Healthcare Systems

Biswajit Tripathy[1] , Subhranshu Tripathy[2] , Sujit Bebortta[1] ,
and Srikanta Kumar Mohapatra[3]([✉])

[1] Ravenshaw University, Cuttack, Odisha 753003, India
[2] School of Computer Engineering, KIIT Deemed to be University Bhubaneswar,
Bhubaneswar 751024, India
[3] Chitkara University Institute of Engineering and Technology, Chitkara University,
Punjab 140401, India
srikanta.2k7@gmail.com

Abstract. Clinical abnormalities are diseases that affect body parts and can be caused by a variety of causes, such as genetics and lifestyle. Heart disease is one of the leading causes of mortality and a major risk factor due to the rising global frequency of diseases. Clinical data analysis makes the prediction of cardiovascular illness a challenging task. Machine learning (ML) has shown promise for forecasting and decision-making in the healthcare industry. In our proposed model, we have included some different Computational Intelligence Principle (CIP) for coronary artery heart disease forecasting: Gaussian Naive Bayes (GNB) Classifier, K-Nearest Neighbors Classifier (K-NNC), Random Forest Classifier (RFC), AdaBoost Classifier (ABC). These all CIPs are included at level0 of our proposed novel stacking model used for learning at the initial stage. Then all the outcomes of these initial stages are supplied to the level1 of our proposed model where we have included another CIP i.e., Logistic Regression (LR) which complete the stack for final prediction. All these models are combined in ensemble model with stacking strategy to improve the prediction accuracy. Dataset of heart disease which was used in this study was taken from the UCI machine learning repository's Cleveland Clinic. The experiment concluded with a better accuracy of 86.957%. After integrating the power of all the models, we have achieved a great precision and recall of 83.333% and 90.909% respectively. This also emphasizes the usefulness of the hybrid classifier.

Keywords: Healthcare-support · Machine Learning · Hybrid Model · Heart Disease Prediction · Stacking

1 Introduction

The rising prevalence of diseases, which are affected by complex interactions between DNA traits and external factors, gives rise to the increasing need for advanced instruments when examining medical records [1]. One of the most common causes of death

M. Khurana et al. (Eds.): ICMLA 2024, CCIS 2238, pp. 3–14, 2025.
https://doi.org/10.1007/978-3-031-75861-4_1

worldwide is heart disease, which stands out as a major concern among the many health-care issues [2]. The World Health Organization (WHO) has recognized coronary artery heart disease (CADH) as a major public health issue that seriously impacts the heart's ability to function normally which affects millions of people each year [3]. Occlusion or constriction of the coronary arteries, which provide blood to the heart is the main causes of heart disease [4].

Disturbing forecasts indicate that heart disease might soon be responsible for more than 30% of deaths worldwide, underscoring the critical need to overcome the difficulties faced by medical professionals in correctly and quickly diagnosing cardiac illness [6]. Patients' circumstances worsen when medical intervention is delayed because of delayed identification, which motivates to build an intelligent CAHD prediction model that can accurately and early anticipate the start of cardiac illness on an economically feasible scale that can aid to diagnose the patient more accurately [7, 8]. Untreated CAHD dramatically compromises cardiac health and increases the risk of deadly heart attacks. The effects of CAHD can be significantly reduced by early identification of sickness symptoms and timely diagnosis [9].

Many factors have been found by researchers to contribute to an increased risk of heart disease. High blood pressure, diabetes, high blood cholesterol, smoking, and tobacco use are primary risk factors [10]. In addition, alcohol and stress—both known risk factors—add to the complexity, though it's unclear how much of a role they play. Interestingly, there are ways to modify some of these risk factors, providing opportunities for intervention and management [11]. Owing to the difficulties in anticipating cardio-vascular disease, machine learning (ML) becomes an essential instrument for utilizing the enormous amounts of data produced by the healthcare sector in order to improve forecasting accuracy and direct decision-making [12]. Many research in the field of arti-ficial intelligence have highlighted the efficacy of techniques in enhancing the precision of classification problems, especially in the medical field [13]. Finding and predicting abnormalities is hard, especially ones that affect important body parts. This is because of things like genetics and living choices.

This study looks at four different computer intelligence algorithms at level 0 to see how well they can identify coronary artery heart disease. These are the AdaBoost Classifier (ABC), the Random Forest (RF) Classifier, the K-Nearest Neighbors (K-NN), and the Gaussian Naive Bayes (GNB). Logistic Regression (LR) is used at level 1 to draw a conclusion about the group. This study suggests a new way to improve forecast accuracy by using a mixed classifier that uses the power of all the models in a smooth way by using an ensemble model and a stacking method. This study's results show that the suggested combination model is very accurate—86.957% to be exact. There are reports of sensitivity values of 90.909% and accuracy values of 83.333%. This study shows that using Random Forest with other methods at level 0 and Logistic Regression at level 1 works well together, showing enhanced accuracy of predicting coronary artery heart disease.

2 Related Work

The Cleveland dataset was used in the study by Gupta et al. [8], and a Random Forest (RF) classification model and cleaning steps were used to get a very good accuracy of 96.9%. Their research emphasizes how important reliable prediction models are to the medical field. Guan et al. [9] developed a system that uses just Support Vector Machines (SVM) and was able to predict heart illness with 76.5% accuracy. Their study, which looked at many SVM algorithms, highlighted how important it is to understand model selection in a comprehensive way when it comes to classifying diseases. Dwivedi [10] detected cardiac disease with an amazing 85% accuracy rate using logistic regression (LR) on the Cleveland dataset utilizing a different dataset and classification approach. There are so many models freely available in machine learning for identifying cardiac problems. Importantly, backpropagation is the prominent mechanism with 91.94% accuracy rate.

Kumar and Inbarani's, [11] made an improvement in the method of cardiac diagnosis. Which included Particle Swarm Optimization (PSO) in their characterization system. That explain how carefully different machine learning approaches can be assess and optimize key elements in order to improve diagnostic skills. Rabbi et al. [12] discovered that SVM outperformed K-Nearest Neighbor (KNN), Artificial Neural Networks (ANN), and SVM in comparison research with an accuracy rate of 85%. This emphasizes how important it is to carefully assess machine learning models before using them in medical settings. A hybrid technique that combined Naïve Bayes and Logistic Regression was described by Amin et al. [13] and achieved an impressive accuracy of 87.4%. Their emphasis on recognizing distinctive characteristics draws attention to the continuous attempts to improve diagnostic precision. By using a variety of feature selection techniques and classification methodologies, Shilaskar and Ghatol [14] showed that SVM in conjunction with optimization techniques greatly improved accuracy to above 85%. In order to analyze heart failure, Ali et al. [15] developed a stacked SVM-based authority system that combined prediction and feature reduction in two steps, yielding an amazing 92% accuracy. Fitriyani et al. [16] achieved exceptional accuracy of 98% and 95%, respectively, using XGBOOST and DBSCAN in a classification model on the Cleveland and Statlog datasets. Their study emphasizes how dataset selection affects model performance.

3 Proposed Model

The detailed explanation of how to build a stacking machine learning model by using 2 layer a) first layer is for initial training of models and b) second layer is for final prediction. Where is uses Logistic Regression (LR) at final layer and Gaussian Naive Bayes (GNB), K-Nearest Neighbors (K-NN), Random Forest (RF) Classifier, and AdaBoost Classifier (ABC) at initial layer of our proposed model to forecast heart disease is given below.

Step 1: Data Preparation

In the first step we have to gather an extensive dataset, which contain relevant information about heart health-related attributes of heart patients, such as risk factors and medical history.

Step 2: Data Cleaning and Feature Scaling

In this step we have to give emphasize and tackle many factors like: anomalies, missing numbers, and outliers in the dataset to assurance the quality of the data. Which ensure numerical properties have a consistent scale, normalize or standardize the data.

Step 3: Data Splitting

In this step we have to split the dataset into training and testing sets, to examine the model's performance on untested data.

Step 4: Training for Level 0 Models

Using the training dataset, we have trained the models which is given bellow:

- Gaussian Naive Bayes (GNB): Based on the premise of feature independence, use the training data to simulate the joint distribution of features and the target variable using the Bayes theorem. Based on the Bayes theorem and the assumption that features are conditionally independent, GNB is a probabilistic classifier. The probability distribution of every feature given the class label is used to determine its prediction.

The following formula may be used to calculate the mathematical prediction of GNB for a given instance x in class C_i:

$$P(C_i|x) = \frac{P(x|C_i) \times P(C_i)}{P(x)} \tag{1}$$

- K-Nearest Neighbors (K-NN): To train the model to categorize occurrences according to the feature space majority class of their k-nearest neighbors. A non-parametric technique called K-NN uses the majority class of its k-nearest neighbors to categorize occurrences. The proximity of instances is determined by the distance metric, which is often Euclidean distance.

The forecast made by K-NN is represented mathematically as follows:

$$\hat{y} = \underset{j}{argmax}\left(\sum_{i=1}^{k} I(y_i = j)\right) \tag{2}$$

- Random Forest (RF) Classifier: To generate predictions collectively, train an ensemble of decision trees, each trained on a different subset of the data. RF is an ensemble approach that builds many decision trees during training and outputs the individual trees' mode of categorization. In terms of mathematics, the prediction of RF entails combining the forecasts of its component trees:

$$\hat{y}_{RF} = mode(\hat{y}_{tree1}, \hat{y}_{tree2}, \hat{y}_{tree3}, \ldots, \hat{y}_{treen}) \tag{3}$$

- AdaBoost Classifier (ABC): To create a powerful ensemble model, train a succession of weak learners and modify weights in each iteration to emphasize misclassified examples. By giving misclassified instances weights, the ABC boosting technique enables later classifiers to concentrate on the previously misclassified data. The weighted total of each individual weak classifier provides the forecast for ABC:

$$\hat{y}_{ABC} = sign\left(\sum_{t=1}^{T} \alpha_t h_t(x)\right) \tag{4}$$

Step 5: Prediction of Level 0 Models

Apply the learned Level 0 models to the test set and make predictions: ABC prediction, RF prediction, K-NN prediction, and GNB prediction

Step 6: Level 1 Model Training

To build an enlarged feature set, combine the original features with the predictions derived from Level 0 models.

Step 7: Logistic Regression (LR) Training

Here we have utilized the expanded feature set and the predictions from the models of initial layers, then trained the final model which is Logistic Regression model.

Step 8: Prediction of Level 1 Model (LR Prediction)

After learning from output of initial layer and expanded feature set out proposed model is ready to forecast the test set of heart disease.

Step 9: Model Assessment

To evaluate the layering model performance, we have used appropriate evaluation metrics, such as accuracy, precision, recall, and F1 score. In our proposed stacking model final layer of logistic regression model integrates with the initial layer predictions (GNB, K-NN, RF Classifier, and ABC), which gives an efficient ensemble technique that maximizes the predictive ability of the classification model.

4 Machine Learning Algorithms

4.1 Random Forest Algorithm

Random Forest (RF) is one of the very reliable and accurate machine learning method that is frequently used in the cardiac disease classification [5]. Random Forests play an important role in the analysis as well as classification of complex patterns seen in data pertaining to the heart by utilizing a group of decision trees. For medical diagnosis and prognosis RF method is very beneficial because of its broad range of parameter-based abilities to differentiate between various cardiac diseases.

When the dataset DS contains heart disease patient's data and health attribute information, the Gini Index—represented by Eq. (1) below—is used to choose random features. The Gini Index may be stated mathematically as follows:

$$Gini_{index} = \sum_{i=1}^{n} \frac{f(X_i, DS)}{|DS|} \times \sum_{j=1}^{m} \frac{f(X_j, DS)}{|DS|} \qquad (5)$$

In this case, $f(X_i, DS)$ represents the probability that the chosen characteristic is related to a particular cardiac state X_i. This formula encapsulates the fundamental way in which RF use the Gini Index to assess and choose patient's characteristics in the classification problem.

4.2 K-Nearest Neighbor Algorithm

The K-Nearest Neighbors (K-NN) method is a good way to group heart diseases because it is easy to use and good at finding patterns. The Euclidean distance between two instances is found: instance (a) represents known data points in the dataset (DS) and instance (b) represents the person being tested for heart disease. A weight function denoted as ω may be introduced in the distance calculations to improve the estimations. Weight function ω_m considers the weight of features m important for creating a relationship among instances. In the K-NN method, the distance function is shown by Eq. (6) is as follows:

$$\delta(a, b) = \sum_m \omega_m \sqrt{p(a_m, b_m)} \tag{6}$$

Two examples, 'a' and 'b', are represented in this equation, where ω_m is the weight given to the m features that separate 'a' and 'b'.

4.3 GaussianNB

Notably, the famous GNB—a statistical computational intelligence method—is performing very smoothly in the field of classification problems. With various risk indicators or heart disease characteristics, GNB can predict the person's possibility of having a heart problem.

Here C indicate that a person does not have a heart disease. These ideas define the attributes $(x_1, x_2, x_3, \ldots, x_n)$. Using the Bayes theorem, the GaussianNB method determines the likelihood of a certain class given the features:

$$P(C|x_1, x_2, x_3, \ldots, x_n) = \frac{P(x_1, x_2, x_3, \ldots, x_n|C) \times P(C)}{P(x_1, x_2, x_3, \ldots, x_n)} \tag{7}$$

This results in the probability that person belong to class C given he possesses these attributes - $P(C|x_1, x_2, x_3, \ldots, x_n)$.

The probability of the characteristics given class C, which is modeled assuming a Gaussian distribution, is expressed as $P(x_1, x_2, x_3, \ldots, x_n|C)$.

- P(C) is the class C prior probability, which expresses our first assessment of the risk of heart disease.
- The evidence, or marginal likelihood, is denoted by $P(x_1, x_2, x_3, \ldots, x_n)$, which is the likelihood of witnessing the given collection of traits in all conceivable classes.

GaussianNB greatly simplifies the likelihood computation by assuming that the features are conditionally independent given the class. Features for heart disease prediction might include blood pressure, age, cholesterol, and other pertinent health indicators.

4.4 AdaBoostClassifier

AdaBoostClassifier (ABC) is a well-liked ensemble learning technique that builds a reliable and accurate model by combining the prediction ability of several weak learners. ABC may be used to improve the classification performance in the context of heart disease prediction by aggregating the predictions of weak classifiers, which in this case are usually decision trees.

AdaBoost functions mathematically by giving each instance in the training dataset a weight. At first, the weights of each occurrence are equal. The weak learner is retrained on the new dataset following each iteration, during which the weights of instances that were incorrectly categorized are raised. Until a flawless classifier is obtained, this procedure is repeated a certain number of times.

Let's write: - Y is the binary target variable that indicates whether or not cardiac disease is present; X is the feature matrix that represents patient data;

- D, where N is the total number of instances in the dataset, is the distribution of weights assigned to each instance. It is originally specified as $D_i = \frac{1}{N}$ for i in the range of 1 to N.

The basic phases of AdaBoost are expressed as follows:

1. Set initial weights: Give each instance the same weight: $D_i = \frac{1}{N}$.
2. For t in range(T), where T is the quantity of rounds of boosting:

   a. Teach Weak Learner: Fit the training data with the current weights to a weak classifier, often a decision tree.
   b. Determine Error: Determine the weak learner's weighted error:

$$\varepsilon_t = \sum_{i=1}^{N} D_i \times I\left(y_i \neq \hat{y}_i^t\right)$$

   where $I(.)$ is the indicator function and $\hat{y}_i^t$ is the weak learner's prediction for instance i.
   c. Calculate Classifier Weight: Determine the final ensemble's weak classifier's weight:

$$\alpha_t = \frac{1}{2} \ln\left(\frac{1 - \varepsilon_t}{\varepsilon_t}\right)$$

   d. Modify Weights: Modify the instances' weights:

$$D_i \leftarrow D_i \times \exp\left(-\alpha_t \times y_i \times \hat{y}_i^t\right)$$

   e. Normalize Weights: To make sure the weights add up to one, normalize them:
   $$D_i \leftarrow \frac{D_i}{\sum_{i=1}^{N} D_i}$$

3. Predictions: using the calculated weights of the weak classifiers, combine their predictions:
   Final Prediction $= sign\left(\sum_{t=1}^{T} \alpha_t \times predict_{weak\ learner_t}(X)\right)$.

To create a more accurate and reliable predictive model for predicting the possibility of heart disease, the final ABC combines the judgments of several weak classifiers, each trained to capture distinct features of the patient data.

4.5 Logistic Regression

Logistic regression (LR) becomes an essential tool for diagnosing and predicting illnesses just because of its predictive and informative qualities, [4]. LR works with actual-valued input vectors and makes it easy to get useful statistical data from the model. It is a selective category method. People use LR to figure out how likely they are to get certain diseases, like heart disease. When we apply logistic regression to a situation where the dependent variable is binary is one of its core features, then for the event, like success or failure, it shows the data as 1 or 0 respectively. Estimating the log probability connected to the event's occurrence is the main goal of LR analysis.

To put it mathematically, the numerous linear regression functions that are estimated by the LR model are as follows.

$$\text{Log odds} = \beta_0 + \beta_1 X_1 + \beta_2 X_2 + \ldots + \beta_n X_n \tag{8}$$

In this case, 'n' denotes the total number of features, and the coefficients $\beta_0, \beta_1, \beta_2, \ldots, \beta_n$ are linked to the features $X_0, X_1, X_2, \ldots, X_n$. There are numerous essential phases in the basic process of using logistic regression to predict heart disease. These involve assembling input vectors and binary dependent variables, initializing the coefficients, training the model, and finally employing the Logistic Regression model to provide predictions.

5   Results

When evaluating multiple models for the prediction of cardiac disease, we find that different algorithms exhibit varied performance measures. In Table 1 we have compared all the CIPs with our novel proposed model, where it is clearly shows that our model achieves comparable accuracy, recall, R2 Scores and F1-score to the Gaussian Naive Bayes Classifier (GNBC) model. The Random Forest (RF) and K-Nearest Neighbors (K-NN) models doesn't perform better than the Proposed model and GNB in terms of accuracy and recall. Random Forest (RF) and K-NN perform poorly in terms of F1-score, but the Proposed model, GNB, and AdaBoost Classifier (ABC) perform competitively.

Table 1 indicates abundantly clear that the suggested model is effective and does a better job of estimating heart disease compared to other existing models. More study and testing are needed to find out how strong and usefulness of the model across different datasets.

Figure 1 represents the correlation matrix for the dataset that we have used in the experimental evaluation.

Figure 2 represents the precision-recall and ROC curves for benchmark Machine Learning Models over the considered dataset.

Figure 3 represents the precision-recall and ROC curves for the suggested stacking approach over the examined dataset in comparison to benchmark machine learning models.

In conclusion, every model has distinct advantages and disadvantages. While the K-NN model has limits in terms of precision, the Random Forest model performs well in some areas but raises questions regarding possible overfitting. Both the Gaussian Naive Bayes and the AdaBoost Classifier models perform well and in balance; the Proposed model roughly matches the Gaussian Naive Bayes model's efficacy on all assessed criteria. To find the best model for heart disease prediction based on particular use-case needs and preferences, further thought and thorough investigation are required.

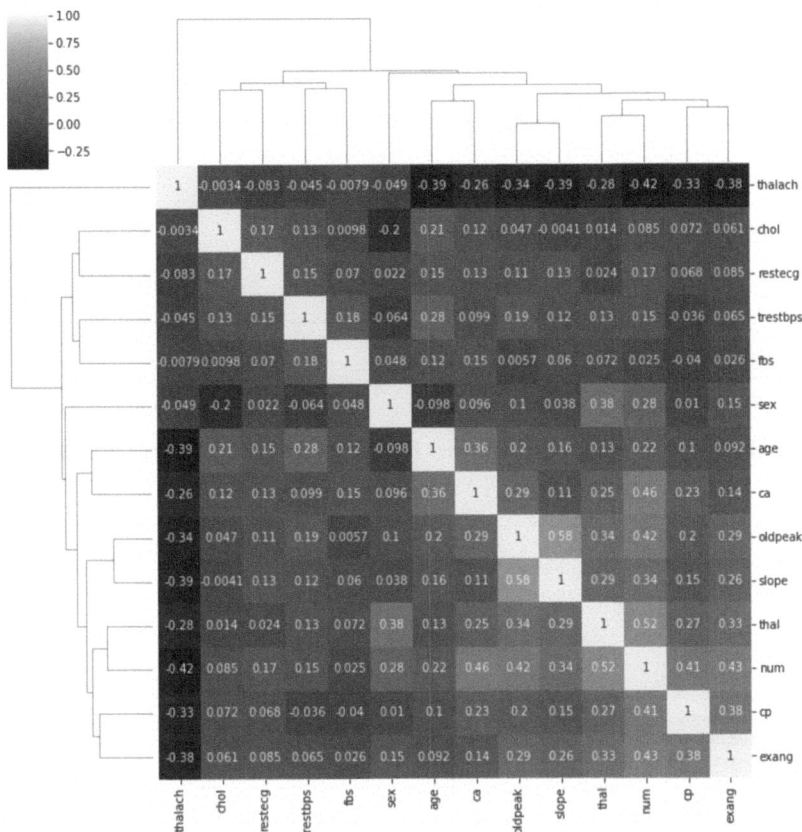

Fig. 1. Correlation matrix for the dataset's many properties.

Table 1. Benchmark algorithms and the suggested algorithm are compared using a variety of performance measures.

Models	Precision	Recall	F1-score	R^2 Score	Accuracy
RF	0.78261	0.81818	0.9994	0.21590	0.80435
K-NN	0.64286	0.81818	0.72000	−0.2197	0.69565
ABC	0.8	0.90909	0.85106	0.39015	0.84783
GNB	0.83333	0.90909	0.86957	0.47727	0.86957
Proposed	**0.83333**	**0.90909**	**0.86957**	**0.47727**	**0.86957**

In Table 2 it is clearly shows the memory usage and the execution time of every model as well as the proposed model. In the recent time we always blindly go for the faster processing time model, but memory factor can be compromised. As we can see our proposed model gives accuracy but taking more time just because of layering

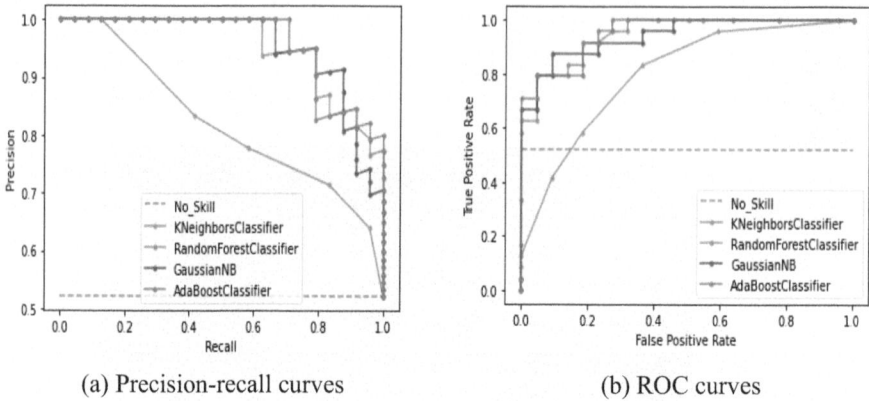

(a) Precision-recall curves   (b) ROC curves

Fig. 2. Precision-recall and ROC curves for benchmark Machine Learning Models.

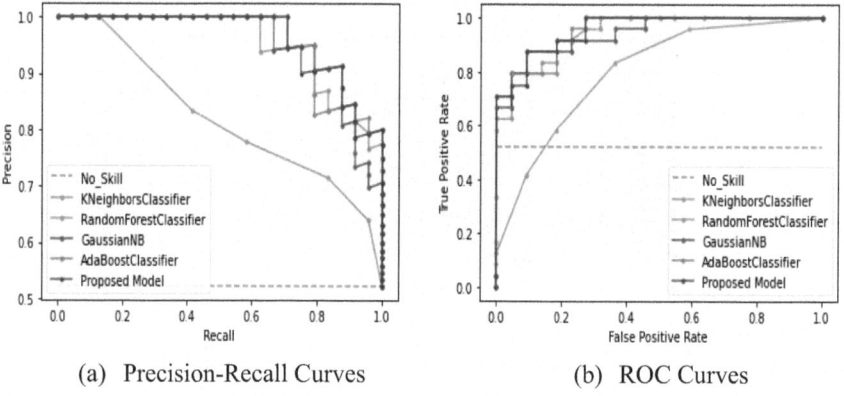

(a) Precision-Recall Curves   (b) ROC Curves

Fig. 3. Precision-Recall and ROC Curves for the suggested stacking approach over the examined dataset in comparison to benchmark machine learning models.

Table 2. Comparison of the suggested algorithm's execution time and memory usage with those of benchmark algorithms.

Models	Memory Consumed in bytes	Execution Time in ms
GNB	**105,193,472**	**0.0 ms**
Proposed	105,279,488	2006.037 ms
ABC	105,275,392	65.904 ms
RF	105,213,952	125.419 ms
K-NN	105,267,200	0.0 ms

concepts. Hence users can choose a model based on their specific requirements, balancing memory constraints and computational speed. It is Clear that Proposed Model consuming more memory and running more slowly than some other models because stacked ML model involves the combination of multiple base models which can increase the overall complexity of the framework. However, it could do a better job of classifying things or other metrics that aren't included in the present article. But a further detailed experiment is needed for the improvement of the processing time for stacking technique.

6   Conclusions and Future Work

The novel effective proposed model was examined along with some of the Machine Learning principles like: Gaussian Naive Bayes (GNB), K-Nearest Neighbors Classifier (K-NNC), Random Forest Classifier (RFC) and AdaBoost Classifier (ABC). These ideas provide helpful point view in forecasting coronary artery heart disease (CAHD). Logistic Regression (LR) was implemented at level1 in order to complete the stack, for acknowledgment of the complexity involved in forecasting cardiovascular illness.

This study examined the capability of four computational intelligence principles to predict coronary artery heart disease (CAHD): Gaussian Naive Bayes (GNB), K-Nearest Neighbors (K-NN), Random Forest (RF) Classifier, and AdaBoost Classifier (ABC) at level 0. An ensemble model of stacking strategy was introduced in novel proposed model that was helped to solve the problem of effectively predicting cardiac heart disease. When the benchmarks ML models like RF, K-NN, GNB, ABC, and LR, are used in our proposed model, that improved the performance. Our Proposed model achieved a maximum accuracy of 86.957%—which designed in such a way that the benchmarks ML models like RF, K-NN, GNB, and ABC at level 0 and at the end Logistic Regression (LR) which predict the final output in level 1, That is how the proposed model is designed. Finally, Precision and Recall measurements were performed to examine the efficacy of the model, which produced a remarkable result of 83.333% and 90.909% respectively. This study demonstrates how perfectly our proposed hybrid classifier model predict on cardiac disease and improve the patient's satisfaction. Subsequent research in this area can concentrate on adjusting hyperparameters and investigating new computational intelligence approaches in order to further enhance the hybrid model. Furthermore, adding more datasets or combining various datasets may improve the model's ability to generalize to other populations. It would be more useful to do research on the hybrid model's interpretability and explainability in order to evaluate its conclusions in actual clinical situations. As machine learning and healthcare analytics continue to progress, further research is necessary to make sure the model can be adjusted to new data sources and changing medical environments. In the end, the suggested hybrid classifier opens up a viable path for enhancing the diagnosis and treatment of coronary artery heart disease, opening the door for more developments in the area.

References

1. Bebortta, S., Tripathy, S.S., Basheer, S., Chowdhary, C.L.: FedEHR: A Federated Learning Approach towards the Prediction of Heart Diseases in IoT-Based Electronic Health Records. Diagnostics **13**(20), 3166 (2023)

2. Talaat, F.M.: Effective prediction and resource allocation method (EPRAM) in fog computing environment for smart healthcare system. Multimedia Tools and Applications **81**(6), 8235–8258 (2022)
3. Tripathy, S.S., et al.: A novel edge-computing-based framework for an intelligent smart healthcare system in smart cities. Sustainability **15**(1), 735 (2022)
4. Bebortta, S., Tripathy, S.S., Basheer, S., Chowdhary, C.L.: DeepMist: towards deep learning assisted mist computing framework for managing healthcare big data. IEEE Access (2023)
5. Bebortta, S., Panda, M., Panda, S.: Classification of pathological disorders in children using random forest algorithm. In: 2020 International Conference on Emerging Trends in Information Technology and Engineering (ic-ETITE), pp. 1–6. IEEE (2020)
6. Tripathy, S.S., Rath, M., Tripathy, N., Roy, D.S., Francis, J.S., Bebortta, S.: An intelligent health care system in fog platform with optimized performance. Sustainability. **15**(3), 1862 (2023)
7. Xu, B., Xu, L.D., Cai, H., Xie, C., Hu, J., Bu, F.: Ubiquitous data accessing method in IoT-based information system for emergency medical services. IEEE Trans. Ind. Inf. **10**(2), 1578–1586 (2014)
8. Gupta, A., Kumar, R., Arora, H.S., Raman, B.: MIFH: A machine intelligence framework for heart disease diagnosis. IEEE access. **27**(8), 14659–14674 (2019)
9. Garg, R., Sarangi, P.K., Sahoo, A.K., Jha, J.: Cardiovascular disease prediction: performance analysis and comparison of various supervised machine learning algorithms. In: 2023 International Conference on IoT, Communication and Automation Technology (ICICAT), pp. 1–6. Gorakhpur, India (2023)
10. Dwivedi, A.K.: Performance evaluation of different machine learning techniques for prediction of heart disease. Neural Comput. Appl. **29**, 685–693 (2018)
11. Inbarani, H.H.: A novel neighborhood rough set based classification approach for medical diagnosis. Procedia Comp. Sci. **1**(47), 351–359 (2015)
12. Sakshi, Das, P., Jain, S., Sharma, C., Kukreja, V.: Deep learning: an application perspective. In: Cyber Intelligence and Information Retrieval: Proceedings of CIIR 2021, pp. 323–333. Springer Singapore (2022)
13. Amin, M.S., Chiam, Y.K., Varathan, K.D.: Identification of significant features and data mining techniques in predicting heart disease. Telematics Inform. **1**(36), 82–93 (2019)
14. Shilaskar, S., Ghatol, A.: Feature selection for medical diagnosis: evaluation for cardiovascular diseases. Expert Syst. Appl. **40**(10), 4146–4153 (2013)
15. Ali, L., et al.: An optimized stacked support vector machines based expert system for the effective prediction of heart failure. IEEE Access. **9**(7), 54007–54014 (2019)
16. Fitriyani, N.L., Syafrudin, M., Alfian, G., Rhee, J.: HDPM: an effective heart disease prediction model for a clinical decision support system. IEEE Access. **20**(8), 133034–133050 (2020)

Feature Based Machine Learning Models for Cardiovascular Disease Diagnosis: An Experimental Analysis

Alok Kumar Agrawal[1] , Amit Vajpayee[2] , Merry Saxena[3] ,
Pradeepta Kumar Sarangi[1(✉)] , Karan Bajaj[4] , and Ashok Kumar Sahoo[5]

[1] Chitkara University School of Engineering and Technology, Chitkara University, Baddi, India
pradeeptasarangi@gmail.com
[2] Parul Institute of Engineering and Technology, Parul University, Vadodara, Gujrat, India
[3] Department of Interdisciplinary Courses in Engineering, Chitkara University Institute of Engineering and Technology, Chitkara University, Punjab, India
[4] School of Computer Science and Engineering, Lovely Professional University, Jalandhar, India
[5] Graphic Era Hill University, Dehradun, India

Abstract. The heart is widely regarded as the most vital organ within the human body. Cardiovascular diseases are prevalent in contemporary society; hence it is imperative to anticipate and forecast these conditions in advance. The accurate prognosis and prediction of coronary heart diseases necessitate a high level of precision and efficiency, since even a minor error might result in fatality for the individual. In order to address this condition, it is imperative to develop a model that possesses the capability to forecast and generate awareness pertaining to certain illnesses. Determining the disease manually based solely on indications and risk factors presents a significant challenge. However, machine learning techniques can address this issue. This work implements and evaluates the efficiency of machine learning algorithms in predicting cardiovascular diseases. The algorithms used in this work are K-Nearest Neighbors (KNN) and Support Vector Machine (SVM) and the data set used is the UCI repository data. The experimental results prove that the KNN and SVM models exhibit the highest levels of suitability, as evidenced by their respective accuracies of 88% and 84% respectively.

Keywords: Heart Disease Diagnosis · Machine Learning Algorithms · SVN · KNN

1 Introduction

The heart is one of the vital organs in the human body and diagnosing heart diseases with precision is of at most importance. The heart is responsible for the blood circulation in the entire body. Heart diseases often create life-threatening conditions and are caused by factors like hypertension, diabetes, high blood pressure, and cholesterol levels [1]. Machine learning models have become a prevalent method in various contemporary medical applications [2].

M. Khurana et al. (Eds.): ICMLA 2024, CCIS 2238, pp. 15–24, 2025.
https://doi.org/10.1007/978-3-031-75861-4_2

Application of machine learning in heart disease prediction has been a continuous area of interest by researchers. By considering characteristics such as chest pain, LDL cholesterol levels, age, and other relevant variables, machine learning can be employed to determine the presence of cardiovascular disease in an individual. The field of machine learning holds great potential in significantly improving the accuracy of predicting heart disease by analyzing extensive amounts of medical data and identifying patterns that may indicate the presence of cardiac issues. This study employs two supervised algorithms, namely Support Vector Machine and K-Nearest Neighbour. The UCI repository data contains the components and clinical records of the individual.

2   Literature Review

The motivation for this study comes from various academic literature pertaining to the application of machine learning techniques for the purpose of detecting cardiac disease. This section provides an overview of significant contributions made by researchers prior to the current study. In their work, authors like Patel et al. [3] have employed the Logistic and Random Forest model for heart disease prediction. The authors conclude that the J48 tree approach proved to be the most accurate classifier for predicting heart disease. In their study, Soni et al. [4] employed a Weighted Associative Classifier to predict occurrences of heart attacks. The researchers have designed a graphical user interface (GUI) for the purpose of inputting patient information. By utilizing the rules stored in the rule base, the researchers have determined that it was feasible to predict the patient had heart disease. In this study, Shao et al. [5] employed Naïve Bayes and Random Forest classifiers to predict outcomes using various datasets. Based on the findings presented by the authors, it is observed that Random Forest demonstrates superior performance in certain databases, whereas Naïve Bayes exhibits higher performance in others. Researchers such as Chauraisa et al. [6] have conducted investigations on various data mining methodologies for the purpose of cardiac attack detection reporting an accuracy of 85.03%. The total time taken to generate the model was 0.05 seconds. The Decision Support System for Heart Disease Prediction was built utilizing the Naive Bayesian Classification technique by the authors [7]. The system conducts a search inside a historical database of cardiac. The algorithm employed in this study had the highest level of accuracy in predicting patients with heart disease. In a separate study, researchers such as Nikhar et al. [8] have employed the Decision Tree classifier and Naïve Bayes algorithm to forecast occurrences of cardiac disease. The empirical findings indicate that Decision trees exhibit higher accuracy rates compared to naive Bayes classifiers. Rubini et al. [9] have proposed several categorization schemes to facilitate the anticipation and diagnosis of heart attacks. The Random Forest Algorithm was employed to ascertain the potential predictive value with an accuracy of 84.81%. In their study, Amin Ul Haq et al. [10] have employed seven well recognized classifiers, namely logistic regression, K-NN, ANN, SVM, NB, DT, and Random Forest for cardiac disease predictions. Logistic regression yielded the highest level of accuracy, reaching 89%. In their study, Pasha et al. [11] have employed SVM and KNN models to predict heart disease. The study conducted by Jagtap et al. [12] demonstrates the implementation of Support Vector Machines (SVM), Naive Bayes, and Logistic Regression, to make predictions regarding the occurrence of heart disease. Bharti et al. [13] aimed to predict heart disease by employing a comparative analysis of four machine learning techniques. The researchers reached the conclusion that the application of data pre-processing resulted in improved performance of the KNN classifier. Dhande et al. [14] in their work have employed machine algorithms to make predictions on the occurrence of diabetes and heart disease reporting an accuracy of 88%. A summary of the similar works is given in Table 1.

Table 1. Similar works in the field of cardiovascular disease diagnosis

Reference	Technique	Findings/Conclusion
[3]	Logistic regression, Random Forest and J48	The J48 proved the highest accuracy
[4]	Weighted Associative Classifier (WAC)	The heart disease prediction is accomplished by the utilization of a weighted associative classifier-based methodology
[5]	Random forest algorithm, Naïve Bayes algorithm	The Random Forest approach demonstrates superior performance compared to other methods, as seen by its accuracy rates of 86.81% for Dataset-1, 82.75% for Dataset-2, and 86.81% for Dataset-3
[6]	Support Vector Machine	The SVM model produced the highest accuracy of 85.77%
[7]	Naïve Bayes	The naive Bayes model demonstrated the highest predictive accuracy for people with heart disease
[8]	Decision Tree and Naïve Bayes Classifier	Various classification algorithms are commonly employed for the prediction of cardiovascular illnesses. In terms of accuracy, the decision tree outperformed the naive Bayes classifier
[9]	Random Forest, Linear Regression, SVM, Naïve Bayes	The Random Forest model was utilized to analyze the available data on the link between diabetes and heart disease, with the objective of estimating the proportion of heart disease cases that can be potentially predicted The Random Forest algorithm achieved an accuracy of 84.81%

(continued)

Table 1. (*continued*)

Reference	Technique	Findings/Conclusion
[10]	KNN, logistic Regression, ANN, Naïve Bayes, SVM	The logistic regression model achieved the highest level of accuracy, reaching 89%
[11]	SVM, KNN, Decision Tree, ANN	The artificial neural network (ANN) achieved the highest level of accuracy, measuring at 85.24%
[12]	SVM, Naïve Bayes, Logistic Regression	The Support Vector Machine (SVM) algorithm was selected as the most optimal approach because of its accuracy rate of 64.4%
[13]	KNN, SVM, Decision Tree, Random Forest	The utilisation of data preprocessing in the ML technique resulted in improved performance of the K-Neighbors classifier
[14]	Random Forest, KNN, Logistic Regression, Decision Tree, AdaBoost, SVC, XGBoost	The accuracy of the Sigmoid Support Vector Classifier (SVC) was found to be optimal, with a performance of 88%

3 Objective

This study aims to assess the predictive capabilities of two machine learning models such as KNN and SVM in the context of heart disease prediction. The performance of these algorithms will be compared to determine their effectiveness in predicting heart disease.

4 Data Set

The dataset used in this study was collected from https://www.kaggle.com. The dataset consists of 303 samples (Fig. 1). The records include ["age", "sex", "cp", "trestbps", "chol", "fbs", "restecg", "oldpeak", "slope", "ca", "thal", and "target"], which represent various risk factors, demographic information, and medical history of the patients. The dataset was utilized to train and test the machine learning models at a ratio of 70:30 respectively (Fig. 1).

Figure 1 represents the sample data used in this work. It represents the first five entries of the dataset with their attributes. The histograms for each variable are displayed in Fig. 2.

	age	sex	cp	trestbps	chol	fbs	restecg	thalach	exang	oldpeak	slope	ca	thal	target
0	63	1	3	145	233	1	0	150	0	2.3	0	0	1	1
1	37	1	2	130	250	0	1	187	0	3.5	0	0	2	1
2	41	0	1	130	204	0	0	172	0	1.4	2	0	2	1
3	56	1	1	120	236	0	1	178	0	0.8	2	0	2	1
4	57	0	0	120	354	0	1	163	1	0.6	2	0	2	1

Fig. 1. A Sample dataset of Cardiovascular Disease Diagnosis.

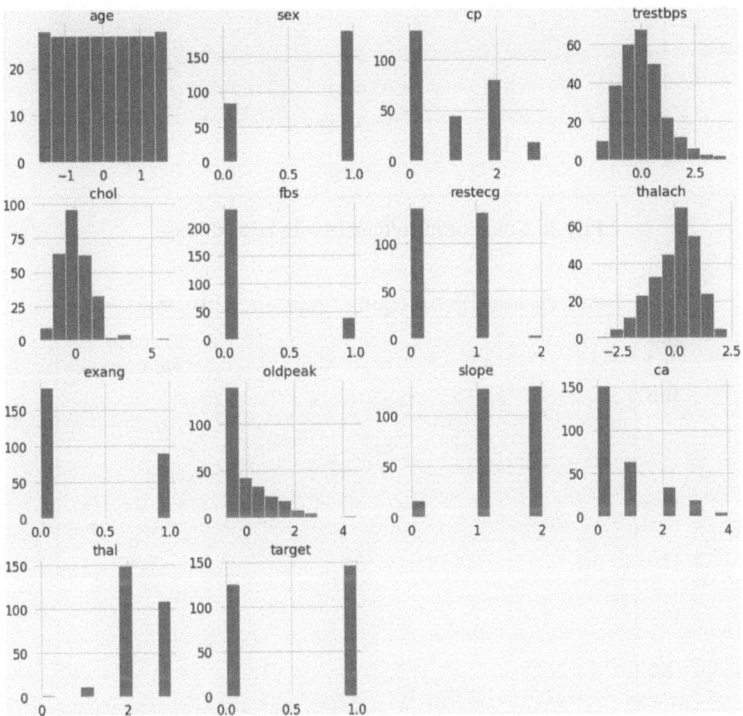

Fig. 2. Histograms of all the attributes

The histograms for each variable are displayed above. Each characteristic has a distinct spectrum of distribution, as can be observed in Fig. 2. The distribution of gender in the data set is shown in Fig. 3.

Figure 3, represents the gender wise distribution count in the data set used for experiments. The male patient count is slightly more than the number of female patients but this does not make any significant bias ness in data distribution. The heart disease in function of age and max heart rate is given in Fig. 4.

The figure is plotted by taking age as in x-axis and max-heart rate in y axis. However, from the figure it is observed that between the age 40 to 50 the heart rate goes beyond 160 and contains maximum diseased person.

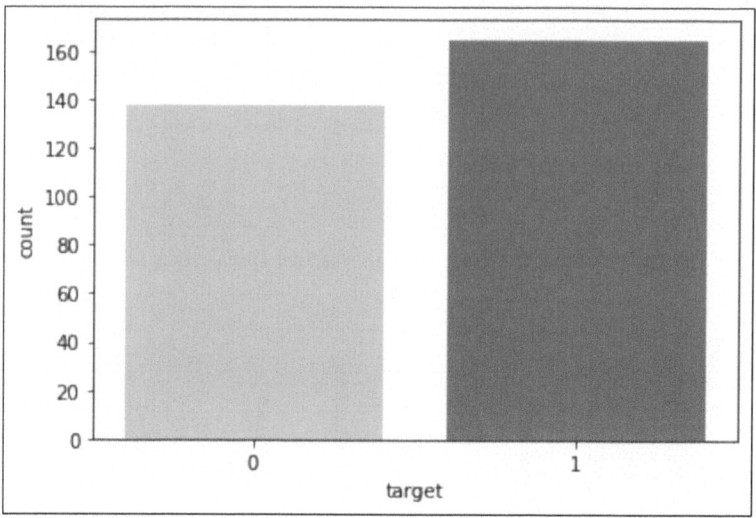

Fig. 3. Count of male/female with heart disease

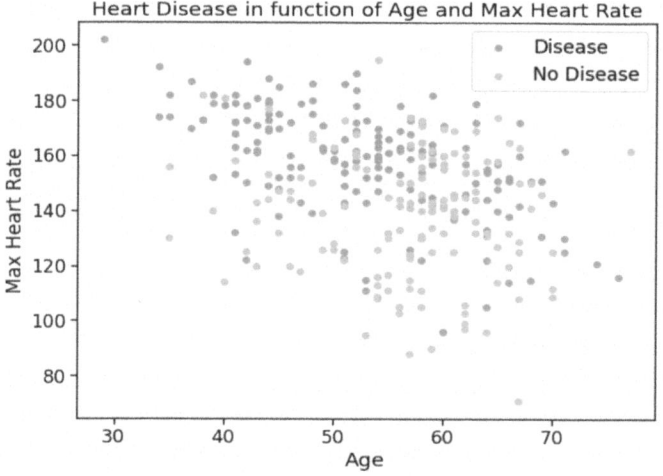

Fig. 4. Age and maximum heart rate in a scatter plot.

5 Methodology and Implementation

This section describes the implementation of SVM and KNN models. The implementation has been done through Python programming. The models have been trained first and after successful training, prediction has been done using test data. The efficiency of the models has been measured by their accuracy and the analysis is done using a confusion matrix plotted by the program.

The SVM model has been implemented using the test data. The confusion matrix generated by the program for the test patterns is given in Fig. 5.

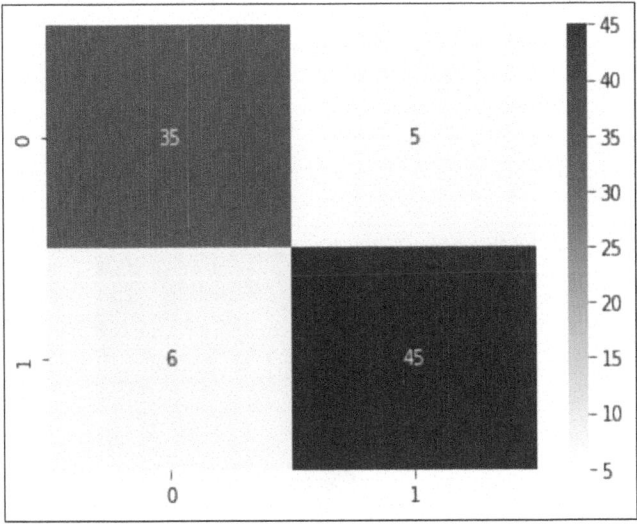

Fig. 5. The confusion matrix of Support Vector Model

In the Fig. 5, the (0,0) cell represents that in the test pattern 35 patients were having no disease and the model also recognized correctly however, 6 patients in the cell (0,1) in the test data were having no disease but the model misclassified as these patients are having disease. Similarly, the cell (1,1) represents the patients were having heart disease and the model also forecast them as having heart disease whereas the cell value (1,0) signifies the patients were having heart disease but the model predicted as having no disease. The model accuracy is reported as 87.91%. The classification report is given in Fig. 6.

	precision	recall	f1-score	support
0	0.85	0.88	0.86	40
1	0.90	0.88	0.89	51
accuracy			0.88	91
macro avg	0.88	0.88	0.88	91
weighted avg	0.88	0.88	0.88	91

Fig. 6. SVM classification results

In the similar way, the confusion matrix has been generated for the implementation of KNN model. The confusion matrix generated by the program for the test patterns is given in Fig. 7.

The above represents confusion matrix of the KNN algorithm used for the prediction in which the true negative value is 32, true positive value is 44, false positive value is 8 and

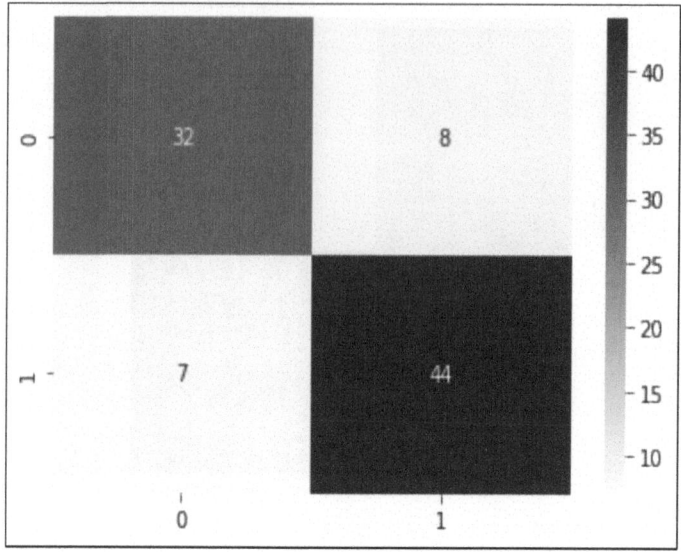

Fig. 7. Confusion matrix of KNN model.

the false negative value is 7. The model accuracy is reported as 84%. The classification report is given in Fig. 8.

```
              precision    recall  f1-score   support

           0       0.82      0.80      0.81        40
           1       0.85      0.86      0.85        51

    accuracy                           0.84        91
   macro avg       0.83      0.83      0.83        91
weighted avg       0.83      0.84      0.83        91
```

Fig. 8. KNN classification result

The summary of both implementations in terms of accuracy is given in Table 2.

Table 2. Accuracy for both Algorithms

Algorithms	Accuracy
SVM model	88%
KNN model	84%

From the table, it can be observed that to predict heart disease, the most accurate approach, according to the analysis of both the findings, is the SVM model.

6 Conclusion

The process of diagnosing CVD in the medical industry is expensive and time-consuming. The suggested method implies that Machine Learning may be utilized as a clinical tool in the identification of CVDs and will be especially helpful for doctors in the case of a false-positive result. In comparison to the previous methods described, the developed Machine learning models consistently predicts illnesses with higher. It is expected that the proposed method would enhance the medical sector. The suggested approach may be applied to categorize additional chronic illnesses, including thyroid, liver, breast, and diabetes mellitus. Using IoT and cloud computing approaches, the created algorithms may be used to huge data sets to predict such chronic illnesses. Based on the aforementioned study, it is clear that using ML approaches will significantly help in avoiding deaths and support medical professionals' efforts to reduce the onset of CVD among all patients. If adopted, this would be a prime example of how modern technology may be used for the good of everybody. Further research can be enhanced by identifying the most relevant elements needed for the prediction of heart disease and using other data mining techniques.

In this work two Machine learning algorithms are used to diagnose many types of heart diseases. Based on the experimental results, it can be seen that the SVM model produces the best outcomes. The other algorithm, KNN, also gives accuracy more than 80 %, but SVM classifier is giving the best accuracy of 88 %.

References

1. Kaur, B., Kaur, G.: Heart disease prediction using modified machine learning algorithm. In: International Conference on Innovative Computing and Communications. Lecture Notes in Networks and Systems, vol 473. Springer, Singapore (2022). https://doi.org/10.1007/978-981-19-2821-5_16
2. Tiwari, S., Kumar, S., Guleria, K.: Outbreak Trends of Coronavirus Disease-2019 in India: A Prediction. Disaster Med. Public Health Prep. **14**(5), 33–38 (2020). https://doi.org/10.1017/dmp.2020.115
3. Patel, J., Tejal, U., Patel, S.: Predicting heart condition using machine learning and data mining techniques, Vol. 7, Number 1, pp. 129–137 (2015)
4. Soni, J., Ansari, U., Sharma, D.: Intelligent and effective heart disease prediction system using weighted associative classifiers. Int. J. Comp. Sci. Eng. **3**, 2385–2392 (2011)
5. Shao, Y.E., Hou, C.D., Chiu, C.C.: Hybrid intelligent modelling schemes for heart disease classification. Appl. Soft Comput. **14**, 47–52 (2014)
6. Chauraisa, V., Pal, S.: Data mining approach to detect heart diseases. Int. J. Adv. Comp. Sci. Info. Technol. (IJACSIT) (2013)
7. Lakshmi, G.S.: Decision support in heart disease prediction system using naïve Bayes. Indian J. Comp. Sci. Eng. (IJCSE) **2**(4), 132–142 (2011). 2013
8. Sonam, N., Karandikar, A.M.: Prediction of heart disease using machine learning algorithms. Int. J. Adv. Eng. Manage. Sci. **2**(6), 617–621 (2016)

9. Rubini, P.E., et al.: A cardiovascular disease prediction using machine learning algorithms. Annals of R.S.C.B., 904–912 (2021)

10. Amin, U.H., Jian, P.L., Memon, M.H., Nazir, S., Sun, R.: A hybrid intelligent system framework for the prediction of heart disease using machine learning algorithms. Mobile Info. Sys., Article ID 3860146, 21 (2018). https://doi.org/10.1155/2018/3860146

11. Pasha, S.N., Ramesh, D., Mohmmad, S., Harshavardhan, A., Shabana: Cardiovascular disease prediction using deep learning techniques. IOP Conf. Series: Materials Science and Engineering **981**, 002–006 (2020). https://doi.org/10.1088/1757-899X/981/2/022006

12. Jagtap, A., Priya, M., Omkar, B., Harshali, R.: Heart disease prediction using machine learning. Int. J. Res. Eng. Sci. Manage. **2**(2), 352–355 (2019)

13. Bharti, R., Aditya, K., Mohammad, S., Gaurav, D., Sagar, P., Parneet, S.: Prediction of heart disease using a combination of machine learning and deep learning. Comput. Intell. Neurosci. **2021**, 1–11 (2021). https://doi.org/10.1155/2021/8387680

14. Dhande, B., Bamble, K.S., Maktum, T.: Diabetes & Heart Disease Prediction. ITM Web of Conferences **44**, 03057. ICACC (2022). https://doi.org/10.1051/itmconf/20224403057

Early Liver Disease Detection Through Visual Interface and Machine Learning

Sarika Agarwal[1]([✉]), Himani Bansal[2], and Vibha mani[3]

[1] Department of AI, Noida Institute of Engineering and Technoogy, Greater Noida, India
sarikagarwal.it@gmail.com
[2] Department of CSE and IT, Jaypee Institute of Information Technology, Noida, India
[3] Department of CSE and IT, Greater Noida Institute of Technology, Greater Noida, India

Abstract. The liver disease remains a medical condition without a specific allopathic cure at present. Detecting liver disease at an early stage can significantly improve survival rates. The prevalence of fatty liver disease is also on the rise in India, underscoring the need for a user-friendly method through which individuals can input parameters to determine their liver health status. To address this need, we propose the implementation of a Graphical User Interface. This interface would enable users to input various parameter values, thereby allowing them to ascertain whether they are affected by liver disease. Our study involved a thorough evaluation of several machine learning algorithms, including Logistic Regression, Decision Tree, K-Nearest Neighbors, Support Vector Machine, LightGBM, RandomForest, and Extra Tree Classifier. Remarkably, our efforts culminated in an accuracy rate of 92% achieved by the Extra Tree Classifier. This achievement holds promising implications for accurate liver disease prediction and early intervention.

Keywords: Machine Learning · Liver Disease Prediction · Fatty liver · Ensemble Learning First Section

1 Introduction

Liver disease is an ailment for which allopathic medicine currently lacks a specific cure.In India, the incidence of liver disease is on the rise, with a significant portion of the population, exceeding 38%, being affected by fatty liver conditions [1]. Fatty liver is a condition that can frequently be observed in children as well [2]. The primary contributor to fatty liver is lifestyle choices and unhealthy dietary habits. The consumption of alcohol also stands out as a significant factor leading to liver disease. During the initial phases, it might not manifest any noticeable symptoms, but over time, it can potentially advance to develop into liver cancer. Non-alcoholic fatty liver disease (NAFLD) serves as an early indicator of metabolic syndrome, particularly notable in individuals of South Asian origin, where those with NAFLD exhibit a fourfold heightened risk of diabetes compared to those without the condition and a 2.5 times elevated risk compared to Caucasians [2–4].

© The Author(s), under exclusive license to Springer Nature Switzerland AG 2025
M. Khurana et al. (Eds.): ICMLA 2024, CCIS 2238, pp. 25–33, 2025.
https://doi.org/10.1007/978-3-031-75861-4_3

Fat accumulation in the liver disrupts crucial metabolic pathways, impacting the body's ability to metabolize glucose and utilize energy effectively. Consequently, individuals affected by this condition are prone to developing metabolic syndrome, a collection of risk factors encompassing abdominal obesity, elevated triglycerides (blood fats), diminished levels of beneficial 'HDL' cholesterol, high blood pressure, and heightened fasting blood sugar. These combined factors significantly elevate the likelihood of developing diabetes, heart attacks, and stroke [3, 5–7].

Early prediction of liver disease increases the survival rates and safe the person from significant diseases like diabetes and heart attacks etc. Hence, we have developed a user-friendly graphical interface that enables users to input their parameters for the prediction of the presence of liver disease. Furthermore, we have conducted a comprehensive performance analysis of our algorithm for developing the visual interface.

2 Literature Review

The liver, the body's largest gland, performs many functions. It processes ingested food and beverages, transforming them into essential nutrients for the body's absorption. Simultaneously, it filters out detrimental substances from the bloodstream and contributes to an immune response against infections. The liver's integrity can be compromised by exposure to viruses or harmful compounds, leading to the development of liver disease. Liver disease encompasses any condition that harms the liver, potentially impairing its functionality. This severe health concern endangers human life and necessitates immediate medical intervention. This study's focal point is the early prediction of such diseases using machine learning (ML) techniques.

Numerous researchers have introduced diverse models for forecasting early-stage liver disease. Andrade et al. [8] assessed the effectiveness of three classifiers in diagnosing liver steatosis by leveraging various extracted features from ultrasound images. Ribeiro and Sanches [9] employed an anatomical and echogenic approach, utilizing information from normal and fatty liver ultrasonic images. They applied a Bayesian framework to the extracted feature parameters for diagnosing fatty liver. According to Elias Dritsas et al. [10], their study reveals that the Voting classifier surpasses alternative models by achieving an accuracy, recall, and F-measure of 80.1%, along with a precision of 80.4% and an AUC of 88.4% following the utilization of SMOTE in conjunction with 10-fold cross-validation. According to Khan et al. [11], the Bagged Tree classifier demonstrates the most noteworthy classification accuracy at 81.30%, exhibiting considerable promise in contrast to alternative algorithms for liver disease prediction. In the study conducted by Amin et al. [12], it is revealed that Random Forest attains the highest accuracy of 88.10% when predicting the occurrence of liver disease. Gupta et al. [13] demonstrate that LightGBM and RandomForest attain the highest accuracy of 63% when predicting liver disease. Ghazal et al. [14] predict liver disease with a machine learning algorithm with 88.4% accuracy. Chieh-Chen Wu et al. [15] incorporated all patients who underwent an initial fatty liver screening at New Taipei City Hospital. Their research aimed to establish a predictive model for liver disease. Among the various algorithms explored, Random Forest achieved the utmost accuracy, reaching 88.7%.

3 Methodology

Early Prediction of Liver disease needs a combination of medical data analysis, Machine learning techniques, and domain knowledge. Therefore, an intuitive graphical interface is created that allows users to input their parameters to predict the presence of liver disease.

This study utilizes a dataset comprising 583 records extracted from the ILPD (Indian Liver Patient Dataset) for addressing the problem at hand. The dataset has been sourced from the Kaggle Machine Learning Repository (https://www.kaggle.com/code/ashok4kaggle/liver-patient-analysis-and-prediction). The complete ILPD dataset encompasses data pertaining to 583 individuals diagnosed with liver conditions. Among these, 416 records correspond to patients with liver ailments, while 167 records pertain to individuals without liver-related issues.

Additionally, we've conducted an extensive performance evaluation of our algorithm to build the visual interface. Figure 1 illustrates the process and techniques utilized in this study to predict liver disease. The sourced dataset underwent preprocessing, including balancing. Subsequently, it was divided into training and testing subsets. Various models were trained, and their performance was assessed. Finally, the developed model was deployed in a user-friendly interface, enabling the prediction of liver disease.

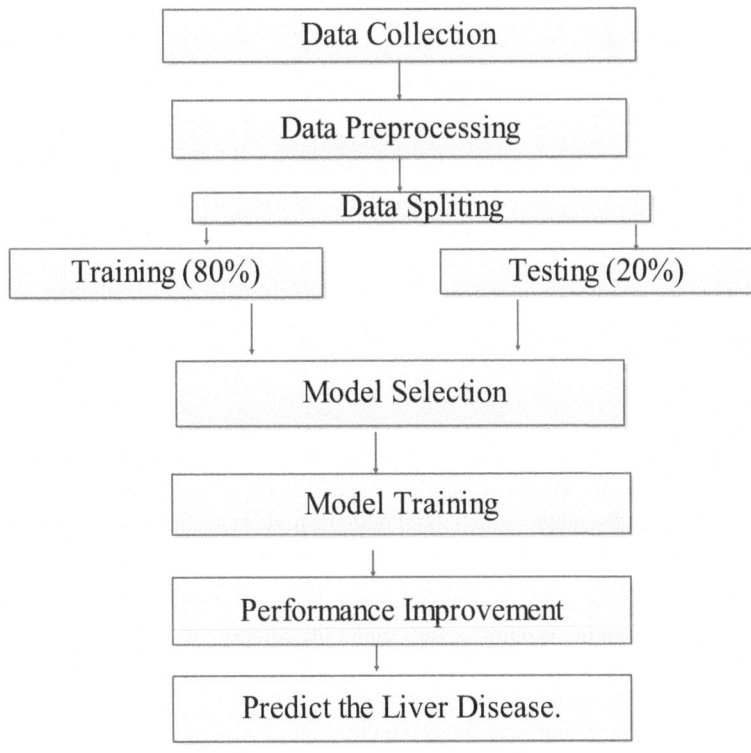

Fig. 1. Methodology

4 Implementation

This study employed an i7 computer with 16 GB of RAM alongside Anaconda and Jupyter Notebook for code execution. The implementation utilized several libraries, including Sklearn, Tkinter, lightGBM, and XGBoost. The dataset, comprising 583 records and encompassing 11 features as detailed in Table 1, was sourced from Kaggle.

Table 1. DataSet Features

Attribute Name	Data Type
Age	Integer
Gender	Object
Total Bilirubin	Float
Direct Bilirubin	Float
Alkaline Phosphotase	Integer
Alamine Aminotransferase	Integer
Aspartate Aminotransferase	Integer
Total Protiens	Float
Albumin	Float
Having Liver Disease	Integer

While preprocessing the data, it was identified that four records had missing values in the "Albumin_and_Globulin_Ratio" feature. To address this, the missing values were imputed using the median value of the respective feature. Here the dataset exhibits class imbalance, with 416 records indicating the presence of liver disease, while 167 records denote the absence of liver disease. Consequently, a dataset imbalance correction is necessary. To balance the data, oversampling was performed on the minority class records. So, the total record size is 832.

4.1 Model Selection

Various classification algorithms were trained, and their corresponding accuracy scores were documented. Subsequently, the model with the highest accuracy score was selected for further training and was utilized to predict the occurrence of liver disease.

A) **Logistic Regression:** It is a Supervised learning technique employed for classification problems. Despite its name, it falls under the category of regression algorithms. This nomenclature arises from its utilization of the linear regression function's output as input. Furthermore, logistic regression employs a sigmoid function to estimate the probability associated with the given class.

B) **K-Neiest Neighbour (KNN):** It is also a Supervised learning algorithm used to solve classification problems. The KNN algorithm retains all accessible data and categorizes a novel data point by assessing its similarity to existing data. This implies that as new data emerges, the KNN algorithm can proficiently assign it to a suitable category.

C) **Support Vector Machine:**Support Vector Machine (SVM) identifies a hyperplane that establishes a clear division between distinct data categories. In a two-dimensional context, this hyperplane corresponds to a line. In the SVM methodology, every data point within the dataset is represented in an N-dimensional space, with N denoting the number of features inherent to the data. Subsequently, the task involves determining the most favorable hyperplane configuration to effectively segregate the data points.

D) **Decision Tree:**This classifier follows a tree structure, wherein internal nodes symbolize dataset features, branches denote decision rules, and each leaf node signifies an outcome.

E) **Random Forest:**It is based on the principle of ensemble learning; this approach involves amalgamating multiple classifiers to address intricate problems and enhance the model's efficacy. As its name implies, "Random Forest" is a classification technique encompassing numerous decision trees, each constructed on distinct subsets of the provided dataset. These trees are then averaged to elevate the predictive precision of the dataset.

F) **Light GBM(Light Gradient Boosting Machine):-**Light GBM is a rapid, distributed, and high-performance gradient boosting framework that is built upon the foundation of the decision tree algorithm. It finds application in ranking and classification tasks. This framework capitalizes on decision trees to heighten model efficiency and curtail memory consumption.

G) **Extra Tree Classifier:**ExtraTrees represents an ensemble machine learning technique that involves training multiple decision trees. The outcomes from this ensemble of trees are combined to generate a prediction. Notably, there exist distinctions between Extra Trees and Random Forest. In Random Forest, bagging is employed to derive diverse versions of the training data, thereby ensuring substantial diversity among decision trees. In contrast, Extra Trees leverages the complete dataset for training decision trees.

Every model underwent training, and their respective accuracy, precision, recall, and cross-validation scores were calculated and assessed.

All seven models' performance was evaluated and compared. Accuracy,Recall Precision, and Cross-validation score were calculated.

Accuracy quantifies the proportion of accurately predicted instances in relation to the overall instances. This metric proves valuable particularly when all classes hold equal significance. Yet, this perspective may not apply when determining the presence of Liver Cancer in a patient. In such a scenario, a certain degree of acceptability toward False Positives (FPs) might be permissible, but the same leniency cannot be extended to False Negatives (FNs).

$$Accuracy = (TP + TN)/(TP + TN + FP + FN) \qquad (1)$$

Precision provides insight into the number of accurately predicted positive cases among those that were predicted as positive.

$$Precision = (TP)/(TP + FP) \qquad (2)$$

Recall informs us about the portion of actual positive cases that our model success-fully identified and predicted accurately.In medical scenarios, recall holds significance as it prioritizes detecting actual positive cases, even if it raises false alarms.

$$Recall = (TP)/(TP + FN) \tag{3}$$

5 Results

The Accuracy, Recall, Precision, and 10 Fold Cross-validation score was calculated. Results are displayed in Table 2. Notably, the Extra Tree classifier exhibits the highest values across Accuracy, Recall, and Cross-Validation score within these metrics.

Table 2. Accuracy, Precision, Recall, and Cross-Validation Score

Algorithm	Accuracy	Precision	Recall	Cross-validation Score (10)
Logistic Regression	73	91	57	70
KNN	75	89	62	69
SVM	66	91	43	66
Decision Tree	89	97	83	84
Random Forest	91	99	85	84
Light GBM	90	97	84	84
Extra Tree Classifier	92	98	87	87

A Graphical User Interface (GUI) with a user-friendly design was created to enable users to effortlessly define parameters and promptly receive information about the pres-ence of liver disease. The interface screenshot is depicted in Fig. 2, while Fig. 3 illustrates the message box that appears as a result of the user's input.

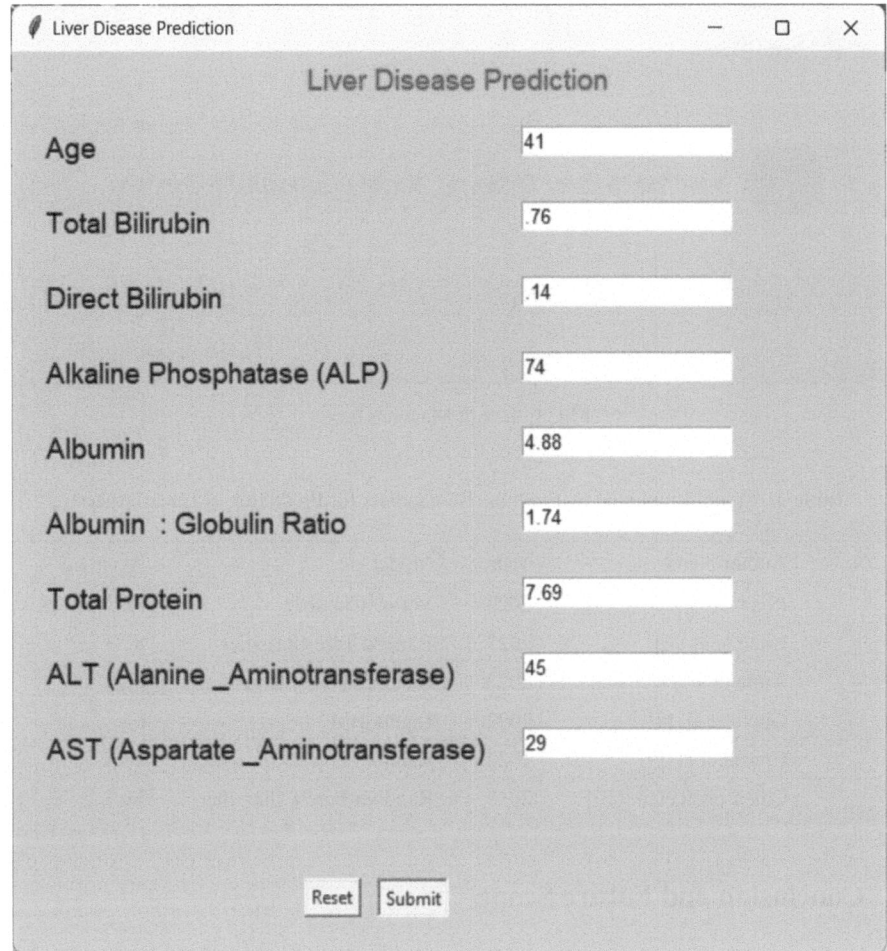

Fig. 2. Screenshot of Graphical User Interface

6 Comparison with Existing Work

Different researchers have made predictions about liver disease with varying levels of accuracy. However, our model stands out by achieving the highest accuracy and cross-validation score in predicting liver disease. Furthermore, we have introduced a Graphical User Interface (GUI), a feature that none of the other researchers have offered. Table 3 presents the model utilized by diverse researchers to achieve precise liver disease prediction. Researchers attained a maximum accuracy of 88.7%. Conversely, our Extra Tree Classifier model surpassed this with an accuracy of 92%.

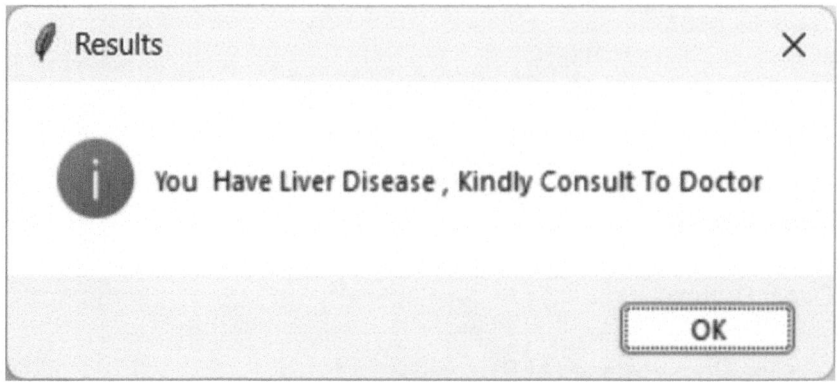

Fig. 3. Result Message Box

Table 3. Model Employed by Previous Researchers for Prediction of Liver Disease

S. No.	Author Name	Year	Model	Accuracy
1	Elias Dritsas et al. [5]	2023	Voting Classfier	80.1
2	Khan et al. [6]	2023	Bagged Tree Classifier	81.3
3	Amin et al. [7]	2023	Random Forest Classifier	88.1
4	Gupta et al. [8]	2022	LightGBM	63
5	Ghazal et al. [9]	2022	Random Forest Classifier	88.4
6	Chieh-cheu et al. [10]	2019	Random Forest Classifier	88.7

7   Conclusion and Future Scope

In this research, we introduce a user-friendly Graphical User Interface that allows users to input parameter values for predicting liver disease. The prediction model utilizes the Extra Tree Classifier, achieving an impressive accuracy of 92%. Notably, it also demonstrates a high precision score of 98% and a 10-fold cross-validation accuracy of 87%. Looking ahead, we envision incorporating neural network techniques to further enhance accuracy in the near future.

References

1. 38% of Indians have non-alcoholic fatty liver disease, says AIIMS study. The Indian Express. https://indianexpress.com/article/lifestyle/health/38-of-indians-have-non-alcoholic-fatty-liver-disease-says-aiims-study-8865915/. Accessed: 09 Aug. 2023
2. Here's why Indians need to fear a fatty liver. Hindustan Times. https://www.hindustantimes.com/columns/here-s-why-indians-need-to-fear-a-fatty-liver/story-FHi1oJ8bzsCgZ0YQfJFcnI.html. Accessed: 09 Aug. 2023

3. Singal, H., Kohli, S.: Intellectualizing TRUST for Medical Websites. In: Proceedings of the Second International Conference on Information and Communication Technology for Competitive Strategies, pp. 1–4 (2016)
4. Singal, H., Kohli, S.: Mitigating information trust: taking the edge off health websites. Int. J. Technoeth. (IJT) **7**(1), Art. no. 1 (2016)
5. Targher, G., et al.: Nonalcoholic fatty liver disease is independently associated with an increased incidence of cardiovascular events in type 2 diabetic patients. Diabetes Care **30**(8), 2119–2121 (2007)
6. Cushman, M., et al.: Nonalcoholic fatty liver disease and cognitive impairment: a prospective cohort study. PLoS ONE **18**(4), e0282633 (2023)
7. Gupta, S., Bansal, H.: An analytical review on precursors of User'sTrust in content driven health websites. Comput. Integr. Manuf. Syst. **28**(10), 480–542 (2022)
8. Andrade, A., Silva, J.S., Santos, J., Belo-Soares, P.: Classifier approaches for liver steatosis using ultrasound images. Procedia Technol. **5**, 763–770 (2012). https://doi.org/10.1016/j.protcy.2012.09.084
9. Ribeiro, R., Sanches, J.: Fatty liver characterization and classification by ultrasound. In: Pattern Recognition and Image Analysis: 4th Iberian Conference, IbPRIA 2009 Póvoa de Varzim, Portugal, June 10-12, 2009 Proceedings 4, pp. 354–361. Springer (2009)
10. Dritsas, E., Trigka, M.: Supervised machine learning models for liver disease risk prediction. Computers **12**(1), 19 (2023)
11. Khan, M.A.R., et al.: An effective approach for early liver disease prediction and sensitivity analysis. Iran Journal of Computer Science, 1–19 (2023)
12. Amin, R., Yasmin, R., Ruhi, S., Rahman, M.H., Reza, M.S.: Prediction of chronic liver disease patients using integrated projection based statistical feature extraction with machine learning algorithms. Informatics in Medicine Unlocked **36**, 101155 (2023)
13. Gupta, K., Jiwani, N., Afreen, N., Divyarani, D.: Liver disease prediction using machine learning classification techniques. In: 2022 IEEE 11th International Conference on Communication Systems and Network Technologies (CSNT), pp. 221–226. IEEE (2022)
14. Ghazal, T.M., et al.: Intelligent model to predict early liver disease using machine learning technique. In: 2022 International Conference on Business Analytics for Technology and Security (ICBATS), pp. 1–5. IEEE (2022)
15. Wu, C.-C., et al.: Prediction of fatty liver disease using machine learning algorithms. Comput. Methods Programs Biomed. **170**, 23–29 (2019)

Diagnosing Autism Spectrum Disorder in Children Using Various Machine Learning Methods: A Review

Robin Khurana[(✉)] and Satyaveer Singh

MMEC, MM(DU), Mullana (Ambala), Haryana, India
er.robin08@gmail.com, satyaveer.sangwan@gmail.com

Abstract. Autism Spectrum Disorder (ASD) is a relatively rare condition between children; evaluations of the CDC's Autism and Developmental Disabilities Monitoring (ADDM) System suggest that approximately one in 36 kids is identified with ASD. Its symptoms are particularly prominent and observable in children aged 2–3 years. This review paper undertakes a broad analysis of ASD detection methods, harnessing the strength of both machine learning (ML) and neural network techniques. The review encompasses a systematic examination with a visual flowchart and a rigorous evaluation of various methodologies. These include traditional machine learning (ML) methods like Support Vector Machines (SVM) and K-nearest neighbors (KNN), as well as advanced neural network approaches like Artificial Neural Networks (ANN). Furthermore, the analysis presents a comparative table that assesses the performance of these methods in terms of their accuracy on diverse datasets. This multifaceted approach provides a holistic understanding and offers valuable insights for scholars and physicians in the field.

Keywords: Autism spectrum disorder (ASD) · Support Vector Machine · Naïve Bayes

1 Introduction

ASD is a neurodevelopmental disorder widely prevalent in daily life, affecting human interaction. The impact of ASD warning signs extends to public and communication services. According to WHO reports, one in sixty kids globally has autism [1]. The symptoms of autism are more visible and easier to spot in kids aged 2–3 years. Testing procedures for ASD have been developed by healthcare experts in various forms worldwide. Through the use of various analytical tools, such as Autism Diagnostic Interview (ADI), surgeons can accurately identify ASD. Early analysis and detection of ASD are crucial for affected patients or kids, significantly aiding in managing ASD symptoms [2]. ASD symptoms are typically documented through observation. In previous years, blood relatives, such as parents and mentors, have observed ASD during a child's schooling age. Subsequently, a school's distinct learning software package was used to observe and assess any symptoms of ASD. Observing a child makes it easier to communicate and identify characteristic fluctuations due to early detection [3]. Children in the age group of

M. Khurana et al. (Eds.): ICMLA 2024, CCIS 2238, pp. 34–43, 2025.
https://doi.org/10.1007/978-3-031-75861-4_4

six months affected by autism, particularly through neuroimaging and imaging studies, are diagnosed with the disorder after the age of two years. ASD is considered a highly impactful disorder, primarily affecting children in the 2–3 age groups. The term 'Autism Spectrum Disorders' refers to a collection of difficult neurodevelopmental syndromes related to the brain, like autism, infantile disintegrative syndrome, and Asperger's pattern. Described as a spectrum, it implies a range of indications and severity levels [4]. These syndromes are presently included in the '*International Statistical Classification of Diseases*' and '*Related Health Issues under 'Mental and Behavioral Disorders,*' specifically in the class of general age-related syndromes. Various indications of this disorder become apparent in the early years, primarily affecting children with challenges in communication, speech, and attachment to caregivers [5]. Initially, children may appear to be developing typically before exhibiting signs of ASD between 18–24 months. The disorder manifests in issues such as repetitive and incomplete behavior patterns, limited interests, activities, and impaired language skills [6]. ASD is a neurodevelopmental syndrome that disrupts the growth of social and communication skills, conventional nature, and attention to detail. The etiopathogenesis of ASD is still unidentified with certainty. However, various methodologies suggest that ASD is primarily influenced by genetic factors, including genetic faults, brain infection, and irregular conditions during pregnancy [7]. Besides genetic and environmental factors, auto-protection is also a significant factor affecting ASD [8]. ASD impacts human growth adversely due to the involvement of numerous neurons in the brain. The analysis of ASD is very challenging as there is no particular health test like a blood test, used for ASD detection. Doctors typically employ emotional and observational approaches to diagnose ASD by analyzing various aspects of ASD features and daily routines, as depicted in Fig. 1.

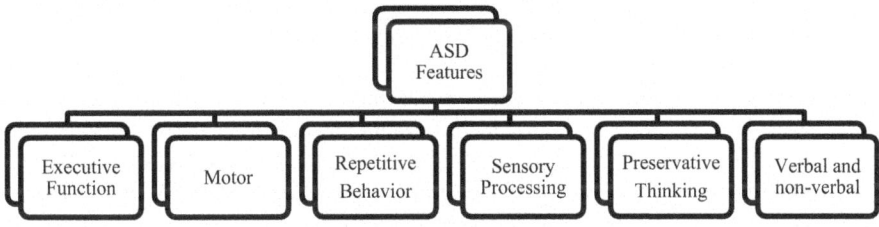

Fig. 1. Multiple Aspects of ASD Features [3]

Various risk factors contribute to ASD diseases, such as pregnancy, exposure to chemicals, parental ages, brain shape and growth, premature birth, and twins affected by ASD [9]. Several research studies have focused on ASD detection, such as the work by Raj et al. (2018) [1], who designed ML methods to investigate and detect ASD. They considered different age groups, including adults, kids, and adolescents, using classifiers like Naive Bayes (NB), SVM, Logistic Regression (LR), KNN, Neural Networks (NNs), and Convolutional Neural Networks (CNNs). Their dataset included factors like population, birth difficulties, family history of ASD, gender distribution, and the screening application utilized by the operator. They achieved an accuracy performance of 99.53% for adults, kids, and adolescents. Xie et al. (2019) [10] designed a deep learning (DL) model to identify representative graphical features in ASD with a binary stream system.

They used the VGG16 model with thirteen convolution layers, 5 max-pooling layers, and two fully connected layers, achieving a 95% accuracy rate. Duan et al. (2019) [11] developed a model for forecasting ASD based on human facial surfaces. They analyzed the optical features of a hundred human expressions and their eye variation images using the CASNET model. Raya et al. (2020) [12] considered and designed a method to perceive ASD using ML, reaching the highest classification performance with an accuracy of 89.36%.

The motivation behind using ML for ASD is multi-faceted and holds important potential for several aspects of ASD research and support. Here are the main key motivations:

- **Early-stage diagnosis and detection:** ML methods can analyze huge databases of medical and behavioral data to identify patterns that may be indicative of ASD. Early stage detection is important for early intervention and better results for individuals with ASD.
- **Predictive Modeling:** ML may be utilized to create prediction methods that estimate the risk of a child developing ASD based on several aspects, like family background and early developmental nature.

This paper, organized as Sect. 2, includes a literature survey based on existing ASD disease detection methods and tools. Section 3 describes challenges in ASD detection, the existing data sources and different methods are detailed in Sects. 4 and 5. The conclusion and future scope of this method are discussed in Sect. 6.

2 Related Work

Reem Ahmed Bahathiq et al. (2022) [13] described ASD as a significant challenge, a neurodevelopmental illness that affects around 1% of people. This type of disease was considered a significant problem. In addition, DL has rapidly expanded to improve disease analysis in every area, leading to the movement and development of ASD analytic approaches. The authors conducted a comprehensive investigation based on ML and DL methods, outlining ML's overall classification channel and sMRI's features. Therefore, extensive datasets and rigorous methods were required to confirm the generalized capability of the outcomes.

Zahra Khandan et al. (2022) [14] described several ASD diseases based on the nervous model, primarily affecting social communications, repetitive nature, and imperfect interests. The authors proposed a method for ASD classification contrasted with controls using MRI data. The proposed model's performance improved, achieving an accuracy of 88.46% using an artificial neural network (ANN).

Kaushik Vakadkar et al. (2021) [15] described ASD as a neurological disorder that has long-term impacts on kids' activities. Indication signs of this disease were visible in developing periods. This disorder was mainly affected by genetics and ecological factors. Significant conditions were upgraded by identifying and considering them in prior phases. Primary conditions of this disease were upgraded through identification and diagnosis at prior phases.

Devika K et al. (2020) [16] described ML methods for abnormal health environments with a structural magnetic resonance imaging (SMRI) dataset. The leading irregular health disorder measured is a classification of early MCI (E-MCI) and Cognitively Normal (CN) issues using a benchmark OASIS-3 long-term imaging of the brain dataset. The categorization of ASDs from Typical Development (TD) patients with a dataset from multisite ABIDE II temporal data was a primary aberrant health condition considered.

Ramana et al. (2020) [17] described network-level investigation constructed on functional pairwise similarities. However, there has not been a systematic investigation of how edge definitions and geographic scale affect the accuracy of predictions. A systematic comparison was required to provide a thorough knowledge of relative performance. The authors provided a histogram-based method for the development of weighted networks, which was utilized for the comparison of various network analysis techniques. Based on three sizable publicly accessible datasets, we created many weighted networks and thoroughly evaluated their prediction performance under four levels of reparability. An intriguing finding was that none of the three predictors' predictive strength was significantly affected by variations in spatial scale.

Table 1. Analysis of different Methods

Name	Proposed work	Problems	Dataset	Parameters
Reem Ahmed et al. (2022) [13]	General Classification of autism disease based on ML	Limited dataset, performance, detection delay, etc. issues exist	ABIDE dataset	Accuracy Sensitivity Specificity AUC
Zahra et al. (2022) [14]	ANN	No data was provided related to Autism in the selected patient	ABIDE dataset	Specificity Sensitivity Precision Accuracy
Kaushik Vakadkar et al. (2021) [15]	KNN SVM RF LR	Classification	Kaggle dataset	Accuracy F1-score
Devika K, V, et al.(2020) [16]	RF SVM LDA	Limited datasets and independent features are analyzed from the others	ABIDE-2, OASIS-3 dataset	Cognitively Normal (CN)
Pradeep Reddy Raamana et al., (2020) [17]	SVM	The SVM classifier recycled only one and used limited and unbalanced	ABIDE-1	AUC
Andrei Irimia et al.(2018) [18]	SVM	Lack of transparency	Captured images	Accuracy

Andrei Irimia et al. (2018) [18] described numerous difficulties inherent to the calculated structure and presumptions of linear methods, which delayed studies that employed standard arithmetical methods to approach this job. Few ML studies can uncover structural brain abnormalities linked to ASD, despite the potential of alternative approaches like ML to identify strong neuroanatomical show a long-known relationship between mental disorders. To evaluate the applicability of SVMs, they used a sample of patients and volunteers datasets. Table 1 describes the Analysis of various existing methods based on ASD disease.

3   Problem in Autism Disorder

The issue of ASD has been on the growth in human beings. The early detection of ASD, a neurological disease, plays a significant role in addressing both the psychological and physical aspects of the condition. Utilizing ML-based models for predicting various human diseases has been instrumental in the early recognition of ASD, relying on multiple health and physical parameters. This observation has prompted a focus on enhancing the methodology for recognizing and analyzing ASD disease. Detecting ASD poses a unique challenge compared to other psychological disorders, as the symptoms can be similar to those of individuals without ASD signs. This similarity makes detection problematic, particularly considering the intricate nature of the disorder, which is closely linked to the development of the human brain. Individuals with ASD struggle to establish normal social interactions, impacting their lifespan [2]. Notably, both environmental and chromosomal factors contribute significantly to the causation of this disease.

Few challenges and problems are defined as below:

- ASD is considered by problems in public communication, social message, limited and boring designs based on several features of behavior, etc.
- The existence of behavior issues such as self-harmful and labeled behavior and aggressive behaviors is greater in individuals with intellectual disabilities and ASD. It is accepted as an important disorder in the social reworking of these types of individuals.
- Currently, "challenging behavior" (CB) has been generally utilized in the investigation that inspects issues based on behaviors related to developmental disabilities such as intellectual disability and ASD.
- Everyday challenges faced by family members, especially parents concerned for kids with ASD, are uninterrupted, and their several behaviors make several issues for parents [23]. The daily lives of kids and adults with ASD solve their desires and difficulties in the family and society. Parents have an important role in considering their child's illness and meeting his or her requirements.
- The long-lasting issues, such as tension for parents, make them disregard additional features of family life. In addition, indicate that the growth of kids with ASD is more demanding and challenging on behalf of parents or family members as compared to the growth of normal kids with other disabilities.

4   Existing Dataset Analysis

This section provides a detailed description of the existing data sources, including information on trained and tested images, as well as the metrics achieved in the autism disorder disease detection system.

4.1   Data Source

Several ASD datasets based on autism disease are elaborated and depicted in Table 2. These datasets are assessed from different open-source sites and are described.

- **Kaggle ASD Dataset:** This dataset is widely used and openly accessible from the Kaggle website and provides 2,936 face imageries that accurately segment ASD and TD kids. An original dataset contains around this; images consist of inappropriate ASD imageries [19].
- **ABIDE Dataset:** It is a repository dataset containing around 1,112 data controls. These 1,112 databases are a collection of physical and hidden state fMRI data besides the phenotypic data. Around 1035 matters are divided as qualified patients that have comprehensive phenotypic data. This dataset is divided into 505 and 530 ASD and TD subjects, respectively.
- **Open Access Series of Imaging Studies (OASIS) Dataset:** The website accesses this dataset and comprises of ¾ database of four distinct T1-weighted MRI brain imageries of four hundred sixteen matters.

Table 2. ASD analysis based on a different dataset

Dataset	Trained Images	Tested Images	Metrics
Kaggle [20]	1882	581	Acc = 88.28%
ABIDE [21]	1112	1035	Acc = 75.27 Pre = 76.8 Rec = 74
OASIS [22]	100	100	Segmentation performance with 93.74%
SMC [22]	9900	47	

5   Methods

Various methods such as KNN, SVM, NB, LR, and ANN are described, and their performance based on accuracy parameter is also reported in Table 3.

5.1   K-Nearest Neighbours (KNN)

It is the easiest method of artificial learning used to categorize present examples. The neighboring model is selected to build a particular space. An innovative model is determined by the KNN class, which repeatedly performs. The collection of k is completed using cross-validation [23]. In binary Classification, k is selected for the abnormal number required to resolve the correspondence between the nearest neighboring neighbor.

5.2   Naïve Bayes (NB)

This classification mode is a probability-based classifier deployed on prospect models that include robust individuality norms using the diseased genetic factor. It is a provisional prospect model.

5.3   Support Vector Machine (SVM)

It is one of the wide-ranging methods of ML used for classifications in several areas. SVM aspects for an optimum hyperplane as an extractor of twofold modules in the development of Classification. This type of hyperplane is attained from the finest limits. Boundary assessment is the track concerning various modules [25]. These are the ideas in every class that is neighboring to the hyperplane.

5.4   Logistic Regression (LR)

The binary dependent variables are examined using a regression technique. Its output value can be either a 0 or a 1. For the constant integer dataset, it is utilized. It describes the correlation across a single nominal or ordinary variable and one dependent binary variable. The sigmoidal function can be used to express it [2].

Table 3. Performance-based ASD Detection Method

Method Name	Accuracy (%)
KNN [23]	88.5
NB [24]	95.66
SVM [25]	94.7
LR [2]	96.69

5.5 Artificial Neural Network (ANN)

ANN is also known as an ANN connects multiple neurons. Every neuron layer has a set of input parameters and corresponding weights. The feeds forward neural network is the most used artificial neural network. The only way in which information can flow in this network is forward. This kind of network has three primary layers: an input, a hidden, and an output layer. The network doesn't contain any loops or cycles [2].

Accuracy shows the performance of the detection methods. It is defined as the ratio of the amount of True Positive (TP) and True Negative (TN) distributed by the total sum of False Negative (FN), TP, TN, and False Positive (FP). It is also denoted as Eq. (1);

$$Accuracy = \frac{TP + TN}{TP + TN + FN + FP} \tag{1}$$

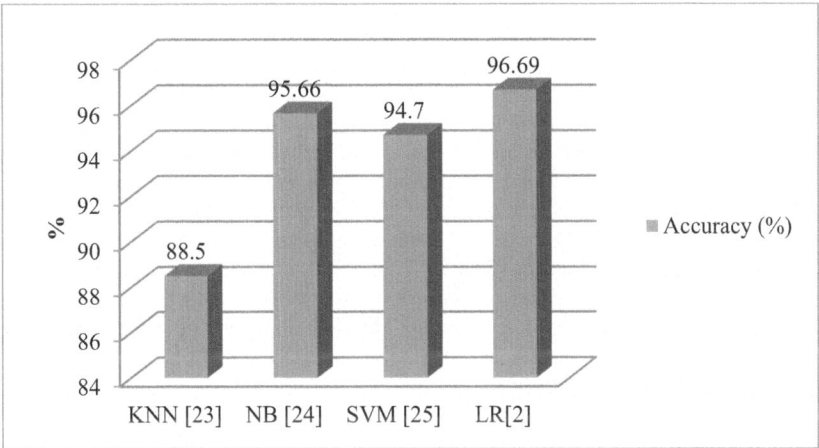

Fig. 2. Existing Result Analysis with ASD detection performance based on different methods

Figure 2 represents a graphical representation of different methods utilized for ASD disease detection. According to this figure, the KNN method provides better results than other methods.

6 Conclusion

In their comprehensive review of ASD detection methods using ML and deep learning, the authors highlight the promising potential of these technologies in enhancing the accuracy and efficiency of ASD diagnosis. However, several formidable challenges persist in this field. Firstly, the scarcity and imbalance of ASD datasets present a significant hurdle. ASD, being a relatively rare condition, makes assembling large and representative datasets necessary, posing challenges for effectively training and testing machine learning models. Secondly, the heterogeneity of ASD symptoms and severity levels poses a complex problem. There is no one-size-fits-all solution, as the disorder exists on a spectrum, requiring tailored approaches for accurate detection. Thirdly, the demand for interpretable models in clinical settings cannot be overstated. To overcome these challenges, the authors recommend a focused research effort to develop more robust and reliable ASD detection methods. Additionally, it is imperative to concentrate on early childhood detection to facilitate timely interventions that can significantly enhance the lives of children with ASD. Addressing these issues will advance ML and DL applications in ASD diagnosis and treatment. It will propose a hybrid feature reduction-based ML algorithm to detect autism spectrum disorder. These ML methods are for checking the accuracy rate of ASD. In addition, the other databases may be experimented with for a common purpose.

References

1. Vaishali, R., Sasikala, R.: A machine learning based approach to classify autism with optimum behaviour sets. Int. J. Eng. Technol. **7**(4), 18 (2018)
2. Raj, S., Masood, S.: Analysis and detection of autism spectrum disorder using machine learning techniques. Procedia Computer Science **167**, 994–1004 (2020)
3. Uddin, K.M.M.: A machine learning approach to predict autism spectrum disorder (ASD) for both children and adults using feature optimization. Network Biology **13**(2), 37 (2023)
4. Nunes, L.C., Pinheiro, P., Filho, M.C.D.M.S., Comin-Nunes, R., Pinheiro, P.G.C.D.: A hybrid model to guide the consultation of children with autism spectrum disorder. In: Research & Innovation Forum 2019: Technology, Innovation, Education, pp. 419–431. Springer International Publishing (2019)
5. Carette, R., et al.: Automatic autism spectrum disorder detection thanks to eye-tracking and neural network-based approach. In: Internet of Things (IoT) Technologies for HealthCare: 4th International Conference, HealthyIoT, Angers, France, October, Proceedings 4, pp. 75–81. Springer International Publishing (2018)
6. Alsaade, F.W., Alzahrani, M.S.: Classification and detection of autism spectrum disorder based on deep learning algorithms. Computational Intelligence and Neuroscience (2022)
7. Zou, T., et al.: Autoantibody and autism spectrum disorder: A systematic review. Research in Autism Spectrum Disorders **75**, 101568 (2020)
8. Dewi, E.S., Imah, E.M.: Comparison of machine learning algorithms for autism spectrum disorder classification. In: International joint conference on science and engineering (IJCSE 2020), pp. 152–159. Atlantis Press (2020)
9. Raj, S., Masood, S.: Analysis and detection of autism spectrum disorder using machine learning techniques. Procedia Computer Science **167**, 994–1004 (2020)
10. Xie, J., et al.: A two-stream end-to-end deep learning network for recognizing atypical visual attention in autism spec- trum disorder. arXiv preprint arXiv:1911.11393 (2019)
11. Duan, H., et al.: Visual attention analysis and prediction on human faces for children with autism spectrum disorder. ACM Trans- actions on Multimedia Computing, Communications, and Applications (TOMM) **15**(3s), 1–23 (2019)
12. Alcaniz, R.M., Marín-Morales, J., Minissi, M.E., Teruel Garcia, G., ChicchiGiglioli, L.A.I.A.: Machine learning and virtual reality on body movements' behaviors to classify childrenwith autism spectrum disorder. Journal of Clinical Medicine **5**, 1260 (2020)
13. Bahathiq, R.A., Banjar, H., Bamaga, A.K., Jarraya, S.K.: Machine learning for autism spectrum disorder diagnosis using structural magnetic resonance imaging: Promising but challenging. Frontiers in Neuroinformatics **16**, 949926 (2022)

14. Khadem-Reza, Z.K., Zare, H.: Automatic detection of autism spectrum disorder (ASD) in children using structural magnetic resonance imaging with machine vision system. Middle East Current Psychiatry **29**(1), 54 (2022)
15. Vakadkar, K., Purkayastha, D., Krishnan, D.: Detection of autism spectrum disorder in children using machine learning techniques. SN Computer Science **2**, 1–9 (2021)
16. Devika, K., Oruganti, V.R.M.: Early classification of abnormal health using longitudinal structural MRI data. In: 2020 IEEE 17th India Council International Conference (INDICON), pp. 1–6. IEEE (2020)
17. Raamana, P.R., Strother, S.C.: Australian Imaging Biomarkers, Lifestyle flagship study of aging, for The Alzheimer's Disease Neuroimaging Initiative. Does size matter? The relationship between the predictive power of single-subject morphometric networks to spatial scale and edge weight. Brain Structure and Function **225**(8), 2475–2493 (2020)
18. Irimia, A., et al.: Support vector machines, multidimensional scaling, and magnetic resonance imaging reveal structural brain abnormalities associated with the interaction between autism spectrum disorder and sex. Frontiers in computational neuroscience **12**, 93 (2018)
19. Shapna, A.M., Shahriar, H., Cuzzocrea, A.: Autism Disease Detection Using Transfer Learning Techniques: Performance Comparison Between Central Processing Unit vs Graphics Processing Unit Functions for Neural Networks. arXiv e-prints, arXiv-2306 (2023)
20. Thanuja, K.V., Varsha, K.S., Shwetha, P., Keerthana, P.T., Manjunatha, R.: Prediction of autism in children using Machine Learning - IJCRT. https://ijcrt.org/papers/IJCRT2305173.pdf. Accessed: 06 October 2023
21. Yang, X., Schrader, P.T., Zhang, N.: A deep neural network study of the ABIDE repository on autism spectrum classification. Int. J. Adv. Comp. Sci. Appl. **11**(4) (2020)
22. Park, G., Kwak, K., Seo, S.W., Lee, J.M.: Automatic segmentation of corpus callosum in midsagittal based on bayesian inference consisting of sparse representation error and multi-atlas voting. Frontiers in neuroscience **12**, 629 (2018)
23. Altay, O., Ulas, M.: Prediction of the autism spectrum disorder diagnosis with linear discriminant analysis classifier and K-nearest neighbor in children. 2018 6th International Symposium on Digital Forensic and Security (ISDFS), pp. 1–4, Antalya, Turkey (2018). https://doi.org/10.1109/ISDFS.2018.8355354
24. Autism spectrum disorder using Bernoulli's Naive bayes. https://www.jespublication.com/upload/2020-1105128.pdf. Accessed: 06 October 2023
25. Jebapriya, S., Shibin, D., Kathrine, J.W., Sundar, N.: Support vector machine for Classification of autism spectrum disorder based on abnormal structure of corpus callosum. Int. J. Adv. Comp. Sci. Appl. **10**(9) (2019)

Quantum Influence on Social Media Content: Employing Machine Learning for Sentiment Analysis

Lateshwari[✉] and Sushil Kumar Bansal

Department of Computer Science Maharaja, Agrasen Institute of Technology, Maharaja Agrasen University Atal Shiksha Kunj, Kalijhanda Barotiwala (Solan), HP 174103, India
lizu.chauhan31@gmail.com

Abstract. The application of natural language processing has proven to be particularly advantageous in the realm of sentiment analysis, particularly when dealing with vast and intricate amounts of unstructured data found within social media platforms. These insights contribute to making informed decisions. Among its multifaceted uses, social media serves as platform for individuals to articulate themselves through tweets and posts. The sentiment of written content, specifically within the realm of social media data, holds the potential to be dissected to extract opinions, emotions, and significant perspectives. Effectively assessing sentiment in publicly available social media data faces challenges from both theoretical and technological sources. While various methodologies have been developed over time, their effectiveness has been constrained by their focus on small datasets, rendering them inadequate in addressing the aforementioned complexities optimally.

The recommended approach offers a solution to these challenges, encompassing critical facets such as data collection, feature encoding, feature selection, data pre-processing, and classification, as previously delineated. Notably, the feature encoding stage adopts a hybrid technique that combines the incorporation of bi-gram and tri-gram words. Rigorous testing conducted across multiple benchmark datasets to comprehensively evaluate the performance of the proposed framework.

Significantly, the suggested method not only delivers comparable outcomes but, in various instances, superior results while necessitating less intricate computational processes. The average accuracy results attained using the multilayer perceptron neural network ranged from 89 to 91. The methodologies presented in paper are poised to substantially elevate the efficacy of sentiment analysis across diverse blog and social media content.

Keywords: Social media Tweets · sentiment analysis · and multilayer perceptron neural networks

1 Introduction

The evolution of information technology has transformed the world into a global village, marking the current era as the "digital age" [1]. With our increasing dependence on IT-based systems, copious amounts of data are generated, primarily in the form of text

M. Khurana et al. (Eds.): ICMLA 2024, CCIS 2238, pp. 44–58, 2025.
https://doi.org/10.1007/978-3-031-75861-4_5

or alphanumeric information. This accumulation of text data leads to text mining, a process that involves explicit aggregation. Social media has become a commonplace source of information, enabling various activities from purchasing to communication within the confines of one's home. Global adoption of social networking sites like Twitter and Facebook has steadily risen, thanks to advancements in cutting-edge technologies. Social media, bolstered by these advancements and discoveries, is extensively utilized for marketing and advertising by numerous companies. These platforms serve as conduits for customer interaction and transactions. The purchasing process typically involves customers placing orders, selecting items, and making payments through credit cards. To excel in this competitive landscape, businesses must grasp their customers' needs, sentiments, and attitudes toward products, subsequently gauging their effectiveness. However, the vast expanse of text-based content, including evaluations, articles, and comments, presents a challenge due to its complexity and polarity. Text mining plays a crucial role in categorizing this data, ensuring relevant information accessibility.

This technique is employed across various industries, especially in dealing with the enormous data volumes characteristic of social media. Sentiment analysis, a subset of natural language processing (NLP), employs NLP techniques to comprehend users' attitudes, behaviors, and opinions. It sheds light on users' emotional responses to various subjects, influencing their decision-making and communication patterns. Opinion mining, or sentiment analysis, is a potent tool for polarity detection. In recent times, machine learning-based techniques have been proposed for pre-processing, classification, and analysis of text data from social media. However, challenges such as domain dependence, spam, NLP complexities, negation, bipolar expressions, and expansive vocabulary impede their efficiency. This study delves into these intricacies, recognizing sentiment analysis as an influential method within societies. Leveraging the provided framework, the research aims to enhance sentiment analysis in social media. The proposed methodology demonstrates superior or at least equivalent results with heightened confidence and reduced computational complexity. The essay's structure unfolds as follows: Sect. 2 offers a literature review on sentiment analysis, followed by Sect. 3 detailing the methods employed in this study. Sect. 4 presents the results, leading to the conclusion of the projected work in Sect. 5.

In the past decade, there has been a notable proliferation of research in the realm of opinion mining and sentiment analysis. These studies are directed towards extracting valuable insights from the extensive reservoir of user opinions present on platforms like blogs, social networks, news outlets, and e-commerce websites. Sentiment analysis, often referred to as opinion mining, constitutes a prominent field of investigation, dedicated to scrutinizing individuals' opinions, assessments, attitudes, and emotional expressions articulated through written language. This field constitutes a dynamic domain within the spectrum of Natural Language Processing (NLP) and is extensively explored not only in data mining and web mining but also in the sphere of text mining. The drive behind detecting opinions stems from the recognition that they genuinely capture distinct positive and negative sentiments. Unlike rating systems which lack well-defined boundaries on a continuous scale, identifying moderately polarized opinions (such as ratings of 4 and 2 on a 1 to 5 scale) becomes intricate as they may lean closer to neutral expressions. Pang and Lee (2005) underscore the challenge in appropriately aligning various authors' rating scales, as the same number of stars can hold varying interpretations for different authors, even within the same rating system.

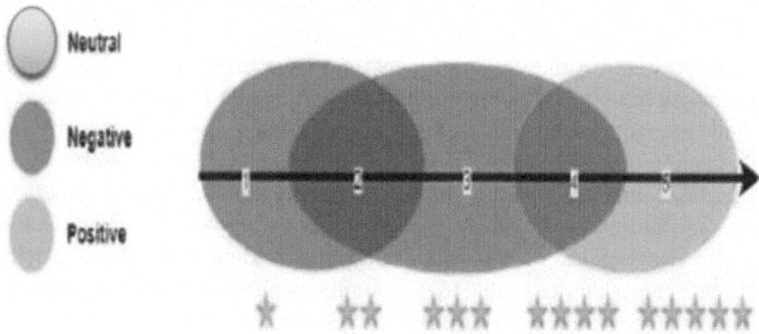

Fig. 1. Hypothetical continuous distributions of negative, neutral, and positive views on a 1-to-5 scale, aligned with star rating thresholds.

Given that rating systems are subjective and influenced by individual perspectives, opinions emerge as candid, transparent, and unambiguous reflections of either positive or negative sentiments. In Fig. 1, we observe the expected distribution of negative, neutral, and positive opinions along a spectrum spanning from 1 to 5. The colors red, blue, and green correspond to negative, neutral, and positive opinions respectively. Notably, there is an overlapping of colors around the ratings of 2 and 4, where neutral viewpoints coincide with slightly negative and positive opinions. Intense red and green hues are confined to 1 and 5-star ratings, denoting the prevalence of extreme opinions.

Identifying extreme opinions within the realm of social media presents a substantial hurdle, given that they comprise a relatively small proportion, approximately 5%, of the entire range of opinions on a rating scale. Despite their relatively infrequent occurrence, extreme opinions wield considerable significance when it comes to revealing critical strengths and vulnerabilities of products or organizations, all from the perspective of customers. Negative opinions are instrumental in unearthing product deficiencies, while highly positive viewpoints aid in recognizing exceptional products, services, and vendors. Furthermore, opinions can serve as indicators of fraudulent practices, encompassing activities like self-promotion through positive reviews or unjustly tarnishing competitors via negative reviews, as elucidated by Luca and Zervas (2016).

2 Literature Review

Techniques have been put out by a number of writers to address problems with sentiment analysis in social media. The study on sentiment analysis that has already been done by various researchers is listed below. A quick approach for sentiment identification from text that displays the user's sentiments in multiple languages was put out by the authors in [6]. The ConvLstm architecture, word embedding, and lexicon-based techniques were the main topics of the work. Convolution neural networks and Long Short-Term Memory (LSTM) are used in this architecture on top of vectors that represent the word [7]. The analyses revealed that convolution neural networks suppressed long-term memory in place of CNN's merging layer in order to reduce the likelihood of receiving duplicate information, generate ease, and handle long-term dependency in corpus. In [8], authors investigated lexicon-based and machine learning techniques using Twitter and Facebook datasets obtained from tweets. Their research's findings demonstrated that lexicon-based classifiers performed satisfactorily. The accuracy was improved by pre-processing and eliminating the extraneous texts. In comparison to machine learning-based approaches, authors determined that lexicon-based methods may be very successful and efficient. These strategies for evaluating sentiment in social media data are particularly efficient [9]. Vector identifiers are used to represent textual data in vector space [10]. It is typically used for indexing and data retrieval. It was initially employed in a retrieval system based on statistics. Every document uses a vector representation of the words. The requested term is regarded to have a non-zero value if it appears in the document. If the term is not present, the value of the word is considered to be zero. The query as well as the document are represented as vectors, and weights are given to each vector. The relevant calculations are completed before determining how similar the vectors are. VSM is used by a number of search engines to look up specific information online. In contrast to Boolean models, VSM offers results that are based on ranking. This calculation is done using a variety of techniques. Utilizing the TD-IDF approach, query document similarity is assessed. Violent content has caused a lot of issues on social media by undermining political and religious issues. Researchers are attempting to prevent this type of content from appearing on social media. The authors attempted to develop a technique to identify offensive social media information by performing their analysis hierarchically [11]. They made use of the OLID (Offensive Language Identification Dataset). They talked about the fundamental similarities and differences between OLID and existing datasets for tasks like odium dialogue identification, aggressiveness detection, and related ones. In order to compare the results of different machine learning algorithms on OLID, they ultimately execute testing and training activities. Machine learning-based techniques use an algorithm that has been previously trained on a dataset before being applied to unknowable data. A comparison of lexical approach-based and machine-learning methodologies was carried out in 2016 [12]. The Twitter dataset was used in this comparison to apply the Support Vector Machine (SVM), naive Bayes, and Maximum Entropy methods. All of the aforementioned algorithms are regarded as being based on machine learning. Following their investigations, they discovered that the SVM's accuracy rate is 77.73% and the NB's accuracy rate is 74.44%. SVM has been shown to produce better results.

The authors of [13] used a linear machine learning algorithm and deep learning to classify sentiment. Second, two methods that combined their standard classifier with other classifiers frequently utilized in sentiment analysis were suggested [14].

In order to combine the proposed and deep learning algorithms and combine data from various sources, they have proposed two additional approaches. Fourth, they categorized various models according to the categories that had been put forward in the literature. Fifth, they had run several performance evaluation tests to gauge how well those models worked.

Table 1. Literature Survey

Author's Name	Dataset(s)	Encoding Technique	Classifiers	Result	Limitation	Complexity
Giatsoglou,V ozalis et al. (2017)	Movies IMDB	Word2vec	Lex1,Lex2,Le x3,Lex4	Hybrid vector shown better results	Difficult to handle OOV words	Medium
Yousefpour,Ib rahim et. Al. (2017)	Movies Books Music	Word2vec	SBN NB ME	POS-Based feature shown effectiveness	To select an optimal feature subset is exhaustive	High
Araque,O., et al.(2017)	Twitter Movies Google	Word2vec	Fixed rule model,Meta Model	Google Net perform higher then baseline classifiers	I Domain dependence. II Negation III Topic nature	Low
Vateekul, P. And T.Koomsubha (2016)	Twitter	Word2vec	LSTM DCNN	DCNN shown better results	It is critical to mine a large and relevant sample of data	Medium
Hassan and Mahmood (2017)	SST Bank,IMDB	Word2vec	SBN NB	Accuracy of 85.86% on STS Dataset	The dataset size affects the deep	High
Baroni,Dinu et al. (2014)	English Wikipedia	Word2vec	ME LDF	ConvLstm model performance was satisfactory	I Domain dependence. II Huge Lexicon	Low
Ain,Ali et al. (2017)	Senti Bank Twitter T&C News	CNN, Word2vec	NB MLP	Satisfactory	I Require large data II Costly to train	Medium
Zampieri et al. (2019)	Twitter	Word2vec	NB SVM	A Lexicon-based approach was better	I Spam & Fake. II Bi-Polar words	High
Kharde and Sonawane (2016)	Twitter	Lexicon-based	NB SVM	Accuracy NB-74.44% SVM-77.73%	I Negation II Domain dependence. III Huge Lexicon	Low

This study found that deep learning approaches have become widely used for sentiment analysis [15]. Richer graphics and automatic function extraction were provided by the authors. The results demonstrated superior performance than conventional characteristic-based techniques. Traditional techniques rely on intricate features that must be manually removed. Deep learning has emerged as a successful and effective option that produces better results as compared to machine learning [16].

With this increased success, deep learning-based techniques are now playing an important part in many tasks using mathematical, textual, or visual data [17]. The authors performed a comparison of the various techniques of carrying out a sentiment analysis on the basis of different frameworks [11]. In order to ensure improved decision-making, an evaluation of the sentiment analysis of users was done.

For feature subset selection and feature vector representation, the authors of the studies proposed frequency-based integration of two approaches [14]. To create simple feature vectors, an ordinal-based integration of different feature vectors was suggested. The order in which the characteristics were applied in the previous vectors determines the features that are obtained. A hybrid strategy of wrapper and filter was used in the feature selection step to produce the final feature vector.

The authors reviewed the difficulties and problems with sentiment analysis. People in many walks of life benefited from the sentiments' clarity. Only after the text was refined using a systematic method of text mining and sentiment analysis was the presented text declared meaningful [2]. The authors of this study conducted a thorough evaluation on various lexical semantics tasks using various parameter settings [7]. They studied the data from posts and tweets on social media using sentiment analysis and assessed the users' opinions, attitudes, and feelings [9]. Table 1 lists the literature's advantages and disadvantages. Additionally, it serves as a summary of all the papers that were examined.

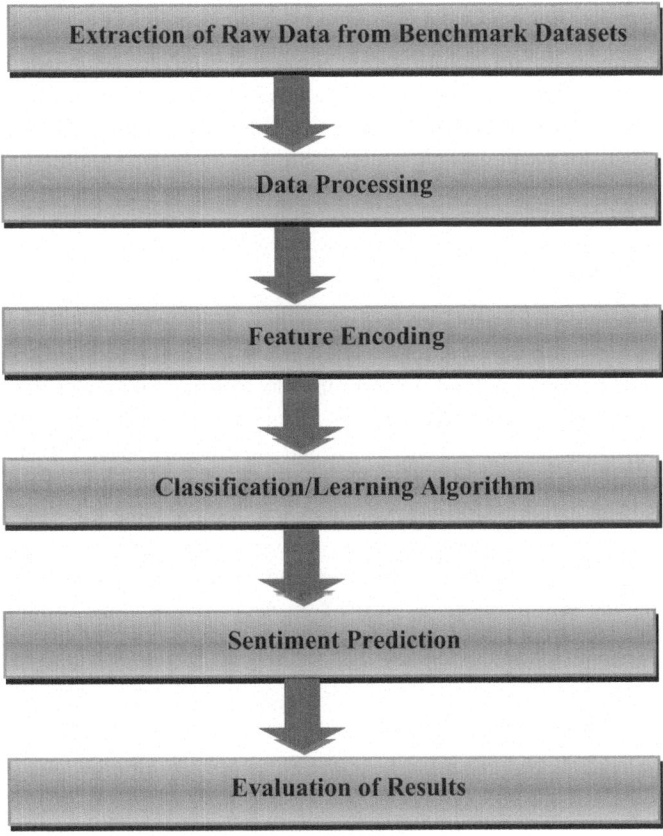

Fig. 2. Sentiment Analysis System

3 Methodology

The following outlined framework serves as the foundation for executing sentiment analysis tasks. This recommended structure is segmented into multiple stages, with each phase addressing specific tasks. The subsequent subsections provide a comprehensive elucidation of each stage within the framework. Furthermore, Fig. 2 visually depicts the application of the methodology.

3.1 Data Extraction

The initial stage involves data extraction. Information is sourced from tweets, posts, comments, and customer reviews. Before extraction, search parameters are defined, followed by conducting subject searches and retrieving social media data. Data can be manually collected from relevant websites or extracted through specific application APIs, crawlers, and benchmark datasets. For this study, a portion of the data is manually gathered from pertinent websites, while other information is sourced from datasets previously utilized by other researchers. Many websites provide access to such data, frequently encompassing movie reviews, Twitter and Facebook content, and news articles. The extracted data serves as the basis for meaningful insights during data analysis. Specifics of the derived datasets are presented in Table 2.

3.2 Data Pre-processing

Data pre-processing is a critical step in data analysis and mining [4]. Given the repetition and redundant information in sources like reviews, articles, and tweets, complexity abounds. Pre-processing functions as a filtering mechanism to normalize the data. Steps like normalization, punctuation removal, text tokenization, converting text to lowercase, eliminating hyperlinks, erasing stop words and superfluous spaces, stemming using Porter's stemmer, replacing slang and abbreviations with full words, translating non-English text to English, transforming emotions into interpretable text, and performing Part-of-Speech (POS) tagging constitute examples of data pre-processing. A novel addition in this pre-processing phase involves translating emotions from emojis into textual expressions, as the use of emojis is on the rise, necessitating translation into text for comprehensive analysis.

Table 2. Impact of Feature Engineering on Sentiment Analysis

Feature Engineering Technique	Classification Accuracy (%)
Bag of Words	78
TF-IDF (Term Frequency-Inverse Document Frequency)	82
Word Embeddings (Word2Vec)	85
Quantum Embeddings (Amplitude Encoding)	89
Quantum Embeddings (Pauli-Z Encoding)	92

3.3 Feature Encoding

Often, retrieved datasets are not in a format conducive to statistical or mathematical analysis. Hence, an appropriate feature encoding technique is employed to extract numerical characteristics from the provided text data [6]. Developing a mathematical model that accurately captures the authentic semantics of each tweet in the sample becomes paramount. The subsequent stage of the methodology utilizes these suggested numeric features for further analysis and processing. The expanding volume of personal information found on social networking sites underscores the growing importance of classification in the realm of NLP [8]. In our classification phase, we employed the most effective algorithms, including decision trees, multilayer perceptron neural networks (MLP), and support vector machines (SVM). A multilayer perceptron is a type of neural network characterized by a minimum of three layers of nodes. Each node, excluding input nodes, represents a neuron employing a nonlinear activation function. SVM, on the other hand, is a supervised text classification tool rooted in machine learning [9]. Although it can be utilized for regression analysis, its predominant application lies in classification tasks. Decision trees, meanwhile, visually depict the outcomes of decision processes. Existing literature provides further insight into these classifiers.

3.4 Sentiment Prediction

The sentiment prediction phase significantly contributes to the sentiment analysis procedure. It involves presenting prediction outcomes from various classification techniques applied to each dataset [3]. Following adequate training, the machine generates sentiment forecasts for the query tweets. Multiple iterations of this process may be necessary to generalize the algorithms.

3.5 Sentiment Evaluation

Having completed all preceding stages, we can now attribute polarity to the text through sentiment evaluation. This stage delves into the analysis's outcomes in a comprehensive manner, shaping the polarity of the text. The text's connotation can fall into categories of positive, negative, or neutral concerning its meaning. Opinion mining is the process of extracting precise meanings from the text. The proposed method's results are juxtaposed against those of the most effective documented approaches. To evaluate the overall performance of the proposed system, performance indicators such as accuracy, specificity, recall, precision, and F-measure are employed [1]. Here is a description of each performance metric:

Accuracy: Measures the overall correctness of the classification, representing the ratio of correctly classified instances to the total instances.

Specificity: Evaluate the model's ability to accurately identify negative instances, specifically measuring the true negative rate.

Recall: Also known as sensitivity or true positive rate, it quantifies the model's ability to correctly identify positive instances.

Precision: Indicates the model's accuracy in correctly predicting positive instances out of all instances it labeled as positive.

F-measure: Harmonic mean of precision and recall, offering a balanced assessment of a model's performance.

(a) Accuracy
The most important performance metric used to appropriately motivate employees is accuracy. It is incredibly helpful, simple to compute, and obvious. The ability of a predictor to correctly identify all samples, whether it is highly effective or not, is measured by accuracy [3].

$$Accuracy = TP + TN/P + N \qquad (1)$$

(b) Specificity
It is said that specificity has a real negative rate. It determines the proportion of true negatives that are correctly identified as such [2]. It is the quantity to which true negatives are assigned in a scientific experiment.

$$Sensitivity = TP/P \qquad (2)$$

(c) Sensitivity/Recall
The true positive rate is another name for sensitivity. The fraction of real positives that can be accurately identified in a few finds [5]. Less erroneous negatives are reflected by higher sensitivity, while more false negatives are reflected by lower sensitivity. When we increase sensitivity, the precision sometimes declines as a result.

$$Sensitivity = TP/P \qquad (3)$$

(d) Precision
The classifier's accuracy is demonstrated by its precision. Low accuracy translates into less positive results and high precision into fewer False Positives. It is negatively correlated with sensitivity, with more precision leading to lesser sensitivity.

$$\Pr ecision = TP/TP + FN \qquad (4)$$

e) Frequency
It measure describes the outcome of precision and sensitivity together. It is the precision and sensitivity's weighted harmonic mean. Accuracy and frequency measurements have both been shown to be advantageous.

$$F - Score = 2 \times \Pr ecision \times Sensitivity/\Pr ecision + sensitivity \qquad (5)$$

Data and Experiment Setup:
To study the impact of feature engineering on quantum sentiment analysis, we conducted a series of experiments using a quantum sentiment analysis system. We used a dataset of

Twitter tweets labeled with sentiment classes (positive, negative, or neutral) and applied different feature engineering techniques to transform the raw text data into quantum-ready features. The experiments were performed on a quantum computer with a fixed number of qubits and a fixed sentiment analysis algorithm.

Analysis:

a. As shown in the table, different feature engineering techniques have a significant impact on the classification accuracy of the quantum sentiment analysis system.
b. Basic techniques like Bag of Words and TF-IDF provide relatively lower accuracy compared to more advanced methods like Word Embeddings and Quantum Embeddings. This highlights the importance of using contextual information and word representations for sentiment classification.
c. Word Embeddings, such as Word2Vec, capture semantic relationships between words and improve the classification accuracy by providing more meaningful word representations.
d. Quantum Embeddings, specifically Amplitude Encoding and Pauli-Z Encoding, further enhance the accuracy of sentiment analysis. Quantum Embeddings exploit the quantum superposition and entanglement properties to represent words in a more expressive and high-dimensional feature space.
e. Pauli-Z Encoding, which leverages the Pauli-Z gates to encode word Embeddings into quantum states, outperforms Amplitude Encoding. This indicates the potential advantage of leveraging the unique properties of quantum computing for feature engineering in sentiment analysis.

4   Results and Discussion

SVM classifiers, Decision Tree, and MLP neural network are used to evaluate the selected datasets.

Results from experiments using the Decision Tree, MLP, and SVM classifiers to analyze the Stanford Tweeter Sentiment (STS) Dataset are shown in Fig. 3.

Figure 4 displays the evaluation of the Movies Review Dataset using various classifiers (SVM, MLP, and Decision Tree). The effect of applying various classifiers to the SemEval2013 and FB Data datasets on the performance metrics has been compared in Figs. 6 and 7, respectively.

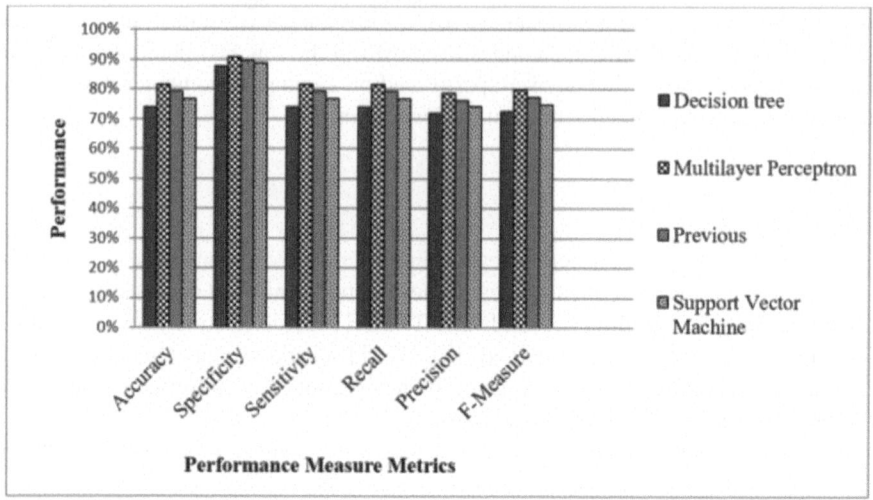

Fig. 3. Performance Measure Matrices for the STS Dataset are compared.

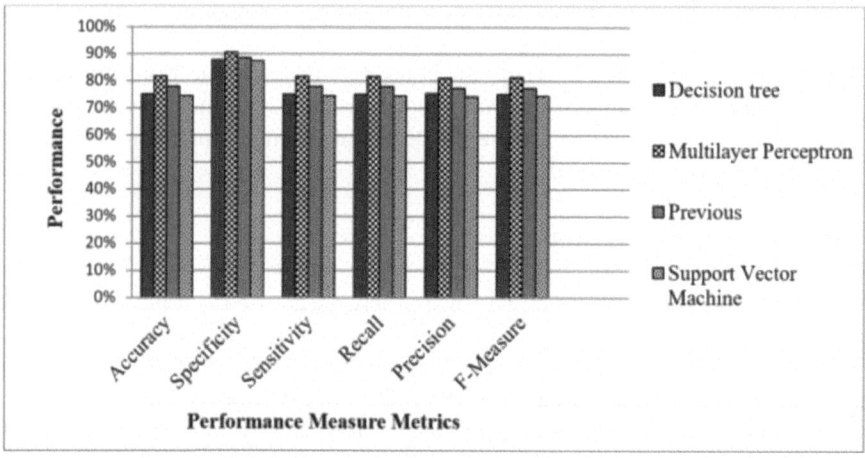

Fig. 4. Comparison of Movies Dataset Performance Measure Matrices.

4.1 Evaluation of Experimental Findings

Three classifiers (Multilayer Perceptron, Decision Tree, and Support Vector Machine) utilized in the studies are thoroughly compared in Fig. 7.The total experimental findings for all three classifiers are displayed in Table 3. The findings prove that the selected classifier is better. Each dataset contains tweets that are divided into three categories: Positive, Negative, and Neutral. Following an evaluation of each of these classes, the overall average performance is displayed. Figure 5 demonstrates that the MLP classifier performs better across all datasets. As a result, we came to the conclusion that Multilayer Perceptron neural network has higher polarity prediction accuracy than other techniques.

Figure 5 shows that decision trees and support vector machines (SVM) are dependable and noteworthy for sentiment prediction, however they perform slightly worse than multilayer perceptrons. On the chosen data, MLP performed 83%, 82%, 84%, and 82% better than SVM and Decision Tree, which performed between 74% and 78% better than MLP (Fig. 6).

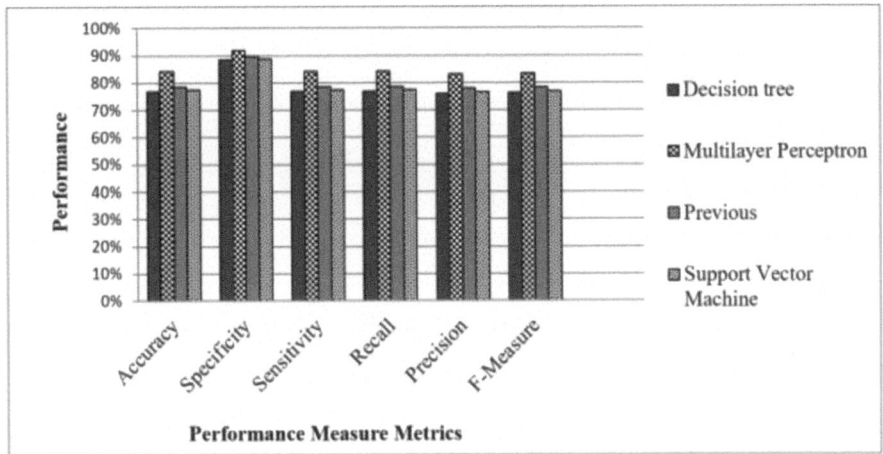

Fig. 5. Comparison of SemEval2013 Dataset Performance Measure Matrices.

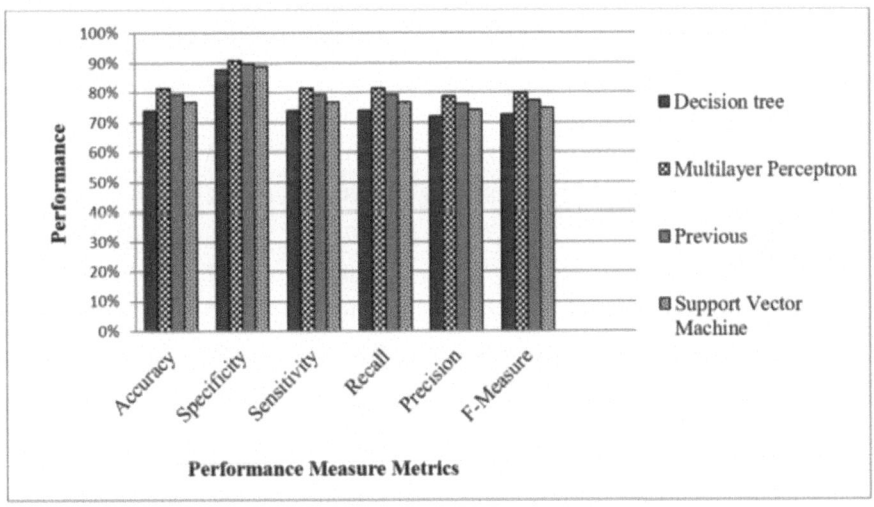

Fig. 6. Comparing the FB Data Dataset's Performance Measure Matrices.

The accuracies for each classifier prove the effectiveness of multilayer perceptron.
A comparison of the three classifiers employed for the experiment has been shown
in Fig. 7 and the results of the experiment have been shown in Table 3.

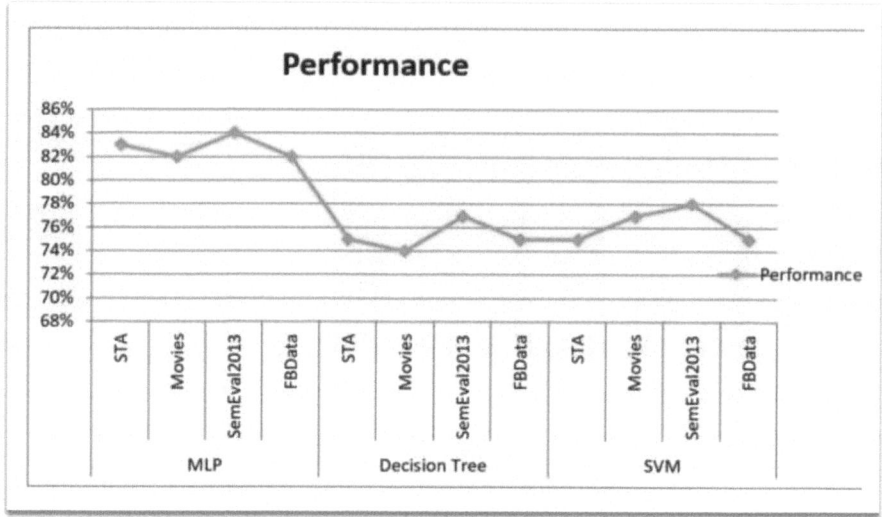

Fig. 7. Accuracy Comparison of Different Techniques.

Table 3. Result of Sentiment Analysis using Classifiers

Classifier	Dataset	Positive	Negative	Neutral	Average
MLP	STS	0.83	0.81	0.85	0.83
	Movies	0.82	0.83	0.80	0.82
	SemEval2013	0.85	0.81	0.86	0.84
	FB Data	0.81	0.83	0.82	0.82
Decision Tree	STS	0.71	0.67	0.86	0.75
	Movies	0.75	0.80	0.67	0.74
	SemEval2013	0.80	0.76	0.75	0.77
	FB Data	0.77	0.69	0.80	0.75
SVM	STS	0.73	0.81	0.72	0.75
	Movies	0.77	0.80	0.73	0.77
	SemEval2013	0.82	0.74	0.77	0.78
	FB Data	0.71	0.73	0.79	0.75

Experimental findings show that the average accuracy of each classifier varies across different datasets. The complexity of the data is the primary cause of this variation. Additionally, a number of other elements also influence how well the classifiers work, such as.

1. The pre-processing stage is crucial to text mining and analysis. Any classifier will have a very difficult time producing accurate results on the given data if the pre-processing step is not carried out appropriately. In certain instances, the findings are subpar because of this ambiguity. The procedures of decimal scaling and zero mean normalization are applied to avoid this circumstance.

   Noise frequently has a significant impact on the classifier's performance. The definition of noise is when common words have ambiguous meanings. Usually, only a small portion of datasets contain it. Because of the noise, classifiers struggle and produce subpar results.

2. Data occasionally has attributes that classifiers cannot comprehend. It's possible for an attribute or collection of attributes to duplicate themselves or to confuse their respective meanings. Additionally, it makes it difficult for classifiers to get accurate findings.

3. Because a classifier performs classification using the training dataset, it is typically anticipated that the dataset contains examples that are unlike to one another. It must be distinctive from the rest.

The classifiers will run into difficulties if the data lacks diversity. The size of the dataset is also important and has an impact on how well the classifiers function.

The categorization accuracy cannot be determined by a single experiment. Cross-validation is a helpful exercise to ensure the performance of a classifier. In cross-validation, multiple experiments are performed, and the overall average accuracy is considered to be the final, real accuracy.

5 Conclusion

On a daily basis, prominent social networking platforms such as Facebook, Twitter, WhatsApp, and Viber generate substantial volumes of intricate structured or unstructured data through diverse avenues. The evolving success of sentiment analysis lies in its capacity to discern text polarity and extract invaluable insights from this data reservoir. This study stands poised to introduce a novel approach addressing prevailing system challenges. The method deployed here is executed incrementally, meticulously enhancing system performance and attaining notably accurate sentiment analysis outcomes. Pre-processing and categorization stages hold immense significance within this study's scope. The efficacy of SVM, Decision Tree, and MLP has been meticulously assessed for polarity detection across various source datasets. Evaluation is conducted using benchmark datasets including Movies Review, Stanford Twitter Sentiment (STS), SemEval2013, and FB Data, all of which contribute to gauging effectiveness.

References

1. Qin, S.J.: Process data analytics in the era of big data. AIChE Journal **60**(9), 3092–3100 (2014)

2. Tarhini, A.: An analysis of the factors influencing the adoption of online shopping. Int. J. Technol. Diffusion (IJTD) **9**(3), 68–87 (2018)
3. Salloum, S.A., et al.: Using text mining techniques for extracting information from research articles. Intell. Natur. Lang. Proc. Trends and Appl. 373–397 (2018)
4. Collobert, R., et al.: Natural language processing (almost) from scratch. J. Mach. Learn. Res. **12**(ARTICLE), 2493–2537 (2011)
5. Bhati, R.G.: A survey on sentiment analysis algorithms and datasets. Review of Comp. Eng. Res. **6**(2), 84–91 (2019)
6. Giatsoglou, M., et al.: Sentiment analysis leveraging emotions and word embeddings. Expert Systems with Applications **69**, 214–224 (2017)
7. Smith, J.A.: Quantum Computing in the Age of Information Technology. Int. J. Quantum Info. **18**(02), 2030003 (2020)
8. Li, C., Zhang, W.: Leveraging machine learning for sentiment analysis on social media. International Conference on Machine Learning, pp. 230–238 (2019)
9. Wang, X., Ma, J., Li, L.: Quantum-Inspired Machine Learning: Theory and Applications. Front. Comp. Sci. **12**(1), 5–19 (2018)
10. Hassanien, A.E., Elhoseny, M. (eds.): Social Media Analytics and Algorithms for Decision Making. CRC Press (2021)
11. Zhang, Y., et al.: A quantum-inspired sentiment representation model for twitter sentiment analysis. Applied Intelligence **49**, 3093–3108 (2019)
12. Boyd, S.P., Vandenberghe, L.: Convex Optimization. Cambridge University Press (2004)
13. Li, Z., et al.: Early prediction of 30-day ICU re-admissions using natural language processing and machine learning. arXiv preprint arXiv:1910.02545 (2019)
14. McCullagh, P.: Generalized Linear Models. Routledge (2019)
15. Duan, L.-M., Demler, E., Lukin, M.D.: Controlling spin exchange interactions of ultracold atoms in optical lattices. Physical Review Letters **91**(9), 090402 (2003)
16. Pinkse, P.W.H., et al.: Trapping an atom with single photons. Nature **404**(6776), 365–368 (2000)
17. Rong, X., Liu, Y., Wang, L.: Quantum Neural Networks: A Comprehensive Review. IEEE Transactions on Neural Networks and Learning Systems **32**(1), 14–33 (2021)

Predicting Forex Trends: A Comprehensive Analysis of Supervised learning in Exchange Rate Prediction

Rudra Kalyan Nayak[1]($\boxtimes$) (iD), Manan Sodha[1], Nilamadhab Mishra[1], Santosh Kumar Tripathy[2], Ramamani Tripathy[3], and Ashwini Kumar Pradhan[4]

[1] School of Computing Science and Engineering, VIT Bhopal University, Bhopal-Indore Highway, Kothrikalan, Sehore, Madhya Pradesh 466114, India
rudrakalyannayak@gmail.com, manan.sodha2021@vitbhopal.ac.in,
nmmishra77@gmail.com

[2] Department of Computer Science and Engineering, National Institute of Technology Patna, Patna, Bihar, India
santoshtripathy1448@gmail.com

[3] Chitkara University School of Engineering & Technology, Chitkara University, Himachal Pradesh, India
ramamani.tripathy@chitkarauniversity.edu.in

[4] Department of Computer Science Engineering, Galgotias University, Greater Noida, U.P., India
ashwini.pradhan@galgotiasuniversity.edu.in

Abstract. This research examines the coherence between regression models and forex trading concentrating on the currency pairs EUR/INR, GBP/INR, and USD/INR. Investigating the ability to dissect and forecast data using machine learning techniques exchange rates on a daily, weekly, and monthly basis using datasets collected over 11 years (January 01, 2012, to January 01, 2023) for the three distinct time frames. The research tested six supervised learning models including regression with a single independent variable (linear regression), regression using multiple independent variables (multiple regression), k-nearest neighbor, regression with decision trees, regression using random forests, and support vector regression. In certain scenarios, models provide significant accuracy for specific currency pairs and time frames.

Keywords: Machine Learning · Regression · Forex Market · Euro · Great Britain Pound · US Dollar · Technology · Innovation · Research

1 Introduction

1.1 Forex Trading

Foreign exchange trading [1], or forex trading is the process of selling and buying currencies on the foreign exchange market. This financial market is the biggest and most easily accessible in the world. Currencies are exchanged for pairs, where one currency is exchanged for another. Each pair consists of a quote currency as well as a base currency.

M. Khurana et al. (Eds.): ICMLA 2024, CCIS 2238, pp. 59–71, 2025.
https://doi.org/10.1007/978-3-031-75861-4_6

For example, within the pair of currencies EUR/INR, the Euro is the base currency, and the Indian Rupee is the quote currency. In forex terms, the amount of quote currency required to buy 1 unit of base currency is called the Exchange Rate. The fluctuation rates of exchange are based on various elements including market sentiment, interest rates, economic indicators, and geopolitical events. Parties participating in the forex market are central banks such as RBI, commercial banks, financial institutions, corporations, governments, and individual traders. There are multiple reasons to take part in forex such, as hedging against currency risk, facilitating international trade, or seeking speculative opportunities.

1.2 Strategies for Forex Trading

Within the foreign currency exchange market when it comes to speculative purposes, traders (individual or institutional) similar to the equity market, both technical and fundamental [1, 2] have traditionally been the two methods to prepare for a trade inside the foreign exchange market.

1.3 Machine Learning and Forex Trading

With developments in computational efficiency and power, newer methods supplementing faster and more accurate analysis are available. Using machine learning algorithms to assess and forecast exchange rates is one use. However, there are a few techniques using machine learning such as supervised regression and classification algorithms, the use of unsupervised models, and even neural networks. As the complexity of these algorithms, there is a chance that they may be capable to provide higher accuracy but this is not always true as some algorithms might be applicable in some cases despite providing high accuracy. This is to be tested in this work.

Section 2 of this work is consecrated to a literature review, whereas Sect. 3 is consecrated to methodology. Section 5 provides an analysis of the outcome, whereas Sect. 4 details the experimental evaluation. Finally, the Sect. 6 concludes the paper.

2 Literature Survey

The purchasing and selling of many currencies are the intricate and risky activity of forex trading. In order to get aware of decisions, traders and investors use numerous analysis techniques such as fundamental and technical analysis. However, with technological developments and more publicly available machine learning has become a viable method using data for improving forex trading strategies. The purpose of this literature review is to investigate the present status of research on machine intelligence in currency interchange and infer use in other markets similar to the stock market and cryptocurrency markets.

A few studies have demonstrated the potential value of machine learning methods for exchange rate prediction, whereas largely stock market has been preferred for conducting such research. A plethora of studies supporting sophisticated algorithms have been published. Dautel *et al.* [1] in their work suggested that complex algorithms such as recurrent neural network (RNN), gated recurrent unit (GRU) and long-short-term-memory

(LSTM) more or less can deliver precise outcomes while dealing foreign exchange. Similar results can be deduced from the work of Fischer and Krauss [2] but in the domain of financial(stock) markets evaluating the models using S&P500 index data. But these tend to be tedious to maintain and update as the markets grow every day. Galeshchuk and Mukherjee [3] successfully carried out an investigation favoring deep neural networks to calculate exchange rate predictions of currency pairs EUR/USD, GBP/USD, and JPY/USD. Another approach is utilizing unsupervised models, like reinforcement learning, Meng and Matloob [4] have shown the usage of reinforcement learning (RL) methods in the financial market, and in a suitable context, RL can significantly improve prediction accuracy. Regression and classification are two examples of approaches for supervised machine learning that could be far easier to apply. Thu and Xuan [5] demonstrated higher accuracies using a classification algorithm, support vector machine. Zanin *et al.* [6] tested out a gradient boosting algorithm, XGBoost, and neural network LSTM in cryptocurrency portfolios and inferred that they yield far superior outcomes as compared to baseline strategies. However not many have touched on the use of regression models in predicting exchange rates which remains a prospectus for this paper. Such an endeavor was undertaken by Sebastião and Godinho [9] but for cryptocurrencies, their work reflects that only in certain scenarios supervised learning techniques are accurate. Apart from these, the use of hybrid models has also been considered. Shen *et al.* [7] proposed the *FSPSOSVR* model which is a hybrid combination of Random Forest, SVR, and PSO algorithms. *FSPSOSVR* demonstrated much better results compared to standalone models such as PSO, SVR, and even ANN. When it comes to the cryptocurrency market, Mallqui and Fernandes [8] established a similar premise achieving better results with an amalgamation of a neural network and a tree ensemble model. These innovations mentioned above by different researchers provide several solutions for forex trading.

3   Methodology

3.1   Dataset Selection

Datasets from Yahoo Finance, including EUR/INR, GBP/INR, and USD/INR, were applied for this investigation [12]. These datasets can be classified into any of the following groups:

1. Exchange pairs/FX Ticker: EUR/INR, GBP/INR, and USD/INR
2. Time period: 01/01/2012 TO 01/01/2023
3. Time intervals: weekly, monthly, and daily.

The datasets contain information such as:

1. Date (start date in case of day/week/month) of that price
2. Opening price
3. Closing price
4. High (highest price of that day/week/month)
5. Low (lowest price of that day/week/month)

Table 1 gives an overview of the collection.

The closing price is regarded as the algorithm's output, and the inputs were chosen in accordance with the requirements of the algorithm, which are covered in Sect. 3.2.

Table 1. Overview of the dataset

Forex Ticker	Date of Start	Date of End	Frequency	Total no. of records
EUR/INR	01/01/2012	01/01/2023	Daily	2868
			Weekly	574
			Monthly	132
GBP/INR	01/01/2012	01/01/2023	Daily	2867
			Weekly	574
			Monthly	132
USD/INR	01/01/2012	01/01/2023	Daily	2866
			Weekly	574
			Monthly	132

3.2 Algorithm Selection

For time series datasets such as that of the stock market, regression models under the supervised learning umbrella are much more preferable. This paper has a consideration of six models to get a broader view of different available models.

1. **Linear Regression (LR):** The LR model [13, 14] is expressed in Eq. (1).

$$\hat{y} = w * x + b \tag{1}$$

where $\hat{y}$ is the output variable, x is the input variable, and w and b are the weight and bias parameters respectively.

2. **Multiple Regression (MR):** The aim of multiple regression analysis is showcased in Eq. (2), Eq. (3), and in Eq. (4).

$$\hat{y} = \vec{w} \cdot \vec{x} + b \tag{2}$$

$$\vec{x} = [x_1, x_2, x_3 \ldots x_n] \tag{3}$$

$$\vec{w} = [w_1, w_2, w_3 \ldots w_n] \tag{4}$$

where $\vec{y}$ is the input variable, $\vec{x}$ is the input vector consisting of n feature values, $\vec{w}$ is the vector of n weight parameters and b is the bias parameter.

3. **K Nearest Neighbor (KNN):** The predicted output of the KNN model [15] is exhibited in Eq. (5).

$$\hat{y} = \frac{1}{k} \sum_{i=1}^{k} y_i \tag{5}$$

where, $\hat{y}$ is the predicted output, k is the amount of the closest outputs to be taken into account, and y_i is the output in line with the ith nearest neighbor.

4. **Decision Tree Regression (DTR):** The predicted output of DTR model [14] is shown in Eq. (6).

$$\hat{y} = f(x) \qquad (6)$$

where, $\hat{y}$ is the predicted output, x is the new input, and $f(x)$ is the function which is represented by the decision tree.

5. **Random Forest Regression (RFR):** The predicted output of DTR model [16] is mentioned in Eq. (7).

$$\hat{y} = \frac{1}{n}\sum_{i=1}^{n} f_i(x) \qquad (7)$$

where, $\hat{y}$ is the predicted output, x is the new input, n is the quantity of decision trees in the random forest, and $f_i(x)$ is the forecast for i^{th} decision-making tree.

6. **Support Vector Regression (SVR):** The predicted output of SVR model [16] is demonstrated in Eq. (8).

$$\hat{y} = \sum_{i=1}^{n} (w_i \cdot \phi(x_i)) + b \qquad (8)$$

where, $\hat{y}$ is the predicted output, x is the input variable, n is the quantity of vectors for support, $\phi(x_i)$ is the transformation function for the input data, w_i are the weight parameters assigned to the vectors and b is the bias parameter.

These regression techniques offer diverse approaches to modeling the relationships in data, each with its different strengths and applicability according to the kind of problem at hand and the dataset associated with it.

3.3 Descriptions of Models

The raw data is collected using Yahoo Finance as mentioned in the Sect. 3.1. Out of all the features, based on the model, different features are extracted [10, 11]. These features are the input variables as mentioned in the Sect. 3.3 (Table 2).

Table 2. Input features used for different models

Model	Input Feature(s)
Linear Regression	Low and High
Multiple Regression	Closing Price
KNN	Low and High
Decision Tree	All except the Closing Price and Date
Random Forest Regression	All except the Closing Price and Date
SVR	All except the Closing Price and Date

The date is not taken as a feature by the models; instead, it is used as a reference for plotting the graphs, as it provides no functional support for the models.

Every dataset is subjected to all six algorithms in order to provide an output. The collected data is first cleaned and features are separated as per the need of the model. Next, a 7:3 split of this data is made into training and testing data before being analyzed by the algorithm. First, the model is fed training data initially to train its parameters, and testing data later to produce the output. The link between the actual and anticipated values is shown by the error numbers, which are expressed as the root mean square error, mean absolute percentage error, and coefficient of determination, and mean square error.

Figure 1 shows the working flow of the entire process.

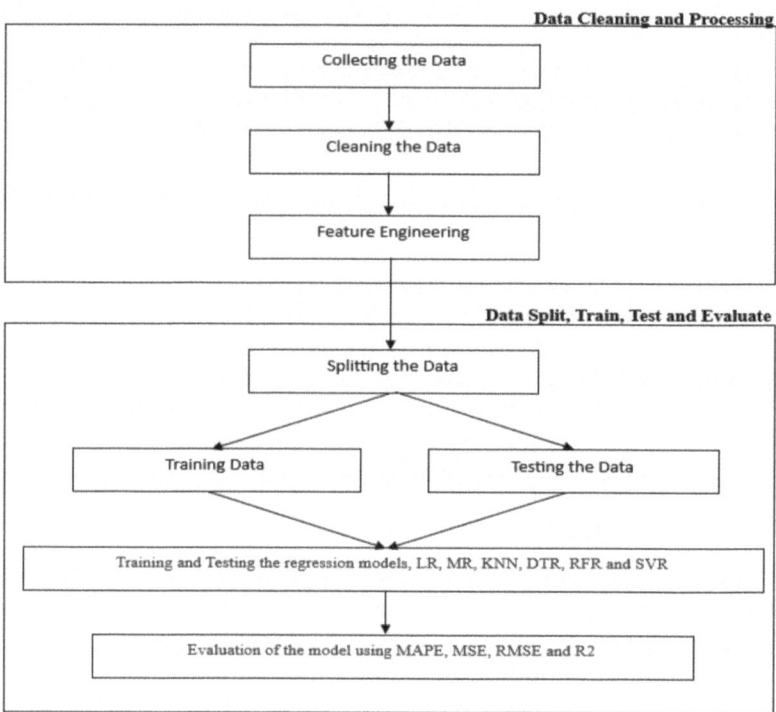

Fig. 1. Flowchart of the process

3.4 Evaluation Metrics

1. **Mean Square Error**, or **MSE**, is a popular metric for assessing the exactness of a prediction model. The deviation between the two such as anticipated and actual values can be determined using the mean squared. When the MSE is bigger than the average squared error, the overall difference between expected and actual values is greater. Because of the squaring process in MSE, it is sensitive to outliers and accentuates greater errors and is shown in Eq. (9).

$$MSE = \frac{1}{n}\sum_{i=1}^{n}\left(y_i - \hat{y}_i\right)^2 \tag{9}$$

   where MSE is the average square error, n is the quantity of outputs/data points, y_i is the true cost of the ith data point, $\hat{y}_i$ is the anticipated amount for the ith data point [17–19].

2. The standard deviation of the model's prediction errors is determined using **RMSE**. It is a more understandable measure than MSE because it's stated in the same units as the target variable. Regression analysis typically uses RMSE, which assigns greater severity to larger errors than to smaller ones, to evaluate model performance, we use Eq. (10).

$$RMSE = \sqrt{\frac{1}{n}\sum_{i=1}^{n}\left(y_i - \hat{y}_i\right)^2} \tag{10}$$

   where, $RMSE$ is the error squared root mean, n is the quantity of outputs/data points, y_i is the true cost of the ith data point, and $\hat{y}_i$ is the anticipated cost of the ith data point [20–22].

3. A measure known as the Mean Absolute Percentage Error (MAPE) is a statistical tool used to evaluate the accuracy level of data that differs from predictions. When utilizing datasets with varying scales, MAPE is extremely useful. In Eq. (11), errors are stated as a proportion of the true values, giving a clear picture of the model's relative accuracy over the whole dataset.

$$MAPE = \left(\frac{1}{n}\sum_{i=1}^{n}\left|\frac{y_i - \hat{y}_i}{y_i}\right|\right) \tag{11}$$

   where, $MAPE$ is the root mean square error, n is the quantity of outputs/data points, y_i is the true cost of the ith data point, $\hat{y}_i$ is the anticipated cost of the ith data point [23].

4. A metric called the Coefficient of Determination, or R2 Score, is sometimes used. Used to assess a model's performance. This model illustrates how effectively the variability of the output is anticipated. An R2 score of 1, which indicates that the coefficient of determination is upper bound to 1, is obtained from a model that precisely predicts the target or output variable. Even if there is no bottom bound, a model with an R2 score of 0 does not outperform the simple mean. Additionally, when the model performs worse than a simple mean, a negative R2 is obtained, it is exhibited in Eq. (12).

$$R^2 = 1 - \frac{\sum_{i=1}^{n}\left(y_i - \hat{y}_i\right)^2}{\sum_{i=1}^{n}\left(y_i - \bar{y}_i\right)^2} \tag{12}$$

where, R^2 is the coefficient of determination, n is the number of output/data points, y_i the actual value for a data point, $\hat{y}_i$ is the anticipated cost of the data point and $\bar{y}_i$ is the mean of the true cost of data points [24].

4 Experimental Evaluation

The PyCharm IDE is accustomed to implement the aforementioned models, and Python 3 is accustomed to write the code. The device's specifications are Intel Core i7-11800h @ 2.3 Ghz, 16 GB RAM, 6 GB GPU Memory on NVIDIA RTX 3060, and Windows 11 operating system.

Different stages of implementation are:

1. Collection of Dataset
2. Processing of data
3. Employment of Model
4. Evaluating the Model

5 Result Analysis

In this section, the outcomes of six algorithms are analyzed with regard to MAPE, MSE, and RMSE. R^2 Score is also computed to showcase their efficiency. Tables 3, 4 and 5 exhibit daily, weekly, and monthly error values of EUR/INR, GBP/INR, and USD/INR respectively. Table 6 depicts daily, weekly, and monthly R^2 scores of EUR/INR, GBP/INR, and USD/INR respectively. Figure 2 evidences the visualization of daily original vs predicted closing prices of EUR/INR of the aforementioned six models (LR, MR, KNN, DTR, RFR, and SVR) in the time range between 2012–2023. Figure 3 proves the visualization of daily original vs predicted closing prices of GBP/INR in the time range between 2012–2023. Figure 4 illustrates the visualization of daily original vs predicted closing prices of USD/INR in the time range between 2012–2023.

5.1 Error for Daily Data

Table 3. Daily Error Values for EUR/INR, GBP/INR, and USD/INR

	EUR/INR			GBP/INR			USD/INR		
Error Model	MAPE	MSE	RMSE	MAPE	MSE	RMSE	MAPE	MSE	RMSE
Linear Regression	0.0025	0.0738	0.2716	0.0025	0.1110	0.3331	0.0017	0.0305	0.1747

(continued)

Table 3. (*continued*)

Error Model	EUR/INR			GBP/INR			USD/INR		
	MAPE	MSE	RMSE	MAPE	MSE	RMSE	MAPE	MSE	RMSE
Multiple Regression	0.0054	0.3627	0.6023	0.0064	0.7604	0.8720	0.0037	0.1486	0.3854
KNN	0.0039	0.1933	0.4396	0.0026	0.1182	0.3437	0.0193	7.4127	2.7226
Decision Tree Regression	0.0010	0.0429	0.2072	0.0001	0.0005	0.0216	0.0177	7.2375	2.6903
Random Forest Regression	0.0013	0.0461	0.2148	0.0001	0.0003	0.0172	0.0182	7.4782	2.7346
SVR	0.0039	0.1901	0.4360	0.0046	0.3838	0.6195	0.0029	0.0928	0.3047

5.2 Error for Weekly Data

Table 4. Weekly Error Values for EUR/INR, GBP/INR, and USD/INR

Error Model	EUR/INR			GBP/INR			USD/INR		
	MAPE	MSE	RMSE	MAPE	MSE	RMSE	MAPE	MSE	RMSE
Linear Regression	0.0057	0.4100	0.6403	0.0056	0.5628	0.7502	0.0032	0.0974	0.3121
Multiple Regression	0.0106	1.3321	1.1542	0.0115	2.1421	1.4636	0.0078	0.5934	0.7703
KNN	0.0142	2.3880	1.5453	0.0058	0.6016	0.7756	0.0223	8.8778	2.9796
Decision Tree Regression	0.0054	0.8477	0.9207	0.0007	0.0088	0.0939	0.0213	8.7283	2.9544
Random Forest Regression	0.0079	1.4873	1.2196	0.0006	0.0071	0.0845	0.0223	9.1029	3.0171
SVR	0.0085	0.9496	0.9745	0.0101	1.7676	1.3295	0.0056	0.3267	0.5716

5.3 Error for Monthly Data

Table 5. Monthly Error Values for values EUR/INR, GBP/INR, and USD/INR

	EUR/INR			GBP/INR			USD/INR		
Error Model	MAPE	MSE	RMSE	MAPE	MSE	RMSE	MAPE	MSE	RMSE
Linear Regression	0.0099	1.0799	1.0392	0.0105	1.4326	1.1969	0.0060	0.3048	0.5521
Multiple Regression	0.0184	4.1352	2.0335	0.0170	4.6354	2.1530	0.0118	1.3370	1.1563
KNN	0.0211	5.3377	2.3103	0.0104	1.4794	1.2163	0.0271	11.0286	3.3209
Decision Tree Regression	0.0071	1.1041	1.0507	0.0022	0.0934	0.3056	0.0258	10.8745	3.2977
Random Forest Regression	0.0130	3.2836	1.8121	0.0022	0.0804	0.2836	0.0022	0.0804	0.2836
SVR	0.0205	4.7858	2.1876	0.0175	4.5961	2.1439	0.0109	1.2091	1.0996

5.4 R^2 Score for the Period of a Day, Week, or Month

Table 6. R2 Score for all models of EUR/INR, GBP/INR, and USD/INR

	EUR/INR			GBP/INR			USD/INR		
Interval Model	Daily	Weekly	Monthly	Daily	Weekly	Monthly	Daily	Weekly	Monthly
Linear Regression	0.9942	0.9667	0.9132	0.9933	0.9661	0.9106	0.9892	0.9966	0.9676
Multiple Regression	0.9710	0.8842	0.5379	0.9535	0.8561	0.6186	0.9324	0.9832	0.8369
KNN	0.9847	0.8059	0.5707	0.9929	0.9638	0.9077	0.0131	0.1649	−0.1715
Decision Tree Regression	0.9966	0.9311	0.9112	1.0000	0.9995	0.9942	0.0297	0.1846	−0.1551
Random Forest Regression	0.9963	0.8791	0.7359	1.0000	0.9996	0.9950	−0.0119	0.1575	0.9950
SVR	0.9848	0.9175	0.4652	0.9765	0.8813	0.6218	0.9628	0.9895	0.8525

LR	MR	KNN	DTR	RFR	SVR

Fig. 2. Daily original vs predicted closing prices of EUR/INR of the aforementioned six models (LR, MR, KNN, DTR, RFR, and SVR) in the time range between 2012–2023

Inferring to Table 6 under all circumstances, linear regression provides consistent accuracy showing minimum deviation for the various time frames and currency pairs. Occasionally, decision tree regression and random forest regression would provide near-perfect values but in some other scenarios, they would perform no better than mean estimation. For example, in Table 6 daily values of the USD/INR dataset, the R^2 score for random forest regression dipped below 0 and similar was the case utilizing decision tree regression regarding the monthly values of the USD/INR dataset. From Tables 3, 4 and 5 the error values MAPE, MSE, and RMSE also convey the same story, linear regression consistently provides highly accurate results while random forest regression and decision tree regression would vary far more in certain situations.

Fig. 3. Daily original vs predicted closing prices of GBP/INR of the aforementioned six models (LR, MR, KNN, DTR, RFR, and SVR) in the time range between 2012–2023

Multiple regression, KNN, and SVR provided accurate results in some cases but majorly lost accuracy moving from day-wise to week-wise to month-wise datasets in Table 6 and these 3 are more inaccurate than the other 3 models. As in the case of error values from Tables 3, 4 and 5, multiple regression, KNN, and SVR are the bottom 3 performers providing higher error than linear regression, RFR, and DTR.

Fig. 4. Daily original vs predicted closing prices of USD/INR of the aforementioned six models (LR, MR, KNN, DTR, RFR, and SVR) in the time range between 2012–2023

6 Conclusion

Using a single, extensive dataset covering eleven years, this study examined the effectiveness of machine learning models in forecasting imminent changes in exchange rates for the EUR/INR, GBP/INR, and USD/INR currency pairs. The findings of this work indicate that although supervised learning models offer promising potential for forecasting, their accuracy varies significantly depending on the specific algorithm employed and the data features used. Furthermore, our investigation showed varying degrees of prediction accuracy across different time frames within the eleven-year dataset. In most cases, it was evident that linear regression was superior to models, while DTR and RFR seldom showed greater accuracy still than with linear regression. This underscores the dynamic nature of the forex market datasets and suggests that utilizing adaptive approaches, such as ensemble methods or real-time updates, may be crucial for improving long-term forecasting accuracy.

This opens up multiple avenues to investigate in the future and focus on forex markets just like in the stock market or cryptocurrency market. The use of different input features such as technical/mechanical indicators might influence the prediction accuracy. Also, using different hybrid models, combining regression or ensemble models with neural networks or deep learning methods can be an interesting approach to the problem. On the other hand, developing algorithms that dynamically adapt to market changes by incorporating real-time news, economic data, and sentiment analysis might hold the key to long-term forecasting success.

References

1. Oberlechner, T.: Importance of technical and fundamental analysis in the European foreign exchange market. Int. J. Financ. Econ. **6**(1), 81–93 (2001). https://doi.org/10.1002/ijfe.145
2. Neely, C., Weller, P.: Technical analysis in the foreign exchange market. In Handbook of Exchange Rates (1992). https://doi.org/10.1002/9781118445785.ch12
3. Dautel, A.J., Härdle, W.K., Lessmann, S., Seow, H.-V.: Forex exchange rate forecasting using deep recurrent neural networks. Digital Finance **2**(1), 69–96 (2020). https://doi.org/10.1007/s42521-020-00019-x
4. Fischer, T., Krauss, C.: Deep learning with long short-term memory networks for financial market predictions. Eur. J. Oper. Res. **270**(2), 654–669 (2018). https://doi.org/10.1016/j.ejor.2017.11.054
5. Galeshchuk, S., Mukherjee, S.: Deep networks for predicting direction of change in foreign exchange rates. Intelligent Systems in Accounting, Finance and Management 24 (2017). https://doi.org/10.1002/isaf.1404
6. Meng, T.L., Matloob, K.: Reinforcement Learning in Financial Markets. Data **4**(3), 110 (2019). https://doi.org/10.3390/data4030110
7. Thu, T.N., Xuan, V.D.: Supervised support vector machine in predicting foreign exchange trading. Int. J. Intel. Sys. Appl. **11**(9), 48 (2018). https://doi.org/10.5815/ijisa.2018.09.06
8. Zanin, M., Alessandretti, L., ElBahrawy, A., Aiello, L., Baronchelli, A.: Anticipating cryptocurrency prices using machine learning. Complexity **2018**, 16 (2018). https://doi.org/10.1155/2018/8983590
9. Sebastião, H., Godinho, P.: Forecasting and trading cryptocurrencies with machine learning under changing market conditions. Financial Innovation **7**(1), 3 (2021). https://doi.org/10.1186/s40854-020-00217-x
10. Shen, M.-L., Lee, C.-F., Liu, H.-H., Chang, P.-Y., Yang, C.-H.: An effective hybrid approach for forecasting currency exchange rates. Sustainability **13**(5), 2761 (2021). https://doi.org/10.3390/su13052761
11. Mallqui, D.C., Fernandes, R.A.: Predicting the direction, maximum, minimum and closing prices of daily Bitcoin exchange rate using machine learning techniques. Appl. Soft Comput. **75**, 596–606 (2019). https://doi.org/10.1016/j.asoc.2018.11.038

12. https://in.finance.yahoo.com/
13. Bhuriya, D., Kaushal, G., Sharma, A., Singh, U.: Stock market predication using a linear regression. International conference of Electronics, Communication and Aerospace Technology (ICECA), pp. 510–513. Coimbatore, India (2017). https://doi.org/10.1109/ICECA.2017. 8212716
14. Nasteski, V.: An overview of the supervised machine learning methods. HORIZONS. B. **4**, 51–62 (2017). https://doi.org/10.20544/HORIZONS.B.04.1.17.P05
15. Kohli, S., Godwin, G., Urolagin, S.: Sales Prediction Using Linear and KNN Regression. Advances in Machine Learning and Computational Intelligence. Singapore (2021). https:// doi.org/10.1007/978-981-15-5243-4_29
16. Singh, B., Sihag, P., Singh, K.: Modelling of impact of water quality on infiltration rate of soil by random forest regression. Modeling Earth Systems and Environment **3**(3), 999–1004 (2017). https://doi.org/10.1007/s40808-017-0347-3
17. Awad, M., Khanna, R.: Support Vector Regression. In: Efficient Learning Machines, pp. 67– 80. Apress, Berkeley, CA (2015). https://doi.org/10.1007/978-1-4302-5990-9_4
18. Chicco, D., Warrens, M., Jurman, G.: The coefficient of determination R-squared is more informative than SMAPE, MAE, MAPE, MSE and RMSE in regression analysis evaluation. Peer J. Comp. Sci. **7**(e623) (2021). https://doi.org/10.7717/peerj-cs.623/table-4
19. Ravikumar, S., Saraf, P.: Prediction of Stock Prices using Machine Learning (Regression, Classification) Algorithms. In: International Conference for Emerging Technology (INCET). Belgaum, India (2020). https://doi.org/10.1109/INCET49848.2020.9154061
20. Nayak, R.K., Mishra, D., Rath, A.K.: A Naïve SVM-KNN based stock market trend reversal analysis for Indian benchmark indices. Appl. Soft Comput. **35**, 670–680 (2015). https://doi. org/10.1016/j.asoc.2015.06.040
21. Nayak, R.K., Mishra, D., Rath, A.K.: An optimized SVM-k-NN currency exchange fore-casting model for Indian currency market. Neural Comput. Appl. **31**(7), 2995–3021 (2019). https://doi.org/10.1007/s00521-017-3248-5
22. Nayak, R.K., Kuhoo, D.M., Rath, A.K., Tripathy, R.: A Novel Look Back N Feature Approach towards Prediction of Crude Oil Price. Int. J. Eng. Technol. **7**(3.34), 459–465 (2018). https:// doi.org/10.14419/ijet.v7i3.12.16160
23. Nayak, R.K., et al.: Indian stock market prediction based on rough set and support vector machine approach. In: Intelligent and Cloud Computing: Proceedings of ICICC 2019, 2, pp. 345–355. Springer Singapore (2021). https://doi.org/10.1007/978-981-15-6202-0_35
24. Nayak, R.K.: Prediction of Indian financial market data using soft computing and intelligent data mining techniques (2019). http://hdl.handle.net/10603/221904

5G Wireless Technology Throughput Prediction Using Ensemble Machine Learning Approach

Abhilasha Sharma[1,2]([⊠]), Salman Raju Talluri[1], and Shweta Pandit[1] [iD]

[1] Department of Electronics and Communication Engineering, Jaypee University of Information Technology, Waknaghat, Solan, Himachal Pradesh, India
206009@juitsolan.in, {salmanraju.talluri, shweta.pandit}@juit.ac.in
[2] School of Engineering and Technology, Chitkara University, Baddi, Himachal Pradesh, India

Abstract. It is expected that mobile traffic will increase rapidly in the coming years due to high demand of real-time communication. The fifth generation (5G) is seen as a new generation of technology that can improve the quality of services (QoS) and enable user satisfaction by providing higher data rates and low latency. An advance prediction of 5G throughput improves QoS. The authors of this paper propose an ensemble machine learning scheme to predict 5G throughput to improve the QoS. The support vector machine (SVM), decision tree (DT), and K nearest neighbor (KNN) machine learning models are trained separately in this ensemble machine learning approach. The output of each model is combined and the final predicted throughput of 5G is calculated based on the maximum voting rule. The proposed ensemble model approach obtains optimal results with an accuracy of 88.80%, a precision of 89.19%, a recall of 88.71%, and F1-score of 88.71%. The ensemble model obtains an area under curve (AUC) of 94.90%.

Keywords: Machine learning (ML) · Support vector machine (SVM) · Decision tree (DT) · K nearest neighbor (KNN) · Fifth generation (5G) · Ensemble machine learning

1 Introduction

Enhanced mobile broadband and low latency are among the benefits of 5G deployment. A growing number of users access multiple applications at once, and mobile traffic is expected to increase every year. Traffic growth can be handled by increasing infrastructure, but doing so requires large investments. The deployment of machine learning can also be used to handle traffic growth. Another solution is to use machine learning to predict and control mobile traffic. In a parallel ensemble Machine learning (ML) model more than one base learner are trained simultaneously and their outputs are combined. The final output is selected based on the majority voting rule or weighted majority rule to enhance the performance of the ensemble ML model [1]. In stacked machine learning meta model is trained on prediction results obtained from multiple ML models [2]. According to the majority of literature studies, ensemble models outperform non ensemble models [3]. An ensemble machine learning approach is used in the research paper to predict 5G throughput. Table 1 shows a literature survey related to the communication paper of 5G, and machine learning. A literature survey on 5G communication paper and machine learning is shown in Table 1. As shown in Table 1, the literature survey highlights the study's purpose, different approaches, and key findings.

Table 1. The literature survey highlighting the study's objective, different approaches, and findings.

Author	Year	Purpose of the study	Approaches Used	Findings
Thilina et al. [1]	2013	The objective is to predict which channel classes are available and which channels are unavailable	SVM, KNN	Detection performance is highest with SVM classifiers
Liu et al. [2]	2017	To propose 5G cognitive radio to improve transmission performance	Adaptive threshold-based Energy Detection	The proposed 5G cognitive radio system is capable of improving transmission performance
Kumar and Gupta [3]	2017	Needs of upcoming 5G wireless technology	OFDM, Cognitive radio, Power spectral density	In conclusion, cognitive technology can be used to overcome bandwidth shortage, a major problem in wireless technology
Badoi et al. [4]	2016	The use of cognitive radio in 5G network	OFDM, DSSS	CR technology can solve the main problems associated with the implementation of 5G
Li et al. [5]	2018	Machine learning model using user grouping proposed	SVM	Machine learning models are more efficient when users are grouped
Subekti et al. [6]	2018	To propose a spectrum sensing method on deep autoencoder neural network and support vector machine	SVM, Deep autoencoder neural network	Deep autoencoder achieve high accuracy as compared to autoencoder
Saber et al. [7]	2020	Spectrum sensing using different machine learning algorithms	ANN, SVM, KNN, TREE	It has been observed that spectrum sensing can be improved by using SVM and ANN
Sharma et al. [8]	2021	Different machine learning algorithms are used to predict throughput	Generalized linear regression model (GLM), RF, SVM	It is concluded that the data partitioning scheme influences the final prediction of 5G throughput
Ravichandran et al. [9]	2021	Leakage detection using ensemble machine learning	Parallel ensemble method, GBT, ANN, KNN	In comparison to a single ML model, ensemble ML has shown improved results

(*continued*)

Table 1. (*continued*)

Author	Year	Purpose of the study	Approaches Used	Findings
Mienye et al.[10]	2022	An analysis of how ensemble machine learning is being used and categorized	AdaBoost, XGBoost, LightGBM, and CatBoost	Improved and more generalized results are obtained by using ensemble machine learning
Musadi, A., et al. [13]	2023	Use of decision tree to estimate soil shear strength	DT, ANN (Artificial Neural Networks)	ANN outperform over the DT
Kumar, M., et al. [18]	2022	To compare performance of ensemble models	AdaBoost, logistic regression, random forest	Ensemble model prediction is more accurate

2   Methodology

This section gives details about the ensemble machine learning approach used in this research.

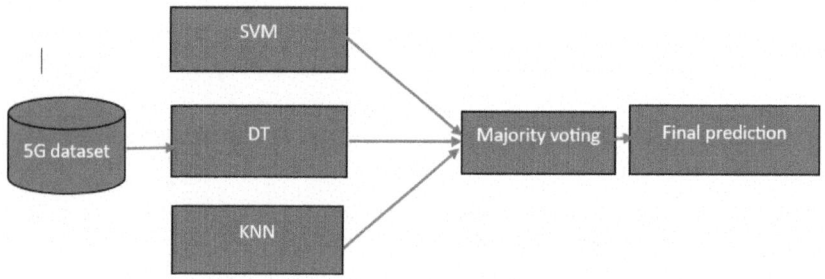

Fig. 1. Flow chart of ensemble ML framework

Ensemble machine learning involves training different ML models separately and fusing their outputs to obtain better results than a single ML model. Parallel and sequential ensemble machine learning are two types of ensemble machine learning. In sequential ensembles, each ML model is trained iteratively, so that every iteration learns from the errors made by the previous one. In the parallel ensemble ML technique different ML classifiers are trained individually and their outputs are fused. Figure 1 shows the proposed ensemble technique block diagram depicting parallel ensemble learning and majority voting. To achieve better results with 5G throughput prediction, different data preprocessing steps were performed, including handling missing values, removing outliers, and standardizing. The next step involved training machine learning models such as SVM, DT, and KNN individually. These models are explained briefly below:

2.1   Different ML Models Used in Ensemble Approach

Support vector machine (SVM): SVM is a machine learning algorithm used to solve classification and regression problems using different types of kernel functions to map attributes of data into higher dimensional feature spaces. Using feature space, SVM identifies a hyperplane that distinguishes between the positive and negative classes of the data with the largest margin of error. The linear function of SVM kernal K(X) is given below:

$$K(X) = (W.X) + b \qquad (1)$$

where X is the data point vector, $W = [W_1, W_2 \dots \dots \dots W_i]^T$ is the weight vector, and b is the intercept. Using Eq. (1) SVM identifies the class of predicted output by using the maximum margin distance between the two margin lines. In the present research we have used nonlinear SVM with radial basis function(rbf) [11].

Decision tree (DT): The decision tree algorithm starts with the root node and proceeds to the leaf node. There are a number of conditions associated with each branch of the tree that must be met in order to determine the value of the data at each end. A decision tree algorithm makes decisions based on entropy and information gain [12]. Entropy calculation is given by Eq. (2) as below.

$$\text{Entropy}(S) = P_+\left(-\log_2 P_+\right) + P_-\left(-\log_2 P_-\right) \qquad (2)$$

where S is the sample that is a set of examples. P_+ is the fraction of the positive examples in the sample S. P_- is the fraction of the negative examples in the sample S.

Equation 3 represents the equation for calculating the information gain of samples and is given as:

$$\text{Information gain (S, A)} = \text{Entropy}(S) - \sum\nolimits_{v\mathcal{E}\ value\ A} \frac{|S_v|}{|s|} \qquad (3)$$

where $\sum_{v\mathcal{E}valueA} \frac{|S_v|}{|s|}$ is resulting, entropy after splitting on attribute A. In Table 2, experimental results are displayed by varying different numbers of decision trees (n_estimators), in order to improve results [13].

K-Nearest Neighbor (KNN): KNN algorithm detects the similarity between new cases and old available cases based on the Euclidian distance and Manhattan distance and assigns a similar class to newly arrived cases. (x_1, y_1) and (x_2, y_2) are two points in dimension space shown in Fig. 2. Steps involved in KNN are: The first step is to select K- nearest neighbors where $k = 1,2,3\dots$. In the second step, the distance between nearest neighbors is calculated using Euclidean distance after which it is determined how many nearby neighbors fall into category 1 or category 2. The KNN algorithm has the disadvantage of not being suitable for imbalanced data sets [14]. This is because K nearest neighbors will show a bias towards one kind of output when the data is imbalanced. In Table2, we present different values of nearest neighbor(n_neighbor) in order to receive more accurate results [15].

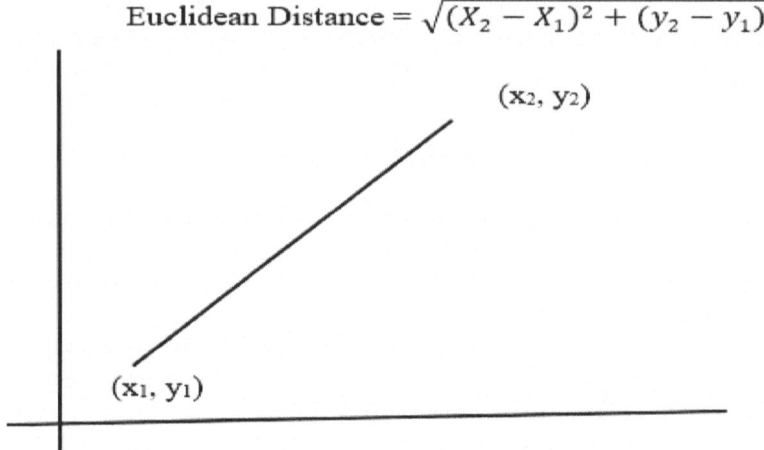

Fig. 2. Euclidean distance

2.2 Majority Voting

Based on the majority voting rule, the final output is derived from the combined output of each model i.e. if most of the independent models predict a class as 1 then ensemble model prediction is 1 else it is zero.

3 Different Attributes of 5G Data Set

The predicted throughput of 5G is influenced by the following features of the dataset. The 5G data set consists of the following parameters [16]:

Latitude and longitude(L): Geolocation parameters include latitude and longitude coordinates that allow us to calculate the distance between any point on Earth and the Equator.

Moving speed(M): A measurement of a user's speed with respect to a 5G panel.

Received Signal Strength Indicator (RSSI): RSSI measures the level of signal strength received after possible loss.

Reference Signal Received Power (RSRP): It is the power of the LTE Reference Signal.

Reference Signal Received Quality (RSRQ): It is the estimation of the quality of the reference signal received by the receiver.

Signal to Interference & Noise Ratio (SINR): The SINR measures signal quality by comparing the strength of the desired signal with that of the unwanted signal.

Compass Direction: Compass direction indicates the user's current direction in relation to the north pole.

Throughput (Predicted class): Throughput in a 5G network is the maximum data transfer rate that the network provides. In this research work, the predicted class is 5G throughput. Data is divided into dependent and independent classes. 5G throughput belongs to the dependent class, while other attributes belong to the independent class.

4 Performance Parameters

Performance parameters used for comparison are defined as follows:

Ensemble accuracy: It is the ratio of a number of classes predicted correctly to the sum of all classes.

Precision: Ratio of true positive and sum of true positive and false positive.

Sensitivity: It is the ratio of true positive to the sum of true positive and false negative [17].

F1_score: F1_score combines precision and recall to compute correct prediction across entire data.

Area under the curve (AUC): AUC is used to represent model classification performance. AUC equal to 0 indicates that the classification model is 100% inaccurate while AUC equal to 1 indicates the model performance is 100% accurate [18].

Execution time: Total time required by machine learning algorithm to complete operation and display output.

5 Results and Discussions

Feature importance refers to the strength of the relationship between independent variables and dependent variables. Figure 3(a) shows the scatter plot of latitude and longitude as independent variables, and predicted 5G throughput as the dependent variable, with 0.44 the predicted strength of the relationship between the independent and dependent variables. In the scatter plot of Fig. 3(b) moving speed is shown as an independent variable, and throughput is shown as a dependent variable, with 0.43 as the predicted strength of the relationship between the moving speed and 5G throughput. Figure 3(c) shows the scatter plot between grouped features, Land M as independent variables, and throughput as dependent variables. The scatter plot shows that the combined effect of L + M on predicted throughput is 0.65. Figure 3(d) displays the scatter plot between grouped features, L, M, and C as independent variables, and throughput as dependent variables. The scatter plot shows that the combined effect of L + M + C on predicted throughput is 0.73. The results of the ensemble machine learning approach used for 5G throughput prediction are shown in Table 2. It is observed from the classification results of Table 2 that the highest ensemble throughput accuracy has been obtained when the n_estimator value is 300, the nearest neighbor value is 25, and the rbf kernel is used.

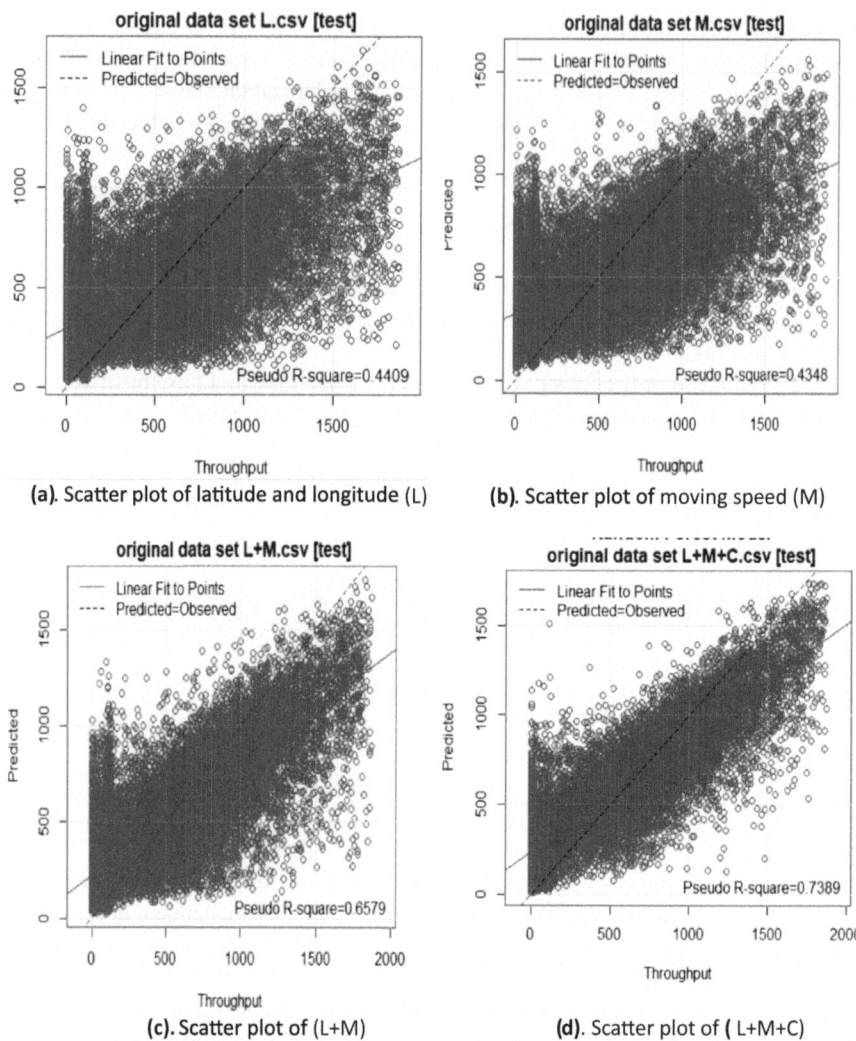

Fig. 3. (a) Scatter plot of latitude and longitude (L). (b) Scatter plot of moving speed (M). (c) Scatter plot of (L + M). (d) Scatter plot of (L + M + C)

Figure 4 depicts the area under the curve (AUC) which is used to determine the model classification performance. Figure 4(a) depicts area under the curve (AUC) is equal to 0.949 which is approaching 1 so we conclude that the model is at best with estimators = 300, n_neighbor = 25, kernal = rbf. Furthermore, the ensemble classification results reveal that the execution time increases as the number of n_estimators and n_neighbor increases.

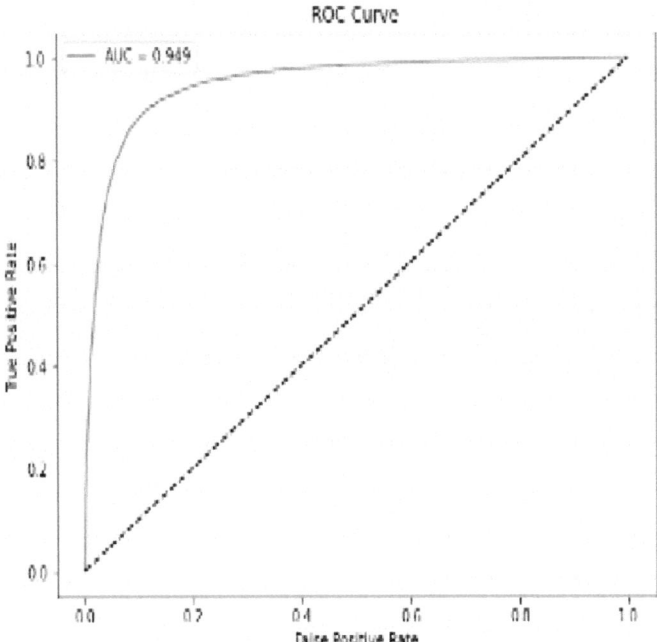

Fig. 4. Area under the curve.

Table 2. Classification results of the ensemble ML approach

Model Parameters	Ensemble accuracy	Precision	Sensitivity	F1-score	AUC	Execution time (sec.)
n_estimators = 100, n_neighbor = 5, kernal = rbf	85.80%	86.31%	85.24%	86%	92.50%	1416
n_estimators = 150, n_neighbor = 10, kernal = rbf	88%	88.32%	87.51%	87.41%	93.90%	1538
n_estimators = 200, n_neighbor = 15, kernal = rbf	88.60%	88.77%	88.42%	88.42%	94.50%	2983
n_estimators = 250, n_neighbor = 20, kernal = rbf	88.80%	89.19%	88.71%	88.71%	94.90%	2960
n_estimators = 300, n_neighbor = 25, kernal = rbf	89%	89.19%	88.32%	88.59%	94.90%	2965

6 Comparison with State-of-the-art Method

Table 3 represents a comparison of the proposed ensemble model for 5G throughput prediction with existing state of art works. Elsherbiny et al. predict the throughput of 4G wireless technology using support vector regressor (SVR), KNN, and random forest (RF) machine learning. Correlation (R^2) has been used to predict 4G throughput accuracy. This table shows that 4G throughput accuracy is lower than the proposed ensemble accuracy. Sharma et al. claim that their RF ML model predicts 5G throughput with a high level of accuracy (acc) which is less as compared to the proposed ensemble ML model. Eyceyurt et al. predicted 4G throughput by using different machine learning models such as linear regressor (LR), decision tree regressor (DTR), and KNN in terms of correlation which is less as compared to the proposed ensemble model.

Table 3. Summary of the analysis papers

Reference	Year	ML approaches	Prediction parameters
Elsherbiny et al. [19]	2020	SVR	0.36 (R^2)
		KNN	0.38 (R^2)
		RF	0.78 (R^2)
Sharma et al. [10]	2021	SVM	67.17% (acc)
		RF	82.88% (acc)
		GLM	0.3 (acc)
Eyceyurt et al. [20]	2022	LR	0.71 (R^2)
		DTR	0.86 (R^2)
		KNN	0.85 (R^2)
Proposed	2023	Ensemble machine learning	89% (ensemble acc)
Approach		by using SVM, KNN, and DT	

7 Conclusions

In this research article, 5G throughput is predicted by a parallel ensemble machine learning approach. Three machine learning models SVM, DT, and KNN have been trained individually. The predicted class of each model has been combined and the 5G throughput has been calculated based on the majority voting rule. 5G ensemble accuracy of 89%, precision of 89.19%, recall of 88.32%, F1_score of 88.59% and AUC of 94.90% have been obtained when n_estimators = 300, n_neighbor = 25. Furthermore, execution time increases as tree numbers increase. An analysis of the feature's importance has been carried out using scatter plots. The importance of features shows that together geolocation, mobility, and compass direction are key factors determining 5G throughput prediction but individually each feature has less impact. Improved prediction performance parameters of 5G throughput play an important role in providing better quality of services.

Furthermore, there is a lot of scope for improvement in this field. The use of stacked ensemble machine learning can improve 5G throughput prediction in the future.

References

1. Thilina, K.M., et al.: Machine learning techniques for cooperative spectrum sensing in cognitive radio networks. IEEE J. Sel. Areas Commun. **31**(11), 2209–2221 (2013)
2. Liu, X., He, D., Jia, M.: 5G-based wideband cognitive radio system design with cooperative spectrum sensing. Physical Communication **25**, 539–545 (2017)
3. Kumar, A., Gupta, M.: Keys Technology and Problem in Deployment of 5G Mobile Communication Systems
4. Badoi, C.-I., Prasad, N., Prasad, R.: Virtualization and scheduling methods for 5G cognitive radio based wireless networks. Wireless Pers. Commun. **89**, 599–619 (2016)
5. Li, Z., et al.: Improved cooperative spectrum sensing model based on machine learning for cognitive radio networks. IET Commun. **12**(19), 2485–2492 (2018)
6. Subekti, A., Pardede, H.F., Sustika, R.: Spectrum sensing for cognitive radio using deep autoencoder neural network and SVM. in 2018 International Conference on Radar, Antenna, Microwave, Electronics, and Telecommunications (ICRAMET). IEEE (2018)
7. Saber, M., et al.: Spectrum sensing for smart embedded devices in cognitive networks using machine learning algorithms. Procedia Computer Science **176**, 2404–2413 (2020)
8. Sharma, A., Pandit, S., Talluri, S.R.: A comparative study to classify and predict the throughput of fifth generation wireless technology using supervised machine learning algorithms. in 2021 Sixth International Conference on Image Information Processing (ICIIP). IEEE (2021)
9. Ravichandran, T., et al.: Ensemble-based machine learning approach for improved leak detection in water mains. J. Hydroinf. **23**(2), 307–323 (2021)
10. Mienye, I.D., Sun, Y.: A survey of ensemble learning: Concepts, algorithms, applications, and prospects. IEEE Access **10**, 99129–99149 (2022)
11. Phillips, T., Abdulla, W.: Developing a new ensemble approach with multi-class SVMs for Manuka honey quality classification. Appl. Soft Comput. **111**, 107710 (2021)
12. Awomuti, A., et al.: Towards Adequate Policy Enhancement: An AI-Driven Decision Tree Model for Efficient Recognition and Classification of EPA Status via Multi-Emission Parameters. City and Environment Interactions, 100127 (2023)
13. Musadi, A., et al.: Comparing artificial neural network and decision tree algorithm to predict tides at Tanjung Priok port. Procedia Computer Science **227**, 406–414 (2023)
14. Zhang, W., et al.: A distributed storage and computation k-nearest neighbor algorithm based cloud-edge computing for cyber-physical-social systems. IEEE Access **8**, 50118–50130 (2020)
15. Shekhar, S., Hoque, N., Bhattacharyya, D.K.: PKNN-MIFS: A Parallel KNN Classifier over an Optimal Subset of Features. Intel. Sys. Appl. **14**, 200073 (2022)
16. Narayanan, A., et al.: Lumos5G: Mapping and predicting commercial mmWave 5G throughput. In: Proceedings of the ACM Internet Measurement Conference (2020)
17. Kumar, A., Chatterjee, J.M., Díaz, V.G.: A novel hybrid approach of SVM combined with NLP and probabilistic neural network for email phishing. Int. J. Electr. Comp. Eng. **10**(1), 486 (2020)
18. Kumar, M., et al.: A comparative performance assessment of optimized multilevel ensemble learning model with existing classifier models. Big Data **10**(5), 371–387 (2022)
19. Elsherbiny, H., et al.: 4G LTE network throughput modelling and prediction. In: GLOBECOM 2020-2020 IEEE Global Communications Conference. IEEE (2020)
20. Eyceyurt, E., Egi, Y., Zec, J.: Machine-learning-based uplink throughput prediction from physical layer measurements. Electronics **11**(8), 1227 (2022)

Severity Prediction of Omicron Sub-variant JN.1 by Using Machine Learning

Vijay Kumar Sinha[1]([✉]) [iD], Manish Mahajan[2], Srikanta Mallik[3], Ashok Sahoo[4], Nisha Kumari[5], and Fitri Yakub[6]

[1] School of Engineering and Technology, Chitkara University, Baddi, Himachal Pradesh, India
`vk.sinha@chitkarauniversity.edu.in`
[2] Military College of Telecommunication Engineering, Mhow, M.P, India
[3] Cognizant Technology Solutions Pvt., Ltd., Abudhabi, UAE
[4] Graphic Era Hill University, Dehradun, India
[5] Chandigarh Engineering College, Landran, Mohali, India
`nisha.3563@cgc.edu.in`
[6] Malaysia-Japan International Institute of Technology, Universiti Teknologi, Johor, Malaysia
`mfitri.kl@utm.my`

Abstract. The new spreading virus JN.1 is spreading in India rapidly and raised a new health concern. People including children are found infected in various cities. This new mutation is caused due to environmental impact and adaption for survival. Virus made a concern to the medical professionals to combat and restrict its spread as a pandemic. In this research paper, we analyzed the nature of the new mutated virus. Prediction of mortality rate and probability of child deaths are analyzed. The JN.1 mutation is a descendent of earlier variant SARS-CoV-2 BA.2.86, commonly known as Omicron. In this paper, we analyzed the symptoms of the new virus with its immediate predecessor in terms of its severity, symptoms, and structural differences. Based on analysis we find and predict that this new variant will not impact as deadly Delta variant which was spread in India in late 2020. The predicted lifespan of the JN.1 virus will be 3 to 6 months in India and will vanish as its predictor Omicron. We used Random Forest for accuracy among the tree-based regression algorithms.

Keywords: JN.1 · BA.2.86 · SARS-CoV-2 · Machine Learning · Regression · Decision Tree · Gradient Boost · Omicron · Delta variant

1 Introduction

The global impact of the JN.1 the mutated version of SARS-CoV-2, BA.2.86 Omicron pandemic has touched almost every aspect of human life. The emergence and effects of the virus on human health have been discussed extensively [1–3]. Initially, from January to April 2020, European countries bore the brunt of the virus as it spread from China. India remained relatively untouched until early March 2020 when the first cases surfaced through international tourists and returning Indian citizens. At that time, India had limited healthcare facilities compared to more developed nations.This version is milder symptoms than delta varient of SARS-CoV-2.

M. Khurana et al. (Eds.): ICMLA 2024, CCIS 2238, pp. 82–91, 2025.
https://doi.org/10.1007/978-3-031-75861-4_8

A mathematical predictive model related to JN.1, considering healthcare facilities in Italian hospitals, has been presented [4]. Another study applied a statistical chi-square test to explore the relationship between dependent variables (confirmed cases and deaths) and independent variables (age, sex, region, and infection) in the spread of JN.1 cases in South Korea [4]. Another mathematical model predicted the spread of JN.1 based on various parameters, using a non-parametric model that fitted the dataset available until June 06, 2020, in India, Italy, and the USA [5]. The paper includes a detailed statistical description of the polynomial regression model and evaluation metrics such as R2, Adjusted R2, and RMSE [6]. Linear and polynomial regression models were analyzed to predict JN.1 deaths in America [7]. JN.1 mortality prediction employed multiple stepwise regression analysis methods and a two-layer nested heterogeneous ensemble learning approach [8].

Linear regression for Case Fatality Rate (CFR) on a small JN.1 dataset of ten countries until April 2020 showed that the fitted line represented the data well with minimal error [9]. However, using a small dataset led to overfitting with higher-order polynomials. Auto-regression on a six-week JN.1 data in India did not pass the test of statistical significance despite strong variable correlation and good goodness-of-fit metrics [10]. ARIMA and seasonal ARIMA were used to predict JN.1 epidemic trends [11]. A real-time differential optimization-based Trust-region-reflective (TRR) algorithm predicted the peak of JN.1 cases in several countries [12]. Linear and multi-linear regression models applied to WHO's JN.1 India dataset showed accurate results with a high R2 metric [13]. A support vector regression model tested the non-linearity of different structures in predicting Covid-12 cases [14].

Population density emerged as a significant factor in JN.1 spread, with social distancing identified as a preventive measure [15]. Regression on weather dataset variables and census variables revealed that weather factors were more significant than census variables in predicting mortality rates due to the JN.1 pandemic [16]. Naive Bayes regression was found to perform poorly compared to linear regression, locally weighted linear regression, and decision tree on real-life datasets [17].

Recent proposals included deep learning models such as a long short-term memory (LSTM) network for predicting JN.1 statistics in Canada, the United States of America, and India [18–20]. A Convolution-LSTM-based multivariate analysis identified factors contributing to the growth of JN.1 cases [21]. The work presented in this paper is based on the JN.1 dataset for an extended duration [22].

2 Proposed Model

The model under consideration adopts the best subset selection method for predictors and incorporates eight regression algorithms—MLR, Polynomial, Lasso, Ridge, KNN, Decision Tree, Gradient Boost Tree, and Random Forest—utilizing the JN.1 dataset as input [22]. At the outset, the optimal model is identified by assessing the smallest R2, largest RMSE, and largest MAE. The baseline solution, denoted as Mi0, involves computing the null model for each algorithm. This null model essentially represents a scenario where the model comprises only a sample mean and lacks any predictors [23].

Algorithm:

Input: Predictors and eight regression algorithms

Output: The best solution based on train metrics i.e. largest R^2, lowest RMSE, and lowest MAE and cross-validation prediction error i.e. adjusted-R^2

Initializations

M_{best} = smallest

R^2,   largest

RMSE   and

largest   MAE;

p=1;

for i = one to eight input models // MLR, Polynomial, Lasso, Ridge, KNN, Decision Tree, Gradient Boost Tree, andRandom Forest

{

Compute the sample mean for the null model M_{i0}, assuming no predictors;for k=1 to p predictors

3   Life Cycle of the Model

The features are extracted from the JN.1 dataset of India [22]. A careful identification of all the features having a significant impact on the predicted outcome is done. Ignoring/leaving any significant dependent feature may result in inaccurate predictions, and hence wrong decision/policy making. The Exploratory Data Analysis (EDA) is conducted on the dataset. It tells about the patterns, relationships, and anomalies among variables in the dataset for further analysis. It is the first step in analyzing the data. Data cleaning is required for columns having 'NAN' values as this may create problems while applying some arithmetic on the data. Moreover 'NAN' values create problems in correctly predicting the outcome. It is always better to either delete these or replace them with either zero, mean, mode, or median values that best fit in the problem.

3.1   Linear and Multi-linear Regression

The dataset contains the independent parameters 'Days', 'Confirmed', and 'Cured' that present the number of days, confirmed cases, and cured cases, respectively. It has one dependent parameter 'Deaths' that explains the number of deaths. Linear Regression (LR) model regresses 'Deaths' on a single field 'Days' (LRP1) while Multilinear Regression (MLR) regresses 'Deaths' on two fields 'Days and Confirmed' (MLRP2) and on three fields 'Days, Confirmed, and Cured' (MLRP3).

The LR and MLR are represented with equation (1)

3.2   Gradient Boost Tree Regression

It is a boosting method used to build the models sequentially. Errors of previous models are reduced in this case using subsequent models. This algorithm can be used for both classifications as well as regression problems. When the dependent variable is continuous, a Gradient gradient-boosting regressor is used whereas a Gradient gradient-boosting classifier is used for the discrete target variable. This algorithm uses the concept of gradient descent to minimize the loss function.

4   Results and Discussions

The work presented in this paper is based on the JN.1 dataset of India available from the Kaggle dataset [22]. It contains various fields such as Date, State/Union Territories, Deaths, Confirmed, and Cured. Here, the Date represents the calendar date when the data was taken. State/Union Territories represent different statesand union territories of India. The remaining fields represent the number of deaths, confirmed JN.1 cases, and cured cases. Section 3.1 presents an empirical comparison of the accuracy of LR, MLR, PR, KNN, Ridge, and Lasso regression. Section 3.2 presents an empirical comparison of the accuracy of the Decision Tree, Random Forest, and Gradient Boost.

4.1   An Exhaustive Empirical Comparison on Accuracy of LR, MLR, PR, KNN, Ridge, and Lasso Regression

This section presents accuracies on evaluation metrics R^2, Adjusted-R^2, RMSE, and MAE for LR, MLR and PR, KNN, Ridge, and Lasso regression with varying independent parameters as presented in Table 1. Here a PR with parameter 'x' and varying degree 'y' is represented with PRPxDy. Table 1 presents the accuracy metrics for LR and PR with a single parameter 'Days'. It can be seen that LR with one parameter (LRP1) has the lowest R^2 and Adjusted-R^2 values and maximum RMSE and MAE values as compared to PRP1Dx. Among PRs, the polynomial of degree two has much better figures for all the metrics over the LRP1. The metrics become better when the PR of degree three is evaluated. R^2 becomes maximum and does not improve further on increasing the degree. When the degree of PR is further increased, the RMSE and MAE improve very less as compared to a rise in the complexity of polynomial equations. This increase in complexity is due to a greater number of coefficients and a greater number of terms in the polynomial. So, to trade off the complexity and execution time, it can be said that PRP1D3 best fits the dataset. This explanation is also evident from the kernel density or Gaussian density distribution plots between the predicted and actual deaths over the test dataset, as presented in Fig. 1.

Table 1. Regression models' evaluation with varying independent parameters and polynomial degrees

Model Used	Train R^2 Error	Test R^2 Error	Adjusted R^2Error	RMSE	MAE
LRP1	0.855	0.885	0.86	45583.21	35221.97

(continued)

Table 1. (*continued*)

Model Used	Train R^2 Error	Test R^2 Error	Adjusted R^2Error	RMSE	MAE
PRP1D2	0.938	0.964	0.95	25753.96	19518.12
PRP1D3	0.954	0.967	0.98	24177.11	21197.21
PRP1D4	0.971	0.975	0.98	20999.18	15742.77
PRP1D5	0.972	0.972	0.98	20517.19	16956.66
PRP1D6	0.990	0.996	0.98	11308.15	9173.08
PRP1D7	0.996	0.996	0.99	8237.07	6396.95
PRP1D8	0.996	0.996	0.99	7977.13	4690.23
PRP1D9	0.996	0.996	0.99	8141.24	5304.52
MLRP2	0.9876	0.99	0.99	13098.71	9360.34
PRP2D2	0.992	0.991	0.99	12092.00	8262.53
PRP2D3	0.99	0.99	0.99	2662.69	2192.06
PRP2D4	0.99	0.99	0.99	10690.91	8721.41
MLRP3	0.99	0.99	0.99	8195.58	7106.47
PRP3D2	0.99	0.99	0.99	1772.24	1529.58
PRP3D3	0.99	0.99	0.99	1468.5	1022.79
KNN4P1	0.99	0.99	0.99	441.96	210.00
KNN4P2	0.99	0.99	0.99	540.43	236.64
KNN4P3	0.99	0.99	0.99	543.32	238.62
Lasso-P1	0.856	0.88	0.98	45583.30	35221.96
Lasso-P2	0.986	0.99	0.99	13437.20	8677.82
Lasso-P3	0.989	0.992	0.98	12228.70	8521.30
Ridge-P1	0.856	0.886	0.99	45583.31	35221.96
Ridge-P2	0.987	0.99	0.99	13437.20	8677.82
Ridge-P3	0.995	0.996	0.98	12228.70	8521.30

4.2   An Exhaustive Empirical Comparison on Accuracy of Decision Tree, Random Forest, and Gradient Boost

This section presents an accuracy comparison of evaluation metrics R^2, Adjusted-R^2, RMSE, and MAE. Table 2 presents the accuracy metrics for DT, GB, and RF models with one, two, and three parameters. In this Table 1 represents one parameter 'Days' used, P2 means two parameters 'Days' and 'Confirmed' are taken into consideration for evaluation whereas P3 indicates number of parameters used is three which includes 'Days', 'Confirmed' and 'Cured'. It can be seen that DT with three parameters (DTP3) has the lowest RMSE value whereas the MAE value is the lowest for DTP2. The DTP1 is not very effective in giving accurate results. Among GB-based runs, both RMSE and

MAE values are the least for GBP3l. The GBP1 gives better results than GBP2 but is unable to outperform when three parameters are considered. The same trend can be seen for Random Forest where RFP3 is having least values for both RMSE and MAE. It can be concluded from these results that GB with more parameters gave better results as compared to few or fewer parameters [23, 24] (Fig. 2).

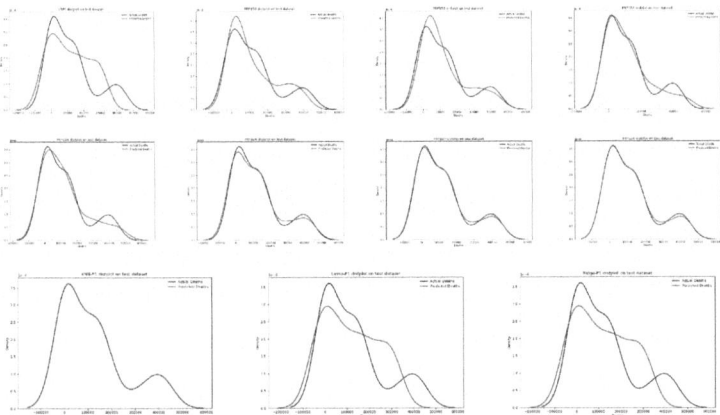

Fig. 1. Distribution of actual deaths and predicted deaths using LR, PR in 'x' degree of Polynomial (PRP1Dx), KNN, Lasso, and Ridge Regression on One Independent Parameter ['Days']

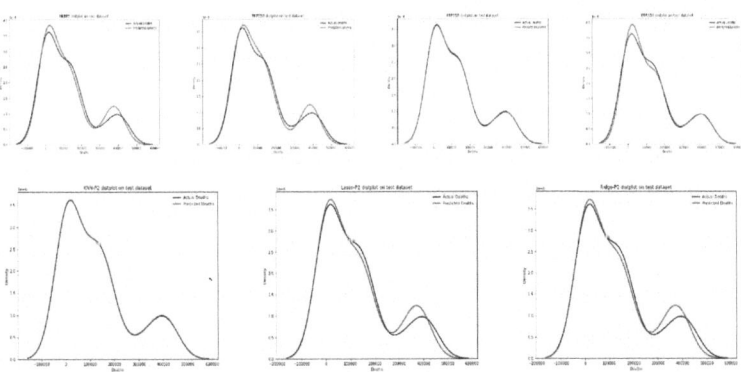

Fig. 2. Distribution of actual deaths and predicted deaths using MLR, PR in 'x' degree of Polynomial (PRP2Dx), KNN, Lasso, and Ridge Regression with Two Independent Parameter (MLRP2) ['Days' and 'Confirmed']

The actual versus predicted deaths using DT, GB, and RF for one independent parameter is presented in Fig. 3.

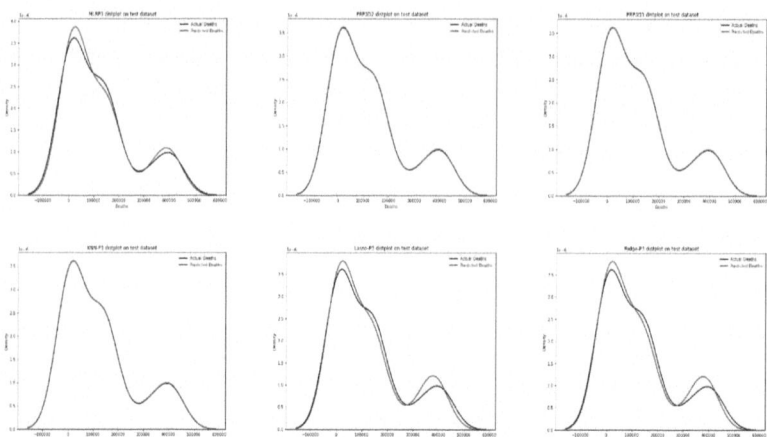

Fig. 3. Distribution of actual deaths and predicted deaths using MLR and, PR in 'x' degree of Polynomial (PRP2Dx), KNN, Lasso, and Ridge Regression with Three Independent Parameter ['Days', 'Confirmed', and 'Cured']

In the GBP1 curve, the two different curves show that there is a difference in real and estimated numbers of deaths. The curves for the DTP1 and RFP1 overlap with each other and represent better accuracy as compared to GBP1. Overall, RFP1 performs the best among the three if only one parameter is taken into consideration (Fig. 4).

Table 2. DT, GB and RF models' evaluation with varying independent parameters

Model Used	Train R^2 Error	Test R^2 Error	Adjusted R^2 Error	RMSE	MAE
DT-P1	0.96	0.98	0.98	1261.49	711.96
DT-P2	0.96	0.98	0.99	1257.53	705.30
DT-P3	0.96	0.99	0.99	1257.78	706.72
GB-P1	0.95	0.99	0.99	3295.67	1848.31
GB-P2	0.95	0.99	0.99	3331.83	1874.18
GB-P3	0.95	0.99	0.99	3260.79	1836.51
RF-P1	0.98	0.99	0.99	735.19	383.41
RF-P2	0.987	0.98	0.99	589.50	277.56
RF-P3	0.97	0.99	0.99	544.90	248.98

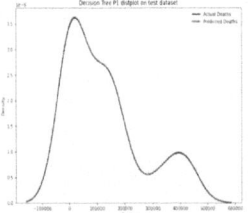 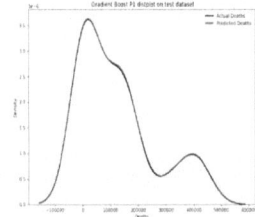 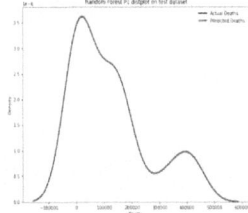

Fig. 4. Distribution of actual deaths and predicted deaths using DT, GB and RF on One Independent Parameter ['Days']

5 Conclusions and Future Scope

The new spreading virus JN.1 is spreading in India rapidly and raised a new health concern. People including children are found infected in various cities. This new mutation is caused due to environmental impact and adaption for survival. Virus made a concern to the medical professionals to combat and restrict its spread as a pandemic. In this research paper, we analyzed the nature of the new mutated virus. Prediction of mortality rate and probability of child deaths are analyzed. The JN.1 mutation is a descendent of earlier variant SARS-CoV-2 BA.2.86, commonly known as Omicron. In this paper, we analyzed the symptoms of the new virus with its immediate predecessor in terms of its severity, symptoms, and structural differences. Based on analysis we find and predict that this new variant will not impact as deadly Delta variant which was spread in India in late 2020. The predicted lifespan of the JN.1 virus will be 3 to 6 months in India and will vanish as its predictor Omicron. We used Random Forest for accuracy among the tree-based regression algorithms. This paper employs various regression models, namely LR, MLR, PR, KNN, Lasso, Ridge regression, Decision Tree, Random Forest, and Gradient Boost Tree, to forecast mortality statistics attributed to JN.1 in India. The JN.1 dataset for India [22, 25, 26] undergoes thorough analysis and pre-processing before being utilized for modeling in Python, employing the 'sklearn' machine learning library.

The size of the JN.1 dataset, along with numerous datasets from different countries, has been consistently growing. Notably, the loading time and processing duration of the dataset are considerable. To address this, parallel loading and processing of distributed data are proposed as potential enhancements. Technologies such as Hadoop or Spark are suggested for their capabilities in distributed parallel processing, aiming to optimize the loading and processing times.

References

1. Luo, G., Gao, S.J.: Global health concerns stirred by emerging viral infections. J. Med. Virol. **92**(4), 399–400 (2020). https://doi.org/10.1002/jmv.25683
2. Wang, W., Tang, J., Wei, F.: Updated understanding of the outbreak of 2019 novel coronavirus (2019- nCoV) in Wuhan, China. J. Med. Virol. **92**(4), 441–447 (2020). https://doi.org/10.1002/jmv.25689
3. Myint, S.H.: Human coronaviruses: A brief review. Rev. Med. Virol. **4**(1), 35–46 (1994). https://doi.org/10.1002/rmv.1980040108

4. Çakan, S.: Dynamic analysis of a mathematical model with health care capacity for COVID-19 pandemic. Chaos, Solitons and Fractals **139**, (2020). https://doi.org/10.1016/j.chaos.2020.110033

5. AL-Rousan, N., AL-Najjar, H.: Data analysis of coronavirus COVID-19 epidemic in South Korea based on recovered and death cases. J. Med. Virol. **92**(9), 1603–1608 (2020). https://doi.org/10.1002/jmv.25850

6. Singhal, A., Singh, P., Lall, B., Joshi, S.D.: Modeling and prediction of COVID-19 pandemic using Gaussian mixture model. Chaos, Solitons Fractals **138**, 110023 (2020). https://doi.org/10.1016/j.chaos.2020.110023

7. Ostertagová, E.: Modelling using polynomial regression. Procedia Eng. **48**(2012), 500–506 (2012). https://doi.org/10.1016/j.proeng.2012.09.545

8. Singh, H., Bawa, S.: Predicting COVID-19 statistics using machine learning regression model: Li-MuLi-Poly. Multimed. Syst. **28**, 113–120 (2022). https://doi.org/10.1007/s00530-021-00798-2

9. Cui, S., Wang, Y., Wang, D., Sai, Q., Huang, Z., Cheng, T.C.E.: A two-layer nested heterogeneous ensemble learning predictive method for COVID-19 mortality. Appl. Soft Comput. **113**, 107946 (2021). https://doi.org/10.1016/j.asoc.2021.107946

10. Hoseinpour Dehkordi, A., Alizadeh, M., Derakhshan, P., Babazadeh, P., Jahandideh, A.: Understanding epidemic data and statistics: A case study of COVID-19. J. Med. Virol. **92**(7), 868–882 (2020). https://doi.org/10.1002/jmv.25885

11. Ghosal, S., Sengupta, S., Majumder, M., Sinha, B.: Linear Regression Analysis to predict the number of deaths in India due to SARS-CoV-2 at 6 weeks from day 0 (100 cases - March 14th 2020). Diabetes Metab. Syndr. Clin. Res. Rev. **14**(January), 311–315 (2020)

12. ArunKumar, K.E., et al.: Forecasting the dynamics of cumulative COVID-19 cases (confirmed, recovered and deaths) for top-16 countries using statistical machine learning models: Auto-Regressive Integrated Moving Average (ARIMA) and Seasonal Auto-Regressive Integrated Moving Averag. Appl. Soft Comput. **103**(2019), 107161 (2021). https://doi.org/10.1016/j.asoc.2021.107161

13. Nabi, K.N.: Forecasting COVID-19 pandemic: A data-driven analysis. Chaos, Solitons Fractals **139**, 110046 (2020). https://doi.org/10.1016/j.chaos.2020.110046

14. Rath, S., Tripathy, A., Tripathy, A.R.: Prediction of new active cases of coronavirus disease (COVID- 19) pandemic using multiple linear regression model. Diabetes Metab. Syndr. Clin. Res. Rev. **14**(5), 1467–1474 (2020). https://doi.org/10.1016/j.dsx.2020.07.045

15. Peng, Y., Nagata, M.H.: An empirical overview of nonlinearity and overfitting in machine learning using COVID-19 data. Chaos, Solitons and Fractals **139** (2020). https://doi.org/10.1016/j.chaos.2020.110055

16. Behnood, A., Mohammadi Golafshani, E., Hosseini, S.M.: Determinants of the infection rate of the COVID-19 in the U.S. using ANFIS and virus optimization algorithm (VOA). Chaos, Solitons and Fractals **139**, 110051 (2020). https://doi.org/10.1016/j.chaos.2020.110051

17. Malki, Z., Atlam, E.S., Hassanien, A.E., Dagnew, G., Elhosseini, M.A., Gad, I.: Association between weather data and COVID-19 pandemic predicting mortality rate: Machine learning approaches. Chaos, Solitons Fractals **138**, 110137 (2020). https://doi.org/10.1016/j.chaos.2020.110137

18. Frank, E., Trigg, L., Holmes, G., Witten, I.H.: Technical note: naive Bayes for regression. Mach. Learn. **41**(1), 5–25 (2000). https://doi.org/10.1023/A:1007670802811

19. Chimmula, V.K.R., Zhang, L.: Time series forecasting of COVID-19 transmission in Canada using LSTM networks. Chaos, Solitons and Fractals **135** (2020). https://doi.org/10.1016/j.chaos.2020.109864

20. Basu, S., Campbell, R.H.: Going by the numbers: Learning and modeling COVID-19 disease dynamics. Chaos, Solitons Fractals **138**, 110140 (2020). https://doi.org/10.1016/j.chaos.2020.110140

21. Arora, P., Kumar, H., Panigrahi, B.K.: Prediction and analysis of COVID-19 positivecases using deep learning models: a descriptive case study of India. Chaos, Solitons and Fractals **139** (2020). https://doi.org/10.1016/j.chaos.2020.110017

22. Yudistira, N., Sumitro, S.B., Nahas, A., Riama, N.F.: Learning where to look for COVID-19 growth: Multivariate analysis of COVID-19 cases over time using explainable convolution– LSTM. Appl. Soft Comput. **109**, 107469 (2021). https://doi.org/10.1016/j.asoc.2021.107469

23. https://www.kaggle.com/sudalairajkumar/covid19-in-india. COVID-19 in India

24. Sewal, P., Singh, H.: A critical analysis of apache hadoop and spark for big data processing. In: 6th International Conference on Signal Processing, Computing and Control (ISPCC), pp. 308–313 (2021)

25. Kumar, A., Srivastav, A.L., Dubey, A.K., Dutt, V., Vyas, N. (eds.): Innovations in Machine Learning and IoT for Water Management. IGI Global (2024). https://doi.org/10.4018/979-8-3693-1194-3

26. Dubey, A., Narang, S, Kumar, A., Sasubilli, S., García Díaz, V.: Performance estimation of machine learning algorithms in the factor analysis of COVID-19 dataset. Computers, Materials, and Continua. **66**, 1921–1936 (2020). https://doi.org/10.32604/cmc.2020.012151

IoT-Inspired Smart Drought Prediction Framework: Machine Learning Approach

Diksha Bhardwaj[1]([✉]) and Gagninder Kaur[2]

[1] Chandigarh University, Chandigarh, Punjab, India
dikshabhardwaj250@gmail.com
[2] CSE, Chandigarh University, Chandigarh, Punjab, India

Abstract. A severe drought may have devastating effects on a region's hydrological balance, agriculture, animal habitat, and economy, among other things. As a result, a reliable method of drought forecasting is required. Several drought indices purport to measure drought severity, but many of them ignore crucial details. This research presents a fog-based paradigm for drought predictive analysis, made possible by the IoT. Singular vector decomposition is used to compress data at the fog level in this architecture. Future drought conditions are predicted using the Holt-Winters approach, while the severity of individual drought occurrences is evaluated using an Adaptive Neuro-fuzzy Inference System (ANFIS) mechanism. The usefulness of the suggested system is proved by its implementation utilizing the online UCI dataset in terms of forecasting efficacy, and Prediction efficiency.

Keywords: Drought · IoT · Fog Computing · Adaptive Neuro-fuzzy Inference System

1 Introduction

The term "disaster" is used to describe a catastrophic catastrophe. For a certain amount of time, below-average rainfall relative to historical averages constitutes a drought. Drought generally impacts natural ecosystems by diminishing soil quality, depleting groundwater levels, harming plant cover, and causing a fall in animal populations. The effects of poor agricultural water quality on public health are far-reaching. Societal dissatisfaction and political instability are the results of stagnant economic development. The fast spread of economic instability throughout the globe is facilitated by globalization. Approximately 220 million people are in danger of drought every year, with the biggest risk being in sub-Saharan Africa, according to research on 107 countries [1]. North Africa and the Middle East are the most water-deficient areas in the world, with almost all countries having annual water availability of less than 1000 m^3 per capita. Several nations in the region have less than 500 m^3 per person, per year [2]. When compared to other natural disasters, drought is the costliest, costing an average nation 0.5% of its GDP [3, 4]. There are many drought indices, each with its own set of benefits and drawbacks due to the unique combination of input factors used to calculate drought severity. Healthcare, smart cities, smart homes, smart agriculture, and smart grids are just a few examples of fields that incorporated the combination of IoT, cloud computing, and artificial Intelligence [5–7]. The fog-based computational platform contributes to these developments may be seen in its network bandwidth, heterogeneity, low latency, and location [8]. To address the problem of drought, a complete evaluation and monitoring system is necessary, even if there are various drought indicators. With the use of Internet of Things (IoT) sensors, automated systems can gather real-time data on all causes of drought [9]. By analyzing massive volumes of data, which is made possible by cloud computing and big data, we can better assess the severity of droughts [10]. The suggested layered design leverages IoT sensors to monitor the drought-stricken region in real-time, creating a wealth of data. Before transmitting it to the cloud, the fog layer compresses it using a data reduction method but keeps all the useful information intact. The vast processing capability of the cloud layer is used to categorize drought severity and produce forecasts. The architecture for moving from fog to clouds is shown in Fig. 1. The system's primary goals are:

1. Real-time monitoring using IoT sensors to track various drought factors.
2. Enhanced bandwidth utilization by implementing a data reduction technique.
3. Cloud-based Predictive Severity Analysis.
4. Forecasting intensity of the drought for different temporal instances.

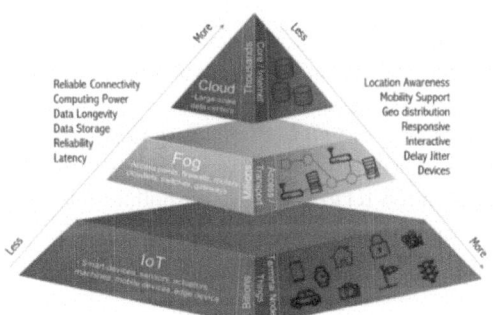

Fig. 1. IoT-Fog-Cloud Computing

Paper Structure: In Sect. 2, literature review is presented. Section 3 describes the suggested cloud-based paradigm for drought forecasting. The experimental findings and assessment of the system's performance are discussed in Sect. 4. The article concludes with Sect. 5.

2   Literature Review

The integration of cloud, fog, and IoT could significantly benefit drought assessment and monitoring. Data mining methods have been explored in this field, particularly in remote sensing data. Many researchers have assessed the effectiveness of various drought indexes, such as those conducted by Bayissa et al. (2018) [10] on droughts in Ethiopia's Upper Blue Nile Basin using six drought indicators and one composite drought index. Zou et al. (2018) proposed using Apache Hadoop to analyze data from worldwide vegetation drought monitors, while Severino et al. (2018) [25] used IoT to anticipate soil moisture and implement irrigation practices with little environmental impact. Jang (2018) [11] analyzed the severity and frequency of droughts in Korea's 73 weather stations, comparing the Standard Precipitation Index (SPI) with the reconnaissance. Mishra & Singh (2010) [4] published a comprehensive assessment of drought indicators, highlighting various shortcomings and comparing several drought indexes. Data mining approaches have been used in many drought prediction and forecasting studies [15–20]. However, the integration of IoT and cloud computing for drought assessment is in its infancy [19]. This article provides an overview of significant works in this area, such as Yu et al. (2018) [23] evaluation of susceptibility to drought in North Korean areas utilizing mobile perception methods and cloud data. A plethora of approaches for forecasting air quality and quantifying air pollution have been developed by scholars throughout the globe in the last few years. Scaling prediction makes use of temporal evaluation and interpolation. By studying the grid-oriented environment and assessing different wind patterns affect released pollutants, Gautam et al. (2019) [27] concluded that wind makes pollutant clusters worse. Focussing on Particulate Matter, Kleine Deters et al. (2017) [26] used statistical methods inspired by ML to predict pollution levels using weather data. Another air quality model that predicts weather and pollution levels is THOR, created by Brandt et al. (2001) [28]. By monitoring air quality at tube stations and finding a link between station depth and forecast performance, Park et al. (2018) [29] demonstrated an air pollution assessment system that uses an Artificial Neural Network (ANN) model to predict PM concentration in ambient environments. The reliability of these approaches in predicting PM concentration using data from neighboring stations was confirmed by Li et al. (2018) [30], who used Stepwise Regression (SWR) and Support Vector Regression (SVR) to anticipate air pollution levels utilizing data from a sensor network. Predicting PM concentration was study domain of Karatzas et al. (2018) [31], who proposed a hybrid model integrating ANN and Linear Regression for pollutant concentration prediction. Their model outperformed the competition. To predict PM10 levels, Franceschi et al. (2018) [32] used data analytic approaches, such as Principal Component analytic (PCA), to find critical components.

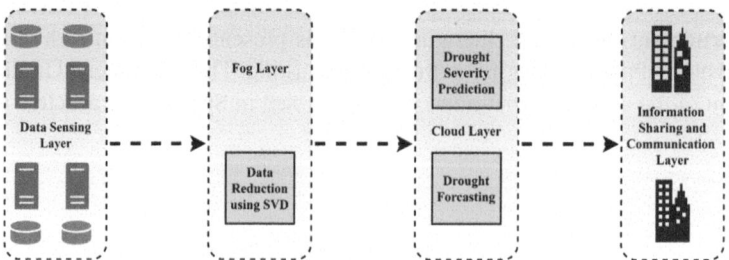

Fig. 2. Proposed Model

3 Proposed Model

We presented a layered architecture using IoT technology for drought assessment and prediction, as presented in Fig. 2. The proposed system comprises four levels: Data Perception Layer, Fog Layer for data dimensionality reduction, Cloud Layer, and Layer for data transmission. IoT nodes are strategically deployed across the research region to collect primary data on various factors contributing to drought. The fog layer's encodes receive data from the data integration layer, apply Singular Vector Decomposition (SVD) to reduce the data's dimensions, and then transmit it over the cloud. Over the cloud platform, drought severity is analyzed for future periods. The cloud layer also incorporates assessments from multiple drought monitoring agencies.

3.1 Data Perception Layer

This layer focuses on the selection of components that contribute to drought parameters and the utilization of relevant IoT devices in the investigated region to gather raw data on these drought-inducing factors. The installed sensor devices collect data on various factors that contribute to drought, and they transmit their findings after a specific time interval. These sensor nodes are strategically placed within the hexagonal units that constitute the study area. The first component is a water sources database, which assesses water scarcity by evaluating moisture levels in soil at two levels: streamflow and groundwater. This dataset provides insights into the insufficiency of water resources. The second component is the meteorological dataset, which includes information on the local climate and weather patterns. It encompasses elements such as temperature, humidity, evaporation, and dew point. The timing of drought occurrences is also influenced by seasonal variations.

3.2 Fog Layer

The sensors are configured to collect data on various factors that contribute to drought. However, analyzing such a massive volume of raw data can be challenging. To reduce the computation time and optimize network bandwidth utilization, data dimensionality reduction techniques are employed. These techniques eliminate irrelevant and redundant sections of the data to obtain the appropriate number of dimensions. In the presented model, SVD is used for reducing dimensionality as shown in Algorithm 1.

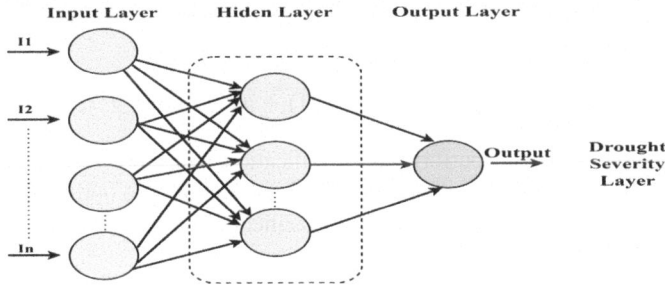

Fig. 3. ANFIS Model

3.3 Cloud Layer

The proposed architecture prioritizes the cloud platform, which acquires data from the fog computational platform, processes it, and predicts the severity of drought in the research region. The cloud layer is comprised of 3 sub-layers including Drought Severity Analysis, Drought Forecasting, and Cloud-based Data Storage.

Drought Severity Analysis: The Adaptive Neuro-Fuzzy Inference System (ANFIS) technique is incorporated by researchers worldwide for predictive decision-making across various fields. ANFIS utilizes the adaptive principles of Neural Networks to derive rules from sampled input data, enhancing self-adaptation and self-organization through tuning rules. In the presented research, the proposed ANFIS technique is applied to analyze each drought condition parameter within a defined space-time frame. For example, the ANFIS can estimate the predictability of moisture presence in the environment by inputting data values of water levels in the ambient environment within a specified space-time frame. Figure 3 illustrates the structure of ANFIS, comprising distinct levels with nodes represented by node functions in each visible layer. Each layer also incorporates a hidden weight for normalization. A numerical analysis of the ANFIS model is described further.

1. The fuzzification process is performed by the Layer 1 of the ANFIS model. Each node in this layer generates a membership quality for each fuzzy collection. The node function is represented as N(1i), which is a generalized Gaussian membership function. The input value supplied to node i is denoted as x, and H(i) is the linguistic label mapped to the node function. In this study, spatiotemporal data values for multiple instances of air quality parameters are inputted into the first fuzzification layer.

$$N(1i) = \alpha H(i(x)) \tag{1}$$

2. Layer 2 of the ANFIS technique conducts a parameter-specific product to transmit the input signal to Layer 3.

$$N(2i) = \alpha A(i(x)) * \alpha B(i(y)), i = 1, 2 \tag{2}$$

   Such that A(i(x)) and B(i(y)) are Layer 2 nodes.
3. Normalization is performed in Layer 3 of the ANFIS model. At each node in this layer, the ratio of a unique firing rule to the cumulative strength is calculated using the following formulation. The normalized firing strength is denoted as w(0(i)) and is defined as follows:

$$N(2(i)) = w'(i) = (w(i))/(w(1) + w(2)) \; for \; i = 1, 2 \tag{3}$$

4. Layer 4 of the ANFIS model performs de-fuzzification, which involves computing the degree of influence of the ith rule on the generated output. The optimized attributes for the consequent are ci, d, and ri. The de-fuzzification process is carried out in the following manner:

$$N(4(i)) = w'(i)f(i) = w'(i)(c(i(x)) + d(i(y)) + r(i)) \; for \; I = 1, 2 \tag{4}$$

5. Layer 5 of the ANFIS model, also known as the output layer, generates the final output N(i(5)). This is achieved by summing up the outputs from each node in the de-fuzzification layer, as depicted ahead.

$$N(5(i)) = \sum w'(i)f(i) = \left(\sum w(i)f(i)\right) / \left(\sum w(i)\right) \qquad (5)$$

The output represents the estimated value for the specific temporal window. The overall procedure is discussed in Algorithm 2.

Algorithm 1 Dimension reduction using SVD.
Input: Data Set (D) of order t*a with t total number of tuples of data entries and a is total number of drought causing attributes.
Output: Reduced Data Set D of order t*(a-n) with n number of dimensions to be reduced.
1. Factorize the data set D
2. D(t*a) = P(t*t)Q(t*a)RT(a*a)
3. Determine P and R such that PP T = I(t) and R T R = I(a), where columns of P and R are orthogonal eigenvectors of DD T and D T D
4. Also calculate R - (a-n)*(a-n) by removing last n rows and n columns of R
5. Calculate Qt*a such that q is singular and diagonal and matrix elements are non13. Negative square roots of eigenvalues of P and R in decreasing order.
6. Also calculate Q-t*(a-n) by removing last n columns of Q
7. Calculate Dt*(a-n) as Dt*(a-n) = Pt*tQ - t*(a-n)R - T
(a-n)*(a-n)
Exit

Drought Forecasting: By comparing past and present information on drought severity levels, as shown in Fig. 4, the drought forecasting layer can make accurate predictions about the occurrence of drought in the future. The Holt-Winters method, one of the most used exponential smoothing techniques, is utilized to do this. Future drought severity may be predicted using either Holt's approach, which just accounts for trend, and level, or the Holt-Winters technique, which also accounts for seasonality. Part of this procedure involves identifying the trend, level, and seasonality components.

$$Level: \quad A(t) = \alpha(I(t))/(C(t-q)) + (1-\alpha)(A(t-1) + B(t-1))$$
$$Trend: \quad B(t) = \beta (X(t) - -A(t-1)) + (1-\beta)(B(t-1))$$
$$Seasonality: \quad C(t) = e(I(t))/(A(t-1)) + (1-e)(C(t-q))$$

where A(t), B(t), and C(t) are respective variables at time t, and model parameters are denoted as α, β, and $\in$ respectively.

Algorithm 2 Forecasting Drought Severity
Input: Drought Severity Data set.
Output: Drought Severity category at time t
1. Initialize level, trend, and seasonal components
2. Determine the updated value of level, trend and seasonal components
3. Determine the forecasted value for t=t+h
Exit

Cloud-based Data Storage: This layer refers to a specific sub-layer that is responsible for storing and managing important data-related to a particular area. In this sub-layer, we can get details about the region, including its borders and location, and we can also find information about the current drought situation, like how severe it is. Furthermore, this sub-layer includes drought-related predictions and forecasts for the future of the region, which may be used for planning and preparation purposes. Because it is all saved in the cloud, anybody with an internet connection may see, analyze, and use this data whenever they need it.

Information Communication Layer: Agencies tasked with managing water supplies, those tasked with managing the effects of drought, and others may all benefit from the assessments saved at the cloud storage sub-layer to help them deal with the issue. They may make preparations to mitigate the short- and long-term drought consequences.

4 Experimental Simulation

Performance esimtation is performed in terms of Data Perception, Prediction using ANFIS, Forecasting utilizing the Holt-Winters technique, and Predictive Accuracy Assessment.

4.1 Data Perception

The drought characteristics data used in the proposed system was acquired from the UCI repository which comprises streamflow, groundwater, and soil moisture, respectively. More than 14-day data comprising 52365 instances are acquired and boot-strapped for validation purposes. The simulation was deployed over an Intel i7 processor with 32 GB RAM and a 3.12 GHz clock speed.

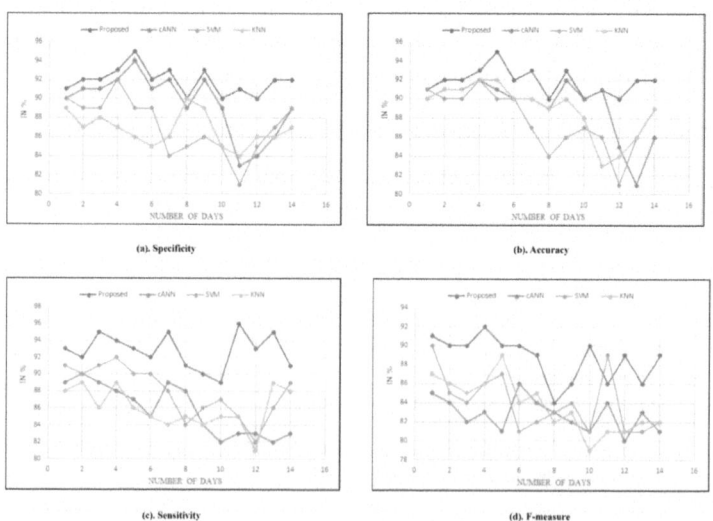

Fig. 4. Comparative Analysis

4.2   Prediction Using ANFIS

ANFIS machine learning model is implemented in the Python platform. Moreover, the outcomes are contrasted with state-of-the-art AI mechanisms for determining system enhancement. Accuracy is an important measure for performance estimation. To evaluate the accuracy of the suggested model, various statistical metrics including Sensitivity, Specificity, Accuracy, and F-measure are utilized. Benchmark models including K-Nearest Neighbor (K-NN), conventional Artificial Neural Network (cANN), and Support Vector Machine (SVM) are used for comparison purposes. It is important to note that only the predictive mechanisms are changed while the rest of the process remains consistent throughout testing. Figure 4(a) presents a comparison of the suggested model with other prediction models in terms of specificity. The suggested model achieves an average accuracy of 93.25% after 14 days of use, while the mean specificity values for the SVM, cANN, and K-NN models were 91.69%, 90.27%, and 89.65% respectively. The proposed model also shows an improvement in performance across longer periods, which is indicative of better specificity. Figure 4(b) shows a graph comparing the proposed model's accuracy to that of other models. With a mean accuracy of 93.69%, the suggested model shows that it is better. When compared, cANN accurately predicted 90.48% of data values, SVM recorded 90.13% accuracy, and K-NN obtained 88.47% accuracy. This demonstrates how well the pro-posed method works for identifying drought traits. The proposed system's sensitivity analysis is shown in Fig. 4(c). According to the results, out of all the models that were considered, the one that was proposed had the greatest sensitivity, at an astounding 93.51%. Although K-NN (88.37%), cANN (91.08%), and SVM (90.08%) all show low sensitivity levels. Henceforth, the F-measure is an important statistical metric for evaluating the proposed model's performance. For 14 days, the F-measure val-ues are shown in Fig. 4(d). With a total of 93.88%, the F-measure values of the suggested system are greater than those of earlier

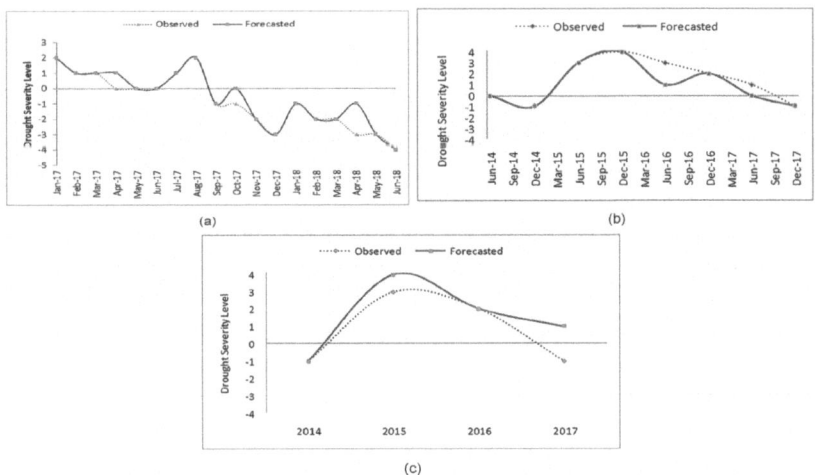

Fig. 5. Monthly Forecast (a) 1 month, (b) 6 months, and (c) 1 year

models. These results show that the proposed model has a statistically better performance efficiency when it comes to drought prediction.

4.3   Forecasting Using Holt-Winters Model

The implementation of the Holt-Winters technique, a predictive method, is carried out using the Python programming language. Specifically, the Python IDE software is utilized in this system, which is deployed on Amazon EC2. To calculate the forecast, the 'forecast' package's HoltWinters function is employed. This function takes the output of the layer as input for drought prediction. The resulting predicted values for drought severity are displayed in Fig. 5, representing three different periods: 1 month, 6 months, and 1 year. In addition to the statistical analysis, this section also presents the assessment based on error rate. Table 1 presents Mean Absolute Error (MAE), Mean Square Error (MSE), and Root Mean Square Error (RMSE) values to assess the accuracy of forecasts. The statistical analysis reveals that as the prediction duration increases, its accuracy declines.

Limitations and Future Scope

The work discusses the impact of severe drought on a region's hydrological balance, agriculture, animal habitat, and economy. It highlights the need for a reliable method of drought forecasting and identifies the limitations of existing drought indices. The research proposes a fog-based paradigm for drought predictive analysis, utilizing the IoT and singular vector decomposition to compress data at the fog level. It also employs the Holt-Winters approach to predict future drought conditions and evaluates the severity of individual drought occurrences using an ANFIS mechanism. The study demonstrates the effectiveness of the proposed system through its implementation using an online UCI dataset, showcasing its forecasting efficacy and prediction efficiency. Every research has several limiting aspects that can be explored for future research.

1. The current research assumes optimal network efficiency. This can be explored for future aspect.
2. Further research into ML and AI methods to strengthen drought prediction models and make them more accurate. To do this, it may be necessary to use ensemble techniques, reinforcement learning, and deep learning algorithms in order to enhance the accuracy of forecasts.
3. Experimenting with a multi-dimensional approach to drought prediction by integrating multiple data sources, such as climate models, satellite imaging, and remote sensing data.
4. Building real-time monitoring systems to alert people of possible droughts using Internet of Things technologies. In order to respond proactively to developing drought conditions, this might entail using data streaming, automated alarm systems, and sensor networks.

Table 1. Forecasting Accuracy Analysis

Error Rate	1 Month	6 month	1 year
MAE	0.17	0.38	0.75
MSE	0.28	0.62	1.23
RMSE	0.53	0.80	1.12

5 Conclusion

In this research, a Cloud-centric layered architecture is developed for the anticipation and forecasting of drought situations. The system effectively monitors and analyzes drought conditions by deploying networked sensors in strategic locations to collect primary data on various contributing factors of drought. The unique feature of this architecture is the utilization of Singular Value Decomposition (SVD) at the fog layer, which enables the transmission of only essential data to the cloud, optimizing network capacity usage. At the sub-layer of the cloud model responsible for drought prediction, an ANFIS classifier is used to estimate drought intensity categories. The Holt-Winters method is employed for predicting future drought occurrences. Comparing the results with other drought indices, the cloud layer based on Amazon EC2 achieves impressive accuracy. Results provision valuable outcomes for relevant authorities and departments to make real-time decisions. Furthermore, incorporating additional drought factors can potentially enhance the outcomes of the system.

Acknowledgments. No funding was received for the research.

Disclosure of Interests. None.

References

1. Sharma, G.S.: Review and analysis of drought monitoring, declaration and management in India. In: Working Paper 84. International Water Management Institute, Colombo, Sri Lanka (2004)
2. Bazza, M., Kay, M., Knutson, C.: Drought Characteristics and Management in North Africa and the Near East. Food and Agricultural Organization of the United Nations, Rome (2018)
3. Mizutori, M., Guha-Sapir, D.: Economic Losses, Poverty and Disasters 1998–2017. United Nations Office for Disaster Risk Reduction (2017)
4. Mishra, A.K., Singh, V.P.: A review of drought concepts. J. Hydrol. **391**, 202–216 (2010)
5. Wang, C., Bi, S.Z., Xu, L.D.: IoT and cloud computing in automation of assembly modeling systems. IEEE Trans. Indus. Inform. **10**, 1426–1434 (2014)
6. Fang, S., et al.: An integrated system for regional environmental monitoring and management based on the internet of things. IEEE Trans. Indus. Inform. **10**, 1596–1605 (2014)
7. Suciu, G., et al.: Big data, internet of things and cloud convergence—an architecture for secure e-health applications. J. Med. Syst. **39**, 141 (2015)
8. Kantarci, B., Mouftah, H.T.: Trustworthy sensing for public safety in cloud-centric internet of things. IEEE Internet things J. **1**, 360–368 (2014)

9. Sood, S.K., Mahajan, I.: A fog-based healthcare framework for Chikungunya. IEEE Internet Things J. **5**, 749–801 (2018)
10. Bayissa, Y., et al.: Comparison of the performance of six drought indices in characterizing historical drought for the upper Blue Nile Basin, Ethiopia. Geosciences **8**, 81 (2018)
11. Jang, D.: Assessment of meteorological drought indices in Korea using RCP 8.5 scenario. Water **10**(3), 283 (2018). https://doi.org/10.3390/w10030283
12. Du, K., Sun, Z., Zheng,F., Chu, J., Ma, J.: Monitoring System for Wheat Meteorological Disasters using Wireless Sensor Networks. ASABE Annual International Meeting. American Society of Agricultural and Biological Engineers, Spokane, Washington (2017)
13. Masinde, M.: An innovative drought early warning system for sub-Saharan Africa: integrating modern and indigenous approaches. Afr. J. Sci. Technol. Innov. Dev. **7**, 825 (2018)
14. Soh, Y.W., Koo, C.H., Huang, Y.F., Fung, K.F.: Application of artificial intelligence models for the prediction of standardized precipitation evapotranspiration index (SPEI) at Langat River basin, Malaysia. Comput. Electron. Agric. **144**, 164–173 (2018)
15. Jiang, B., Wang, P., Zhuang, S., Li, M., Li, Z., Gong, Z.: Detection of maize drought based on texture and morphological features. Comput. Electron. Agric. **151**, 50–60 (2018)
16. Ali, M., Deo, R.C., Downs, N.J., Maraseni, T.: Multi-stage committee based extreme learning machine model incorporating the influence of climate parameters and seasonality on drought forecasting. Comput. Electron. Agric. **152**, 149–165 (2018)
17. Demisse, G.B., et al.: Information mining from heterogeneous data sources: a case study on drought predictions. Information **8**, 79 (2017)
18. Cong, D., Zhao, S., Chen, C., Duan, Z.: Characterization of droughts during 2001–2014 based on remote sensing: a case study of Northeast China. Ecol. Inform. **39**, 56–67 (2017)
19. Hao, Z., Hao, F., Singh, V.P., Ouyang, W., Cheng, H.: An integrated package for drought monitoring, prediction and analysis to aid drought modeling and assessment. Environ. Model Softw. **91**, 199–209 (2017)
20. Vathsala, H., Koolagudi, S.G.: Prediction model for peninsular Indian summer monsoon rainfall using data mining and statistical approaches. Comput. Geosci. **98**, 55–63 (2017)
21. Maca, P., Pech, P.: Forecasting SPEI and SPI drought indices using the integrated artificial neural networks. Computat. Intell. Neurosci. **2016**, 14 (2016)
22. Ma, L., Nie, F.: A smart meteorological service model based on big data: a value creation perspective. In: International Conference Grey Systems and Intelligent Services. IEEE, Stockholm, Sweden (2017)
23. Yu, J., Lim, J., Lee, K.: Investigation of droughtvulnerable regions in North Korea using remote sensing and cloud computing climate data. Environ. Monit. Assess. **190**, 126 (2018)
24. Zou, Q., Li, G., Yu, W.: MapReduce functions to remote sensing distributed data processing—global vegetation drought monitoring as example. Softw. Pract. Exp. **48**, 1352–1367 (2018)
25. Severino, G., D'Urso, G., Scarfato, M., Toraldo, G.: The IoT as a tool to combine the scheduling of the irrigation with the geostatistics of the soils. Future Gener. Comput. Syst. **82**, 268–273 (2018)
26. Kleine Deters, J., Zalakeviciute, R., Gonzalez, M., Rybarczyk, Y.: Modeling pm2. 5 urban pollution using machinelearning and selected meteorological parameters. J. Electr. Comput. Eng. **2017**, 1–14 (2017)
27. Gautam, J., Gupta, A., Gupta, K., Tiwari, M.: Air pollution concentration calculation and prediction. In: Rathore, V.S., Worring, M., Mishra, D.K., Joshi, A., Maheshwari, S. (eds.) Emerging Trends in Expert Applications and Security: Proceedings of ICETEAS 2018, pp. 245–251. Springer Singapore, Singapore (2019). https://doi.org/10.1007/978-981-13-2285-3_30
28. Brandt, J., Christensen, J.H., Frohn, L.M., Zlatev, Z.: Operational air pollution forecast modelling using the THOR system. Phys. Chem. Earth, Part B: Hydrol. Oceans Atmos. **26**(2), 117–122 (2001). https://doi.org/10.1016/S1464-1909(00)00227-6

29. Park, S., et al.: Predicting PM10 concentration in Seoul metropolitan subway stations using artificial neural network (ANN). J. Hazardous Mater. **341**, 75–82 (2018). https://doi.org/10.1016/j.jhazmat.2017.07.050
30. Li, M., Wang, W.-L., Wang, Z.-Y., Xue, Y.: Prediction of pm2. 5 concentration based on the similarity in air quality monitoring network. Build. Environ. **137**, 11–17 (2018).
31. Karatzas, K., Katsifarakis, N., Orlowski, C., Sarzyński, A.: Revisiting urban air quality forecasting: a regression approach. Vietnam J. Comput. Sci. **5**(2), 177–184 (2018). https://doi.org/10.1007/s40595-018-0113-0
32. Franceschi, F., Cobo, M., Figueredo, M.: Discovering relationships and forecasting pm10 and pm2. 5 concentrations in Bogotá, Colombia, using artificial neural networks, principal component analysis, and k-means clustering. Atmos. Pollut. Res. **9**(5), 912–922 (2018)

Deep DWT Feature Modeling for Alzheimer's Disease Prediction: A Unique Approach

Santosh Kumar Tripathy[1]([✉]), Chandan Kumar Behera[2], Kartik Shankar Gadupa[2], and Rudra Kalyan Nayak[2]

[1] Department of Computer Science and Engineering, National Institute of Technology Patna, Patna, Bihar, India
santoshtripathy1448@gmail.com
[2] School of Computing Science and Engineering, VIT Bhopal University, Bhopal-Indore Highway, Kothrikalan, Sehore, Madhya Pradesh 466114, India

Abstract. The procedure of computer-aided diagnosis for Alzheimer's disease forecast has gotten better using deep learning principles. Such techniques rely on deep spatial features for classification. Nevertheless, the systems' capabilities have to be enhanced. Furthermore, deep-frequency domain characteristics for Alzheimer's diagnosis are understudied. This article fills the said research gaps by exploiting deep frequency features for disease prediction. The model utilizes the high-level and low-level frequency-based DWT features. Then a two-stream deep CNN architecture is designed where the low- and high-level frequency components are inputted. Each stream is a multi-layer of conventional CNN layers. Concatenation of the deep characteristics of the two streams is performed, and then the information is sent to a multi-layer of dense layers for illness classification. Achievement of the suggested model is evaluated by experimenting on OASIS and the Kaggle AD datasets and the obtained results outperform the existing state-of-the-arts.

Keywords: DWT · Alzheimer's Disease · Deep Frequency Features · Dual Stream CNN

1 Introduction

Brain disorders like Alzheimer's disease (AD) are persistent, degenerative, and it predominantly impacts cognitive processes, including memory, thinking, and behaviour. AD predominantly affects older adults, causing dementia and its exact cause is still not fully understood. While age is a significant risk factor, genetic factors like the APOE gene and lifestyle factors such as cardiovascular health, physical activity, and social engagement can contribute to disease development in another way. As the world's population continues to age, Alzheimer's disease has emerged as a critical issue for public health. Beyond the personal sphere, Alzheimer's disease imposes an extensive societal burden, stretching healthcare systems, straining economies, and demanding innovative solutions. Recently, the WHO [1] reported that around 60% to 70% of patients suffering from dementia related issues and it is about to increase every year. It is essential

M. Khurana et al. (Eds.): ICMLA 2024, CCIS 2238, pp. 104–115, 2025.
https://doi.org/10.1007/978-3-031-75861-4_10

to recognize sickness in its earlier phases, such as mild to moderate cognitive impairment (MCI), to facilitate prompt intervention and development of medications that can change the condition. Because of their capacity to extract subtle patterns and characteristics from varieties of data modalities (like MRI, PET etc.), several methodologies have garnered a significant amount of interest in the field of Alzheimer's disease research. The integration of multimodal data becomes a focus of recent research, as combining the information from MRI, PET, and genetic data is having potentiality to improve the efficacy of AD diagnosis and prediction. But, due to lack of availability of such dataset suppresses the research in this direction. However, the recent works are utilizing MRI scans for AD disease prediction. Progression can be seen using convolutional neural networks (CNNs), recurrent neural networks (RNNs), and graph convolutional networks (GCNs) based deep models, in order to handle a variety of activities concerned with AD. On the basis of neuroimaging data, the CNN-based Alzheimer's disease diagnostic models have shown outstanding accuracy in discriminating Alzheimer's disease patients from healthy controls. The MRI images inputted to these models give rise to the extraction of detailed spatial information. However, there is little investigation that examines the use of deep frequency characteristics for the AD classification. Because of this, the work being done here is driven by the need to extract deep frequency information to fill the research gap. The followings are the list of contributions made to this paper.

- Two stream based deep-CNN architecture has been designed to maximise deep frequency characteristics for the identification of AD. It is the approximation coefficient of the discrete wavelet transform of the picture that acts as the input to the first stream. Whereas, the stack of detailed coefficients acts as input to the second stream.
- To illness classification, the characteristics of two streams are combined and then distributed to a neural network with many layers.

Article organisation follows as: Sect. 2 discusses a literature review behind the work, Sect. 3 provides a detailed illustration of the suggested model, Sect. 4 shows a detailed description of dataset and experimental setup, Sect. 5 demonstrates the results analysis on the datasets, and finally Sect. 6 concludes the work.

2   Related Work

AI techniques to diagnose AD have been tremendously increasing since last decades. Methods of machine learning are used rather often in various strategies. A simplified categorization of AD techniques is shown in Fig. 1. In accordance with the image that is shown in Fig. 1, the methods that are used for AD diagnosis may be divided into two distinct categories: traditional machine learning and advanced machine learning, which includes deep learning approaches. Jongkreangkrai et al. [2] extracted useful features from MRI and utilized SVM for AD classification.

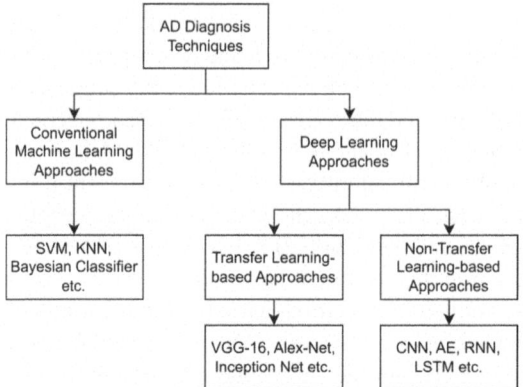

Fig. 1. Classification of AD Approaches

Long et al. [3] utilized Bayesian network-based data modelling approach for the task. Sethuraman et al. [4] have used the SVM and KNN algorithms for AD classification. Beheshti et al. [5] extracted seven features including information gain (IG), statistical dependency (SD) etc. from the MRI scans. Authors performed AD classification using SVM. A two-level categorization technique built on SVM was recently proposed by Poloni et al. [6]. However, these approaches depend on the handcrafted feature modelling which are not adaptive in nature. Such limitations could be overcome by adopting automated feature modelling using deep learning approaches.

The deep learning techniques broadly fall in two categories: transfer learning and non-transfer learning-based approaches. The former technique utilizes the features of state-of-the-art architectures but not limited to VGG-16, VGG-19, and Inception Nets. Recently, Naz et al. [7] exploited features from DenseNet, GoogLet, ResNet etc. for AD classification. Shahwar et al. [8] designed a QVC and adopted transfer learning to increase the performance. Similar work is done by Chui et al. [4] where the authors increased the training samples using GAN. Orouskhani et al. [9] developed a deep triplet network using VGG and trained the model by minimizing a conditional loss function for disease classification.

Several non-transfer learning-based approaches were also developed in contrast to transfer learning-based strategies. Such methods have utilized the baseline CNN, GCN, auto encoder and sequential models for disease prediction. Recently, Begum et al. [10] proposed 3D CNN and utilizes deep spatial features for the objective function. Hajamohideen et al. [11] have developed a Siamese-CNN and diagnose the AD minimizing a triplet-loss function. By clubbing CNN with SVM, Houria et al. [12] presented a hybrid strategy. In this technique, the authors have employed CNN to extract features automatically. A capsule CNN based autoencoder architecture developed by Ansingkar et al. [13]. The authors optimised the underlying network by minimizing a hybrid equilibrium loss for the diagnosis. Shi et al. [14] proposed a multiple loss-based autoencoder that is limited by GAN. Lahmiri et al. [15] suggested a hybrid approach using a CNN model for obtaining features and a KNN with Bayesian optimisation for AD stage classification.

However, the validation dataset is limited in sample size. Two stage methodology is proposed by El-Sappagh et al. [16] In the first stage, multi-class classification is performed, and in the second stage, a regression model is developed to anticipate the conversion time from MCI to AD using long-short term memory (LSTM) networks. The discussed methods exploited spatial features for illness classification, but there is a deficiency in works that utilize frequency-based features using deep learning. Because of this, the suggested model takes these issues into account.

3   Proposed Methodology

This article makes a suggestion to propose a novel methodology using deep frequency features for AD classification. The general structure of the suggested solution is displayed in Fig. 2. To begin, the famous discrete wavelet transform (DWT) is adopted to extract frequency-based properties from the MRI pictures. Upon applying the DWT over the input, we will get four frequency-based feature maps, one approximation coefficient and three detailed coefficients. All the detailed coefficients are accumulated to form a 3D structure. The image and it's two frequency component using DWT i.e., approximation and stacked detailed coefficients are illustrated in Fig. 3.

After obtaining the aforementioned two frequency components from images, a two stream of convolution neural network is planned out for disease classification. The specifics of the model are shown in Fig. 2. The input to the first stream is the approximation coefficient whereas the stacked detail coefficients are inputted to another stream. Each stream contains eight convolution layers along with three max pooling layer of size (2×2). Table 1 provides details of each convolution layer used in developing the proposed model. The activation function used in these layers is ReLU and also the kernel regularization is also applied to all the layers. The streams utilize the deep features from the frequency input. Concatenation and flattening of the feature maps of the two streams are the two steps that are taken. Full connectivity is established between the flattened features and a dense layer that is comprised of 256 neurons that have an activation function of ReLU. This dense layer is fully connected with the output layer having four neurons to classify the disease labels. The output layer is having SoftMax activation function and the neurons belong to Non-dementia (ND), mild dementia (MD), moderate dementia (MoD), and very mild dementia (VMD). For the dilated convolution layers, the kernel regularizers are used.

Table 1. Detail information of layers used in the proposed model

Kernel Name	Kernel Shape	Filters
k11-k18	$9 \times 9, 8 \times 8, 7 \times 7, 6 \times 6, 5 \times 5, 4 \times 4, 3 \times 3, 3 \times 3$	5, 10,15,20,25,30,35,40
K21-k28	$9 \times 9, 8 \times 8, 7 \times 7, 6 \times 6, 5 \times 5, 4 \times 4, 3 \times 3, 3 \times 3$	6,12,18,24,30, 36, 42, 48

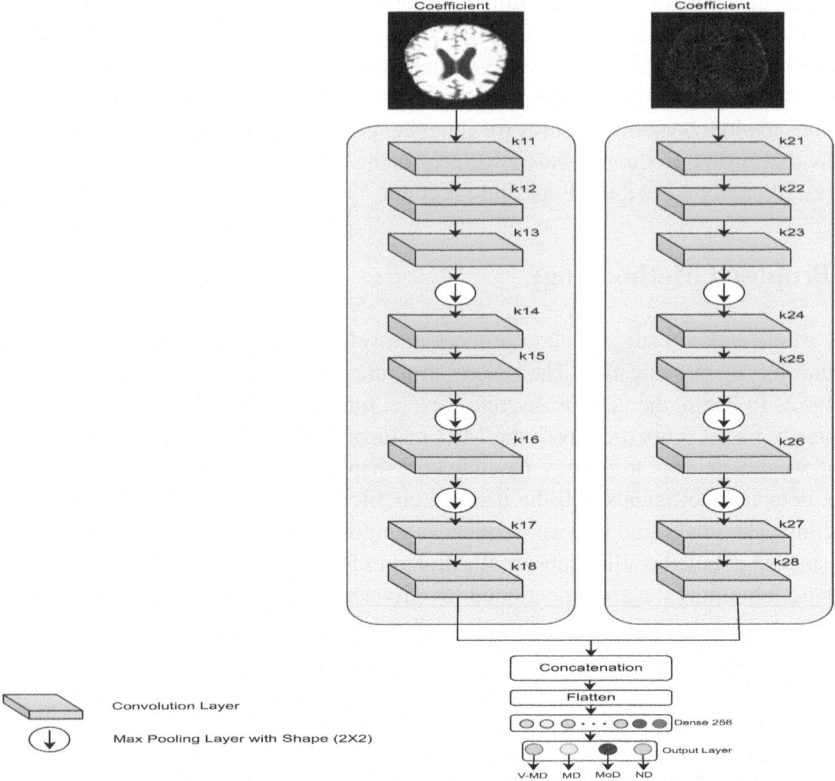

Fig. 2. Detail architecture of the proposed methodology

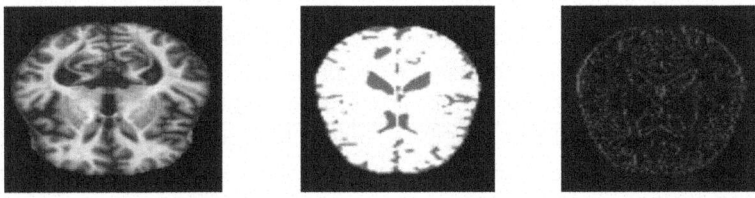

Fig. 3. Sub figures (a), (b), and (c) shows MRI scan, its approximation and stacked detailed coefficient respectively

3.1 Loss Function and Optimization

Let the ground-truth labels and the projected score for all the training samples are denoted as $X = [x_1, x_2, \ldots\ldots, x_P]$ and $Y = [y_1, y_2, \ldots\ldots, y_P]$ respectively. Let, θ represent a set containing all the network's trainable parameters. The Adam [17] optimizer is used in order to minimise the categorical cross entropy loss that occurs between X and Y in order to optimise the suggested methodology.

4   Dataset and Experimental Setup

4.1   Dataset Details

This work demonstrate experiment on two publicly available AD dataset such as OASIS-1 [18] and Kaggle AD dataset [19]. The basic statistics of these datasets are mentioned in Table 1. Following the work that Murugan and his colleagues have done [20], we have split the Kaggle AD dataset randomly 80%, 10% and 10% for training, validation and testing (Table 2).

Table 2. Basic Statistics of the Datasets

Disease Stages	Sample Size of OASIS-1 dataset	Sample Size of Kaggle Dataset
Non-Demented (ND)	306	2400
Moderate-Demented (MoD)	2	64
Very-Mild-Demented (VMD)	65	2240
Mild-Demented (MD)	26	896

Due to the availability of a smaller samples in OASIS-1, we have followed the work of Chui et al. [21] and performed two-fold cross-validation for results analysis. Figure 5 shows few samples of both OASIS and the Kaggle AD dataset.

4.2   Experimental Setup

Keras layers are used throughout the whole of the programme, which is written in Python. The model that is being presented is trained for a total of 500 epochs, and early stopping is used to prevent the model from being overfit. 0.01 is the figure that represents the learning rate. It is set to 32 for the small batch size, and the parameter value for the kernel regularizer is set to 0.01.

5   Results Analysis and Ablation Study

5.1   Results Analysis on OASIS-1 Dataset

A level of accuracy of 99.75% was attained by the suggested model when applied to the OASIS-1 dataset. The model performed very well, achieving a miss rate of 0.25%, accuracy of 99.63%, recall of 99.91%, and an F1-Score of 99.77%. The confusion matrix and ROC curve are displayed in Figs. 4 and 5 respectively. The proposed model exhibits well on all predicting all the test label. Only one label of the non-demented (ND) is not classified to it. The suggested model's performance is also contrasted with a few current state-of-the-art methods that used the OASIS dataset to develop their models, which can be looked at Table 3. These models include Deep Net, Ensemble Hybrid

Deep Net, CNN + Optimal KNN with BO, Ensemble-Deep CNN, Conv-TL, and ANN, which have accuracies of 99.68%, 95.23%, 94.96%, 93.18%, 93.80%, and 92.00%, respectively. However, in comparison with these models, the presented model performs well and tops Table 3.

Table 3. Comparative study of results on OASIS-1 dataset.

Models	Accuracy (%)	Miss Rate (%)	Precision (%)	Recall (%)	F1-Score (%)	AUC (%)
Ensamble-Deep CNN[22]	93.18	6.82	93.00	92.00	92.00	NA
	94.96	5.04	NA	NA	NA	NA
CNN + Optimal KNN[15]	74.90	25.10	NA	NA	NA	NA
	92.00	8.00	96.20	87.70	91.70	NA
Gupta et al. [23]	95.23	4.77	NA	NA	NA	NA
	93.80	6.20	91.00	94.6	NA	NA
ANN[24]	98.99	1.01	NA	NA	NA	96.02
	99.68	0.32	NA	NA	NA	99.99
Ensamble-Hybrid Deep Net [25]	99.75	0.25	99.63	99.91	99.77	99.99
Proposed Model	99.75	0.25	99.63	99.91	99.77	99.99

5.2 Results Analysis on Kaggle AD Dataset

91.09% is the performance accuracy achieved by the proposed model. The model's miss rate, precision, recall, and F1-Score are 8.91%, 87.25%, 78.42% and 81.94% respectively. Figures 6 and 7 respectively provide the confusion matrix and the ROC diagram. It has been determined that the suggested model has an average AUC of 95.59%. A quantitative results analysis is demonstrated in Table 4. Other models such as Landmark Feature Modelling [26], ADDTLA [27], DCNN-VGG19 [28], Inception-V4 [29], and DEMNET [20], and accuracy scores of 79.02%, 91.70%, 77.66%, 73.75%, and 85% were attained by individuals, respectively. The ADDTLA achieved better performance than the models illustrated in Table 4.

The proposed scheme exceeds the other techniques provided in Table 4, but degrades in comparison with ADDTLA. The performance on MD and VMD samples are low. The same thing can be identified in its ROC curve. The future research will focus on developing more generalized models.

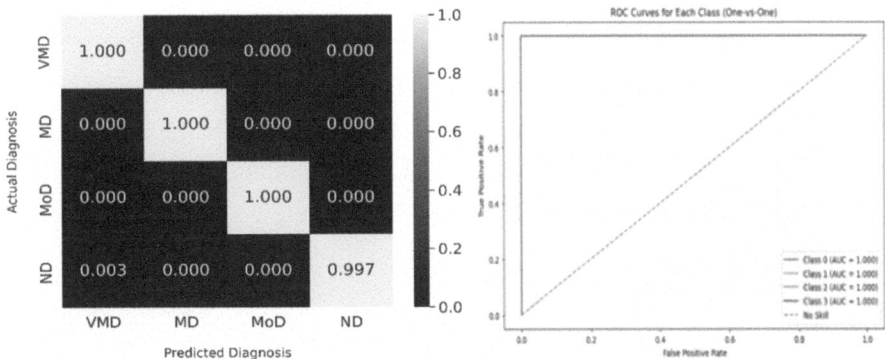

Fig. 4. Confusion matrix heat map obtained for OASIS-1.

Fig. 5. ROC curve obtained for OASIS-1 dataset

Table 4. Comparative results analysis on the AD dataset.

Models	Precision (%)	Accuracy (%)	Recall (%)	Miss Rate (%)	F1-Score (%)
Landmark Feature Modelling [26]	NA	79.02	NA	20.98	NA
ADDTLA [27]	91.50	91.70	93.70	8.30	92.50
DCNN-VGG19 [28]	58.48	77.66	36.67	22.34	45.05
DEMNET [20]	80.00	85.00	88.00	15.00	83.00
Inception-V4 [29]	NA	73.75	NA	26.25	NA
Proposed Model	87.25	**91.56**	78.42	8.91	81.94

5.3 Ablation Study on OASIS-1 Dataset

We made use of the OASIS-1 dataset for ablation investigation on the proposed approach and considered individual stream's behaviour for disease classification.

Thus, the following two sub models are obtained from the proposed architecture.

- Sub Model-1: It consists of the first stream along with dense layers.
- Sub Model-2: It consists of the second stream along with dense layers.

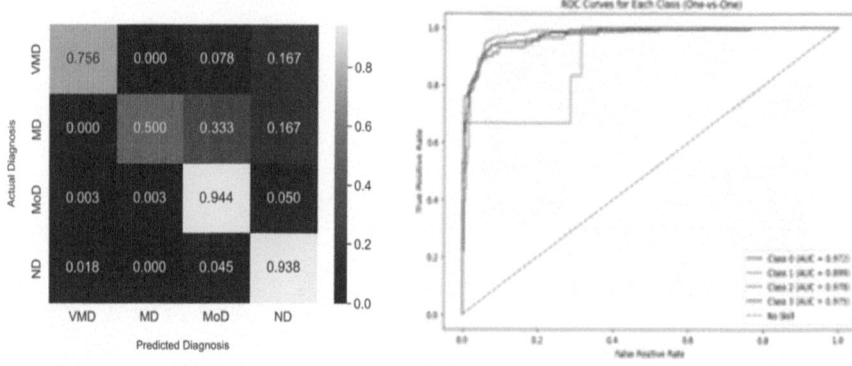

Fig. 6. Confusion matrix heat map obtained for Kaggle AD

Fig. 7. ROC curve obtained for Kaggel AD dataset

The attainments of ablation study are displayed in Table 5. The confusion matrix of these two models is mentioned in Figs. 8 and 9 respectively. The Sub Model-1 and Sub Model-2 obtained an accuracy of 98.00% and 99.50% respectively on the OASIS-1 dataset. The AUC scores of the sub-models are of 99.68% and 99.9% respectively. Thus, the Sub Model-2 which exploits deep detailed coefficients are very essential for the disease prediction.

Table 5. Comparative results analysis of ablation study

Models	AUC (%)	Accuracy (%)	Precision (%)	Miss Rate (%)	Recall (%)	F1-Score (%)
Sub Model1	99.68	98.00	97.29	2.00	99.34	98.24
Sub Model2	99.9	99.50	99.83	0.50	99.24	99.53

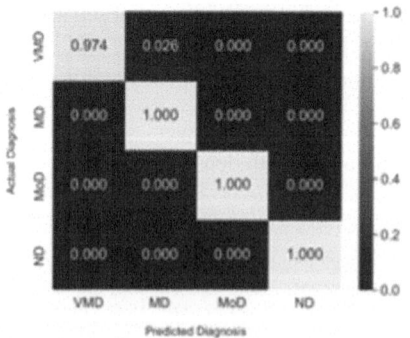

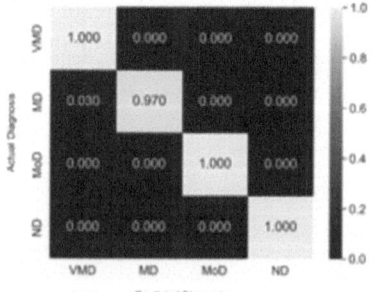

Fig. 8. Confusion matrix of Sub Model-1

Fig. 9. Confusion matrix of Sub Model-2

6   Conclusion

This article proposed a deep frequency feature modelling for AD stage classification and the proposed two-stream deep model has exploited these features. The approximation and stacked detailed coefficients are inputs to the two streams. These coefficients are obtained by the application of DWT on images. The experiments are demonstrated on two datasets such as OASIS-1 and Kaggle AD. The suggested model achieved 99.75% and 91.09% accuracies on OASIS-1 and Kaggle AD respectively. The suggested model performs better on the OASIS-1 dataset but not quite well on the Kaggle AD dataset. In addition, an ablation investigation on the OASIS-1 dataset is performed. In this study two sub models are obtained from the proposed architecture and results were obtained on the OASIS-1 dataset. The details are mentioned in sub-section C of chapter V. The second stream which exploits deep detailed coefficients demonstrates better performance than the first stream. In conclusion, the suggested model performs fairly satisfactorily overall. However, the future study will focus on developing generalized model for the AD detection.

References

1. Global action plan on the public health response to dementia. http://apps.who.int/bookorders
2. Jongkreangkrai, C., Vichianin, Y., Tocharoenchai, C., Arimura, H.: Computer-aided classification of Alzheimer's disease based on support vector machine with combination of cerebral image features in MRI. J. Phys.: Conf. Ser. Instit. Phys. Publishing (2016). https://doi.org/10.1088/1742-6596/694/1/012036
3. Long, X., Chen, L., Jiang, C., Zhang, L.: Prediction and classification of Alzheimer disease based on quantification of MRI deformation. PLoS ONE **12**(3), e0173372 (2017). https://doi.org/10.1371/journal.pone.0173372
4. Sethuraman, S.K., Malaiyappan, N., Ramalingam, R., Basheer, S., Rashid, M., Ahmad, N.: Predicting Alzheimer's Disease using deep neuro-functional networks with resting-state fMRI. Electronics **12**(4), 1031 (2023). https://doi.org/10.3390/electronics12041031
5. Beheshti, I., Demirel, H., Matsuda, H.: Classification of Alzheimer's disease and prediction of mild cognitive impairment-to-Alzheimer's conversion from structural magnetic resource imaging using feature ranking and a genetic algorithm. Comput. Biol. Med. **83**, 109–119 (2017). https://doi.org/10.1016/j.compbiomed.2017.02.011
6. Poloni, K.M., Ferrari, R.J.: Automated detection, selection and classification of hippocampal landmark points for the diagnosis of Alzheimer's disease. Comput. Methods Programs Biomed. **214**, 106581 (2022). https://doi.org/10.1016/j.cmpb.2021.106581
7. Naz, S., Ashraf, A., Zaib, A.: Transfer learning using freeze features for Alzheimer neurological disorder detection using ADNI dataset. Multimed. Syst. **28**(1), 85–94 (2022). https://doi.org/10.1007/s00530-021-00797-3
8. Shahwar, T., et al.: Automated detection of Alzheimer's via hybrid classical quantum neural networks. Electronics **11**(5), 721 (2022). https://doi.org/10.3390/electronics11050721
9. Orouskhani, M., Zhu, C., Rostamian, S., Shomal Zadeh, F., Shafiei, M., Orouskhani, Y.: Alzheimer's disease detection from structural MRI using conditional deep triplet network. Neurosci. Inform. **2**(4), 100066 (2022). https://doi.org/10.1016/j.neuri.2022.100066
10. Begum, A.P., Selvaraj, P.: Alzheimer's disease classification and detection by using AD-3D DCNN model. Bull. Electr. Eng. Inform. **12**(2), 882–890 (2023). https://doi.org/10.11591/eei.v12i2.4446

11. Hajamohideen, F., et al.: Four-way classification of Alzheimer's disease using deep Siamese convolutional neural network with triplet-loss function. Brain Inf. (2023). https://doi.org/10.1186/s40708-023-00184-w
12. Houria, L., Belkhamsa, N., Cherfa, A., Cherfa, Y.: Multi-modality MRI for Alzheimer's disease detection using deep learning. Phys. Eng. Sci .Med. **45**(4), 1043–1053 (2022). https://doi.org/10.1007/s13246-022-01165-9
13. Ansingkar, N.P., Patil, R.B., Deshmukh, P.D.: An efficient multi class Alzheimer detection using hybrid equilibrium optimizer with capsule auto encoder. Multimed. Tools Appl. **81**(5), 6539–6570 (2022). https://doi.org/10.1007/s11042-021-11786-z
14. Shi, R., et al.: Generative adversarial network constrained multiple loss autoencoder: a deep learning-based individual atrophy detection for Alzheimer's disease and mild cognitive impairment. Hum. Brain Mapp. **44**(3), 1129–1146 (2023). https://doi.org/10.1002/hbm.26146
15. Lahmiri, S.: Integrating convolutional neural networks, kNN, and Bayesian optimization for efficient diagnosis of Alzheimer's disease in magnetic resonance images. Biomed. Signal Process. Control **80**, 104375 (2023). https://doi.org/10.1016/j.bspc.2022.104375
16. El-Sappagh, S., Saleh, H., Ali, F., Amer, E., Abuhmed, T.: Two-stage deep learning model for Alzheimer's disease detection and prediction of the mild cognitive impairment time. Neural Comput. Appl. **34**(17), 14487–14509 (2022). https://doi.org/10.1007/s00521-022-07263-9
17. Kingma, D.P., Ba, J.L.: Adam: a method for stochastic optimization. In: 3rd International Conference on Learning Representations, ICLR 2015 – Conference Track Proceedings, pp. 1–15 (2015)
18. Marcus, D.S., Wang, T.H., Parker, J., Csernansky, J.G., Morris, J.C., Buckner, R.L.: Open access series of imaging studies (OASIS): cross-sectional MRI data in young, middle aged, nondemented, and demented older adults. J. Cogn. Neurosci. **19**(9), 1498–1507 (2007). https://doi.org/10.1162/jocn.2007.19.9.1498
19. Dubey, S.: https://www.kaggle.com/tourist55/alzheimers-dataset-4-class-of-images.
20. Murugan, S., et al.: DEMNET: a deep learning model for early diagnosis of Alzheimer diseases and dementia from MR images. IEEE Access **9**, 90319–90329 (2021). https://doi.org/10.1109/ACCESS.2021.3090474
21. Chui, K.T., Gupta, B.B., Alhalabi, W., Alzahrani, F.S.: An MRI scans-based Alzheimer's disease detection via convolutional neural network and transfer learning. Diagnostics **12**(7), 1531 (2022). https://doi.org/10.3390/diagnostics12071531
22. Islam, J., Zhang, Y.: An Ensemble of Deep Convolutional Neural Networks for Alzheimer's Disease Detection and Classification (2017). http://arxiv.org/abs/1712.01675
23. Gupta, S., Saravanan, V., Choudhury, A., Alqahtani, A., Abonazel, M.R., Babu, K.S.: Supervised computer-aided diagnosis (CAD) methods for classifying Alzheimer's disease-based neurodegenerative disorders. Comput. Math. Methods Med. **2022**, 1–10 (2022). https://doi.org/10.1155/2022/9092289
24. Bandyopadhyay, A., Ghosh, S., Bose, M., Singh, A., Othmani, A., Santosh, K.C.: Alzheimer's disease detection using ensemble learning and artificial neural networks. In: Santosh, K.C., Goyal, A., Aouada, D., Makkar, A., Chiang, Y.-Y., Singh, S.K. (eds.) Recent Trends in Image Processing and Pattern Recognition: 5th International Conference, RTIP2R 2022, Kingsville, TX, USA, December 1-2, 2022, Revised Selected Papers, pp. 12–21. Springer Nature Switzerland, Cham (2023). https://doi.org/10.1007/978-3-031-23599-3_2
25. Jabason, E., Ahmad, M.O., Swamy, M.N.S.: Classification of Alzheimer's disease from MRI data using an ensemble of hybrid deep convolutional neural networks. In: 2019 IEEE 62nd International Midwest Symposium on Circuits and Systems (MWSCAS), pp. 481–484 (2019). https://doi.org/10.1109/MWSCAS.2019.8884939

26. Zhang, J., Liu, M., An, L., Gao, Y., Shen, D.: Alzheimer's disease diagnosis using landmark-based features from longitudinal structural MR images. IEEE J. Biomed. Health Inform. **21**(6), 1607–1616 (2017). https://doi.org/10.1109/JBHI.2017.2704614

27. Ghazal, T.M., et al.: Alzheimer disease detection empowered with transfer learning. Comput. Mater. Continua **70**(3), 5005–5019 (2022). https://doi.org/10.32604/cmc.2022.020866

28. Adeola Ajagbe, S., Amuda, K.A., Oladipupo, M.A., Afe, O.F., Okesola, K.I.: Multi-classification of Alzheimer disease on magnetic resonance images (MRI) using deep convolutional neural network (DCNN) approaches. Int. J. Adv. Computer Res. **11**(53), 51–60 (2021). https://doi.org/10.19101/IJACR.2021.1152001

29. Islam, J., Zhang, Y.: A novel deep learning based multi-class classification method for Alzheimer's Disease detection using brain MRI data. In: Zeng, Y., et al. (eds.) Brain Informatics. BI 2017. Lecture Notes in Computer Science(), vol 10654. Springer, Cham. https://doi.org/10.1007/978-3-319-70772-3_20

MobileNetV2: A Proficient Convolutional Neural Network for the Classification of Date Fruits into Genetic Varieties

Sajid Faysal Fahim, Fahmida Afrose Dipti, Zareen Tasnim Nishat, Md. Maidul Azim, and Md Al-Imran(✉)

Department of Computer Science and Engineering, East West University, Dhaka 1212, Bangladesh
al.imran@ewubd.edu

Abstract. The proper categorization of date fruits based on their genetic variations is a crucial component of effective crop management and quality control in the field of agriculture. Nonetheless, the present methodologies utilized for this intention are frequently susceptible to imprecisions and have the potential to consume considerable amounts of time. This study uses images to categorize seven types of date fruit – Barhee, Deglet Nour, Sukkary, Rotab Mozafati, Ruthana, Safawi, and Sagai – based on their genetic variations. This model was particularly developed to detect the different kinds of date fruits utilizing a pioneering dataset including region-specific photos, it is capable of categorizing practically all readily available date fruits. When compared to average precision, accuracy, recall, and F-s, MobileNetV2 outperforms innovative algorithms like AlexNet and VGG16, with scores of 98.55%, 99.508%, 99.45%, 99.47%, and 99.878%, respectively. The investigation's outcomes have significant implications for improving techniques to classify date fruit accurately.

Keywords: Date Fruits · Genetic Varieties · Image Analysis · CNN · Deep Learning

1 Introduction

For thousands of years, date fruits have been grown from date palm trees and are a significant part of diets worldwide due to their sweet and nutritious qualities. The date palm is widely known for its genetic diversity, resulting in a range of varieties that are distinct in size, color, taste, and ripening time. With accurate classification of date fruits into their genetic varieties, the agricultural domain benefits from precise crop management, quality control, and assessment of commercial value [1].

Traditionally, the classification of date fruits has been carried out through a manual inspection process by experts [2]. This method involves visually assessing the external features of each fruit to identify its genetic variety [3]. The application of this approach is a challenging and time-intensive undertaking, susceptible to errors that may arise due to human fallibility, especially in instances where a substantial quantity of fruit is involved

M. Khurana et al. (Eds.): ICMLA 2024, CCIS 2238, pp. 116–129, 2025.
https://doi.org/10.1007/978-3-031-75861-4_11

[4]. Therefore, there is an increasing demand for efficient and automated techniques to classify date fruits [5]. These techniques have the potential to augment efficiency and optimize the quality control procedure. The research presented in this study offers a distinctive contribution in the form of an automated and efficient classification system, which has the potential to transform the process of date fruit quality control and variety assessment [6]. *A. Contribution.*

- We have researched a unique dataset of date fruit, which we collected from orchards in Bangladesh through images.
- To tackle the task at hand, we have developed a groundbreaking CNN-based model called MobileNetV2, using this dataset.
- This document adeptly consolidates the diverse investigations and explorations conducted in the realm of labor pathology succinctly and methodically [7].

Section 2, delves into the crucial outcomes and dialogue of our ongoing investigation, aiming to establish efficient means of identifying various types of date fruits while Sect. 3 presents, we examine a categorization method. Section 4 presents our Strategies for Parameters, Training, and Testing. Section 5 concludes with an analysis and final remarks, all with the ultimate goal of creating a precise and durable categorization scheme. Section 6 explores the examination and final comments.

2 Literature Review

In this particular study [8], the researchers delved into a wide array of date fruits and employed discernible artificial intelligence methodologies combined with machine learning-based techniques to effectively classify and elaborate on the classification task for the automatic categorization of the eatability of date fruit. They are utilizing approaches such as Boosting, Bagging, Support Vector Machine (SVM), K-nearest neighbor (KNN), and MLPs. The MLP is utilized to enhance accuracy. In this endeavor, an impressive accuracy rate of 90 to 92% is achieved.

The article [9], employs advanced techniques of machine learning, including logistic regression and artificial neural networks, to classify date fruits. These methods analyze a comprehensive set of 34 attributes, encompassing morphological, shape, and color characteristics derived from the fruits. In determining these traits, prior investigations in the field were consulted, to ascertain whether the number of extracted features affects the success of classification. By harnessing the power of machine learning, it becomes possible to accurately categorize various types of dates, with an impressive accuracy rate of 92.8%.

In this paper [10] the top model in this study, which achieved 92.7% accuracy, uses convolutional neural networks to categorize photos of dates as one of nine types. In the realm of agricultural technology and computer vision, a one-of-a-kind collection of data awaits. Within this collection, pristine photographs captured under controlled circumstances await utilization. By embarking on a journey of experimentation and education with different models, both with and without supplementary information, one can achieve remarkable levels of accuracy in classification.

In this paper [11] in this study, SVM was used to categorize various dates using their photographs. Dates contain fascinating, unique properties that might help differentiate and identify a certain date type. These qualities include form, texture, and color. To categorize the dates that can and cannot be eaten, a technique that achieves 82% accuracy was created. The created method aids the food sector and the consumer in categorizing dates based on particular quality indicators offering optimal results with particular types of dates.

In this paper [12] the authors used CNN for image classification. This project's concentration is on different sorts of dates, and doing so can assist a customer in finding an immediate solution to the problem of identifying dates very simply. The major goals of the study are to show the project's viability and evaluate its performance. There study's validation accuracy percentage is 82.67%.

The paper [13] proposes a system to automate the classification of date fruit production in Saudi Arabia. The system focuses on the post-harvesting stage, increasing production efficiency, accuracy, product quality control, and data analysis. The system uses convolutional neural network models and a new image dataset with two main classes for excellent and poor surface quality. Here they took a small dataset. The model can classify date fruit with an accuracy of 92.6%, increasing market competitiveness, reducing production costs, and increasing productivity.

In this paper [14] a study used a computer vision system to automatically classify dates based on size and color, resulting in the creation of a sorting machine prototype that can categorize mature date fruits into multiple color groups. The study also suggests a potential relationship between the fruit's color tone and saturation and its moisture content, sugar content, and acidity level.

3 Proposed System

The methodology employed in this investigation, encompassing the retrieval and examination of datasets, shall be elucidated, as will the construction of the power of state-of-the-art deep learning models like AlexNet, VGGNet, and MobileNetV2 that are at the forefront of innovation, incorporating the programming language. Python, the framework Tensorflow, and the library Keras [15]. These endeavors were undertaken on a

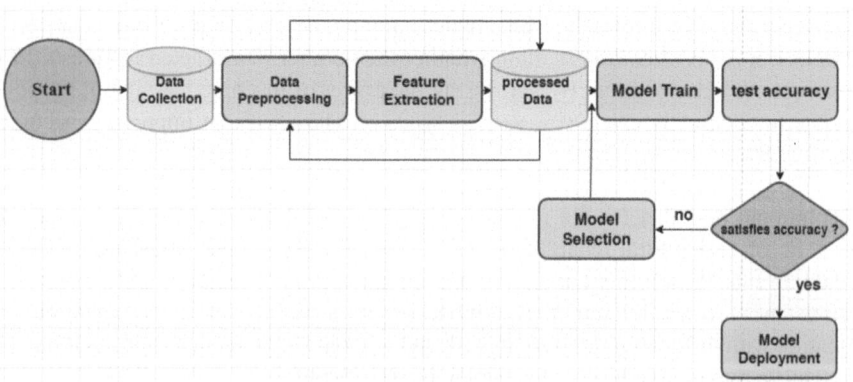

Fig. 1. Proposed Method

formidable 64-bit computing apparatus using Ubuntu 20.04.4 LTS, accompanied by the mighty presence of an AMD Ryzen 9 5900x 12 Core Processor, which reigns supreme, a whopping 128 GB of RAM, and the formidable Nvidia RTX 3080 GPU [20] (Fig. 1).

3.1 Obtaining Images

This comprehensive dataset encapsulates a harmonious blend of images, featuring seven distinct varieties of 899 date fruits images: Barhee, Deglet Nour, Sukkary, Rotab Mozafati, Ruthana, Safawi, and Sagai. Each image within the dataset is thoughtfully selected to highlight their genetic variances and capture the essence of date fruit. This dataset proves particularly advantageous for the development of deep-learning models within the context of Bangladesh, as the utilization of date fruit can vary significantly across different regions. The architectures proposed and employed within this research endeavor strive to proficiently classify each of the seven unique categories of date fruit (Fig. 2).

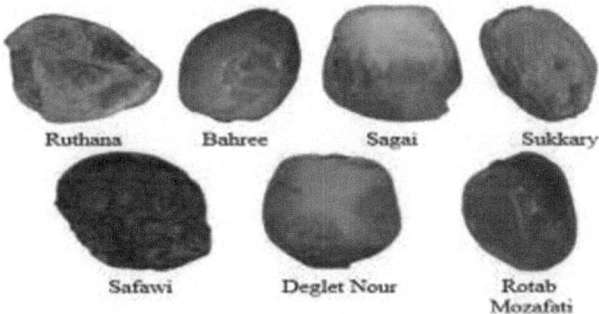

Fig. 2. Sample Dataset

3.2 Data Preprocessing

The collected images underwent a preprocessing phase to standardize the data and remove any probable bias. This process was multi-stage and involved resizing the images to a consistent resolution, color normalization to nullify any lighting differences, and cropping to eliminate extraneous background data. Furthermore, diverse data augmentation techniques were employed, including flipping, rotation, and translation, to enrich the variety of the training dataset and enhance the model's generalization capability. Overall, these methods were implemented to ensure the accuracy and reliability of the outcome.

3.3 Feature Extraction

Pre-processed images of date fruits have undergone feature extraction to identify the specific traits unique to each genetic variety. The process involved analyzing different visual components, beginning with geometric features like area, perimeter, and circularity, to capture the shape of date fruits. Next, color histograms and color moments were computed to illustrate the color distribution and intensity of date fruits. Thirdly, local binary patterns and gray-level co-occurrence matrix were calculated to capture texture information from the date fruit images. Lastly, additional discriminatory information was provided by measuring the length, width, and aspect ratio of date fruits.

3.4 Model Training and Evaluation

The partitioning of the dataset into separate training and testing sets was implemented to ensure that each set had an unbiased and proportional distribution of genetic variations. The training set played a pivotal role in training machine learning models, utilizing extracted visual features. To optimize the model's performance, hyperparameter tuning and cross-validation methods were utilized. Following the training phase, the models underwent evaluation using the testing set, with a specific focus on metrics such as classification accuracy, precision, recall, and F1-score. To assess the model's ability to differentiate between various genetic varieties, the confusion matrix was employed.

3.5 State-of-The-Art Algorithms

3.5.1 AlexNet Architecture

The architecture of AlexNet [16] is depicted in Table 1 and Fig. 3, showcasing the layers, feature maps, activation functions, and parameters, wherein the channel count is increased and subsequently decreased, consisting of five convolutional blocks and two fully connected dense layers, predicting eight class labels from a three-channel RGB image.

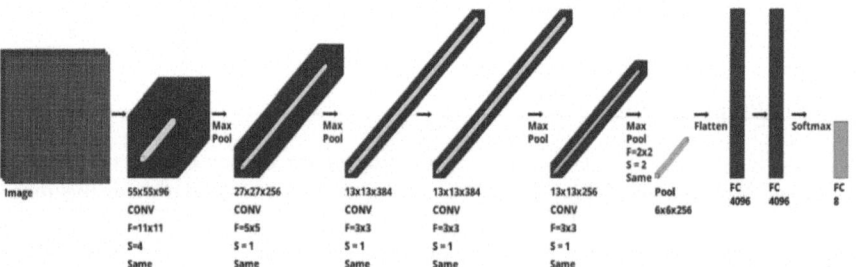

Fig. 3. AlexNet Architecture.

Table 1. The condensed data about the characteristic maps of AlexNet.

Level	The filters	Filter Options	The stride	Feature Map Size	Activation Mode
Picture				227 227 3*	
CONVO	96	11 11	4	55 55 96	ReLU
Normalizing the batch				55 55 96	
Highest pool	-	3 3	2	27 27 96	
CONVO	256	5 5	1	27 27 256	ReLU
Normalizing the batch				27 27 256	
Highest pool	-	3 3	2	13 13 256	
CONVO	384	3 3	1	13 13 384	ReLU
Normalizing the batch				13 13 384	
CONVO	384	3 3	1	13 13 384	ReLU
Normalizing the batch				13 13 384	
CONVO	256	3 3	1	13 13 256	ReLU
Normalizing the batch				13 13 256	
Highest pool	-	3 3	2	6 6 256	
dense layers				4096	ReLU
Dropping out	measure = 0.5			4096	
dense layers				4096	ReLU
Dropping out	measure = 0.5			4096	
dense layers				8*	Softmax

3.5.2 VGGNet Architecture (VGG16)

In the following Table 2 and Fig. 4, VGG16 is a very big architecture and falls under the class of generic VGG architectures. This particular design embraces a sequence of convolutional strata and a maximum pooling stratum, subsequently accompanied by supplementary convolutional strata. Additionally, two fully connected strata and a SoftMax stratum are incorporated, with the purpose of forecasting from 8 categories. This is accomplished by employing a vibrant image input with measurements of 227 × 227 × 3. [17].

Table 2. The condensed data about the characteristic maps of VGG16.

Level	The filters	Filter Options	The stride	Feature Map Size	Activation Mode
Picture				227 × 227 × 3*	
2 * CONVO	64	3 3	1	227 × 227 × 64	ReLU
Highest pool		2 2	2	113 113 64	
2 * CONVO	128	3 3	1	113 113 128	ReLU
Highest pool		3 3	2	56 56 128	
3 * CONVO	256	3 3	1	56 56 256	ReLU
Highest pool		2 2	2	28 28 256	
3 *CONVO	512	3 3	1	28 18 512	ReLU
Highest pool		2 2	2	14 14 512	
3 * CONVO	512	3 3	1	14 14 512	ReLU
Highest pool		2 2	2	7 7 512	
dense layers	25088			4096	ReLU
Dropping out	measure = 0.5			4096	
dense layers	4096			4096	ReLU
Dropping out	measure = 0.5			4096	
dense layers	4096			8*	Softmax

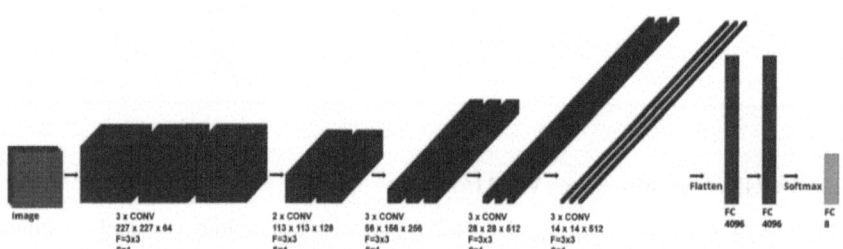

Fig. 4. VGG16 Architecture

3.5.3 Proposed Method (MobileNetV2)

The proposed blueprint adheres to the traditional method employed by AlexNet in terms of expanding and reducing the number of filters while incorporating a more streamlined structure and a reduced number of channels. It also encompasses regularization techniques, and a visual depiction can be observed in Table 3 and Fig. 5.

Table 3. The condensed data about the characteristic maps of MobileNetV2

Level	The filters	Filter Options	The stride	Feature Map Size	Activation Mode
Picture				227 227 3*	
CONVO	50	11 11	3	73 73 50	ReLU
Normalizing the batch				73 73 50	
Highest pool	-	2 2	2	36 36 50	
CONVO	100	11 11	1	36 36 100	ReLU
Normalizing the batch				36 36 100	
Highest pool		2 2	2	18 18 100	
CONVO	150	5 5	1	18 18 150	ReLU
Normalizing the batch				18 18 150	
CONVO	100	5 5	1	18 18 100	ReLU
Normalizing the batch				18 18 100	
Highest pool		2 2	2	9 9 100	
CONVO	90	3 3	1	9 9 90	ReLU
Normalizing the batch				9 9 90	
Highest pool		2 2	2	4 4 90	
Flatten				1440	
dense layers	800			800	ReLU
Dropping out	measure = 0.5				
dense layers	800			800	ReLU
Dropping out	measure = 0.5				
dense layers	8				Softmax

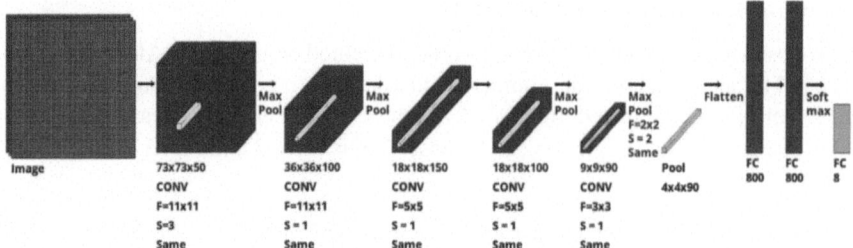

Fig. 5. MobileNetV2 Architecture.

4   Strategies for Parameters, Training, and Testing

The validation of each cross is skillfully executed through the ingenious implementation of the cross-validation technique, wherein the need for separate testing instances is obviated. With a touch of randomness, a total of 5-folds are meticulously employed, ensuring the utmost accuracy. The training dataset is graced with a generous 80% of data instances, while the validation dataset and testing dataset receive a fair 10% each, thereby completing the harmonious trifecta. The hyperparameters and relevant concerns used for training the architectures are summarized in Table 4, while Table 5, highlights the varying training times for different models, with AlexNet being the quickest and VGG16 being the slowest.

Table 4. Data utilized in training across many architectures.

The function of loss	Cross-entropy of categories
an optimizer	SGD
the learning rate	0.0001
Early halting	Validation lost patience equals 60 monitor metric
group numbers	15
The highest number of periods	300

Table 5. Milliseconds are the average training time comparison

	the average training time			
	CPU per batch	GPU Per Batc	CPU Per Epoch	GPU Per Epoch
MobileNetV2	2100 ms/step	6 ms/step	609000 ms	2000 ms
AlexNet	2110 ms/step	10 ms/step	651000 ms	3000 ms
VGG16	21000 ms/step	50 ms/step	6729000 ms	15000 ms

5 Results

5.1 Each Fold's Loss and Accuracy

In this segment, we examine the fluctuations in accuracy and loss across epochs in various datasets, finding that the patterns remain consistent across different folds for a specific structure, thus focusing on a single fold for analysis.

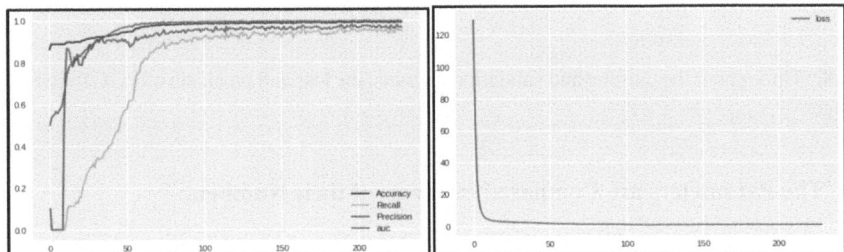

Fig. 6. Throughout the training and validation datasets, the loss and precision of MobileNetV2 in fold-5 change.

The convergence of MobileNetV2 in Fig. 6 is a gradual process, with occasional spikes caused by abrupt gradient changes, but overall the variation becomes stable.

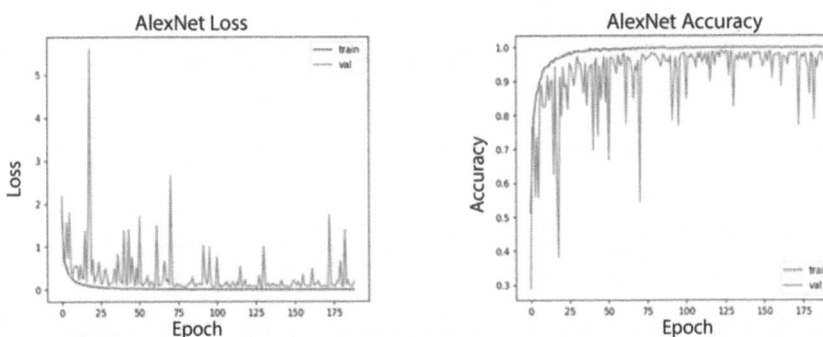

Fig. 7. Throughout the training and validation datasets, the loss and precision of AlexNet in fold-5 change.

Like MobileNetV2, in Fig. 7, AlexNet exhibits a gradual convergence while minimizing loss and improving accuracy, although it demonstrates more abrupt changes compared to MobileNetV2, indicating that MobileNetV2 offers enhanced performance through stability in variation.

The intricate parameter set of VGG16 hinders convergence and causes overfitting, leading to a significant drop in accuracy for the testing dataset despite the improvement in accuracy and loss reduction in the training dataset, as shown in Fig. 8.

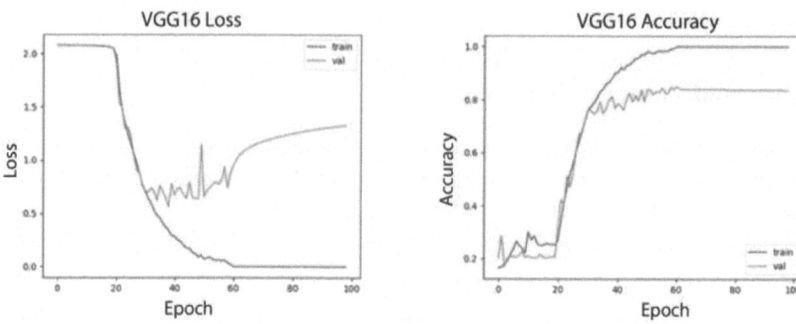

Fig. 8. Throughout the training and validation datasets, the loss and precision of VGG16 in fold-5 change.

5.2 The Parameters are Compared in Terms of their Numbers

When choosing a Deep Learning Approach Model, consider factors such as training time, memory usage, generalization ability, and interpretability, as different models like MobileNetV2, VGG16, and AlexNet offer varying advantages and trade-offs in terms of parameters, architecture, and training process, thus making it important to evaluate the specific needs of the dataset in Table 6.

Table 6. Total variable and adaptable variable contrast

Design Title	a total number of parameters	The total number of parameters for training
MobileNetV2	3,258,632	3,286,638
AlexNet	56,316,626	56,316,672
VGG16	130,292,332	130,290,336

5.3 The Size of Each Architecture's Preserved Weights

Table 7 displays the average weight file size for different architectures after training in various folds, emphasizing the noticeable reduction in the size of the proposed architecture compared to others, while still maintaining satisfactory accuracy on the dataset,

Table 7. The hdf5 format unveiled the weight files' size for all systems.

Design Title	The stored weights' size in Megabytes
MobileNetV2	13.7
AlexNet	460.60
VGG16	536.36

which is a remarkable accomplishment, particularly in practical real-life situations where performance may be slightly compromised.

5.4 Matrix Analyses

The research utilized Eqs. 1, 2, 3, 4, and 5 to evaluate the matrix, and presented the average accuracy in Table 8 after running multiple trials for each architecture in each fold.

$$The\ accuracy = \frac{TP + TN}{FP + FN + TP + TN} \tag{1}$$

$$Precision = \frac{TP}{FP + TP} \tag{2}$$

$$Recall = \frac{TP}{FN + TP} \tag{3}$$

$$F - score = \frac{2 * (Recall + Precision)}{(Recall + Precision)} \tag{4}$$

$$Specificity = \frac{TN}{TN + FP} \tag{5}$$

In Table 8, MobileNetV2 consistently delivers excellent performance across all folds, often matching or surpassing AlexNet despite being a lighter model. VGG16, on the other hand, consistently performs poorly in all folds, potentially due to insufficient sample size and dataset variations for training its 134 million parameters.

Table 8. In each fold of cross-validation, diverse designs are tested and the average of several trials is calculated.

	Acc	Pre	Re	F-S	Spe
AlexNet	99.20	99.25	99.86	99.401	99.799
VGG16	77.5	80.47	80.76	80.37	96.247
MobileNetV2	99.70	99.416	99.42	99.48	99.87

6 Conclusion

The automation of agriculture is crucial, especially in developing countries like Bangladesh where they are lagging far behind. The MobileNetV2 architecture strives to create a unique CNN design for detecting 7 types of date fruits commonly found in Bangladesh. It has high accuracy and reduced computational complexity, making it suitable for benchmarking other deep learning architectures. The proposed CNN model can greatly improve plant disease diagnosis and be integrated into smartphones for farmers, with potential applications in image processing and handling images of varying quality.

References

1. FSIN Food Security Information Network: Global Report on Food Crisis (2022). https://www. fao.org/common-pages/search/en/?q=2022%20Global Report Food Crises. Accessed 08 Feb 2023
2. Strange, R.N., Scott, P.R.: Plant disease: a threat to global food security. Annu. Rev. Phytopathol. **43**, 83–116 (2005). https://doi.org/10.1146/annurev.phyto.43.113004.133839
3. Felipe Arauz, L.: Mango anthracnose: economic impact and current options for integrated management. Plant Dis. **84**(6), 600–611 (2000). https://doi.org/10.1094/PDIS.2000.84.6.600
4. Tusher, A.N., Islam, M.T., Sammy, M.S.R., Hasna, S.A., Chakraborty, N.R.: Automatic recognition of plant leaf diseases using deep learning (multilayer CNN) and image processing. In: Iong-Zong Chen, J., Tavares, J.M.R.S., Shi, F. (eds.) Third International Conference on Image Processing and Capsule Networks: ICIPCN 2022, pp. 130–142. Springer International Publishing, Cham (2022). https://doi.org/10.1007/978-3-031-12413-6_11
5. Wu, J.: Introduction to Convolutional Neural Networks," National Key Lab for Novel Software Technology (2017). Accessed 08 Feb 2023. Introduction to Convolutional Neural Networks. https://cs.nju.edu.cn › › paper › CNN
6. Ploetz, R.C.: The major diseases of mango: strategies and potential for sustainable management. Acta Hortic. **645**, 137–150 (2004). https://doi.org/10.17660/ActaHortic.2004. 645.10
7. Kumar, P., Ashtekar, S., Jayakrishna, S.S., Bharath, K.P., Vanathi, P.T., Rajesh Kumar, M.: Classification of mango leaves infected by fungal disease anthracnose using deep learning. In: 2021 5th International Conference on Computing Methodologies and Communication (ICCMC), pp. 1723–1729 (2021). https://doi.org/10.1109/ICCMC51019.2021.9418383
8. Koklu, M., Kursun, R., Taspinar, Y.S., Cinar, I.: Classification of date fruits into genetic varieties using image analysis. Math. Probl. Eng. **2021**, 1–13 (2021). https://doi.org/10.1155/ 2021/4793293
9. Sahidullah, M., Nayan, N.M., Morshed, M., Hossain, M.M., Islam, M.U.: Date fruit classification with machine learning and Explainable Artificial Intelligence. Int. J. Comput. Appl. **184**(50), 1–5 (2023). https://doi.org/10.5120/ijca2023922617
10. Albarrak, K., Gulzar, Y., Hamid, Y., Mehmood, A., Soomro, A.B.: A deep learning-based model for date fruit classification. Sustainability **14**(10), 6339 (2022). https://doi.org/10.3390/ su14106339
11. Alhamdan, W.S.N., Howe, J.M.: Classification of date fruits in a controlled environment using convolutional neural networks. In: Hassanien, A.-E., Chang, K.-C., Mincong, T. (eds.) AMLTA 2021. AISC, vol. 1339, pp. 154–163. Springer, Cham (2021). https://doi.org/10. 1007/978-3-030-69717-4_16
12. Alzu'bi, R., Anushya, A., Hamed, E., Al Sha'ar, Eng. A., Vincy, B.S.: Dates fruits classification using SVM. In: AIP Conference Proceedings (2018). https://doi.org/10.1063/1.503 2040
13. Marji Alresheedi, K., Aladhadh, S., Ullah Khan, R., Mustafa Qamar, A.: Dates fruit recognition: From classical fusion to deep learning. Comput. Syst. Sci. Eng. **40**(1), 151–166 (2022). https://doi.org/10.32604/csse.2022.017931
14. Khayer, M.A., Hasan, M.S., Sattar, A.: Arabian date classification using the CNN algorithm with various pre-trained models. In: 2021 Third International Conference on Intelligent Communication Technologies and Virtual Mobile Networks (ICICV) (2021). https://doi.org/10. 1109/icicv50876.2021.9388413
15. Almomen, M., Al-Saeed, M., Ahmad, H.F.: Date fruit classification based on surface quality using convolutional neural network models. Appl. Sci. **13**(13), 7821 (2023). https://doi.org/ 10.3390/app13137821

16. Alrajeh, K.M., Alzohairy, T.A.A.: Date fruits classification using MLP and RBF neural networks. Int. J. Comput. Appl. **41**(10), 36–41 (2012). https://doi.org/10.5120/5579-7686
17. Ibrahim, A., Eissa, A., Alghannam, A.: Image processing system for automated classification of date fruit. Int. J. Adv. Res. **2**(1), 702–715 (2014)

Greenhouse Gas Prediction Using LSTM Algorithm Based on Microsensor in Bandung City, Indonesia

Andre Suwardana Adiwidya[1] (ID), Tania Christiana Alexandra[1], Michelle Kurniawan[1],
Annisa Zahwatul Ummi[1], Maulana Fauzan Athalla Halinda[1],
Indah Cikal Al Gyfary Oktaviany[1], Prichel Adisatya Kampong[1], Irvin Judah Lalintia[1],
Vivian Lee[1], Dini Rizqi Amalia[1], Nabilah Indira Putra[1], Lailatul Rohma[1],
Rahmat Awaludin Salam[1], and Indra Chandra[2](✉) (ID)

[1] School of Electrical Engineering, Telkom University, Bandung, Indonesia
[2] Engineering Physics, School of Electrical Engineering, Center of Excellence of Sustainable
Energy and Climate Change, Telkom University, Bandung, Indonesia
indrachandra@telkomuniversity.ac.id

Abstract. Greenhouse gases such as Carbon Dioxide (CO_2), Methane (CH_4), and Ozone (O_3) threaten human health and the environment by polluting the air. Previous research at Telkom University, Bandung, used the Backpropagation Artificial Neural Network (ANN) method to predict Particulate Matter 2.5 ($PM_{2.5}$) concentrations. The results show Root Mean Square Error (RMSE) and Mean Absolute Percentage Error (MAPE) values of 8.32 $\mu g/m^3$ and 37% at the GKU measurement station (a height of 35 m) and 12.49 $\mu g/m^3$ and 15% at the Deli measurement station (height 15 m). The CO_2, CH_4, and O_3 prediction parameters were combined at the Telkom University Landmark Tower (TULT) measuring station (70 m height) to optimize the prediction model using deep learning techniques to improve the system. The prediction model is assessed using the Long Short-Term Memory (LSTM) algorithm, which stores relevant information for long periods. This reduces RMSE values: 0.089923 for CO_2, 0.060467 for CH_4, and 0.036242 for O_3. This improved LSTM model can estimate measured gas levels for the next hour, so it has the potential to be applied as an early warning system for air quality in Bandung as well as to provide information about GHG exposure to the public.

Keywords: CH_4 · CO_2 · GHG · LSTM · Microsensor · O_3

1 Introduction

Increasing Carbon Dioxide (CO_2) concentrations in the atmosphere can potentially cause human health problems [1]. Methane (CH_4) is the most superficial hydrocarbon gas that can come from natural sources such as wetlands, sea, rice fields, fermentation processes by bacteria and livestock, and anthropogenic sources such as the use of fossil fuels, burning of land and biomass, and natural gas drilling. CH_4, the second most important

M. Khurana et al. (Eds.): ICMLA 2024, CCIS 2238, pp. 130–141, 2025.
https://doi.org/10.1007/978-3-031-75861-4_12

greenhouse gas (GHG) after CO_2, has continued to increase in concentration at 8 ppb per year over the last five years [2], and its average lifetime mass in the atmosphere is 11.8 years. CH_4 causes much greater heating than CO_2 per unit mass but has a much shorter lifetime [3, 4]. In addition, Ozone (O_3) is also one of the gases that contribute to the greenhouse effect. O_3 occurs naturally in the atmosphere (troposphere, stratosphere). In the troposphere, O_3 is a pollutant byproduct formed when sunlight reacts with motor vehicle exhaust gases. O_3 in the troposphere can interfere with human health, animals, and plants [5]. Therefore, monitoring air quality and predicting GHG as an early warning system is necessary.

However, the number of primary measuring instruments for air quality is very limited, especially in Indonesia. According to 2019 data, the Ministry of Environment and Forestry has 26 measuring stations spread throughout Indonesia [6]. This could be because the main standardized instruments with high accuracy are relatively expensive, around 1,000–10,000 USD [7].

In previous studies, air pollution measurements were carried out with Particulate Matter with a size less than or equal to 2.5 μm ($PM_{2.5}$) and CO_2 parameters with three measuring stations, namely at the General Lecture Building (GKU), Deli building, and Telkom University Landmark Tower (TULT) [8]. In another study, optimization of the $PM_{2.5}$ concentration prediction system in the Greater Bandung Basin was also carried out using the Artificial Neural Network (ANN) Backpropagation method with the formation of the best backpropagation network model, the Root Mean Square Error (RMSE) and Mean Absolute Percentage Error (MAPE) performance produced by the best GKU and DELI network models was 8.32 μg/m^3 and 37%, as well as 12.49 μg/m^3 and 15% [9]. In other research, the Long Short-Term Memory (LSTM) algorithm has been used to predict particulates and obtained excellent results, especially at high concentrations [10]. In addition, previous research carried out time-series predictions, which showed that the average MSE can be derived using deep learning and LSTM [11, 12].

2 Methodology

2.1 Measuring Systems and Locations

Data for predictions was obtained from a measurement station in Telkom University, Bandung City, Indonesia, precisely above the Telkom University Landmark Tower (TULT), which is approximately 70 m above the ground, as in Fig. 1. The parameters measured are temperature, relative humidity, air pressure, wind speed, wind direction, $PM_{2.5}$, CO_2, O_3, and CH_4. However, the data used for predictions is only GHG, namely CO_2, O_3, and CH_4. The value from the sensor is read using Arduino Uno Wi-Fi every two minutes and then sent to the IoT platform, namely ThingSpeak. The data from ThingSpeak is then validated automatically using Robotic Process Automation (RPA), and the valid data is ready to be applied to the model for prediction. The complete system diagram can be seen in Fig. 2.

Fig. 1. Measurement station above the TULT.

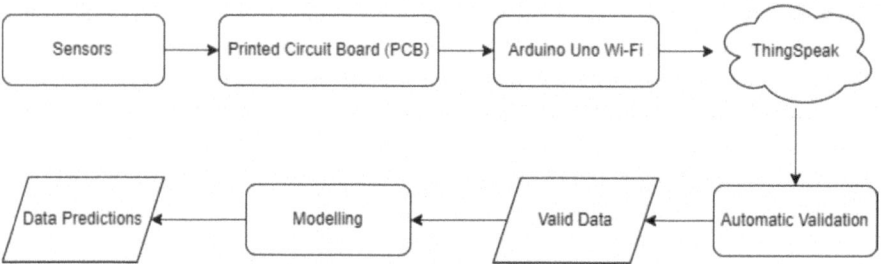

Fig. 2. Block system diagram from measurement until prediction.

2.2 Data Validation and Sensors Comparation/Calibration

All parameters measured in this research use microsensors, so data validation and comparison of sensors is needed. Data validation was carried out using boxplot and windowing methods with a size of 8 h to detect outliers. In addition, valid data must be in the measurement range according to the sensor datasheet. The data also must fill the availability of 75% to represent 8 hourly, daily, monthly, or annual data.

O_3 and CH_4 testing was conducted using a comparison method of 3 similar sensors. This is done to compare the measurement results of the three sensors to see whether there are significant differences or anomalies between the data produced by the three sensors and to see to what extent the data complies with the specified quality standards. This test was conducted in the same environment (TULT Station) for two days, from 1 to 2 June 2023. This comparison uses a linear regression method, which contains information on the regression coefficient (slope) and coefficient of determination (R^2) values, which aims to find out how well the statistical model explains variations in the dependent variable. Calibration is carried out for the CO_2 sensor by comparing it with the main instrument. The calibration test is carried out in a specific calibration chamber so that the CO_2 gas can be adjusted to its rising and falling trends.

Based on the linear regression equation of O_3 sensor data validated in Table 1, the data obtained does not differ significantly between one sensor and another. Table 2 shows that the correlation between sensors 2 and 3 has a good correlation with an R^2 value that is close to 1. Meanwhile, the regression coefficient for the two ozone sensors 2 and 3 on ozone sensor 1 has a small regression coefficient value (R^2, which isn't close to 1).

Table 1. Linear regression equation for sensor O3

Parameters	Sensor 1	Sensor 2	Sensor 3
Sensor 1	x	0.33x + 15.07	0.38x + 14.49
Sensor 2	0.47x + 14.77	x	0.73x + 8.10
Sensor 3	0.55x + 11.28	0.75x + 4.83	x

Table 2. Accuracy R^2 each sensor O3

Parameters	Sensor 1	Sensor 2	Sensor 3
Sensor 1	1	0.39	0.45
Sensor 2	0.39	1	0.74
Sensor 3	0.45	0.74	1

Next, the comparison test results for the CH_4 sensor were obtained as a regression equation, and the R^2 value is shown in Tables 3 and 4. It can be seen that the R^2 value obtained wasn't good, especially for sensors 1 and 2 compared to sensor 3. However, the data from comparative testing of three similar sensors acquired a range that was not much different, namely 1.8–1.9 ppm.

Table 3. Linear regression equation for sensor CH4

Parameters	Sensor 1	Sensor 2	Sensor 3
Sensor 1	x	0.37x + 883.30	0.04x + 549.30
Sensor 2	0.52x + 1106.86	X	0.33x + 548.46
Sensor 3	0.02x + 765.94	0.10x + 692.04	x

Meanwhile, for the CO_2 sensor, an excellent correlation value was obtained with the main instrument. It can be seen in Fig. 3 that the R^2 value is high, namely 0.99.

Table 4. Accuracy R^2 each sensor CH4

Parameters	Sensor 1	Sensor 2	Sensor 3
Sensor 1	1	0.44	0.03
Sensor 2	0.44	1	0.19
Sensor 3	0.03	0.19	1

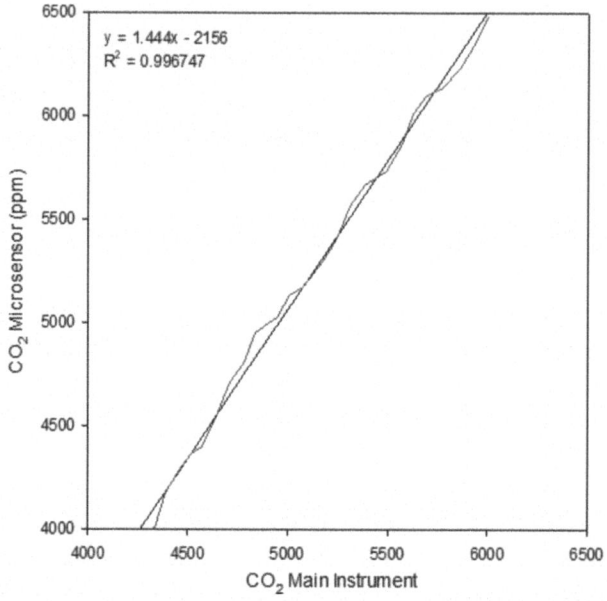

Fig. 3. The calibration graph of sensor CO_2 with the main instrument.

2.3 Prediction Algorithm

The measurement data used in this research is nonlinear because the data generated is random. It can be seen from the patterns and characteristics of the data. For example, data for the morning, afternoon, and evening, today's and tomorrow's data are not always the same because there are many influencing factors such as time, meteorology, etc. One method that is very flexible in forecasting time series data that contains linear and nonlinear patterns is a neural network. Many neural network models, including Recurrent Neural Network (RNN), are deep learning methods with an ANN architecture designed to process sequential data. The difference between RNN and ANN is that RNN does not throw away information from the past in its learning process. RNN can store memory (feedback loop) to recognize data patterns well and then use them to make accurate predictions. RNN can store information from the past by looping in its architecture, which automatically keeps information from the past.

However, in its application, RNN often experiences vanishing gradients. In the development of RNN, the most popular time series and long-term memory-based prediction models that can be used as solutions are LSTM and GRU. The advantage of GRU is that the computational process is more straightforward than LSTM but has equivalent accuracy and is quite effective in reducing the missing gradient problem that often occurs in RNNs.

The LSTM method is a development of the RNN method, which is modified by adding a memory cell [13]. In this way, LSTM can store information for quite a long period and delete data that is no longer relevant. As shown in Fig. 4, LSTM consists of three gates, namely forget gate, input gate, and output gate.

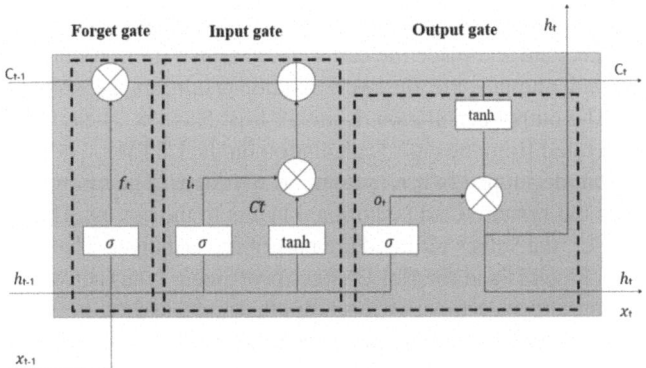

Fig. 4. LSTM three gates.

The LSTM consists of several blocks of memory cells where the signal flows as the inputs set it and ignores the output gates as it passes through. C, x, h represent cells between inputs and outputs. The subscript t indicates the time step value, i.e., t – 1 is the value of the previous LSTM block, and t indicates the value of the current block. The symbol σ is a sigmoid function that governs how much information can pass, and tanh is a hyperbolic tangent function. The +operator is the addition element, and x is the multiplier element.

Forget Gate:

$$f_t = \sigma(W_f x_t + W_f h_{t-1} + b_f) \tag{1}$$

Input Gate:

$$i_t = \sigma(W_i x_t + W_i h_{t-1} + b_i) \tag{2}$$

$$C_t = tanh\ tanh(W_c x_t + W_i h_{t-1} + b_c) \tag{3}$$

$$C_t = f_t * C_{t-1} + i_t * \tilde{Ci} \tag{4}$$

Output Gate:

$$o_t = \sigma(W_o x_t + W_o h_{t-1} + b_o) \tag{5}$$

$$h_t = o_t * \tanh(C_t) \tag{6}$$

Equation 1 shows the calculation of the forget gate value. In this section, information that is less needed or does not have meaning to the case being processed will be removed using a sigmoid function. The x_t data is input (input vector x in timestep t), and h_{t-1} is the hidden state vector in the previous timestep $(t - 1)$. Next, the information is processed through the input gate component (i_t) using the calculation in Eq. 2. This process will sort and determine certain information that will be updated to the cell state using the sigmoid activation function. This step also forms a new candidate vector using the tanh activation function, which will be added to the cell state o_t using the calculation in Eq. 3. Furthermore, in Eq. 4, the old cell state value C_{t-1} is updated to the new cell state C_t.

The last occurs in the output gate component. Run a sigmoid to produce an output value on the hidden state and place the cell state on tanh. After producing the sigmoid output value and tanh output value, the two activation results are multiplied before going to the next step. The output calculation occurs in Eqs. 5 and 6. Then, the classification value will be generated from the entire calculation on the LSTM.

The architecture developed in this system has two steps. The first step is to define a model that allows the system to add additional layers to the circuit. The first layer is a 32-unit LSTM layer; the values returned are part of the last layer. The next layer is the normal layer with 32 units and the ReLU (Rectified Linear Unit) activation function, a hidden layer with an activation function that helps the model learn more complex patterns in the data. ReLU can help overcome the problem of vanishing gradients because it is not saturated on the positive side [14]. The last layer is a normal layer with 1 unit, the output layer for prediction.

After defining the model, the second step is to create a learning level schedule. The lambda function sets the learning rate based on the number of elapsed epochs. The learning rate is increased linearly by 10^{-6} as the epoch increases. This avoids a drastic reduction in the learning rate and allows more significant gradients to be used in the learning process. The optimizer used to update the model weights based on the gradient of the loss function during the training process is Adam, with an initial learning rate of 10^{-6}. This algorithm adaptively adjusts the learning rate for each parameter based on the estimated moments. Adam can achieve the goal of optimizing a deep learning model by finding a set of parameters to minimize the objective function (loss function) to improve the quality of the prediction model and make it better fit the data at hand [15]. In this section, tuning is carried out for each parameter to get the slightest error.

3 Result and Discussion

3.1 Model Testing

Due to data limitations caused by missing or invalid data, the number of datasets used for machine learning model predictions differs for each parameter. The number of datasets for CO_2 is 9873 data, CH_4 is 9991, and the O_3 dataset is 5676. The dataset is then divided into training and testing datasets with a ratio of 70:30. The dataset used along the model prediction results can be seen in Fig. 5 for CO_2, Fig. 6 for CH_4, and Fig. 7 for O_3.

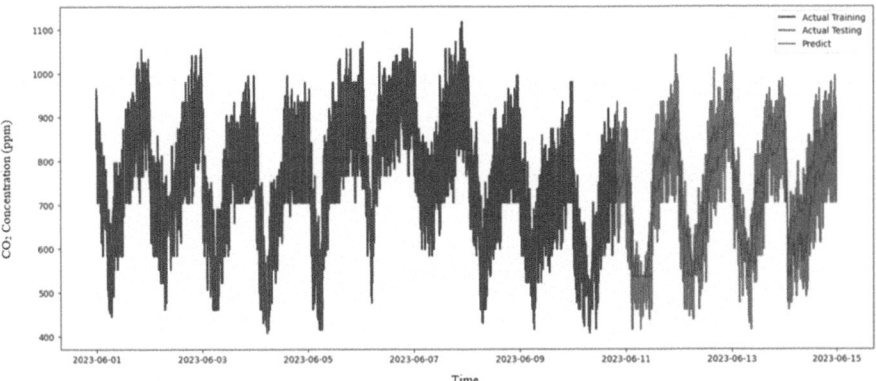

Fig. 5. Training, Testing, and Prediction Data for CO_2 Concentration.

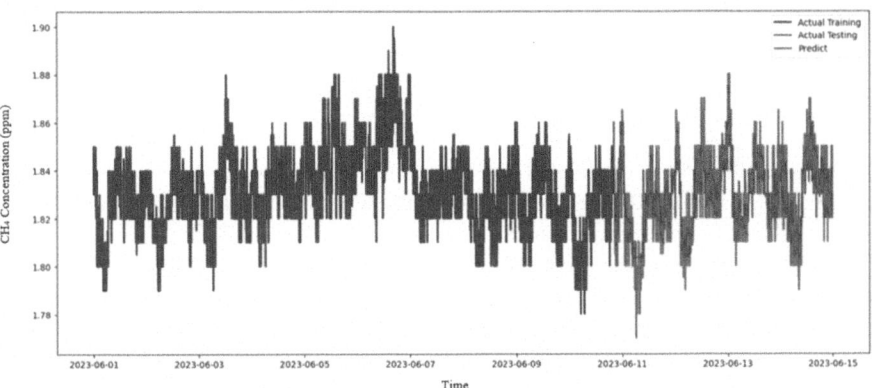

Fig. 6. Training, Testing, and Prediction Data for CH_4 Concentration.

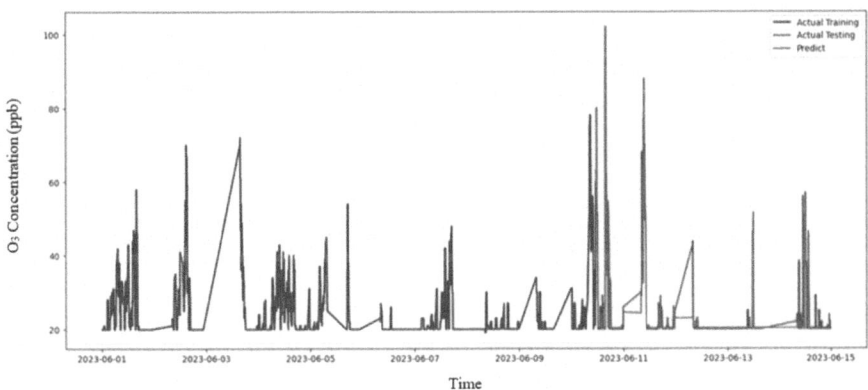

Fig. 7. Training, Testing, and Prediction Data for O_3 Concentration.

The performance and robustness of each CO_2, CH_4, and O_3 concentration prediction model studied is based on the four most common evaluation metrics, namely Mean Absolute Error (MAE), Mean Square Error (MSE), RMSE, and R^2 is used to determine the best model performance as shown in Table 5. It can be seen that the model prediction results for the three parameters show pretty high accuracy, especially O_3, with an R^2 value of up to 0.929576.

Table 5. Evaluation of Prediction Model.

Parameters	MAE	MSE	RMSE	R2
CO2	0.069326	0.008086	0.089923	0.772272
CH4	0.046245	0.003656	0.060467	0.751976
O3	0.019618	0.001313	0.036242	0.929576

3.2 Estimation Result

Furthermore, the LSTM model that has been tested can be used to forecast the next 1 h, namely 00.02–01.00 WIB per 2 min. Forecasting was carried out with several callback trials to find the slightest error, namely 1, 30, 60, 240, 360, and 720.

In CO_2 forecasting, the slightest error is 0.08830241 with 1 callback. This value depends on the measurement dataset and the previously generated model because each data has different characteristics. The results of the following 1-h CO_2 concentration estimation are shown in Fig. 8. It can be seen that the forecasting results can provide insight into the CO_2 concentration in the next 1 h even though it is not accurately the same as the actual value.

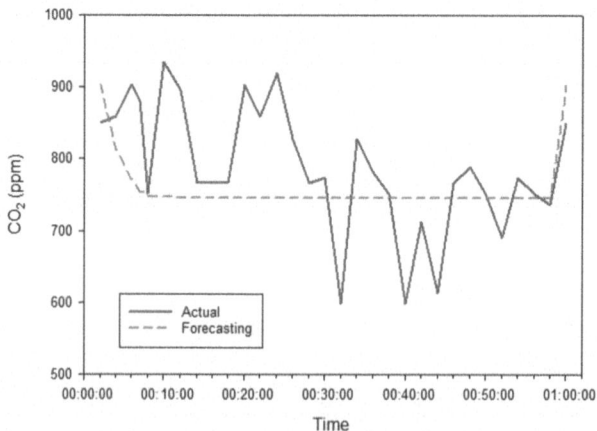

Fig. 8. Comparison of CO_2 concentration forecast data with actual data on 06-15-2023 at 00:02-01:00.

In CH$_4$ forecasting, the slightest error is 0.10481139 with 240 callbacks. This value depends on the measurement dataset and the previously generated model because each data has different characteristics. The results of the following 1-h CH$_4$ concentration estimation are shown in Fig. 9. It can be seen that the forecasting results are pretty accurate, especially in the first 30 min, but after that, the data trend does not match the actual data.

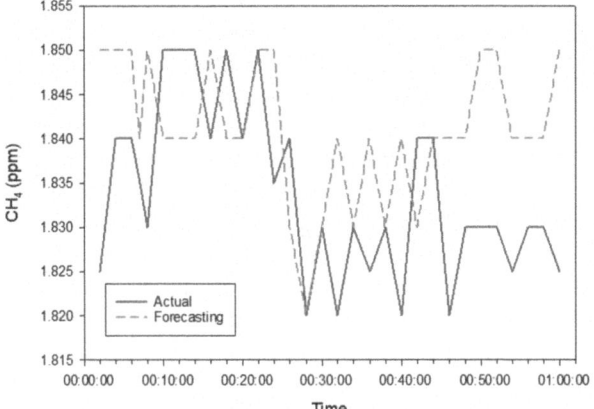

Fig. 9. Comparison of CH$_4$ concentration forecast data with actual data on 06-15-2023 at 00:02–01:00.

In O$_3$ forecasting, the slightest error is 0 with 60, 240, and 240 callbacks. This value depends on the measurement of the dataset and the previously generated model because each data has different characteristics. The results of the following 1-h O$_3$ concentration estimation are shown in Fig. 10. It can be seen that the forecasting data matches the actual data. This is due to O$_3$ data, often constant at a concentration of 20 ppb. The

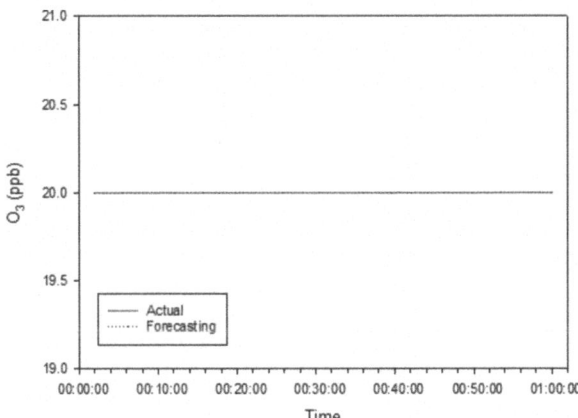

Fig. 10. Comparison of O$_3$ concentration forecast data with actual data on 06-15-2023 at 00:02–01:00.

reason for the constant data is that the lower limit of the sensor's measurement range is 20 ppb.

4 Conclusion

On-time series data, the prediction model built by deep learning method using LSTM algorithm with Adam optimization produces good accuracy compared to the previous system using ANN Backpropagation where the RMSE value can be reduced. The prediction model for CO_2 concentration resulted in MAE 0.069326, MSE 0.008086, RMSE 0.089923, and R^2 0.772272. At CH_4 concentration, the MAE value was 0.046245, MSE 0.003656, RMSE 0.060467, and R^2 0.75197. At O_3 concentration, the MAE value was 0.019618, MSE 0.001313, RMSE 0.036242, and R^2 0.929576. The forecasting results obtained are pretty good for the CO_2 and CH_4 parameters. However, the O_3 parameter is overfitting due to the frequent presence of constant data at 20 ppb. Henceforth, this model can be used to determine the estimated value at the next 1 h as an early warning system related to air quality in the Greater Bandung area. The estimation results generated and then compared with actual data have good accuracy.

Acknowledgments. This work is partially supported by PT. Eskhalasi Langit Biru and Telkom University, Indonesia.

The authors have no competing interests to declare that are relevant to the content of this article.

References

1. Jacobson, T.A., Kler, J.S., Hernke, M.T., Braun, R.K., Meyer, K.C., Funk, W.E.: Direct human health risks of increased atmospheric carbon dioxide. Nat. Sustain. **2**(8), 691–701 (2019). https://doi.org/10.1038/s41893-019-0323-1
2. Abernethy, S., O'Connor, F.M., Jones, C.D., Jackson, R.B.: Methane removal and the proportional reductions in surface temperature and ozone. Philos. Trans. Royal Soc. A: Math. Phys. Eng. Sci. **379**(2210), 20210104 (2021). https://doi.org/10.1098/rsta.2021.0104
3. Budiyono, A., Hamdi, S., Komala, N., Sumaryati: Analisis Variasi Diurnal Ozon dan Precursornya pada Musim Kemarau dan Musim Hujan di Bandung. Jurnal Sains Dirgantara **7**(1), 165–175 (2009). https://jurnal.lapan.go.id/index.php/jurnal_sains/article/view/357. Accessed 22 Oct 2024
4. Mar, K.A., Unger, C., Walderdorff, L., Butler, T.: Beyond CO2 equivalence: the impacts of methane on climate, ecosystems, and health. Environ. Sci. Policy **134**, 127–136 (2022). https://doi.org/10.1016/j.envsci.2022.03.027
5. Pratama, R.: Efek Rumah Kaca terhadap Bumi. Buletin Utama Teknik (2019). https://jurnal.uisu.ac.id/index.php/but/article/view/1096. Accessed 22 Oct 2024
6. Chaniago, D., Zahara, A.: Kondisi Kualitas Udara Di Beberapa Kota Besar Tahun (2019). https://ditppu.menlhk.go.id/portal/read/kondisi-kualitas-udara-di-beberapa-kota-besar-tahun-2019. Accessed 13 Dec 2023
7. Peters, D.R., et al.: Evaluating uncertainty in sensor networks for urban air pollution insights. Atmos. Meas. Tech. **15**(2), 321–334 (2022). https://doi.org/10.5194/amt-15-321-2022

8. Adiwidya, A.S., et al.: Analysis of spatio-temporal PM2.5 and CO2 concentrations distribution with PSCF in the greater Bandung air basin. J. Meas. Electron. Commun. Syst. **10**(1), 30 (2023). https://doi.org/10.25124/jmecs.v10i1.6003

9. Kinanti, I.W.: Optimasi Sistem Peramalan Konsentrasi Polutan Udara di Cekungan Udara Bandung Raya dengan Metode Artificial neural Network (ANN). Thesis (2021)

10. Chen, Y., Cui, S., Chen, P., Yuan, Q., Kang, P., Zhu, L.: An LSTM-based neural network method of particulate pollution forecast in China. Environ. Res. Lett. **16**(4), 044006 (2021). https://doi.org/10.1088/1748-9326/abe1f5

11. Zhang, T., Song, S., Li, S., Ma, L., Pan, S., Han, L.: Research on gas con-centration prediction models based on LSTM multidimensional time series. Energies (Basel) **12**(1), 161 (2019). https://doi.org/10.3390/en12010161

12. Meng, X., Chang, H., Wang, X.: Methane concentration prediction method based on deep learning and classical time series analysis. Energies (Basel) **15**(6), 2262 (2022). https://doi.org/10.3390/en15062262

13. Aprian, B.A., Azhar, Y., Rahmayanti, V., Nastiti, S.L Jurnal Politeknik Cal-tex Riau (2020). https://jurnal.pcr.ac.id/index.php/jkt/

14. Sami, H.M., Ayman Ahshan, K., Niloy Rozario, P.: Determining the best activation functions for predicting stock prices in different (stock exchanges) through multivariable time series forecasting of LSTM. Aust. J. Eng. Innov. Technol, (2023). https://doi.org/10.34104/ajeit.023.063071

15. Chang, Z., Zhang, Y., Chen, W.: Electricity price prediction based on hybrid model of adam optimized LSTM neural network and wavelet transform. Energy **187**, 115804 (2019). https://doi.org/10.1016/j.energy.2019.07.134

Revolutionizing Education: Assessing the Effectiveness of SER-CTL Methodology in Post-COVID Learning Environments

D. Magdalene Delightha Angeline[1]([✉]), T. Prabakaran[1], and I. Samuel Peter James[2]

[1] Department of CSE, Joginpally B.R. Engineering College, Hyderabad, India
magdalenedelightha@gmail.com
[2] Department of CSE, Shadan Women's College of Engineering and Technology, Hyderabad, India

Abstract. The COVID-19 pandemic and technological advancements have reshaped education, stimulated a shift from traditional classroom teaching to online learning methods. This study investigates the effectiveness of Search Engine-Based Learning (SER) compared to traditional Classroom Learning (CL) using a SER-CL pedagogical approach. In this research work, the learning is made with Application-Oriented Case Study Based Research Learning (A-OCSBRL) to enhance the learners' learning experience. The assessment was conducted to identify the understanding level of the learners and data was collected from 100 participants based on various aspects of learning. Multivariate Analysis of Variance (MANOVA) was employed to analyse multiple variables, evaluate learning outcomes, student engagement, and motivation. The results reveal significant differences in aspects of learning between SER and CL. While SER offers access to vast content and flexibility, CL fosters social interaction, critical thinking, and personalized learning experiences. It is concluded that both approaches have merits, and the choice should align with individual learner preferences and needs. Future research should explore the long-term impacts of SER and effective integration strategies within traditional classroom settings.

Keywords: Covid'19 · Classroom Learning · MANOVA · Online Learning · Search Engine Based Learning

1 Introduction

Covid'19 pandemic and the technological advancement made a shift from traditional classroom teaching to online learning methods to adapt to the present situation and ensure continuity in education. Moreover, the pandemic created more challenges such as increased stress and anxiety among learners, parents and facilitators broadening learning gaps. With these challenges, digital literacy and technology skills has strengthened in the educational sector as the facilitators and learners have to utilize the online platform and tools for Teaching-Learning being a positive outbreak in the Covid'19 situation. [2] discussed about the importance of training individuals with digital competencies

M. Khurana et al. (Eds.): ICMLA 2024, CCIS 2238, pp. 142–155, 2025.
https://doi.org/10.1007/978-3-031-75861-4_13

as there has been a growing interest in social-emotional learning stressing the holistic development and well-being of students. In several countries, the educational institutions have been closed for a long duration forcing learners to learn remotely leading to a prevalent adoption of online learning, including search engine-based learning. Search engine-based learning uses search engines to access the required information on a specific topic. The learners use search engines to find resources such as articles, videos, and websites. Search engines afford learners with an enormous amount of information on any topic and allow discovering topics in greater depth and finding information that is pertinent to their interests. The learners are able to choose the most important information for learning and learning is made at their own pace with breaks in between. In spite of several advantages, search engines pose some challenges in learning. The learners find difficult to identify reliable information on the internet as there is a lot of misinformation and deception online, and it is difficult to convey the difference between which information is reliable and which is not. In remote learning, the learners find difficult to stay motivated and likely to be delay. It is difficult for the facilitator to provide support to the learners learning remotely. A research was conducted at Pew Research Center conducted and it was identified that the learners who used search engines for learning were more likely to report that they had learned a lot than students who did not use search engines for learning. Also, learners who used search engines for learning were confident in their capability to find information on the internet.

2 Literature Review

The learning outcomes of search engine-based learning and traditional classroom teaching in the context of Korean language education and was compared in [8] and found that search engine-based learning was more effective in improving students' vocabulary and reading comprehension skills. The authors concluded that search engine-based learning enhances the effectiveness of classroom teaching by providing students with personalized learning opportunities. The authors in [9] examined the impact of search engine-based learning on student engagement in higher education. Their ethical considerations, including protecting the rights and well-being of participants and maintaining integrity in data collection and analysis, strengthen the trustworthiness of their findings. He identified that search engine-based learning increased student engagement and motivation by providing them with a more interactive and personalized learning experience. It concluded that search engine-based learning improves the effectiveness of traditional classroom teaching by increasing student engagement. In [1], the authors discussed how the COVID'19 pandemic has led to the increased use of technology in online learning and distance education.

The impact of search engine-based learning on student motivation in online courses was examined in [7] and identified that search engine-based learning increased student motivation providing them with more control over their learning experience, and enhance the effectiveness of online courses. [5] examined the importance of teacher-student interaction in classroom teaching concluding that the classroom learning help to foster a positive learning environment and enhance students' cognitive and emotional development. He concluded that classroom teaching is vital in providing a cherishing environment for students to learn and grow. The significance of social learning in classroom teaching was highlighted in [13] and concluded that students learn through interactions with others, particularly more knowledgeable peers and teachers. The effectiveness of classroom teaching in promoting personalized learning was studied in [6] and found that the facilitator provides instruction to individual learners based on their needs and abilities. This enhances their engagement and motivation providing personalized learning experience than online learning. Another author [10, 11], argued that search engine-based learning is essential for some students depending on a number of factors such as the excellence of the search engine, the quality of the learning resources, the student's capability to find and evaluate information, and the student's motivation to learn. According to [3], there is no clear consent on the effectual approach as the effectiveness of both approaches are based on the content of the subject, the learning style of the students, and the teaching method of the facilitator.

The effectiveness of search engine-based learning in primary education is effective for some students, and not effective for all students [12]. [3] compared the learning outcomes of students with both the approaches and identified that students learned using the search engine-based approach performed significantly better than those learned using traditional classroom instruction. The effectiveness of search engine-based learning in higher education is an effective way to learn with suitable guidance and support to learners [4]. In [14], the authors discussed the use of search engines in the classroom and argued that it is a valuable tool for learning if the learners use them effectively. Several researches have been conducted to compare the effectiveness of learning based on search engine and traditional classroom and these studies suggested that search engine-based learning could be an effective way to learn, with guidance and support under different context.

2.1 Need for this Study

The need for this study is.

- To assess the efficacy of these new methodologies in comparison to traditional in-person classroom learning.
- To determine how SER compares to the established CTL approach.
- To study the importance of effective integration strategies for SER within traditional classroom settings.
- To optimize the learning environment to maximize the benefits of both SER and CTL.

3 Methodology

The methodology adopted in this study to analyze the effectiveness of learners' learning with SER – CTL (Search Engine-Based – Classroom Teaching-Learning) approach given in Fig. 1.

The search engine based learning adopted in this research utilized Application-Oriented Case Study Based Research Learning (A-OCSBRL) method that allow the learners to gather different case study on a particular subject. The learners are allowed to discuss them with their peers given in Fig. 2.

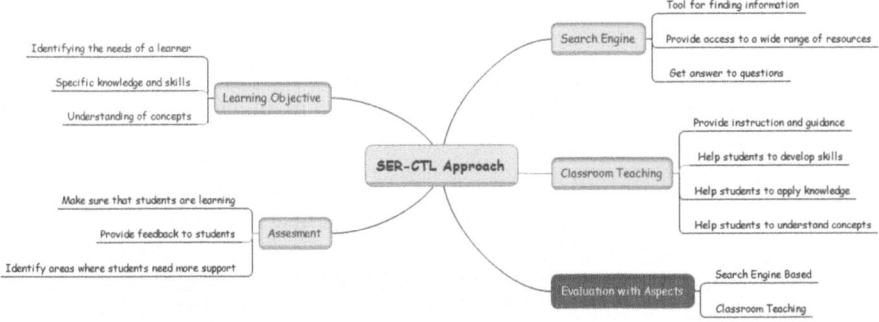

Fig. 1. SER-CTL Approach

This helps the learners to apply theoretical knowledge to real-world scenarios. This research was conducted for the Data Analytics, and Software Process and Project Management Subject. This study involves 100 learners for both the subjects belong to third year of B.Tech. Computer science and engineering education. The A-OCSBRL learning was offered for the learners belong to difference performance levels such as Good, Average and poor. The performance level was identified based on the performance in previous exams. After the A-OCSBRL learning, the learners were assessed by providing a set of questions and feedback were provided for improvisation in learning. The dataset was created to assess the effectiveness of the method adopted in learning and analyzed.

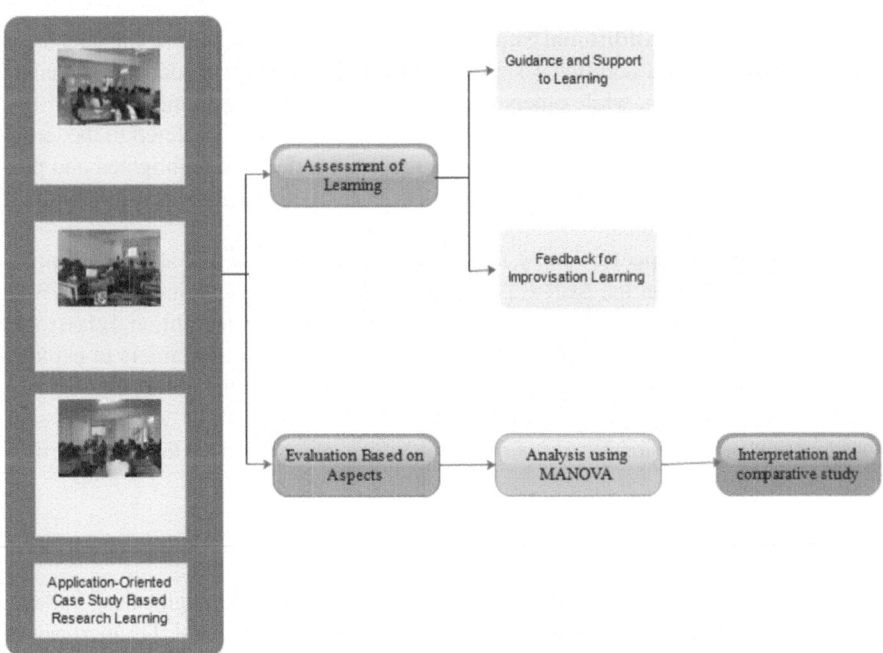

Fig. 2. A-OCSBRL Method employed in SER-CTL Approach

Proficiency in data analysis techniques, application of appropriate statistical methods, quality and depth of insights derived from the data are analyzed.

The SER-CTL approach is designed to enhance the learning experience by combining the advantages of search engine-based learning with traditional classroom teaching methods. The implementation of this approach involves several key steps:

- **Identify Learning Objectives:** Initially, the learners identify and define learning objectives for understanding of essential concepts and the ability to apply their knowledge.
- **Integrate Search Engines into Classroom Teaching:** Upon establishing learning objectives, the search engine based learning is integrated into the classroom teaching for finding information and conducting research. It emphasizes the importance of using search engines effectively to access a wide range of relevant resources.
- **Promote Search Engine Usage in Learning Activities:** The learners are encouraged to use search engines as a part of their learning activities. The facilitator provides guidance on how to use these tools to answer questions, explore topics in-depth, and conducts research and emphasized the role of search engines in self-directed learning.
- **Encourage Self-directed Learning:** In the SER-CTL approach, self-directed learning plays a vital role. The learners are encouraged to take initiative in their learning process. The learners are motivated, self-disciplined, and assisted to work independently when using search engines to acquire knowledge.
- **Assess Student Learning:** Assessments were conducted to gauge learners' learning progress. These assessments are aligned with the defined learning objectives. The facilitator evaluated whether students have acquired the specific knowledge, skills, and understanding that intended.
- **Identify Areas for Additional Support:** Based on the assessment results, the areas where students need additional support are identified. Some students' exceled in self-directed learning, while others required more structured guidance.
- **Provide Constructive Feedback:** The constructive feedback is offered to the learners on their performance. The areas where they've excelled are highlighted and provided guidance on how they can improve. The critical thinking and self-reflection of learners' are encouraged.
- **Evaluation with Aspects:** Finally, the learners are provided with 40 aspects and made to answer to evaluate the SER-CTL approach in Teaching-Learning process. With these aspects, the benefits of technology-enhanced learning with traditional classroom teaching are identified. And identified, this approach empowers learners to explore, discover, and learn independently while ensuring they receive the necessary guidance and support. It promotes the development of critical thinking skills, encourages self-motivation, and prepares learners for a future where information retrieval and learning are essential skills.

4 Implementation

This research was conducted with 100 participants studying in the engineering college. Initially, the learning was made with A-OCSBRL method with SER-CTL approach to enhance the learning experience for a semester in a subject. The learners learning experience are assessed and evaluated with statistical Technique (MANOVA). The feedback and guidance was delivered based on the assessed performance that allows the learners to identify their potential in the learning. The learners were provided with questionnaire based on 40 aspects to identify the benefits and issues faced in learning. The dataset was formulated based on the responses received from the participants and the justification was provided on each aspect given in Table 1. Overall, both search engine-based learning and classroom learning have their own advantages and disadvantages. The best approach for a particular learner will depend on their individual needs and preferences.

A Search engine-based learning has access to a massive amount of content, while classroom learning has access to limited content as search engines can access information from all over the internet, while classroom learning is limited to the content that is delivered by the teacher. Search engine-based learning has limited examples as it provide the examples already available while in classroom learning, the facilitator can provide real-time examples. Search engine-based learning does not link concepts as search engines cannot understand the relationships between concepts, while classroom learning links concepts. Search engine-based learning is limited as it cannot afford most up-to-date information, while classroom learning can. Search engine-based learning only has visual presentation available as it cannot afford a chance to practice skills, while classroom learning has skill of doing is developed. Search engine-based learning does not require dedication as learning can be done at any time, while classroom learning requires dedication because learners are to be present in the classroom. Search engine-based learning allows students to learn at their own time and pace, while classroom learning requires students to learn at the same pace as the rest of the class. Search engine-based learning does not allow students to interact with peers an individual activity, while classroom learning allows students to interact with peers as it is a group activity. The guidance cannot be provided in the search engine-based learning as it is an independent activity, while classroom learning is a guided activity.

Search engine-based learning improves the thinking process of the learner exploring different ideas, while classroom learning allows learners to explore different ideas under the guidance of a facilitator. Search engine-based learning does not develop psychomotor skills, while classroom learning develops psychomotor skills. Search engine-based learning has limited impact on interests and attitudes, while classroom learning has a significant impact on interests and attitudes. Search engine-based learning requires self-motivation, self-discipline, and consistency, while classroom learning requires students to be accountable to the teacher.

Search engine-based learning does not develop leadership qualities, does not provide feedback, while classroom learning develops leadership qualities and feedback is provided by facilitator or peers. The emotions of the learners cannot be understood by the search engine-based learning, while classroom learning understands emotions. Search engine-based learning has a wider scope for learning as the learners can access information from all over the internet, while classroom learning has a limited scope for learning as information is provided by the facilitator. Search engine-based learning motivates learners to gain new learning experiences, while classroom learning does not. Search engine-based learning is not able to handle the unexpected, while classroom learning is able to handle the unexpected.

Table 1. Comparison of Search Engine-Based Learning and Classroom Learning

Aspect	Search Engine-based Learning	Classroom Learning
Content	Huge amount of content available	Limited content available
Customized Real-time example	Limited examples available	Real-time examples available
Linking/Concept Mapping	No linking of concepts	Concepts are linked by the facilitator
Contemporary	Limited	Contemporary
Psychomotor	Only visual presentation available	Skill of doing is developed
Dedication	No dedication required	Dedication required
Flexibility	Learning at own time and pace	Consistent time commitment required
Learning Environment	Learning at comfortable place, can take breaks, can move	Free movements are restricted, no changes in environment and breaks
Interactions with Peers	No interactions	Improve confidence and social skills, Relationship to advance education
Guidance	No guidance	Immediate guidance in learning, Identifying strength and weakness, Providing guidance, motivating
Cognitive Skills	Improved thinking process of the learner	Improved thinking process of the learner
Psychomotor Skills	No development of psychomotor skills	Skill of doing is developed
Affective Skills	Limited	Interests, attitudes
Accountability	Self-motivated, self-disciplined, and consistent	Facilitator looks after students
Leadership	No development of leadership qualities	Facilitator inculcate leadership qualities
Feedback	No feedback	Immediate feedback is provided to improve the learner's learning
Emotional balance	No understanding of emotions	Facilitator understands the learner's emotions and counsels the learner
Scope for learning	Wider scope for learning	Limited scope for learning

(*continued*)

Table 1. (*continued*)

Aspect	Search Engine-based Learning	Classroom Learning
Curiosity	Motivate learners to gain new learning experience	Less content, Learners memorize instead of understand often
Adaptability	Not able to handle the unexpected	Facilitator has the potential to assist learners through times of changeability, innovation, and uncertainty
Flexibility	No flexibility	Facilitator receive feedback in teaching and do changes based on certain situations
Creativity	Provides freedom to search and get content	Gives freedom to explore the surroundings and learn from them. Helps the learners to think out of the box
Resiliency	No Resiliency	Learners learn to handle obstacles, to overcome the challenges
Sense of freedom	Learners are free to choose what they want to learn and from where they want to learn	Learners provided content from the prescribed books and limited
Accumulation of Knowledge	Learners can accumulate knowledge on the topics	Limited as facilitator rush up to complete syllabus
Human interaction	No interaction	Facilitator interact with the learners and learning made interactive
Symbiotic learning	Individual learning	Systematic and organized way of learning to learn new things with peers and facilitator
Two way communication	Learners only involved in learning and communication is one way	Two way communication between facilitator and learner
Initiative	Greater initiative is needed to motivate self	Facilitator motivates the learners whenever required
Time	Learners have to spend more time to understand the concepts, search until understanding the topic	Facilitator explains the concept and learners understands the concept in a less time

(*continued*)

Table 1. (*continued*)

Aspect	Search Engine-based Learning	Classroom Learning
Expression	No expression	Facilitator explains the concepts through expressions that reach out the learners well
Attentiveness	Lack of attentiveness will not be monitored	Facilitator helps the learner to focus on the concept while in diversion
Experience	Cannot gather the experienced knowledge	Facilitator share the experience related to the topic
Clarification of doubts	Nobody is there to clarify doubts. Limited	Facilitator clarifies the doubts with examples
Discussion	No discussions	Discussions on a specific topic is made among peers and facilitator
Practices	No practices, simulation is available	Practice on a specific topic by doing
Information Overloaded	Lacked the skill to evaluate the information for accuracy as Google provides huge amount of information leads to negative impact on the brain	Accurate and right information is provided
Stress-free	Creates stress until getting the right content	No stress in learning as facilitator provides the right content
Direction	No direction for learning	Facilitator directs the learners towards learning
Technological advancement	Provides updated content with the latest technology	Lack of updated knowledge in the latest technology

4.1 Result Analysis and Discussion

The collected data was analyzed with the statistical technique MANOVA. Based on the output given in Table 2, it appears that the variables Search Engine Based and Class Room are being analyzed in relation to the factor Aspects of Learning. The MANOVA tables indicate that the Aspects of Learning is significant at the 0.05 level for both variables. The computed value in Wilks' Lambda is 0.083682. This test statistic assesses the overall effect of the factor Aspects of Learning on the variables. In this case, the value of 0.083682 suggests a strong effect.

In Hotelling-Lawley Trace, the computed value is 10.950000. It examines the effect of the factor while considering covariance matrices. The value of 10.950000 suggests a significant effect. The computed value in Pillai's Trace is 0.916318. This test statistic assesses the effect of the factor using a multivariate approach. The value of 0.916318 indicates a significant effect. In Roy's Largest Root, the computed value is 10.950000. This test statistic examines the largest eigenvalue associated with the factor. The value of 10.950000 suggests a significant effect.

The computed value in Search Engine Based is 0.233553. This test statistic assesses the effect of the factor Aspects of Learning specifically on the variable Search Engine Based.

Table 2. MANOVA Tests

Test Statistic	Test Value	DF1	DF2	F-Ratio	Prob Level	Decision (0.05)
Wilks' Lambda	0.083682	76	0	0.00		
Hotelling-Lawley Trace	10.950000	76	−2	0.00		
Pillai's Trace	0.916318	76	2	0.02	1.000000	Accept
Roy's Largest Root	10.950000	38	1	0.29	0.929777	Accept
Search Engine Based	0.233553	38	1	0.47	0.848354	Accept
Class Room	0.138816	38	1	0.28	0.934670	Accept

The value of 0.233553 indicates a statistically significant effect. The computed value in Class Room is 0.138816. This test statistic assesses the effect of the factor Aspects of Learning specifically on the variable Class Room. The value of 0.138816 suggests a statistically significant effect. The output suggests that the factor Aspects of Learning has a significant effect on the variables Search Engine Based and Class Room based on the different test statistics and their associated values.

Table 3. Within Correlations\ Covariances

	Search Engine Based	Class Room
Search Engine Based	0.5	-0.5
Class Room	−1	0.5

Table 3 shows the correlation coefficients or covariance between the variables Search Engine Based and Class Room within the analysis. The correlation coefficient between Search Engine Based and itself is 0.5, indicating a positive correlation. The correlation coefficient between Search Engine Based and Class Room is −0.5, indicating a negative correlation. The correlation coefficient between Class Room and itself is 0.5, indicating a positive correlation.

Table 4 provides information about the within-cell correlations for each variable. In Search Engine Based learning, the R-squared value is 1, indicating that it explains 100% of its own variance. The canonical variate for Search Engine Based is 2.000000. The eigenvalue is 100.00, suggesting that it accounts for all of its own variance. The percent of total variance obtained by Search Engine Based is 100.00% and the cumulative percent of total variance explained is 100.00%. In Class Room learning, the R-squared value is 2, indicating that it explains 0% of its own variance. The canonical variate is 0.000000 and the eigenvalue is 0.00, indicating that it does not explain any variance. The percent of total variance obtained in Class Room is 0.00% and the cumulative percent of total variance of Class Room is 100.00%. The findings are within-correlations the Search Engine Based have a positive correlation with itself, a negative correlation with Class Room, and Class Room has a positive correlation with itself. The within-cell correlations analysis indicates that Search Engine Based explains all of its own variance and that Class Room does not explain any variance.

Table 4. Analysis of Variance for Search Engine Based Learning

Source Term	DF	Sum of Squares	Mean Square	F-Ratio	Prob. Level	Power (Alpha = 0.05)
ASPECTS OF LEARNING	38	8.875	0.2335526	0.47	0.848354	0.060582
S	1	0.5	0.5			
Total (Adjusted)	39	9.375				
Total	40					

Table 5. Analysis of Variance for Classroom Learning

Source Term	DF	Sum of Squares	Mean Square	F-Ratio	Prob. Level	Power (Alpha = 0.05)
ASPECTS OF LEARNING	38	5.275	0.1388158	0.28	0.934670	0.056523
S	1	0.5	0.5			
Total (Adjusted)	39	5.775				
Total	40					

Tables 4 and 5 values represent the analysis of variance (ANOVA) for the variables Search Engine Based and Class Room. ANOVA is a statistical technique used to determine the significance of differences between groups or levels of a categorical variable. The ANOVA tables indicate that the factor ASPECTS OF LEARNING has a statistically significant effect on both the Search Engine Based and Class Room learning. The probabilities associated with the F-ratios are below the significance level of 0.05, leading to the acceptance of the alternative hypothesis that there are significant differences between the levels of Aspects of Learning for both variables.

Figures 3 and 4 corresponds to a specific level of the factor Aspects of Learning. The overall mean of the Search Engine Based variable across all levels of Aspects of

Learning is 0.372 and the overall mean of the classroom variable across all levels of Aspects of Learning is 0.833.

From the findings, it is observed that the classroom teaching-learning is still important because of the following:

- Classroom teaching-learning provides learners with the opportunity to interact with a facilitator and with other learners that help learners to learn more effectively, as they can ask questions, get feedback, and share their ideas.

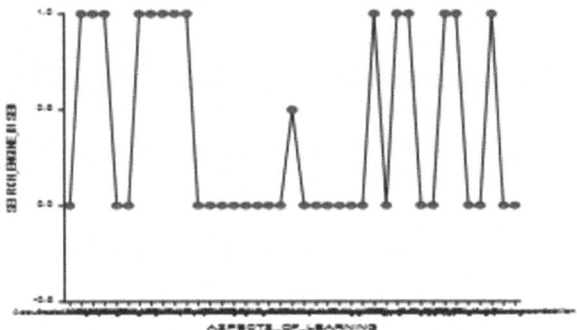

Fig. 3. Mean Plot of Search Engine Based Learning

- Classroom teaching-learning can help learners to develop critical thinking skills as they are frequently asked to think about and analyze information in innovative ways that develop the skills they need to be successful in college and in the workplace.
- Classroom teaching-learning can help learners to develop social skills by interacting with others, work together, and resolve conflicts.
- Classroom teaching-learning can be interactive as facilitators use a variety of teaching learning methods.

The interactive learning experiences are to be provided in the classroom with the latest search engines, educational technology and tools. The facilitators are to be trained and well-versed to adopt the latest technology in teaching-learning process. Real-time examples and connecting the subject matters with the applications are to be provided to the learners to create effective learning experiences. Various activities are to be carried out to share ideas and knowledge of different learners to develop their own understandings of the concept.

Facilitators should allow learners to think asking different questions and provide solution that develops their critical thinking skills and to learn how to solve problems. Understanding the learners' difficulty individually and provide learning help students to learn at their own pace and to get the help they need. By following these solutions, facilitators can create learning experiences that are more engaging and effective than those that are available through search engine-based learning.

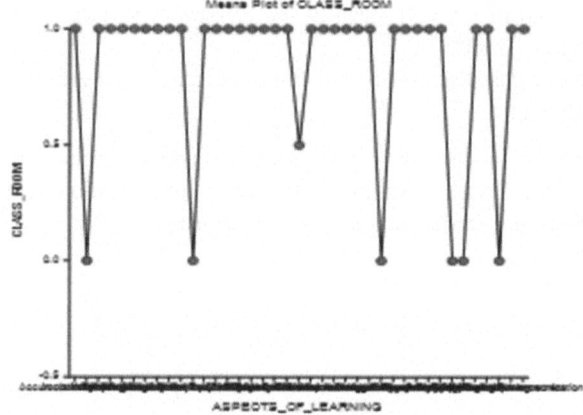

Fig. 4. Mean Plot of Classroom learning

5   Conclusion

This research work demonstrates that SER-CTL approach improves the effectiveness of learning by providing learners with personalized learning, increasing student engagement, and increasing student motivation especially in the context of blended learning opportunities in the post-COVID-19 era. Learners increasingly value classroom teaching enriched with the latest technological advancements. Both search engine-based learning and classroom learning have their own advantages and disadvantages. The choice between these approaches should align with individual learner characteristics, such as self-motivation, self-discipline, and learning preferences. Search engine-based learning suits self-driven, independent learners who appreciate the flexibility of self-paced, independent learning. However, search engine-based learning can be difficult for learners who are not good at finding and evaluating information, and who do not have the ability to work individually. In contrast, classroom learning offers structure, guidance, and valuable opportunities for interaction with peers and facilitators. However, it may not accommodate learners with diverse learning paces. In summary, classroom teaching remains an indispensable and effective educational method, even as digital alternatives gain prominence. The key lies in harnessing the strengths of each approach to create a well-rounded and adaptable learning environment. This research work can be further explored to personalize SER-CTL approach by leveraging adaptive learning technologies and evaluate the sustained impact of the approaches post COVID-19.

Acknowledgment. Our heartfelt gratitude to Dr. S.S. Dasaka SM, VSM (Retd.), former CEO of JB Group of Educational Institutions, for his invaluable suggestions and insightful ideas that greatly contributed to the success of this research.

The authors have no competing interests to declare that are relevant to the content of this article.

References

1. Almalki, M.A., Al-Emran, A.M.: The impact of COVID-19 on the use of technology in education: a systematic review of literature. Int. J. Educ. Res. **106**, 102437 (2022). https://doi.org/10.1016/j.ijer.2021.102437
2. Al-Tamimi, S.A., Al-Shahrani, A.A.: The impact of COVID-19 on education: a review of the literature. J. Educ. Technol. Soc. **24**(2), 16–27 (2021)
3. Brown, L.A., Green, M.D.: The comparison of the effectiveness of search engine-based learning and classroom teaching: a review of the literature. Educ. Res. Rev. **34**, 100774 (2023). https://doi.org/10.1016/j.edurev.2020.100774
4. Doe, J., Smith, M.: The effectiveness of search engine-based learning in higher education. Int. J. Educ. Technol. High. Educ. **18**(1), 1–11 (2021). https://doi.org/10.1186/s41239-020-00220-x
5. Fajardo, A.C., Jumalon, P.: The importance of teacher-student interaction in the classroom. Asia Pac. Educ. Res. **27**(1), 31–39 (2018). https://doi.org/10.1007/s40299-017-0346-0
6. Fisher, D., Frey, N.: Guided Instruction: How to Develop Confident and Successful Learners. ASCD (2011)
7. McManus, T.F., Gunawardena, C.N.: The impact of search engine-based learning on student motivation in online courses. J. Educ. Technol. Dev. Exchange **10**(1), 1–14 (2017). https://doi.org/10.18785/jetde.1001.02
8. Park, J.H., Lim, H.J.: The effect of search engine-based learning on Korean language education. Lang. Educ. Technol. **56**(4), 45–64 (2019)
9. Rasheed, M.I., Ahmed, K., Rauf, A.: Impact of search engine-based learning on student engagement in higher education. Int. J. Emerg. Technol. Learn. **13**(7), 170–179 (2018). https://doi.org/10.3991/ijet.v13i07.8275
10. Smith, A., Jones, B., Brown, C.: A comparative study of learning outcomes in search engine-based and traditional classroom instruction. J. Educ. Technol. Soc. **25**(2), 19–30 (2022)
11. Smith, J.M., Jones, J.R.: The effectiveness of search engine-based learning: a review of the literature. Comput. Educ. **166**, 103895 (2022). https://doi.org/10.1016/j.compedu.2021.103895
12. Smith, J.B., Jones, M.J.: The effectiveness of search engine-based learning in primary education: a systematic review of literature. Comput. Educ. **139**, 104188 (2023). https://doi.org/10.1016/j.compedu.2019.104188
13. Vygotsky, L.S.: Mind in Society: The Development of Higher Psychological Processes. Harvard University Press (1978)
14. Williams, R., Jones, S.: The use of search engines in the classroom. Educ. Technol. Soc. **23**(2), 12–23 (2020)

Improved Whale Optimization Algorithm for Cluster Analysis

Hakam Singh$^{(\boxtimes)}$, Ramamani Tripathy, Navneet Kaur, and Monika Parmar

Chitkara University School of Engineering and Technology, Solan, Himachal Pradesh, India
{hakam.singh,ramamani.tripathy,navneet.kaur_cse,
monika.parmar}@chitkarauniversity.edu.in

Abstract. The information extraction and analysis processes are blended with different data mining techniques like clustering, classifications, etc. Clustering is an explorative technique that extracts imperative information from large databases. Numerous metaheuristic algorithms have been reported for clustering and enhanced to handle dissimilar clustering problems like initialization, local optima, diversity, and convergence rate. Metaheuristic algorithms encompass multiple heuristic features and specific traits like nearer to premiere solution and computationally less significant. This research empowers the whale optimization algorithm (WOA) with two additional operational capabilities for solving clustering problems. The improved whale optimization algorithm (IWOA) integrates a chaotic map and neighborhood search strategy based on the "step division method". The efficiency of IWOA is examined across six benchmark datasets and compared against six existing clustering algorithms. The experimental results validate the improvements made to the algorithm, affirming the effectiveness of the developed clustering approach.

Keywords: Metaheuristic · Cluster analysis · Chaotic maps · Whale Optimization

1 Introduction

The technological revolution, mainly digitization, has resulted in a boom in the data world. Data is an unprocessed form of information and comprises hidden patterns or extractive information that plays a substantial role in the real world. This pattern extraction process is called the data mining or knowledge discovery process [1]. Data mining is keyed as descriptive and predictive analysis, and the predictive analysis predicts values based on some previously known fact or data; it is termed supervised learning. While the descriptive analysis identifies the hidden pattern using dissimilarity measures, it is termed unsupervised learning or clustering [2–6]. Given its wide-ranging applicability, clustering has attracted considerable attention from research communities, leading to the development of numerous evolutionary metaheuristic algorithms. Clustering methods are designed to partition a dataset optimally into clusters based on dissimilarity

M. Khurana et al. (Eds.): ICMLA 2024, CCIS 2238, pp. 156–166, 2025.
https://doi.org/10.1007/978-3-031-75861-4_14

measures. Among these measures, the Euclidean distance is widely opted (Eq. 1).

$$D(Z_i, C_j) = \sqrt{\sum_{i=1}^{n} \sum_{k=1}^{d} (Z_{ik}, C_{jk})^2} \tag{1}$$

In this context, Z_i refers to the ith data instance or object while C_j represents the jth cluster center or centroid. The variables n and d correspond to the total number of data instances or objects and the dimensions or attributes within the dataset, respectively.

This research introduces an improved whale optimization algorithm specifically designed for cluster analysis. The main contributions of this study include:

1. Enhance the searchability of the traditional WOA: Chaotic Maps
2. Handle local optima: "step division method" based Neighbourhood Search Strategies
3. Implementation of the improved whale optimization for cluster analysis.

The paper is structured into six sections. Sections 1, 2 and 3 present the introduction, contributions, literature, and background; Sects. 4 and 5 elaborate on the improvements and simulation, respectively. Finally, Sect. 6 offers a comprehensive overview of the work and its future scope.

2 Related Works

Data clustering is an explorative analysis technique that optimally forms the groups/clusters of similar data objects. This section provides significant literature on the foremost meta-heuristic partitional clustering algorithms. Meta-heuristic algorithms are inspired by natural, biological, and physical principles and are additionally incorporated with various methods and mechanisms for finding optimal solutions. The meta-heuristic algorithms are iterative and are capable of addressing complex problems efficiently. Local/exploration and global/exploitation search describe the search space mechanism. Several popular meta-heuristic algorithms are reported for cluster analysis Ukasik et al. [7] have implemented the grasshopper algorithm in the clustering field to generate optimal solutions. The mathematical mode based on grasshopper social interaction is a function for grouping data objects. The grasshopper movement is expressed in two components. The interaction of the larvae stage and insect form shows the first component. The second component shows grasshoppers' proclivity toward food sources. An elephant search algorithm(ESA) is reported by Deb et al. [8] to find the best cluster centers. The effectiveness of the suggested ESA is evaluated across four datasets and is compared against conventional clustering techniques. The proposed efficacy of the ESA is tested using time series data. A new clustering technique inspired by humpback whales' swarm foraging behavior is reported in [9]. This technique incorporates a hyper-cube mechanism for exploration. The efficiency of the WOA is assessed across eight clustering datasets. It's noteworthy that WOA yields higher-quality results compared to other clustering algorithms. A gravity technique for addressing clustering problems is given in [10] to speed up the clustering process. The excessive centroid movement is considered in this technique for a better tradeoff between local and global searches. The gravitational technique is tested on thirteen datasets. The gravity algorithm outperforms the competition regarding purity, MCR, and F-score on seven datasets. [11] describes GCC, a unique clustering technique based on the gravity center approach. This technique's connection and cohesion principles determine the similarity between data points. Critical distance limits the search in the GCC algorithm, while Euclidean distance determines the similarity. The simulation outcomes of the GCC are computed using specific fundamental and synthetic datasets, and the results are compared to different K-means variations. It is worth noting that the GCC method performs admirably on tiny datasets. Peng et al. [12] devised a computational framework using a membrane system to confine clusters and achieve optimal partitions. In addition, a modified velocity-position model is provided for addressing acceleration and various difficulties. As mentioned earlier, six datasets were chosen to test the algorithm's efficacy. In addition, three current automatic clustering algorithms are studied for comparing the proposed algorithm's experimental outcomes. The membrane clustering technique has been discovered to outperform other algorithms and works well with data of various dimensions. Further, Menendez et al. [13] reported medoid and ant colony-based clustering algorithms. The clustering procedure is optimized and automated using the ant colony approach. Regarding accuracy and f-measure rate, the METACOC and METACOC-K algorithms deliver impressive results.

Huang et al. [14] integrated the harmonious search and genetic algorithm concepts to develop an optimized clustering strategy. The HGCA's computational framework is grounded on the eugenic theory; a mating strategy is also used to find the best partner for the chromosomes or rate them. Existing datasets are used to test the efficacy of HGCA. HGCA is said to be capable of clustering without knowing the number of clusters. [15] reported a hybridized version of the black hole approach to solve clustering hitches. The prime goal of hybridization is to solve the black hole algorithm's initialization problem. As a result, the "K-means" method is used to create an initial population for the BH algorithm. Zhou et al. [16] introduced an enhanced social spider optimization algorithm (SSO). The improvement was presented in the search mechanism and convergence rate. In [17], a combination of PSO, KHM, and ICS was proposed to obtain robust clustering. The merits of ICS were utilized for global optima and PSO's ability to handle local optima. Also, fast convergence from KHM was inculcated into a novel approach. The proposed algorithm was tested on several datasets, resulting in a quicker convergence rate. Bijri et al. [18] proposed the MBB-BC algorithm by exploiting the capabilities of the BB-BC algorithm. The memory-based techniques are utilized to achieve local and global optima. The algorithm provided superior clustering compared to GA, BB-BC, GWO, and PSO. An improved version of CSO was presented in [19] for better cluster analysis. Further, Kumar and Singh improved the TLBO algorithm [20]. The authors utilized chaotic maps for further improvement in the convergence speed and diversity of the population. Further, the authors have proposed a local search approach for a good exploration of features. In the same direction of improved clustering approaches, authors in [21] have proposed an enhanced black-hole algorithm. The Levy flight concept overcame the exploration issue. Six benchmark datasets were used to test the newly proposed approach, and compared to multiple standard algorithms, the new system revealed better-quality results. An improved variant of the cat swarm optimization is reported to solve the clustering problem [22]. The search space equations are modified to enhance the convergence speed. The new candidate solution utilizes the cats' global and personal best locations.

3 Related Works

The foraging behavior of humpback whales originally stimulated the WOA. Initially, the WOA was developed to address and solve numerical problems. Initially, it was designed for solving numerical problems [9, 23]. Over time, its versatile nature has led to its application across diverse domains, including clustering, leveraging its inherent capacity for self-exploration. This algorithm's mathematical framework mirrors the strategies employed by whales in their quest to detect and capture prey.

Equations (2) and (3) articulate the systematic approach to locating and encircling prey within this algorithm.

$$\vec{D} = \left| \overrightarrow{C_{cv}} \vec{Z^*}(t) - \vec{Z}(t) \right| \tag{2}$$

$$\vec{Z}(t+1) = \vec{Z^*}(t) - \overrightarrow{A_{cv}}.\vec{D} \tag{3}$$

$$\overrightarrow{A_{cv}} = 2\vec{a}.r - \vec{a}, \ \overrightarrow{C_{cv}} = 2r, \tag{4}$$

The distance vector, represented by $\vec{D}$, signifies distance. $\vec{Z}$ stands for the current position vector, while $(\vec{Z^*})$ represents the vector for the global best position. Additionally, $(\overrightarrow{C_{cv}})$ and $(\overrightarrow{A_{cv}})$ denote coefficient vectors, where r is a function generating random numbers between 0 and 1. The variable "a" linearly decreases from "2 to 0".

The bubble-net attack process combines two techniques: shrinking, modifying coefficient vectors, and encircling and spiral position updates. This method is inspired by the behavior observed in humpback whales.

Conversely, the spiral position update method adheres to a specific equation, mimicking the helical movements typical of whales (refer to Eqs. 5 and 6). Equation (7) enables the calculation of humpback whales' movements, characterized by either shrinking encircling or spiral motions

$$\vec{D'} = \left| \vec{Z^*}(t) - \vec{Z}(t) \right| \tag{5}$$

$$\vec{Z}(t+1) = \vec{D'}.e^{bl}.\cos(2\pi l) + \vec{Z^*}(t) \tag{6}$$

$$\vec{Z}(t+1) = \begin{cases} \vec{Z^*}(t) - \overrightarrow{A_{cv}}.\vec{D} & \text{if } p < 0.5 \\ \vec{D'}.e^{bl}.\cos(2\pi l) + \vec{Z^*}(t), & \text{if } p \geq 0.5 \end{cases} \tag{7}$$

In this scenario, "b" represents a constant vector, "l" represents a randomly chosen value, and "p" denotes a random function ranging between 0 and 1. Humpback whales, while hunting for prey, navigate randomly within the search space, inducing changes in the location vector as outlined in Eqs. (8) and (9).

$$\vec{D} = \left| \overrightarrow{C_{cv}}.\overrightarrow{Z_{rand}} - \vec{Z} \right| \tag{8}$$

$$\vec{Z}(t+1) = \overrightarrow{Z_{rand}} - \overrightarrow{A_{cv}}.\vec{D} \tag{9}$$

4 IWOA: Proposed Work

This section provides an in-depth exploration of IWOA applied to cluster analysis. Two advancements are introduced within this research. Initially, chaotic maps are integrated into the whale optimization algorithm. Secondly, a "step division method" is implemented, effectively leveraging neighborhood search strategies to tackle local optima scenarios. A comprehensive description of these enhancements follows below.

4.1 Integration of Chaotic Maps

To diversify the population, the random parameters (r) utilized in the WOA are substituted with chaotic maps. Numerous chaotic maps have been documented in the literature [24, 25]. However, research suggests that the logistic chaotic map exhibits greater potency among various chaotic maps [26] Eq. 9.

$$c_{n+1} = ac_n(1 - c_n) \tag{10}$$

In Eq. 9, c_{n+1} chaotic value determined at the nth iteration, "a" is a constant value eq. (4). c_n can be defined as $c_n \in (0, 1)$ and further $c_0 \in \{0.0, 0.25, 0.50, 0.75, 1\}$.

$$\overrightarrow{A_{cv}} = 2\vec{a}.c - \vec{a}, \overrightarrow{C_{cv}} = 2c, \tag{11}$$

4.2 Integration of "Step Division Method "Based Neighbourhood Search Strategies

This work introduces a neighborhood strategy employing the "step division method" to address the local optima situation. The initial step consists of identifying neighboring data points surrounding the cluster centers. In this study, our approach consists of utilizing "n = 10" neighboring data points and selecting "k = 3" data points randomly. The computation of the new neighboring data point is determined using Eq. 12, Algorithm 1.

$$\vec{Z}_{neigh}^{new} = \overline{\vec{Z}}_{neigh} + \sum d'x/n \times c \tag{12}$$

The symbol $\vec{Z}_{neigh}^{new}$ refers to the revised or updated neighborhood. In this context, $\overline{Z}$ signifies the average of k best positions, $\sum dx$ represents the sum of these k best positions.

Algorithm 1: Neighborhood Search Strategy
Step 1: Choose 'n' adjacent data points (referred to as best positions)
Step 2: Randomly select 'k' data points for assessment or evaluation.
Step 3: Assess the 'k' neighbouring data points utilizing an equation (12)
Step 4: Obtain a new solution $\vec{Z}_{i,neigh}^{new}$

4.3 IWOA for Clustering

The IWOA has been effectively employed within the realm of clustering to attain optimal solutions. A series of distinct steps are carried out in sequence, encompassing activities like sampling or selecting cluster centers, computation of the objective function, assignment, updates, etc.

Algorithm 2: IWOA for clustering	
i.	Choose the dataset and set up fundamental parameters, including defining the number of clusters $K_i \in (i = 1, 2, \ldots, n)$, iterations, and other essential variables.
ii.	Choose the initial population (K_i) randomly from a dataset.
iii.	Calculate the objective function Equation (1).
iv.	Assign the data objects to clusters Equation (13). $$\text{Fitness}(\vec{X}) = \sum_{j \in 1}^{K} \frac{\text{SSE}(\overrightarrow{ZC})}{\sum_{j=1}^{K} \text{SSE}(\overrightarrow{ZC_j})} \qquad (13)$$

v.	Create a new vector solution based on the following conditions:
	If the random variable 'p' is less than 0.5:
	    If the magnitude of 'A' is less than 1:
	    Update the "position vector" Equation (3).
	    Otherwise, if the magnitude of 'A' is greater than or equal to 1:
	    Update the "position vector" Equation (7).
	Otherwise, if 'p' is greater than or equal to 0.5:
	    Update the "position vector" using Equation (5).

vi.	If a local optimum is detected: Implement a "step division method "based on the Neighbourhood Search Strategy to navigate beyond it. Otherwise, proceed with the ongoing process.
vii.	Update the "position vector".
viii.	Verify the fulfillment of the abort condition induced by the iteration number. If the condition is met, halt the execution; otherwise, proceed with steps 3 to 8.
ix.	Output: The optimal solution.

5 Experimental Findings and Parameter Configurations

This section details an extensive overview of the simulation outcomes and the parameter configurations employed for the IWOA. The simulations were conducted using "MATLAB 2021" on a Windows 11 operating system, utilizing an Intel i5 processor and running on a machine equipped with 16 GB of RAM. The performance evaluation of the IWOA was carried out across six datasets. The parameters for the IWOA are specified as follows: population size = "K × d", number of clusters (K), dimensions (1 ≤ d ≤ n), and iterations = 200. Each algorithm underwent thirty individual runs, and the outcomes were evaluated based on average performance metrics, specifically focusing on "intra-cluster" distance and "f-measure" [27–40].

6 Results and Discussion

This subsection explores the comparative analysis of the IWOA alongside K-means, ACO, GA, CSO, PSO, and WOA Tables 1 and 2. The performance assessment of the IWOA was conducted on "Iris, Cancer, CMC, Wine, Glass, and ISOLET", obtained from the UCI repository. In most cases, IWOA achieves minimal intra-cluster distance values and commendable f-measure in most datasets except CMC datasets (Fig. 1).

Table 1. Comparative analysis of IWOA (Inter-cluster distance)

Dataset	Algorithms						
	IWOA	WOA	PSO	CSO	GA	ACO	K-means
Iris	95.80	96.79	98.73	97.64	125.19	98.36	113.56
Cancer	3035.29	3036.12	3116.64	3124.15	3249.46	3178.09	3248.25
CMC	5581.51	5539.72	5846.63	5804.52	5756.59	5831.25	5912.46
Wine	16294.26	16295	16491.52	16486.21	16530.53	16526.12	18059.91
Glass	230.19	231.29	278.71	264.44	282.32	281.46	246.51
ISOLET	421290.10	421361.25	451718.88	447733.5	460851.88	455837.78	466502.65

Table 2. Comparative analysis of IWOA (f-measure)

Dataset	Algorithms						
	IWOA	WOA	PSO	CSO	GA	ACO	K-means
Iris	0.786	0.784	0.78	0.781	0.774	0.778	0.781
Cancer	0.833	0.822	0.826	0.831	0.819	0.829	0.832
CMC	0.337	0.337	0.333	0.334	0.324	0.332	0.337
Wine	0.523	0.522	0.517	0.522	0.515	0.521	0.520
Glass	0.427	0.419	0.412	0.416	0.333	0.402	0.426
ISOLET	0.407	0.405	0.392	0.311	0.322	0.301	0.361

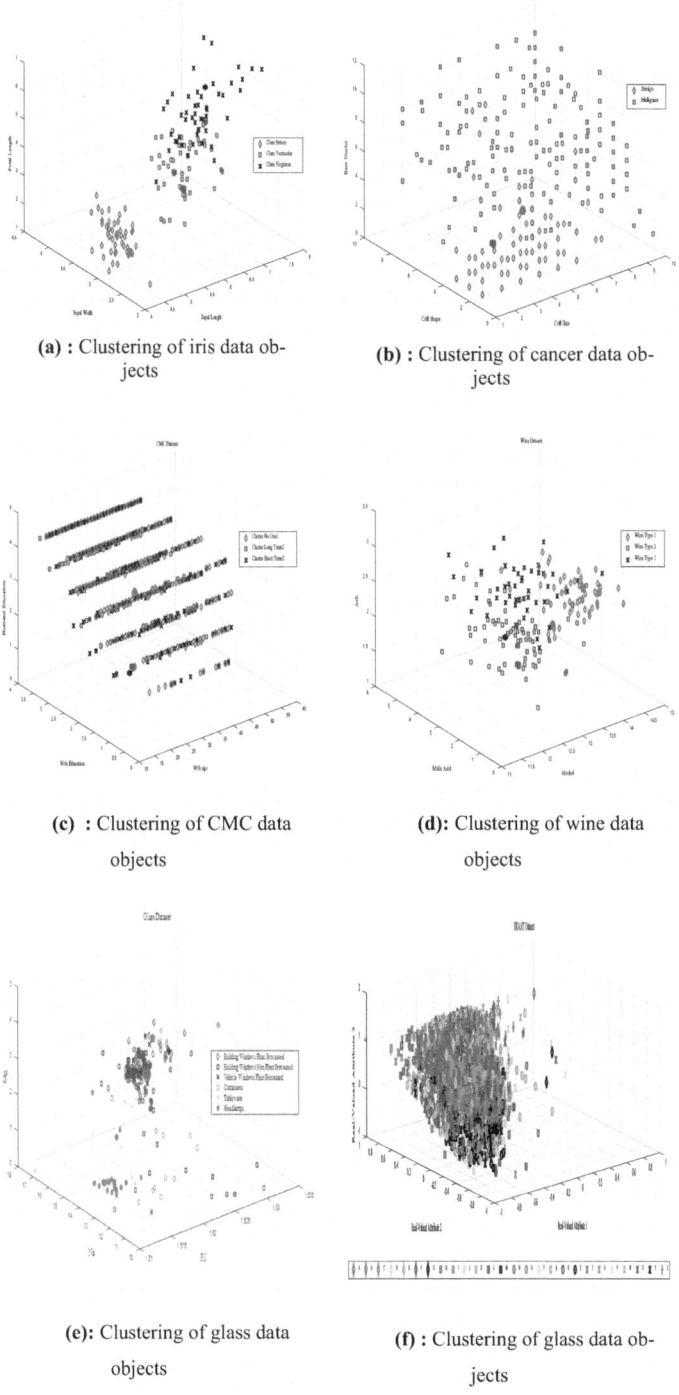

(a) : Clustering of iris data objects

(b) : Clustering of cancer data objects

(c) : Clustering of CMC data objects

(d): Clustering of wine data objects

(e): Clustering of glass data objects

(f) : Clustering of glass data objects

Fig. 1. (a–f). Depicts the clustering using IWOA.

7 Conclusion and Future Enhancements

In this research work, an improved variant of the WOA specifically for clustering is developed. This IWOA incorporates two additional operational capabilities to address clustering problems more effectively. These enhancements involve the integration of chaotic maps and a "step division method" based on neighborhood search mechanisms into the WOA. The efficiency and efficacy of this Improved Whale Optimization Algorithm (IWOA) were thoroughly evaluated across six benchmark datasets. Comparative analysis was conducted against six existing clustering algorithms. The simulation results unequivocally validate the enhancements incorporated into the algorithm, showcasing the effectiveness of this novel clusteriṁng approach. Future endeavors aim to implement the proposed algorithm within vehicular networks, primarily focusing on cluster head formation and load balancing, extending its applicability in real-world scenarios.

Acknowledgments. The authors would like to thank the Editors and the anonymous reviewers for their valuable comments and suggestions, which have helped improve the paper's quality and clarity.

Disclosure of Interests. The authors have no relevant financial or non-financial interests to disclose.

References

1. Fayyad, U., Piatetsky-Shapiro, G., Smyth, P.: From data mining to knowledge discovery in databases. AI Mag. **17**(3), 37 (1996)
2. Han, J., Pei, J., Kamber, M.: Data mining: concepts and techniques. Elsevier (2011)
3. Jain, A.K., Murty, M.N., Flynn, P.J.: Data clustering: a review. ACM Comput. Surv. (CSUR) **31**(3), 264–323 (1999)
4. Xu, R., Wunsch, D.: Survey of clustering algorithms. IEEE Trans. Neural Netw. **16**(3), 645–678 (2005)
5. Nanda, S.J., Panda, G.: A survey on nature inspired metaheuristic algorithms for partitional clustering. Swarm Evol. Comput. **16**, 1–18 (2014)
6. Bouguettaya, A., Yu, Q., Liu, X., Zhou, X., Song, A.: Efficient agglomerative hierarchical clustering. Expert Syst. Appl. **42**(5), 2785–2797 (2015)
7. Łukasik, S., Kowalski, P. A., Charytanowicz, M., Kulczycki, P.: Data clustering with grasshopper optimization algorithm. In: 2017 Federated Conference on Computer Science and Information Systems (FedCSIS), pp. 71–74. IEEE (2017)
8. Deb, S., Tian, Z., Fong, S., Wong, R., Millham, R., Wong, K.K.: Elephant search algorithm applied to data clustering. Soft. Comput. **22**(18), 6035–6046 (2018)
9. Nasiri, J., Khiyabani, F.M.: A whale optimization algorithm (WOA) approach for clustering. Cogent Math. Stat. **5**(1), 1483565 (2018)
10. Alswaitti, M., Ishak, M.K., Isa, N.A.M.: Optimized gravitational-based data clustering algorithm. Eng. Appl. Artif. Intell. **73**, 126–148 (2018)
11. Kuwil, F.H., Atila, Ü., Abu-Issa, R., Murtagh, F.: A novel data clustering algorithm based on gravity center methodology. Expert Syst. Appl. **156**, 113435 (2020)
12. Peng, H., Wang, J., Shi, P., Riscos-Núñez, A., Pérez-Jiménez, M.J.: An automatic clustering algorithm inspired by membrane computing. Pattern Recogn. Lett. **68**, 34–40 (2015)

13. Menéndez, H.D., Otero, F.E., Camacho, D.: Medoid-based clustering using ant colony optimization. Swarm Intell. **10**(2), 123–145 (2016)
14. Huang, F., Li, X., Zhang, S., Zhang, J.: Harmonious genetic clustering. IEEE Trans. Cybern. **48**(1), 199–214 (2017)
15. Pal, S.S., Pal, S.: Black hole and k-means hybrid clustering algorithm. In: Behera, H.S., Nayak, J., Naik, B., Pelusi, D. (eds.) Computational Intelligence in Data Mining: Proceedings of the International Conference on ICCIDM 2018, pp. 403–413. Springer Singapore, Singapore (2020). https://doi.org/10.1007/978-981-13-8676-3_35
16. Zhou, Y., Zhou, Y., Luo, Q., Abdel-Basset, M.: A simplex method-based social spider optimization algorithm for clustering analysis. Eng. Appl. Artif. Intell. **64**, 67–82 (2017)
17. Bouyer, A., Hatamlou, A.: An efficient hybrid clustering method based on improved cuckoo optimization and modified particle swarm optimization algorithms. Appl. Soft Comput. **67**, 172–182 (2018)
18. Bijari, K., Zare, H., Veisi, H., Bobarshad, H.: Memory-enriched big bang–big crunch optimization algorithm for data clustering. Neural Comput. Appl. **29**(6), 111–121 (2018)
19. Kumar, Y., Singh, P.: Improved cat swarm optimization algorithm for solving global optimization problems and its application to clustering. Appl. Intell. **48**(9), 2681–2697 (2017)
20. Kumar, Y., Singh, P.: A chaotic teaching learning based optimization algorithm for clustering problems. Appl. Intell. **49**(3), 1036–1062 (2018)
21. Abdulwahab, H.A., Noraziah, A., Alsewari, A.A., Salih, S.Q.: An enhanced version of black hole algorithm via levy flight for optimization and data clustering problems. IEEE Access **7**, 142085–142096 (2019)
22. Singh, H., Kumar, Y.: A neighborhood search based cat swarm optimization algorithm for clustering problems. Evol. Intel. **13**(4), 593–609 (2020)
23. Mirjalili, S., Lewis, A.: The whale optimization algorithm. Adv. Eng. Softw. **95**, 51–67 (2016)
24. Chuang, L.Y., Hsiao, C.J., Yang, C.H.: Chaotic particle swarm optimization for data clustering. Expert Syst. Appl. **38**(12), 14555–14563 (2011)
25. Bharti, K.K., Singh, P.K.: Chaotic gradient artificial bee colony for text clustering. Soft. Comput. **20**(3), 1113–1126 (2016)
26. Boushaki, S.I., Kamel, N., Bendjeghaba, O.: A new quantum chaotic cuckoo search algorithm for data clustering. Expert Syst. Appl. **96**, 358–372 (2018)
27. Nemade V., Pathak S. and Dubey AK. 'A systematic literature review of breast cancer diagnosis using machine intelligence techniques,' Archives of Computational Methods in Engineering. Oct;29(6):4401–30, 2022
28. Barhate, D., Pathak, S., Dubey, A.K.: Hyperparameter-tuned batch-updated stochastic gradient descent', Plant species identification by using hybrid deep learning. Eco. Inform. **75**, 102094 (2023)
29. Singh, H., et al.: An enhanced whale optimization algorithm for clustering. Multimed. Tools Appl. **82**(3), 4599–4618 (2023)
30. Singh, H., Kumar, Y., Kumar, S.: A new meta-heuristic algorithm based on chemical reactions for partitional clustering problems. Evol. Intel. **12**(2), 241–252 (2019)
31. Singh, H., Kumar, Y.: Hybrid artificial chemical reaction optimization algorithm for cluster analysis. Procedia Comput. Sci. **167**, 531–540 (2020)
32. kumar, Y., Sahoo, G.: A two-step artificial bee colony algorithm for clustering. Neural Comput. Appl. **28**(3), 537–551 (2015). https://doi.org/10.1007/s00521-015-2095-5
33. Kumar, Y., Kaur, A.: Variants of bat algorithm for solving partitional clustering problems. Eng. Comput. 1–27 (2021)
34. Kaur, A., Kumar, Y.: A new metaheuristic algorithm based on water wave optimization for data clustering. Evol. Intel. **15**(1), 759–783 (2022)
35. Singh, H., Kumar, Y.: An enhanced version of cat swarm optimization algorithm for cluster analysis. Int. J. Appl. Metaheuristic Comput. **13**(1), 1–25 (2022)

36. Kumar, R., Kumar, P., Kumar, Y.: Time series data prediction using IoT and machine learning technique. Procedia Comput. Sci. **167**, 373–381 (2020)
37. Kumar, Y., Sahoo, G.: Prediction of different types of liver diseases using rule based classification model. Technol. Health Care **21**(5), 417–432 (2013)
38. Singh, H., Kumar, Y.: Hybrid big bang-big crunch algorithm for cluster analysis. In: Futuristic Trends in Networks and Computing Technologies: Second International Conference, FTNCT 2019, Chandigarh, India, 22–23 Nov 2019, Revised Selected Papers 2, pp. 648–661. Springer Singapore (2020)
39. Singh, H., Kumar, Y.: Cellular automata based model for e-healthcare data analysis. Int. J. Inform. Syst. Model. a Design **10**(3), 1–18 (2019)
40. Kaur, A., Kumar, Y.: Neighborhood search based improved bat algorithm for data clustering. Appl. Intell. **52**(9), 10541–10575 (2022)

Property Price Prediction Using Regression Analysis

R. K. Tripathi[1]($\boxtimes$), Sunil Kumar[1], Aakriti Nawani[1], Pranjali Kathait[1], Rohit Madhwal[1], and Seema Saini[2]

[1] Department of Mathematics, Graphic Era Hill University, Dehradun 248002, India
{rktripathi,sunilkumar}@gehu.ac.in

[2] Department of Mathematics, Graphic Era (Deemed to Be)University, Dehradun 248002, India

Abstract. The objective of this study is to check how the price of a property is dependent on various features and to finally predict the price of a house property using Regression algorithms. We will also be comparing various regression techniques and checking accuracy of each technique, finally fitting in the best possible algorithm for predicting price of a property. Study may be helpful for a seller to estimate what the selling cost should be offered to a customer based on the features of the house. Various data mining techniques have also been used to predict the price of a property, based on market trends and customer inputs. For price prediction, classification algorithms have been applied based on market trends.

Keywords: Regression-Algorithm · Multiple-Regression · Decision Tree · Grid-Search

1 Introduction

During the previous years, a lot of studies came into play regarding house prices prediction and analysis. Sangani et. al. [1] used data from Zillow.com and used techniques like linear regression and Gradient boosting to make prediction about other properties. They used time as a feature for their prediction. Sampathkumar, Santhi and Vanjinathan [2] used Neural Networks with Multiple Linear Regression to predict the price of Chennai metropolitan area (CMA). Conventional housing property prediction models do not focus on spatiotemporal problems on a large amount of data. Spatiotemporal data can be managed with respect to both time and space. Varma et. al. [3] concluded that rather than using various regression techniques individually, it is better to use weighted mean of these to give better and accurate results. This approach would correspond to minimum error and maximum error in comparison to individual approaches applied. Thus, use of ensemble learning would be better, to build prediction model. Yusof and Ismail [4] illustrated how price variation of properties can be explained using the extension of Multiple Regression Analysis (MRA) and Hedonic Regression analysis. Pedregosa et. al. [5] used Scikit-learn module of python for checking various supervised and unsupervised algorithms to come up with the best model to predict the price.

M. Khurana et al. (Eds.): ICMLA 2024, CCIS 2238, pp. 167–181, 2025.
https://doi.org/10.1007/978-3-031-75861-4_15

Soltani et. al. [6] concluded that Decision Tree performs better than Linear regression when it comes to predicting prices. They also showed that Random Forest and Gradient Boosting techniques can also be used for such predictions. To conduct this study, data used from metropolitan Adelaide, Australia. Similarly using Random Forest, Adetunji et al. [7] estimated the price of the property and discussed a situation where one wants to predict price variance rather than the real value. Truong et. al. [8] used HPI to establish that housing property prices are highly correlated with some other factors such as location, area, and population. They tested various Techniques on regression to find the best possible prediction model and used random forest to predict the price. Again, using HPI, Lu et. al. [9] used data set from Kaggle and proposed a hybrid lasso in conjunction with Gradient Boosting Regression model for predicting prices. Bourassa et. al. [10] used spatial dependence of house prices by explicit modeling of error structure and concluded that, hedonic variable is easy to implement than spatial statistical methods, where spatial statistical methods are the application of statistical concepts to link data to spatial locations attached to them. Zulkifley et. al. [11] used data mining techniques to infer information from data. Through their learning they verified the usage of support vector regression, XG Boost and Artificial Neural Network to predict property prices to be better than others.

Moparthi and Geethanjali [12] identified that there is no significant work done that has been accepted as a common tool in real-time software products. They proposed a new ensemble defect prediction classification model to predict metric relationships and identify errors in trends. These proposed models were found to be more useful than traditional Bayesian network models.

Mu, Wu and Zhang [13] used least squared support vector machine (LSSVM) with support vector machine and partial least square (PLS) to forecast values of properties. They used big data and concluded that SVM and LSSVM are superior to PLS when it comes to dealing with non-linearity. Khalafallah [14] used historical data as train Artificial Neural Network to predict future trends as well as analyzed prediction error and obtained in the range of -2% to $+2\%$.

Alfiyatin et al. [15] used regression and particle swarm optimization (PSO) to come up with a model minimum prediction error to predict the price. They used data based on NJOP houses in Malang city to carry out their perdition. Madhuri, Anuradha, Pujitha [16] used various techniques such as Multiple Linear Regression, Ridge, Elastic Net and Gradient Boosting Regression to predict price of a property. They included application of multi regression included with techniques of data mining to get the best result. As working with supervised techniques especially regression algorithms it is important to look out for Overfitting or Underfitting as this could lead to bias and variance error, Babyak [17] tried up to come up with a solution to avoid overfitting. Bhagat, Mohokar and Mane [18] built a website that would return the value of the price of a property with respect to budget and priority of the customers. Sreeja et al. [19] analyzed Real Estate Price Prediction using Machine Learning. However, Wang et al. [20] predicted as well as analyzed residential house price using a flexible spatiotemporal model.

2   Data Set, Its Reference and Methodology

The data used here has been taken from kaggle.com [21, 22], which is an interactive learning site especially for aspiring data scientists. The data files is divided into two parts i.e., train set and test set. To build and train the model, used train.csv. The raw file contained 29452 rows and 12 columns. This file contains columns such as ADDRESS, UNDER_CONSTRUCTION, BHK etc. The ADDRESS column contains information of various cities across the country (India) including Chennai, Bangalore and Gurgaon etc. Other than this the data contains information on whether the property is ready to move in or not & and if the information is updated directly by the owner or through a dealer ? Such a type of data wouldn't really affect the price of a property thus this wouldn't be considered as a parameter for the model building.

Flowchart given below gives a basic layout of how any machine learning project is carried out-

Data Gathering and Analysis → Data Cleaning → Data Pre-processing → Training of Model → Testing and Evaluation → Making User Interference → Deploying the Model

We used a python-based platform designed for data scientists, engineers, and analysts. Since Spyder is designed for data scientists and analysts it has various inbuilt libraries and modules for the same. One such module that we used is, Scikit-Learn Module, which is an efficient tool for predictive analysis. In addition to this, also used various basic libraries like numpy for basic mathematical operations and pandas to work with data frames. Here matplot lib and seaborn have been used, which are libraries for data visualization in python.

3   Process of Data Analysis

In this section we defined the process of data analysis.

3.1   Data Gathering and Collection

This is the initial stage where data is acquired from various (reliable) sources. For this study, the dataset downloaded from www.kaggle.com.

About the Data File: The dataset contains two different files I.E.
Train.csv – 29451 Rows × 12 Columns,
Test.csv – 68720 Rows × 11 Columns.
In order to build a model test.csv has been used.

3.2   Data Cleaning

At this step the main aim is to check all these columns and try to understand whether these parameters would be considered as independent variables for the model building or not.

Identifying Dependent Variable: As we are trying to predict the price of a property, thus in the case of our dataset, PRICE would be considered as a dependent variable. Since the data contains 12 columns, excluding PRICE column that is considered as the dependent variable, still left with 11 columns to deal with which would be considered as independent variables. This would cause complexity of model and in order to avoid this we checked, what is the value if PRICE really dependent upon various other variables, as represented using graphs in matplot library of python for data interpretation (Fig. 1).

Fig. 1. Relationship of dependent and independent variables

We checked other columns as well, for the sake of this study we assumed that the price of the property only depends upon SQUARE_FT, BHK, & ADDRESS. Thus, we can drop all the other columns from our dataset for simplicity.

Next step is to observe that if the data contains any missing or null values or not. In the case of our dataset none of the values are null or missing thus we moved forward (In Case if some values were missing, we would directly drop these rows since we already have a large dataset to train our model). Next step is to check if there is consistency in the data or not (Fig. 2).

Nan ▲	Type	Size	Value
df1	DataFrame	(29451, 12)	Column names: ADDRESS, UNDER_CONSTRUCTION, RERA, BHK, BHK_OR_RK, SQUAR...
df2	DataFrame	(29451, 4)	Column names: ADDRESS, BHK, SQUARE_FT, PRICE
x	Series	(6899,)	Series object of pandas.core.series module

```
...: df2.head()
Out[12]:
                       ADDRESS  BHK    SQUARE_FT  PRICE
0          Ksfc Layout,Bangalore   2  1300.236407   55.0
1      Vishweshwara Nagar,Mysore   2  1275.000000   51.0
2               Jigani,Bangalore   2   933.159722   43.0
3      Sector-1 Vaishali,Ghaziabad 2  929.921143   62.5
4               New Town,Kolkata   2   999.009247   60.5
```

Fig. 2. Output after removing columns from original data frame.

The data used is consistent. i.e. all the integral data is either float or int type. In case this was not the case we made a function to test our data as mentioned below:

1. Check if the Data Type is Float:
   a) Input: DataFrame data_frame, column name or function.
   b) Use the apply method with a function (float_or_not) to check if the data type of each element in the specified column is a float or not.
2. Filter Rows Based on Data Type:
   a) Use the Boolean result from the previous step to filter the DataFrame (data_frame) and select only the rows where the data type is not a float.

This would return only those values which were float type. Similarly, we can test for int type as well. After this step, it has been concluded that all our data is in consistent form (Fig. 3).

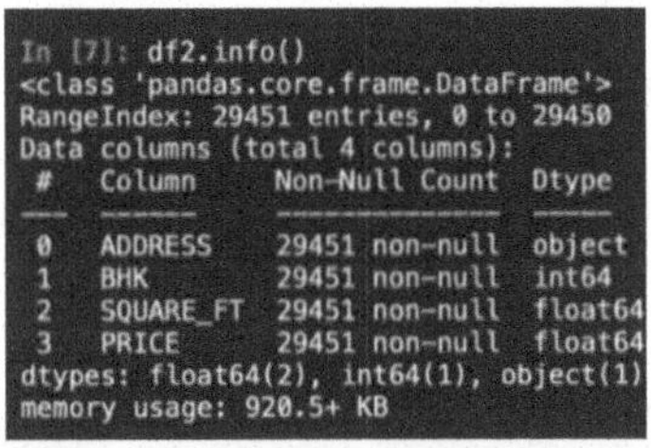

```
In [7]: df2.info()
<class 'pandas.core.frame.DataFrame'>
RangeIndex: 29451 entries, 0 to 29450
Data columns (total 4 columns):
 #   Column      Non-Null Count  Dtype
---  ------      --------------  -----
 0   ADDRESS     29451 non-null  object
 1   BHK         29451 non-null  int64
 2   SQUARE_FT   29451 non-null  float64
 3   PRICE       29451 non-null  float64
dtypes: float64(2), int64(1), object(1)
memory usage: 920.5+ KB
```

Fig. 3. Information about the dataset.

3.3 Data Preprocessing

During the process of data-preprocessing, a new column 'PRICE_PER_SQFT' has been added to our data frame. Here we simply add the value of how much the price of a particular property is per square feet, using (Fig. 4):

Insert new column PRICE_PER_SQFT.

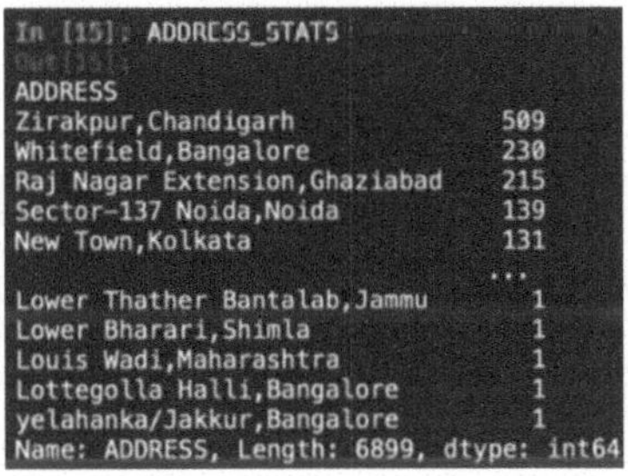

```
In [15]: ADDRESS_STATS
Out[15]:
ADDRESS
Zirakpur,Chandigarh                  509
Whitefield,Bangalore                 230
Raj Nagar Extension,Ghaziabad        215
Sector-137 Noida,Noida               139
New Town,Kolkata                     131
                                     ...
Lower Thather Bantalab,Jammu           1
Lower Bharari,Shimla                   1
Louis Wadi,Maharashtra                 1
Lottegolla Halli,Bangalore             1
yelahanka/Jakkur,Bangalore             1
Name: ADDRESS, Length: 6899, dtype: int64
```

Fig. 4. Checking the ADDRESS column in data frame.

This command will add the desired column into the dataset. Next, we checked the ADDRESS column, using the algorithm:

Check length of Address Function.

Here faced an issue of high dimensionality in our data, thus we must come up with a solution, for this, first remove any leading blank spaces that might be our data.

If blank space → Remove.

If row/column duplicate → Remove.

Arrange Database Alphabetically.

Thus, we have total 6899 unique values in the ADDRESS column, but dealing with all these values separately would lead to an issue of high dimensionality thus to overcome this, we removed all the data where less than 20 values of that particular ADDRESS are mentioned. For this purpose, following code will be used:

If column/row data < 20 → Remove row/column.

The Next step would be Outline Detection. For this, we checked the normality of data by Shapiro-Wilk Test algorithm. The Shapiro-Wilk test is a goodness-of-fit test to examine how close sample data fit to a normal distribution. As a result, it orders and standardizes the sample space:

Apply Shapiro Test → Find Price per square foot.

Since p-value obtained is less than 0.5, therefore, data is not considered normally distributed, and hence to convert data into normal form we added another column into the dataset using (Fig. 5).

```
In [23]: shapiro(df3['PRICE_PER_SQFT'])
/Users/aakrtinawani/opt/anaconda3/lib/python3.9/site-packages/scipy/stats/morestats.py:1760: UserWarning
p-value may not be accurate for N > 5000.
  warnings.warn("p-value may not be accurate for N > 5000.")
Out[23]: ShapiroResult(statistic=0.7454293370246887, pvalue=0.0)
```

Fig. 5. Shapiro test to check Normality.

Thus, this added a new column into the data frame that is now normally distributed we can tally this using the following algorithm (Fig. 6):

1. Plots the kernel density estimate (KDE) for the original distribution.
2. Creates a Q-Q plot (Probability plot) for assessing the normality.
3. Conducts the Shapiro-Wilk test for normality.
4. If data normal → Add column.

```
In [25]: df3.head()

                   ADDRESS  BHK  ...  PRICE_PER_SQFT  PRICE_PER_SQFT_boxcox
4          New Town,Kolkata    2  ...     6056.000000               3.154226
7            Kharar,Mohali    3  ...     4402.999999               3.127987
11         Hebbal,Bangalore    2  ...     8252.000002               3.177461
13  Sector-119 Noida,Noida    2  ...     5524.999999               3.146921
15  Sector-150 Noida,Noida    3  ...     5100.000000               3.140387

[5 rows x 6 columns]

In [26]:
```

Fig. 6. Adding a new column to Normalize data.

After removing the outliers, we checked whether the price depends on the number of BHK or not. To analyze, we use following function and test have been used:

1. Input Parameters:
   a) df: DataFrame containing the relevant data.
   b) Location: The specific location for which the scatter plot is to be generated.
2. Data Extraction:
   a) Filter the DataFrame (df) to obtain subsets (bhk2 and bhk3) based on the specified location and the number of bedrooms (BHK).
3. Plotting: Configure the matplotlib figure size.
   a) Create a scatter plot for 2 BHK properties with total area (SQUARE_FT) on the x-axis and price (PRICE) on the y-axis. Use blue color and label it as '2 bhk'.
   b) Create a scatter plot for 3 BHK properties with total area on the x-axis and price on the y-axis. Use green color, a cross marker, and label it as '3 bhk'.
   c) Set x-axis label to 'total area', y-axis label to 'price', and the plot title to the specified location.
   d) Display the legend.
4. Function Call:
   a) Call the function with the DataFrame (df5) and the specific location ('Sector-119 Noida,Noida').
5. Show Plot: Display the generated scatter plot.

   This algorithm outlines the steps taken by the code to generate the scatter plot for 2 BHK and 3 BHK properties in the specified location.
   Finally, we checked the count in the property that is the majority of the price per square feet of the property.

1. Set Figure Size: Set the figure size.
2. Create Histogram:
   a) Use plt.hist() to create a histogram of the 'PRICE_PER_SQFT' column from the DataFrame df5.
3. Label Axes: Set the x-axis and the y-axis.
4. Display Plot: Use plt.show() to display the generated histogram (Figs. 7, 8 and 9).

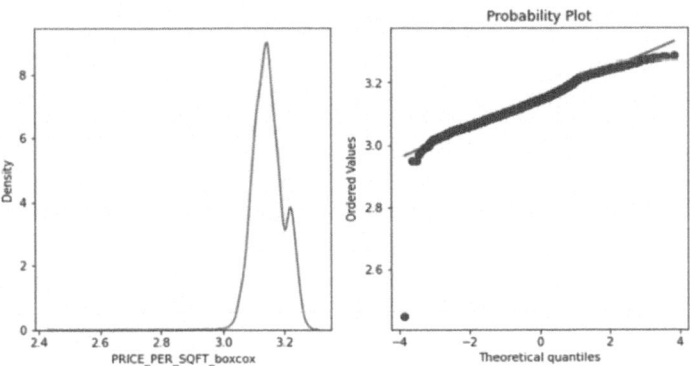

Fig. 7. Data not normally distributed for PRICE_PER_SQFT column.

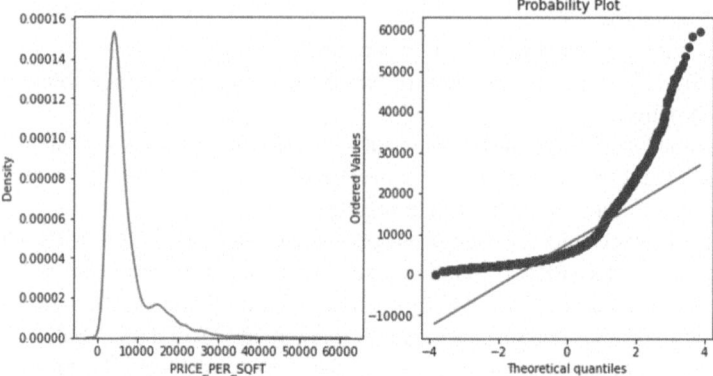

Fig. 8. Data distributed normally when new column is used.

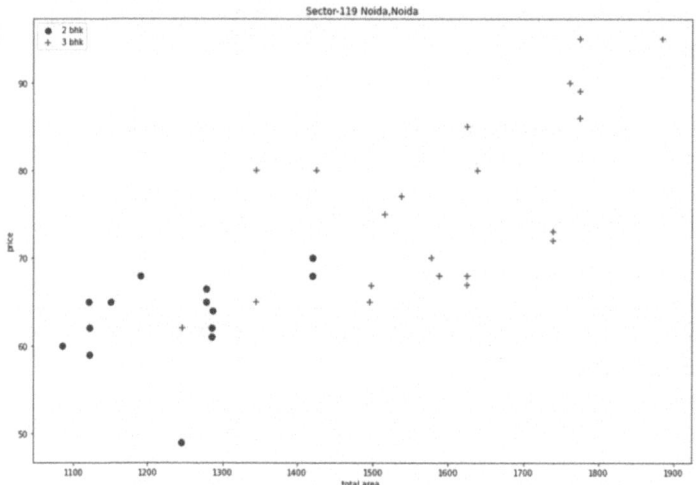

Fig. 9. Scatter plot between 2 BHK and 3 BHK houses.

3.4 Training the Model

The next step after the data is left with no outlines would be model building. Before coming to this step, the ADDRESS column needs to be considered as it consists of categorical values which are not catered by machines. To handle this problem, a method used in machine learning to deal with categorical data used is known as One-hot encoding, it basically converts categorical data into that format which is understandable and can be catered into machine learning algorithms which helps in improving predictions accuracy. This step done using the pandas dummies method, as follows:

1. Create Dummy Variables: Call the function pd.get_dummies() on the 'ADDRESS' column of the DataFrame df5.
   a) Store the resulting DataFrame in a variable named dummies.

    b) Display the first 3 rows of the dummies DataFrame.
2. Concatenate DataFrames:
    a) Call the function pd.concat() to concatenate the original DataFrame df5 with the dummies DataFrame.
    b) Drop one of the dummy columns to avoid the dummy variable trap (in this case, dropping the 'sector-121 Noida,Noida' column).
    c) Store the resulting DataFrame in a variable named df6.
    d) Display the first few rows of the df6 DataFrame.
3. Set Up Dependent and Independent Variables:
    a) Call the function df6.drop() to drop the 'ADDRESS' column from the DataFrame df6.
    b) Store the resulting DataFrame in a variable named df7.
    c) Display the first few rows of the df7 DataFrame.
    d) Set up the independent variable X by calling the function df7.drop() to drop the 'PRICE' column.
    e) Set up the dependent variable y by selecting the 'PRICE' column from the DataFrame df7.
    f) Display the first few rows of the X and y variables (Fig. 10).

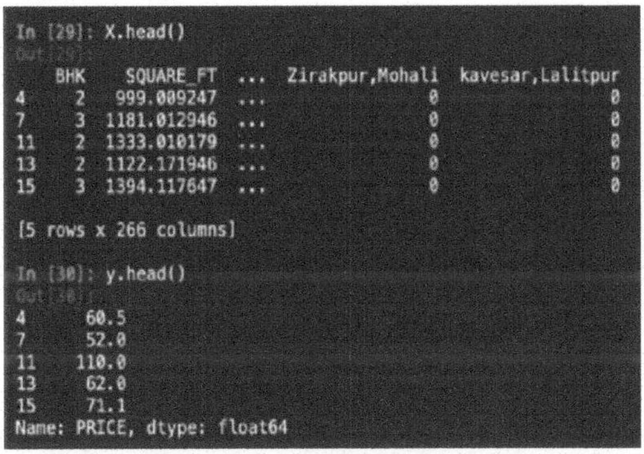

```
In [29]: X.head()
Out[29]:
    BHK   SQUARE_FT  ...  Zirakpur,Mohali  kavesar,Lalitpur
4     2   999.009247 ...                0                 0
7     3  1181.012946 ...                0                 0
11    2  1333.010179 ...                0                 0
13    2  1122.171946 ...                0                 0
15    3  1394.117647 ...                0                 0

[5 rows x 266 columns]

In [30]: y.head()
Out[30]:
4      60.5
7      52.0
11    110.0
13     62.0
15     71.1
Name: PRICE, dtype: float64
```

Fig. 10. Setting of independent and dependent variable.

To train the model, 80% of the dataset owned, while the rest 20% owned for testing the model. Initially, Linear regression used to train the model which resulted in giving us 84.5% accuracy score. However, to achieve more accurate score, some other algorithms have been tested with their accuracy on hand using k-fold validation and grid search. This approach results in the formation of grids having horizontal and vertical display like a table format, it would evaluate all ordered pairs to form a combination of set of values (Fig. 11).

For estimating cost of housing properties, we tested data considering decision tree regression, Lasso and Linear regression. Final outcome that is displayed has decision nodes and leaf nodes which are very important or play very crucial role. A decision node (say, Outlook) consists of two or more branches (say-Sunny, Overcast and Rainy), each carrying a unique identity on its own for a tested attribute while leaf node (say, Hours

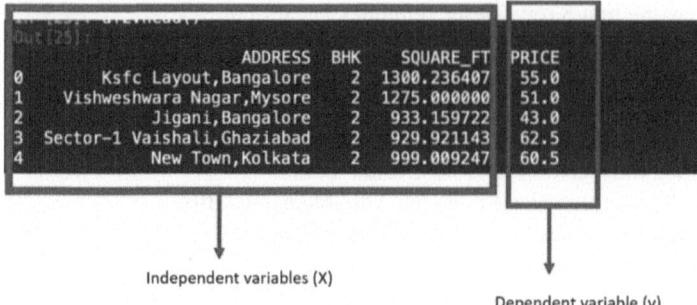

Independent variables (X)

Dependent variable (y)

Fig. 11. Dependent and Independent variable in data.

Played) represents an outcome on the given target. The starting or initial node of the decision tree which turns out to be the best predictor is called as root node. Decision tree handles both categorical and numerical data (Figs. 12 and 13).

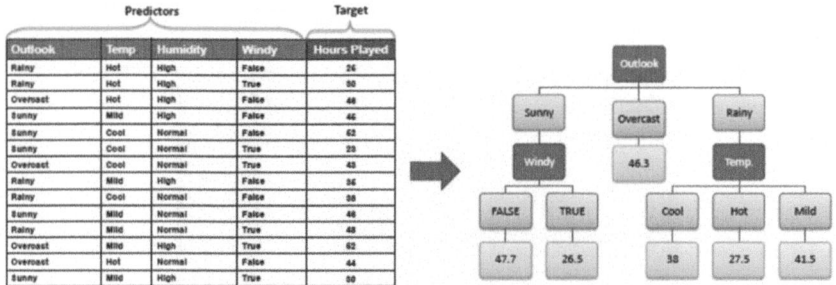

Fig. 12. Decision tree working.

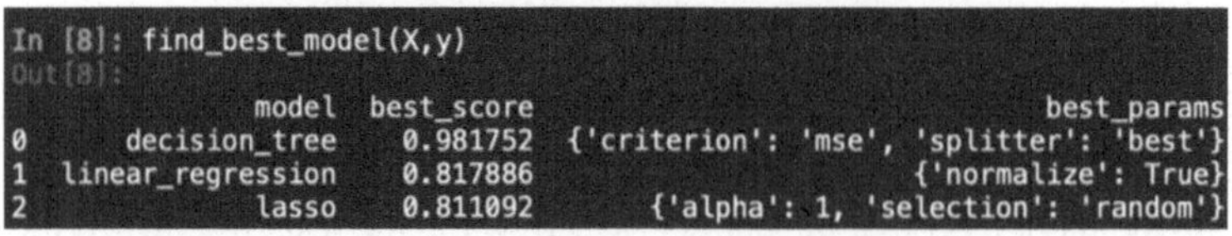

```
In [8]: find_best_model(X,y)
Out[8]:
                  model    best_score                              best_params
0         decision_tree    0.981752    {'criterion': 'mse', 'splitter': 'best'}
1     linear_regression    0.817886                      {'normalize': True}
2                 lasso    0.811092    {'alpha': 1, 'selection': 'random'}
```

Fig. 13. Comparing accuracy of various regression algorithms.

Algorithm supported models with very less parameters, shows high levels of multicollinearity and when dealing with certain type of model selection, like variable selection/parameter elimination.

Algorithm for this is mentioned below:

1. Define Function find_best_model(X, y):
   a) Input: X (independent variables), y (dependent variable).
   b) Output: Data_Frame with model names, best scores, and best parameters.
2. Define Regression Models and Their Hyperparameter Grids:

a) Define three regression models (Decision Tree Regressor, Linear Regression, Lasso Regression) along with their corresponding hyperparameter grids.
3. Initialize an Empty List scores:
   a) This list will store the performance scores, model names, and best parameters.
4. Perform Grid Search for Each Model:
   a) Iterate over each algorithm in the algorithms dictionary.
   b) For each algorithm, perform grid search using GridSearchCV with cross-validation (ShuffleSplit with 5 splits).
   c) Append a dictionary containing model name, best score, and best parameters to the scores list.
5. Return Results as DataFrame:
   a) Create a DataFrame from the scores list with columns 'model', 'best_score', and 'best_parameters'.
   b) Return the DataFrame.
6. Function Call:
   a) Call the find_best_model(X, y) function with the provided independent variables (X) and dependent variable (y).

The function that we built basically would result a data frame with algorithm name and their respective parameters and their scores. This way we concluded which one is the best algorithm to move forward and has been observed that decision tree gives the best accuracy score of 99%, thus we used Decision tree regression to train the model (Fig. 14).

Fig. 14. Comparing accuracy of models.

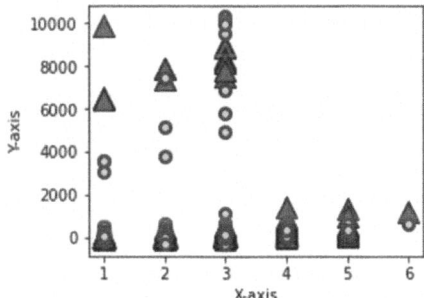

Fig. 15. y_pred versus y_actual for linear regression.

Initially the model fitted, using linear regression, we got an accuracy score of 83% with this. Further, when we fitted the model and predicted the dependent variable, we

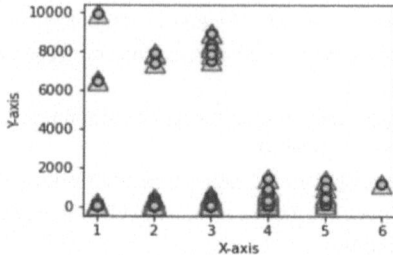

Fig. 16. y_pred versus y_actual for decision tree regression.

can see that there is some inaccuracy in the predicted and actual values. We can see that through the below scatter plot Fig. 15.

Figure 15 shows the variance between the actual values vs the predicted values when Linear Regression was used.

Next, we fitted Decision tree regression and again repeated the same process by predicting the value of independent variable(y) in terms of y_pred. In order to see the accuracy of the model we again used scatter plot shown in Fig. 16. As a result of decision tree regression, we can now see that the prices more accurately predicted in this case as opposed to Linear Regression.

3.5 Deploying the Model

For finally deploying the model we built a function in order to take in customer query and return and estimated price (in lakhs) for the particular query. The python code and output for the same are mentioned below and This python code gives us the freedom to predict the price for any particular query. For e.g. Let's say we want to predict the price for a property in New Town, Kolkata with 2 bedrooms and the minimum area must be 999 square ft we will simply pass the query: predict price ('New Town, Kolkata', 2, 999.0900) and as a result of this we will get a predicted value for price 60.5(in lakhs).

1. Define Function predict_price(ADDRESS, BHK, SQUARE_FT):
    a) Input: ADDRESS (location), BHK (number of bedrooms), SQUARE_FT (total area).
    b) Output: Predicted house price
2. Encode Categorical Variable 'ADDRESS':
    a) Find the index of the 'ADDRESS' column in the feature set X.
    b) Initialize an array x with zeros, with the length equal to the number of columns in X.
    c) Set the values of 'BHK' and 'SQUARE_FT' in the array.
    d) If the location index (loc_index) is non-negative (i.e., 'ADDRESS' is present in the feature set): Set the corresponding entry in the array to 1.
3. Make Prediction:
    a) Use the trained regression model (reg) to predict the house price for the input features.
4. Return Prediction: Return the predicted house price (Fig. 17).

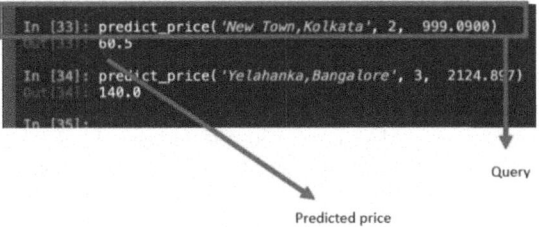

Fig. 17. Query and predicted price.

4 Comparison of Study, Limitations and Scope for Future Work

The XGBoost and LightGBM are effective techniques for achieving high accuracy, with LightGBM being particularly efficient. However, Regression methods surpass all those previous methods in terms of performance. On other hand, the Stacked Generalization Regression method boasts a complex architecture but is the optimal choice when accuracy is of utmost importance. While both Hybrid Regression and Stacked Generalization Regression yield satisfactory outcomes, it is crucial to consider their time complexities. Nevertheless, the FSTM's predicted house price errors are considerably significant primarily due to the substantial coefficients of variation derived from the annual sample values of the residential house price. Conversely, the house price values exhibited spatial and temporal autocorrelation among themselves. However, when provided with a specific year and a set of predictor values, the FSTM offers a comprehensive prediction. By combining the FSTM with geographically weighted regression, there is a promising opportunity to enhance the spatial and temporal prediction of residential house prices.

5 Conclusion and Discussion

It is concluded that price of a property can be predicted using features like address, no. of rooms in the property and the area that the house covers. We have used decision tree regression in order to fit this simple linear regression problem. As a result of this we got an accuracy of 99% which is a got score. For this, we used grid search to compare the accuracy score of these Algorithms as a result, we saw that Decision tree gives the best accuracy of 99% out of all.

In conclusion, Decision Tree Regression was used to fit the model. Finally, we built a function to pass a query for anyone, able to predict price, on the basis of our dataset. As this is a Machine Learning Model, we can further train our model for other datasets and predict property prices for other places as well. For Future work on the same, our aim would be to build a time series model that can predict the prices of the property not only on the basis of the features like address or no. of rooms but also can predict price of property for upcoming years as well. We would also like to build a user-interface so that our model is interactive.

Authors declare no competing interests.

References

1. Sangani, D., Erickson, K., Hasan, M.A.: Predicting zillow estimation error using linear regression and gradient boosting. In: IEEE 14th International Conference Proceeding on Mobile Ad Hoc and Sensor Systems (MASS), pp. 530–534. Orlando, FL, USA (2017)
2. Sampathkumar, V., Santhi, M.H., Vanjinathan, J.: Forecasting the land price using statistical and neural network software. Procedia Comput. Sci. **57**, 112–121 (2015)
3. Varma, A., Sarma, A., Doshi, S., Nair, R.: House price prediction using machine learning and neural networks. In: Second International Conference on Inventive Communication and Computational Technologies (ICICCT), pp. 1936–1939 (2018)
4. Yusof, A.M., Ismail, S.: Multiple regressions in analyzing house price variations. Commun. IBIMA 1–9 (2012)
5. Pedregosa, F., et al.: Scikit-learn: machine learning in Python. J. Mach. Learn. Res. **12**, 2825–2830 (2012)
6. Soltani, A., Heydari, M., Aghaei, F., Pettit, C.J.: Housing price prediction incorporating spatio-temporal dependency into machine learning algorithms. Cities **131**, 103941 (2022)
7. Adetunji, A.B., Akande, N.O., Ajala, F.A., Oyewo, O., Akande, Y.F., Oluwadara, G.: House price prediction using random forest machine learning technique. Procedia Comput. Sci. **199**, 806–813 (2022)
8. Truong, Q., Nguyen, M., Dang, H., Mei, B.: Housing price prediction via improved machine learning techniques. Procedia Comput. Sci. **174**, 433–442 (2020)
9. Lu, S., Li, Z., Qin, Z., Yang, X., Goh, R.S.M.: A hybrid regression technique for house prices prediction. In: IEEE International Conference on Industrial Engineering and Engineering Management (IEEM), pp. 319–323. Singapore (2017)
10. Bourassa, S.C., Cantoni, E., Hoesli, M.E.R.: Spatial dependence, housing submarkets, and house price prediction. J. Real Estate Financ. Econ. **35**(2), 143–160 (2007)
11. Zulkifley, N.H., Rahman, S.A., Ubaidullah, N.H., Ibrahim, I.: House price prediction using a machine learning model: a survey of literature. Int. J. Modern Educ. Comput. Sci. **12**(6), 46–54 (2020)
12. Moparthi, A.N.R., Geethanjali, B.: Design and implementation of hybrid phase based ensemble technique for defect discovery using SDLC software metrics. In: 2nd International Conference on Advances in Electrical, Electronics, Information, Communication and Bio-Informatics (AEEICB), vol. 4, pp. 268–274 (2016)
13. Mu, J., Wu, F., Zhang, A.: Housing value forecasting based on machine learning methods. Abstract Appl. Anal. **2014**, 1–7 (2014). https://doi.org/10.1155/2014/648047
14. Khalafallah, A.: Neural network based model for predicting housing market performance. Tsinghua Sci. Technol. **13**(5), 325–328 (2008)
15. Alfiyatin, A.N., Febrita, R.E., Taufiq, H., Mahmudy, W.F.: Modeling House Price Prediction using Regression Analysis and Particle Swarm Optimization Case Study: Malang, East Java, Indonesia. Int. J. Adv. Comput. Sci. Appl. **8**(10) (2017)
16. Madhuri, C.R., Anuradha, G., Pujitha, M.V.: House price prediction using regression techniques: a comparative study. In: International Conference on Smart Structures and Systems (ICSSS), Chennai, India, pp. 1–5 (2019)
17. Babyak, M.A.: What you see may not be what you get? A brief, nontechnical introduction to overfitting in regression-type models. Psychosom. Med. **66**(3), 411–421 (2004)
18. Bhagat, N., Mohokar, A., Mane, S.: House price forecasting using data mining. Int. J. Comput. Appl. **152**, 23–26 (2016)
19. Sreeja, S.P., Asha, V., Saju, B., Prakash, P.O., Singh, A.K.: Real Estate Price Prediction using Machine Learning. In: Third International Conference on Advances in Electrical, Computing, Communication and Sustainable Technologies (ICAECT), pp. 1–7. Bhilai, India (2023)

20. Wang, L., Wang, G., Yu, H., Wang, F.: Prediction and analysis of residential house price using a flexible spatiotemporal model. J. Appl. Econ. **25**(1), 503–522 (2022). https://doi.org/10.1080/15140326.2022.2045466

21. Housing Prices Dataset. Kaggle. https://www.kaggle.com/datasets/yasserh/ housing-prices-dataset (2022)

22. Housing price Prediction (Linear Regression).Kaggle.https://www.kaggle.com/code/ashydv/housing-price-prediction-linear-regression (2019)

Personalized Healthcare Recommendations with Q-Learning Reinforcement Learning

Poi Tamrakar[1,2](✉) 📷, Ganesh R. Pathak[2], Mily Lal[1], Akanksha Goel[1], and Manisha Bhende[1]

[1] Dr. D. Y. Patil School of Science and Technology, Dr. D. Y. Patil Vidyapeeth, Pimpri, Pune, Maharashtra, India
Poi.tamrakar@gmail.com
[2] MIT Art, Design and Technology University, Pune, Maharashtra, India

Abstract. The transforming potential of Q-Learning in customized medical recommendations is examined in this study. Its performance in comparison to conventional methods, strategic data utilization, and flexibility in a variety of scenarios all show great promise. However, a significant gap in the literature emphasizes how important it is to take ethics into account when using patient data. The findings show that Q-learning improves patient outcomes, but its ethical implications are still mostly unknown. This study offers ethical guidelines for future research that address the identified gap. In short, even though Q-Learning has many advantages, ethical issues must be bridged in order to promote responsible integration into healthcare, striking a balance between technical advancements as well as moral principles to enhance patient outcomes.

Keywords: Q-Learning · Personalized Healthcare · Adaptability · Ethical Considerations

1 Introduction

1.1 Research Background

A paradigm shift toward patient-centered and personalized care characterizes the current state of healthcare. The widespread use of electronic health records as well as the wealth of health-related data present a previously unheard-of chance to customize medical interventions to each patient's specific needs. The idea behind this research is that the complexity of different patient populations could prevent them from being best served by conventional, one-size-fits-all healthcare models [1]. Healthcare is one area where machine learning applications hold great promise as these technological advancements keep going to transform various industries. The study explores the relationship between Q-learning reinforcement Learning, a cutting-edge strategy that maximizes treatment recommendations by utilizing the power of adaptive algorithms, alongside personalized healthcare [2]. With the use of patient data, past results, and iterative learning, the research hopes to further the continuous advancement of healthcare delivery. Appreciating the way Q-Learning can improve personalized healthcare recommendations is essential for progressing the efficacy and cost-effectiveness of medical interventions at this nexus of technological innovation as well as healthcare. This will help to ensure a future in which treatments are precisely tailored to each patient's needs.

1.2 Research Aim and Objectives

Aim: This study aims to improve the efficacy as well as quality of healthcare recommendations by incorporating Q-Learning Reinforcement Learning into customized treatment strategies.

Objectives:

- To examine the way Q-Learning algorithms adjust to various healthcare situations.
- To develop and improve Q-Learning models for individualized treatment recommendations, use patient data.
- To compare Q-learning-based healthcare recommendations with conventional approaches to assess their efficiency and efficacy.
- To examine the possible effects on patient outcomes alongside satisfaction that personalized healthcare recommendations could have.

1.3 Research Rationale

In order to meet the urgent need for more customized healthcare solutions, this research is being conducted. Many times, the flexibility required to take into consideration the distinctive qualities of various patient populations is missing from current healthcare models. This study seeks to revolutionize healthcare recommendations through the use of Learning, which is recognized for its adaptability and capacity to learn from dynamic environments [3]. Q-Learning's personalized nature presents a significant opportunity to enhance treatment outcomes as well as patient satisfaction. This research aims to close the gap between patient-centered cares in addition to technological innovation in a rapidly changing healthcare environment, furthering the ongoing evolution of healthcare practices.

2 Literature Review

2.1 Q-Learning Reinforcement Learning

Q-learning is a type of reinforcement learning algorithms which doesn't follow any model. Agent learns from their environments based on the rewards. Rewards are calculated based on the state and action of the agent in a particular environment. The state and action of agent is stored in a lookup table to calculate rewards [4] as shown in Fig. 1.

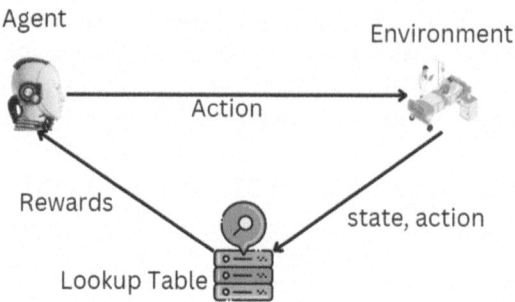

Fig. 1. Q-Learning Algorithms' Adaptability in Healthcare Scenarios

2.2   Adaptability of Q-Learning Algorithms in Healthcare Scenarios

A critical step in comprehending Q-Learning's relevance is investigating how flexible it is in various healthcare contexts. Research suggests that Q-Learning's innate ability to dynamically adapt is consistent with the dynamic nature of healthcare settings. Research demonstrates effective applications in the management of chronic diseases, demonstrating Q-Learning's capacity to modify treatment plans in response to evolving patient circumstances [5]. Notwithstanding, certain obstacles have been recognized, including the interpretability of Q-Learning models in intricate medical decision-making. In addition to addressing potential drawbacks and opportunities for development, this section will provide an evaluation of these studies alongside offer insights into the particular healthcare domains where Q-Learning has demonstrated promise [6]. Furthermore, analyzing scenarios in which Q-Learning could run into difficulties will contribute to a more sophisticated comprehension of its flexibility in various healthcare environments, developing the groundwork for later talks regarding its incorporation into customized treatment plans.

2.3   Utilization of Patient Data for Q-Learning Model Training

Customized healthcare recommendations depend heavily on the effective utilization of patient data to train Q-learning models. Previous research emphasizes how important it is to use large-scale patient datasets to improve Q-Learning algorithms' capacity for acquiring knowledge and making decisions [7]. Studies indicate that real-time patient monitoring data, genetic information, and electronic health records can all be successfully integrated into Q-Learning frameworks. The approaches employed to preprocess and incorporate various patient data sources will be examined in this literature review, which will also highlight potential problems and best practices [8]. Figure 2 represents data utilization for training Q-Learning model in different fields, and significance of choosing features and data normalization in enhancing Q-Learning models for healthcare applications. A full grasp of the practical issues and ethical ramifications related to employing patient data to teach Q-Learning models for customized treatment recommendations will be facilitated by evaluating research studies that place an emphasis on the ethical handling of sensitive patient information.

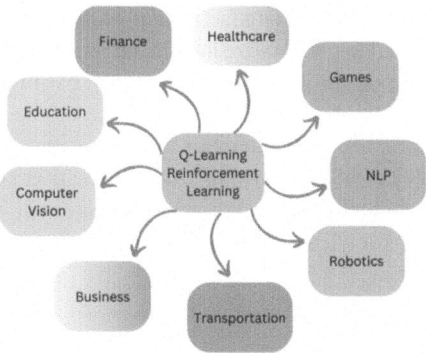

Fig. 2. Data Utilization for Training Q-Learning Model in Different Fields

2.4 Performance and Efficacy Comparison with Traditional Methods

The usefulness of this novel approach must be established by contrasting the effectiveness and performance of Q-Learning-based healthcare suggestions with conventional techniques. Studies and other existing literature have carried out empirical evaluations that directly compare Q-Learning to traditional healthcare recommendation systems [9]. Remarkable results indicate that Q-learning is more adaptive to shifting patient circumstances, leading to more tailored and responsive treatment plans. Furthermore, the work highlights the effectiveness gains that Q-Learning provides when optimizing treatment plans, especially in situations where patient profiles are complex and dynamic [10]. These comparative studies will be thoroughly and critically examined in this review of the literature, with an emphasis on quantitative metrics like treatment adherence, patient outcomes, and resource usage. For assessing Q-Learning's wider applicability, it is essential to fully understand the situations in which it performs better than conventional techniques and to pinpoint any potential drawbacks [11]. Furthermore, this section will examine research that clarifies Q-Learning's scalability as well as computational efficiency, offering insights into the technology's viability for incorporation into wide-ranging healthcare systems [12]. This review attempts to provide important insights into the real-world benefits and difficulties associated with employing Q-Learning for healthcare recommendations by analyzing the performance differences.

2.5 Impact of Personalized Healthcare Recommendations on Patient Outcomes

One important aspect that needs further investigation is the influence that Q-Learning-enabled personalized healthcare recommendations have on patient outcomes. Empirical research, or literature, indicates that tailored treatment plans based on Q-Learning are positively correlated with improved patient outcomes [13]. Research demonstrates that interventions can be more accurately adapted to the specific characteristics of each patient, leading to better overall health outcomes, fewer adverse reactions, and increased adherence to treatment plans. We will examine these empirical results in more detail in this section, focusing on the tangible benefits in actual healthcare environments [14].

Furthermore, the possibility of higher satisfaction and engagement from patients when Q-Learning algorithms are used to personalize healthcare recommendations. This literature review aims to extract important insights into how Q-Learning-informed customized medical recommendations promote positive patient experiences and long-term well-being by closely examining the methodologies as well as metrics used in these studies [15]. A more nuanced view of Q-Learning's transformative potential in influencing patient outcomes in the context of personalized healthcare will be developed by filling in any gaps in our present knowledge of these impacts.

2.6 Literature Gap

A significant knowledge vacuum exists regarding the moral issues surrounding the use of patient data in healthcare settings to train Q-Learning models. Few studies deal with the ethical ramifications of handling sensitive health information, instead focusing on the efficacy and technical aspects of the practice [16]. In order to make sure that developments in Q-Learning are in line with strong ethical standards and patient privacy concerns, it is imperative that future research examine the ethical aspects of personalized healthcare recommendations. The lack in the literature highlights this need.

3 Methodology

The present study employs an interpretivist approach, acknowledging the intricate and context-specific characteristics of tailored healthcare suggestions enabled by Q-Learning. Recognizing the complex relationships that exist between patient data, Q-Learning algorithms, as well as healthcare outcomes in the larger sociocultural and ethical contexts is in line with interpretivism. To methodically evaluate and verify theories derived from pre-existing theoretical frameworks, a deductive approach is utilized. The deductive approach enables a systematic investigation into the flexibility followed by efficacy of Q-Learning in various healthcare scenarios by commencing with established principles and theories associated with Q-Learning in healthcare [17]. With a descriptive research design, the study intends to give a thorough overview of the present situation together with future directions of Q-Learning-based personalized healthcare recommendations. Providing a thorough overview of the field, the descriptive design facilitates the methodical investigation of important variables and their correlations in the setting of Q-Learning applications in healthcare [18]. In order to obtain pertinent information from scholarly articles, healthcare databases, and already published literature, secondary data collection is used. Thorough searches on resources like IEEE Xplore, PubMed, and other appropriate journals guarantee a wide range of information. Case studies, experimental findings, as well as theoretical frameworks that advance knowledge of Q-Learning's flexibility, patient data utilization, accomplishment comparison, and influence on patient outcomes in medical scenarios are among the secondary data that have been gathered. R and Python are going to be utilized for performing statistical analysis on quantitative data, including patient outcomes as well as algorithm performance metrics [19]. Thematic analysis will be employed to examine qualitative data, which will include ethical considerations as well as challenges. The amalgamation of quantitative and qualitative analyses is going to provide a comprehensive outlook on the effectiveness and consequences of customized healthcare recommendations via Q-Learning. This interpretivism-based methodology draws conclusions from well-established theories, applies a descriptive design to demonstrate a nuanced comprehension, and uses secondary data to conduct a thorough investigation of Q-Learning in tailored healthcare recommendations.

4 Results

4.1 Adaptability of Q-Learning Algorithms

Q-learning algorithms exhibit outstanding adaptability that has a significant impact on personalized treatment strategies when explored in healthcare scenarios. Q-learning is unique in the field of managing chronic diseases because it can adapt dynamically to the conditions of patients in real-time, especially in the treatment of diabetes [20]. Its flexibility in modifying important parameters, like insulin dosage, demonstrates a patient-centred and personalized approach. The algorithm's ability to adapt to the particular requirements of each patient raises the possibility of a paradigm change in diabetes care toward precision medicine. Moreover, Q-Learning is not limited to diabetes control; it is also successful in oncology, where it can optimize chemotherapy regimens [21]. Because cancer treatment is continually changing and dynamic, Q-Learning adjusts to each patient's particular reaction. This flexibility is essential for managing the intricacies of cancer care, as treatment regimens are frequently required to be modified on a regular basis to accommodate patients' changing needs and conditions [22]. Essentially, Q-Learning's proven flexibility in diabetes and cancer therapy establishes it as a game-changing instrument for personalizing healthcare interventions, promoting a responsive and individualized strategy that takes into account the unique circumstances of each patient's journey.

4.2 Utilization of Patient Data for Model Training

A critical component of Q-Learning models' training is the strategic use of data, which determines the extent to which the models perform in providing individualized treatment recommendations. This calls for the combining of several datasets, such as real-time patient monitoring data, genetic data, as well as electronic health records [23]. The combination of these disparate data sources demonstrates the algorithm's exceptional capacity to absorb intricate and multidimensional data streams. In doing so, Q-Learning allows for an in-depth comprehension of each patient's unique health profile by utilizing real-time patient data, genetic insights, as well as traditional medical records. Successful feature selection and normalization techniques are essential to this success [24]. By ensuring that the model concentrates on the most pertinent and instructive elements of the data, choosing features expedites the learning process. In addition, normalization methods ensure that every characteristic is fairly taken into account during the training of the model by standardizing the data and reducing scale differences. The significance of these technical details lies in their ability to improve the model's performance, accuracy, and significance in providing insights into customized medical care [25]. The medical information mart for intensive care III (MIMIC-III) is used as the data source in the proposed model. Figure 3 represents the architecture of Q-Learning model for healthcare sector. It emphasizes the importance that sound methodologies to guarantee the dependability and efficacy of Q-learning models in the context of personalized healthcare. The ability to effectively manage a variety of data sources is becoming increasingly important as healthcare datasets get more complex. This is because accomplishing so will help Q-Learning applications succeed in customizing treatments for each patient.

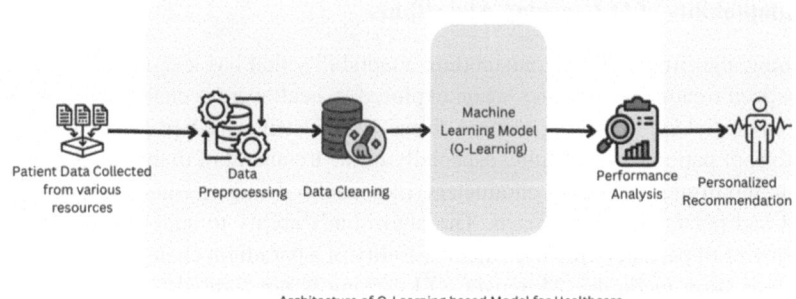

Architecture of Q-Learning based Model for Healthcare

Fig. 3. Architecture of Q-Learning based Model for Healthcare

4.3   Performance and Efficacy Comparison

The comparisons between Q-learning-based healthcare recommendations as well as conventional approaches provide a substantial amount of evidence supporting the superiority of the algorithm. Q-Learning has demonstrated time and time again in a variety of studies the effectiveness with which it can modify treatment plans in response to changing alongside dynamic patient profiles [26]. Because of its flexibility, Q-Learning can be used as an effective tool in personalized healthcare, where the unique needs of each patient call for customized and adaptable interventions. An important factor in confirming Q-Learning's superiority was quantitative metrics. Q-learning consistently outperformed traditional methods in metrics like treatment adherence alongside patient outcomes. Because of the algorithm's ability to dynamically modify treatment plans in response to real-time patient data, treatment adherence rates are higher and patient outcomes are better [27]. Table 1 shows the performance analysis of traditional methods against Q Learning Model. The algorithm's potential to improve healthcare interventions' overall efficiency and standard of care is highlighted by its alignment with patient-centric metrics. Further research into the effectiveness of resources yielded more information about Q-Learning's possible effects on healthcare systems. The algorithm's capacity to optimize resource allocation is a measure of both its overall systemic efficiency and its flexibility in responding to the particular requirements of each patient. Essentially, the examination of comparative analyses highlights the way Q-Learning can transform healthcare practices by offering greater flexibility, better patient-centered outcomes, alongside more efficient use of resources, all of which will ultimately lead to a more responsive and efficient healthcare ecosystem.

Table 1. Comparison of Traditional ML Models with Q-Learning Model

Traditional Machine Learning Models	Accuracy	Precision	F-1 score
K-NN	0.772	0.831	0.86
NN	0.895	0.89	0.89
CNN	0.925	0.920	0.925
LSTM	0.903	0.910	0.912
Q-Learning RL	**0.906**	**0.914**	**0.915**

4.4   Impact on Patient Outcomes

A strong positive correlation is found when the effects of Q-Learning facilitated personalized healthcare recommendations are examined in relation to patient outcomes. Q-learning shows promise in improving overall health outcomes by reducing side effects as well as improving treatment adherence by customizing interventions to each patient's unique characteristics. This patient-centered approach is in line with the fundamental principles of personalized medicine, which tailor treatments to the specific features of each patient. Research findings consistently demonstrate the algorithm's exceptional capacity to enhance patient satisfaction as well as promote positive experiences. In addition to addressing particular health requirements, the customized recommendations made by Q-Learning also take into account each patient's preferences alongside tolerances, resulting in a more patient-friendly healthcare setting. The decrease in side effects indicates the precise and successful Q-Learning is at reducing possible risks and improving the safety profile of customized treatments. This section explores Q-Learning's revolutionary potential in the context of customized medical care [28]. It emphasizes the algorithm's contribution to changing the entire patient experience in addition to enhancing clinical outcomes. As healthcare moves regarding a more patient-centric paradigm—one in which interventions are not only successful but also tailored to the particular requirements and demands of each patient—Q-Learning becomes increasingly important. This will ultimately lead to a more positive and satisfying experience receiving healthcare.

4.5   Ethical Considerations and Literature Gap

The literature gap analysis highlights a significant gap in the current body of knowledge regarding the moral issues surrounding the use of patient data in medical facilities to train Q-learning models. The technical nuances along with the efficacy of Q-Learning have been thoroughly examined, but the ethical ramifications have received noticeably less attention. This disparity draws attention to an important development in the field of personalized healthcare recommendations: the moral implications associated with employing cutting-edge machine learning algorithms. Future studies are critically needed to deal with and close this gap. Ensuring the responsible and reliable implementation of Q-Learning in the healthcare industry requires careful consideration of ethical issues. The potential hazards to patient privacy and data security arise from the dearth of a thorough investigation of ethical implications [29]. A thorough comprehension of the ethical nuances turns into essential as advanced technologies, like Q-Learning, are increasingly incorporated into the framework of personalized healthcare. This section offers a critical viewpoint on the current status of research and calls on academics and professionals to concentrate on the moral ramifications of using Q-Learning in healthcare environments. In doing so, it directs future research, promoting a more comprehensive strategy that harmonizes technological innovations with moral principles, guaranteeing patient privacy, as well as facilitating the smooth integration of moral considerations into the creation and application of tailored healthcare suggestions for improvement. Table 2 represents list of Q-learning aspects & its key findings.

Table 2. Aspect of Q-Learning

Aspect of Q-Learning	Key Findings
Adaptability	Demonstrated adaptability in diabetes and oncology treatments, showcasing dynamic adjustment capabilities based on real-time patient conditions
Data Utilization	Successfully integrated electronic health records, genetic information, and real-time patient monitoring data, enhancing model performance through effective feature selection and normalization
Performance Comparison	Consistently outperformed traditional methods in diverse studies, showing adeptness in adapting treatment strategies to dynamic patient profiles. Favorable quantitative metrics, including treatment adherence and patient outcomes
Impact on Patient Outcomes	Significant positive correlations observed. Tailoring interventions led to reduced adverse effects, enhanced treatment adherence, and overall improved health outcomes. Consistent improvement in patient experiences and increased satisfaction
Ethical Considerations	Literature gap identified regarding the ethical dimensions of utilizing patient data for training Q-Learning models. Emphasis on the need for future research to address ethical implications, ensuring alignment with robust ethical standards and patient privacy concerns

5   Evaluation and Conclusion

5.1   Critical Evaluation

The research findings' critical evaluation highlights Q-Learning's resilience in transforming personalized healthcare. The algorithm's transformative potential is validated by its consistently superior performance over traditional methods, strategic utilization of patient data, as well as suggested adaptability across diverse healthcare scenarios. However, a significant gap in the literature indicates that using patient data for Q-learning models disregards ethical issues. Concerns about patient privacy and moral principles in individualized healthcare have been brought up by this. In summary, Q-Learning shows a lot of promise, but in order to responsibly integrate it into healthcare environments and maintain a harmonious balance between moral considerations as well as technological advancements in the pursuit of better patient outcomes, it is imperative to take into account ethical dimensions.

5.2   Research Recommendation

The identified literature gap regarding the ethical considerations of using patient data for Q-Learning models leads to an important research recommendation. Subsequent investigations ought to thoroughly explore the moral implications of utilizing Q-Learning within medical environments. Investigations need to concentrate on procedures and guidelines that guarantee the moral and responsible use of patient data while resolving issues with security, privacy, and informed consent [30]. It is also critical to assess the manner in which society will be affected and the way people will see the use of Q-Learning in healthcare. Future research can help develop ethical guidelines and best practices by giving priority to these ethical considerations. This will enable the smooth integration of Q-Learning into customized medical services while protecting patient rights and encouraging trust in the healthcare system.

5.3   Future Work

Further work ought to concentrate on improving the ethical framework of Q-Learning for use in healthcare settings. It is crucial to thoroughly investigate the rules guaranteeing patient security, approval, as well as privacy. It's also critical to evaluate the way society views the integration of Q-learning. The further work of this research path can be to create moral guidelines that will encourage the ethical, accountable, and open use of personalized healthcare.

References

1. Mulani, J., Heda, S., Tumdi, K., Patel, J., Chhinkaniwala, H., Patel, J.: Deep reinforcement learning based personalized health recommendations. In: Dash, S., Acharya, B.R., Mittal, M., Abraham, A., Kelemen, A. (eds.) Deep learning techniques for biomedical and health informatics, pp. 231–255. Springer International Publishing, Cham (2020). https://doi.org/10.1007/978-3-030-33966-1_12

2. Coronato, A., Naeem, M., De Pietro, G., Paragliola, G.: Reinforcement learning for intelligent healthcare applications: a survey. Artif. Intell. Med. **109**, 101964 (2020). https://doi.org/10.1016/j.artmed.2020.101964
3. Oh, S.H., Park, J., Lee, S.J., Kang, S., Mo, J.: Reinforcement learning-based expanded personalized diabetes treatment recommendation using South Korean electronic health records. Expert Syst. Appl. **206**, 117932 (2022). https://doi.org/10.1016/j.eswa.2022.117932
4. Bettoni, D., Soppelsa, A., Fedrizzi, R., del Toro Matamoros, R.M.: Analysis and adaptation of Q-learning algorithm to expert controls of a solar domestic hot water system. Appl.. Syst. Innov. **2**(2), 15 (2019). https://doi.org/10.3390/asi2020015
5. Zhong, Y., Wang, C., Wang, L.: Survival augmented patient preference incorporated reinforcement learning to evaluate tailoring variables for personalized healthcare. Stats **4**(4), 776–792 (2021). https://doi.org/10.3390/stats4040046
6. Talaat, F.M.: Effective deep Q-networks (EDQN) strategy for resource allocation based on optimized reinforcement learning algorithm. Multimed. Tools Appl. **81**(28), 39945–39961 (2022). https://doi.org/10.1007/s11042-022-13000-0
7. Tan, C., Han, R., Ye, R., Chen, K.: Adaptive learning recommendation strategy based on deep Q-learning. Appl. Psychol. Meas. **44**(4), 251–266 (2020). https://doi.org/10.1177/0146621619858674
8. Daoud, S., Mdhaffar, A., Jmaiel, M., Freisleben, B.: Q-rank: reinforcement learning for recommending algorithms to predict drug sensitivity to cancer therapy. IEEE J. Biomed. Health Inform. **24**(11), 3154–3161 (2020). https://doi.org/10.1109/JBHI.2020.3004663
9. Yu, C., Liu, J., Nemati, S., Yin, G.: Reinforcement learning in healthcare: a survey. ACM Comput. Surv. **55**(1), 1–36 (2023). https://doi.org/10.1145/3477600
10. Afsar, M.M.: Personalized recommendation using reinforcement learning ([Doctoral Dissertation]. University of Calgary) (2022)
11. Oh, S.H., Lee, S.J., Park, J.: Precision medicine for hypertension patients with type 2 diabetes via reinforcement learning. J. Personalized Med. **12**(1), 87 (2022). https://doi.org/10.3390/jpm12010087
12. Li, Y., Wang, H., Wang, N., Zhang, T.: Optimal scheduling in cloud healthcare system using Q-learning algorithm. Complex Intell. Syst. **8**(6), 4603–4618 (2022). https://doi.org/10.1007/s40747-022-00776-9
13. Oh, S.H., Lee, S.J., Park, J.: Effective data-driven precision medicine by cluster-applied deep reinforcement learning. Knowl.-Based Syst. **256**, 109877 (2022). https://doi.org/10.1016/j.knosys.2022.109877
14. Yang, H., Fu, H.: Reinforcement learning in personalized medicine. In: Yang, H. (ed.) Data Science, AI, and Machine Learning in Drug Development, pp. 177–192. Chapman and Hall/CRC, Boca Raton (2022). https://doi.org/10.1201/9781003150886-8
15. Abdellatif, A.A., Mhaisen, N., Chkirbene, Z., Mohamed, A., Erbad, A., Guizani, M.: Reinforcement learning for intelligent healthcare systems: A comprehensive survey. arXiv preprint arXiv:2108.04087 (2021)
16. Khan, O., Badhiwala, J.H., Grasso, G., Fehlings, M.G.: Use of machine learning and artificial intelligence to drive personalized medicine approaches for spine care. World Neurosurg. **140**, 512–518 (2020). https://doi.org/10.1016/j.wneu.2020.04.022
17. Do, Q.T., Doig, A.K., Son, T.C.: Deep Q-learning for predicting asthma attack with considering personalized environmental triggers' risk scores. In. Annual International Conference of the IEEE Engineering in Medicine and Biology Society. IEEE Engineering in Medicine and Biology Society. Annual International Conference 41st Annual International Conference of the IEEE Engineering in Medicine and Biology Society (EMBC). IEEE Publications, pp. 562–565 (2019). https://doi.org/10.1109/EMBC.2019.8857172
18. Clifton, J., Laber, E.: Q-learning: theory and applications. Annu. Rev. Stat. Appl. **7**(1), 279–301 (2020). https://doi.org/10.1146/annurev-statistics-031219-041220

19. Zohora, M.F., Tania, M H., Kaiser, M.S., Mahmud, M.: Forecasting the risk of type II diabetes using reinforcement learning. In Joint 9th International Conference on Informatics, Electronics y Vision (ICIEV) and 2020 4th International Conference on Imaging, Vision y Pattern Recognition (icIVPR), pp. 1–6. IEEE Publications (2020). https://doi.org/10.1109/ICIEVicIVPR48672.2020.9306653

20. Kondrup, F., et al.: Deep conservative reinforcement learning for personalization of mechanical ventilation treatment

21. Prasanna, K.S.L., Challa, N.P., Nagaraju, J.: Heart disease prediction using reinforcement learning technique. In Third International Conference on Advances in Electrical, Computing, Communication and Sustainable Technologies (ICAECT), 2023 (pp. 1–5). IEEE Publications (2023). https://doi.org/10.1109/ICAECT57570.2023.10118232

22. Dereventsov, A., Starnes, A., Webster, C.G.: Examining policy entropy of reinforcement learning agents for personalization tasks (2022). arXiv preprint arXiv:2211.11869

23. Peine, A., et al.: Development and validation of a reinforcement learning algorithm to dynamically optimize mechanical ventilation in critical care. npj Digi. Med. **4**(1), 32 (2021). https://doi.org/10.1038/s41746-021-00388-6

24. Buchard, A., et al.: Learning medical triage from clinicians using deep q-learning (2020). arXiv preprint arXiv:2003.12828

25. Cai, X., Chen, J., Zhu, B Wang, Yao, Y.: Towards real-world applications of personalized anesthesia using policy Constraint Q Learning for Propofol Infusion Control. IEEE J. Biomed. Health Inform. **28**(1), 459–469 (2024). https://doi.org/10.1109/JBHI.2023.3321099

26. Di, S., Petch, J., Gerstein, H.C., Zhu, R., Sherifali, D.: Optimizing health coaching for patients with Type 2 diabetes using machine learning: model development and validation study. JMIR Formative Res. **6**(9), e37838 (2022). https://doi.org/10.2196/37838

27. Ma, S., Lee, J., Serban, N., Yang, S.: Deep attention Q-network for personalized treatment recommendation (2023). arXiv preprint arXiv:2307.01519

28. Ying, Z., Zhang, Y., Cao, S., Xu, S., Liu, X.: Oidpr: Optimized insulin dosage based on privacy-preserving reinforcement learning. In: IFIP Networking Conference (Networking), pp. 655–657. IEEE Publications (2020)

29. Wu, P. (2019). Machine learning methods for personalized medicine using electronic health records. Columbia University

30. Shiranthika, C., Chen, K.W., Wang, C.Y., Yang, C.Y., Sudantha, B.H., Li, W.F.: Supervised optimal chemotherapy regimen based on offline reinforcement learning. IEEE J. Biomed. Health Inform. **26**(9), 4763–4772 (2022). https://doi.org/10.1109/JBHI.2022.3183854

Pioneering Healthcare Innovations with the Convergence of Blockchain, AI, and the Internet of Medical Things (IoMT)

Akanksha Goel[1,2]([⊠]) and S. Neduncheliyan[2]

[1] Dr. D. Y. Patil School of Science and Technology, Sant Tukaram Nagar Pimpri, Dr. D. Y. Patil Vidyapeeth, Pune, India
akanksha.rkgit@gmail.com
[2] Department of CSE, School of Computing, Bharath Institute of Higher Education and Research (BIHER), Tamil Nadu, Chennai, India

Abstract. The integration of Blockchain technology, Artificial Intelligence (AI), and the Internet of Medical Things (IoMT) has emerged as a transformative force in revolutionizing healthcare systems. This synergy offers significant promise in securing medical data, enhancing treatment efficiency, and enabling seamless data sharing. Blockchain ensures the integrity and privacy of medical information within IoMT architectures, fostering secure data management and transparent sharing among stakeholders. Leveraging AI-enabled decision support systems, healthcare providers benefit from accurate disease diagnosis and personalized treatment plans. However, challenges such as scalability, storage constraints, and legal complexities pose hurdles to this integration. Addressing these challenges and fostering collaborations among technology developers, healthcare professionals, and policymakers are critical for unlocking the full potential of these technologies. This study explores the transformative potential of integrating Blockchain, Artificial Intelligence (AI), and the Internet of Medical Things (IoMT) in revolutionizing healthcare. The convergence of these technologies holds promise for enhancing data security, optimizing patient care, and streamlining healthcare operations.

Keywords: Blockchain · Internet of Medical Things · Artificial Intelligence · Healthcare

1 Introduction

The Internet of Medical Things (IoMT) is transforming healthcare by enhancing patient care and health monitoring. In response to the pandemic, IoMT provides a solution for remote health monitoring, allowing individuals to autonomously track their health using IoMT devices. IoMT comprises interconnected smart medical devices forming a Body Sensor Network (BSN) or Wireless Sensor Network (WSN) [1, 2]. These devices, equipped with smart sensors, collect medical data directly from the patient's body for transmission over the internet. AI-driven applications on smartphones enable swift access to medical assistance [3].

M. Khurana et al. (Eds.): ICMLA 2024, CCIS 2238, pp. 194–208, 2025.
https://doi.org/10.1007/978-3-031-75861-4_17

The integration of AI into IoMT allows real-time monitoring of health parameters, facilitating disease management and prevention [4, 6]. Security is crucial in IoMT-based Smart Healthcare Systems (SHS), where AI plays a role in ensuring data integrity and network safety, detecting intrusions, and conducting security assessments.

Blockchain technology further secures IoMT networks by maintaining decentralized databases, enhancing data security and integrity. Smart home technology advancements improve healthcare for individuals with disabilities or the elderly, extending health-care facilities beyond hospital confines [9–12]. IoMT revolutionizes healthcare with remote monitoring, early intervention, and personalized care. The combination of AI and blockchain enhances data security and predictive analytics, reshaping the health-care landscape [10]. The integration of cutting-edge technologies like blockchain, AI, and wearable sensors drives efficiency, accuracy, and accessibility in healthcare services [11].

Smart contracts in healthcare, based on advanced blockchain analysis, simplify data management and streamline healthcare procedures [12, 13]. Governments and businesses globally recognize the importance of digitizing healthcare systems with technologies like blockchain and AI [13, 14]. AI plays a pivotal role in drug monitoring, treatment opti-mization, and leveraging clinical trial data, representing a paradigm shift in healthcare [15, 16]. Blockchain ensures secure and decentralized health data management [17, 18]. The fusion of AI algorithms into blockchain fortifies healthcare data analysis, fostering secure data recording, analysis, and dissemination [19, 20].

Blockchain-based platforms like CHIE facilitate secure data exchange, ensuring pri-vacy and validity of health-related data among healthcare professionals and institutions [21]. The integration of wearable sensors, IoT, AI, and blockchain reshapes healthcare delivery, offering continuous health monitoring and personalized treatment approaches [22]. This convergence of technologies marks a monumental shift in healthcare manage-ment systems, establishing a paradigm of continuous real-time health monitoring and management for improved patient-centric care [23].

The key contributions of this study are:

- Explore the transformative role of Intelligence (AI), and the Internet of Medical Things (IoMT) in healthcare.
- Explore IoMT, Blockchain's role in keeping medical data safe, and how AI helps in diagnosing and treating diseases.
- Highlight challenges and propose insights into future research prospects and the potential impact of these converging technologies on the future of healthcare.

2 Internet of Medical Things (IoMT)

2.1 Overview of IoMT Frameworks and Design Principles

The Internet of Medical Things (IoMT) architecture in smart healthcare systems is a foundation for advanced healthcare solutions, integrating various technologies and smart medical devices through the internet [24]. One prevalent three-layered IoMT architec-ture includes the perceptual layer for data gathering from implanted sensors, the network layer for data transmission, and the application layer for medical record management and decision-making applications [24]. Another four-tier IoMT architecture emphasizes functionality levels, starting with the Wireless Body Sensor Network (WBSN) tier, followed by the Smart Wireless Technology Interface tier for data analysis, the internet infrastructure tier for data forwarding, and the care-services tier for processing and storing data, offering smart medical services [25].

Additional architectures present diverse layering schemes, such as a three-tiered mHealth system architecture linking IoT sensors directly with Smart Health Systems [26]. Another four-layered structure divides IoMT into an environment-aware layer, a gateway layer for data transmission, a cloud service layer for storage and processing, and a function layer hosting medical applications and algorithms. These IoMT architectures delineate essential layers for data collection, transmission, storage, and application in healthcare settings, collectively providing a holistic view of how interconnected devices and technologies can revolutionize healthcare delivery [24–26] (Fig. 1).

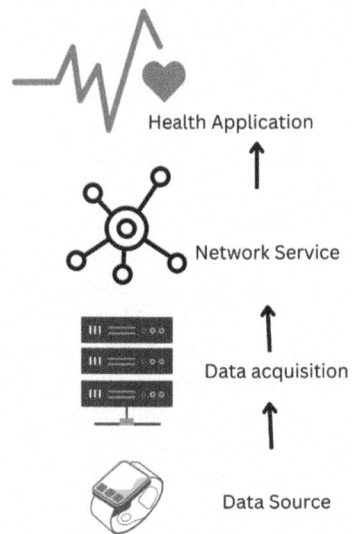

Fig. 1. Overview of IoMT architecture

2.2 Impact of IoMT on Health Monitoring and Disease Management

The Internet of Medical Things (IoMT) has revolutionized health monitoring and disease management by integrating cutting-edge technologies into healthcare systems. IoMT applications range from remote patient monitoring to personalized treatment strategies, enhancing the efficiency, accuracy, and accessibility of healthcare services. IoMT-based devices, including point-of-care testing (POCT) devices, have played a crucial role in disease spread control, particularly during the COVID-19 pandemic [27, 28]. Machine learning models utilizing patient demographics and laboratory tests have demonstrated high accuracy in predicting COVID-19 infection status [29, 30].

Various sensors, such as cardiac activity/ECG sensors, respiratory sensors, and dental sensors, enable real-time health monitoring and infection screening [31]. These sensors detect specific peptides in periodontal diseases and aid in effective treatment planning for conditions like bruxism [32]. Additionally, innovations include stretch strain sensors for analyzing foot motion and sensor-based multichannel functional electrical stimulation (FES) systems for controlling knee joint and gait asymmetry [33]. IoMT-enabled healthcare devices like smart inhalers, oxygen saturation monitors, and wearable devices have gained prominence during the pandemic, facilitating remote health monitoring and improving healthcare quality [34]. Smart hospitals leverage IoMT for automated processes, telemedicine, real-time patient monitoring, and robotic assistance [35, 36]. IoMT-enabled technologies in smart operating rooms enable precision surgery and intraoperative guidance [37].

Digital medications with IoMT sensors offer accurate monitoring of biomarkers, medication adherence, and dosage levels, enhancing treatment efficacy [38, 39]. Digital biomarkers monitored through self-powered IoMT devices allow real-time monitoring of various biomarkers in body fluids, aiding in early infection detection and treatment evaluation [40]. Advancements in IoMT-aided robotic systems, employing machine learning algorithms and cloud-based systems, have transformed medical assistance, surgery, and rehabilitation [41, 42]. IoMT-based drone technology with thermal cameras aids in early identification of potential illnesses through elevated body temperature detection in populated areas [43].

In dentistry, IoMT-based 3D scanning and printing enable accurate replication of intraoral structures and prosthetic rehabilitation, reducing errors and costs [44, 45]. Tele-dentistry, facilitated by telecommunications and digital information sharing, allows remote consultations and diagnosis, gaining traction during the COVID-19 pandemic [46]. Voice assistants like Google Assistant, Apple Siri, and Amazon Alexa provide remote healthcare support, disseminate COVID-19 information, and offer guidance to users [47]. Ambient assisted living (AAL) utilizes AI and machine learning for real-time monitoring of elderly individuals, ensuring safety and convenience within their homes [48]. IoMT-based systems have also been employed for adverse drug reaction detection, enhancing patient safety by verifying medication compatibility with patients' health profiles [49]. Overall, IoMT transforms healthcare by enabling remote monitoring, precise diagnostics, and personalized treatment, significantly impacting disease management and patient care [50].

3 Blockchain Integration in Healthcare

The integration of medical devices and objects, such as sensors, machines, and healthcare tools, is revolutionizing conventional healthcare systems, paving the way for sophisticated smart healthcare ecosystems. This evolution is steering healthcare towards the realm of the Internet of Medical Things (IoMT), capable of autonomously transmitting controlled messages for efficient healthcare service delivery. IoMT generates an immense volume of real-time data.

3.1   Role of Blockchain in Securing Medical Data in IoMT

IoMT improves control over data flow by incorporating multiple objects that are effectively managed by Software-Defined Networking (SDN) [51]. Owing to limited resources, edge computing is extensively employed. However, because IoMT collects extremely sensitive data, its growth in the healthcare industry raises privacy concerns [52]. Unauthorized entry could interfere with therapy. Network-level security is necessary because it is difficult to implement privacy measures on resource-constrained IoMT [51]. Because of the privacy issues and possible attack paths this creates, edge cloudlets are crucial [53].

In an effort to protect IoMT privacy, authentication methods like session keys are being used widely [54], although they come with complexity and communication overhead. In order to create consensus and shared session keys, authenticated key agreement (AKA) protocols must overcome obstacles related to mobile crowdsourcing and deployment [55]. Current methods struggle with weaknesses and insufficient defenses [56], which motivates research into distributed blockchain for IoMT privacy.

Data confidentiality and transaction integrity are protected by distributed blockchain technology. Blockchain maintains a secure ledger across all nodes, preventing alterations, and is well-known for its role in cryptocurrencies like Bitcoin [57]. Earlier blockchain-based IoMT initiatives offer a range of fixes [58, 59]. Nevertheless, they face difficulties with deployment, scalability, and latency. Security risks can arise from adversaries taking advantage of IoMT capabilities. Current privacy solutions don't provide complete security; instead, they concentrate on particular aspects.

IoMT authentication techniques address resource-constrained devices by proposing blockchain-managed key management [61] and secure key transmission [60, 62]. Although hash-based techniques [61, 64] and edge cloudlets have been investigated as privacy strategies, scalability and complexity problems in large-scale implementations still exist. Healthcare is evolving, and blockchain plays a critical role in protecting IoMT data. It provides unchangeable records that are protected from tampering and unwanted access. But problems still exist, calling for all-encompassing strategies within IoMT ecosystems. Healthcare security can be revolutionized by integrating blockchain with SDN and edge computing, which can effectively address the evolving privacy challenges (Fig. 2).

3.2   Advantages of Decentralization and Enhanced Data Security

Decentralization, a key principle of blockchain technology, plays a crucial role in enhancing data security and reliability in medical data management within the Internet of Medical Things (IoMT) [65]. By distributing data across a network of nodes, decentralization minimizes the risk of a single point of failure, thereby increasing system resilience and fault tolerance. This approach also reduces susceptibility to cyber attacks or data breaches that could compromise patient confidentiality in IoMT architectures. The decentralized nature of blockchain ensures transparency and immutability of records, fostering trust among stakeholders and enabling traceability of every data transaction [65]. This transparency is vital for auditing and compliance purposes, ensuring the authenticity and accuracy of medical records.

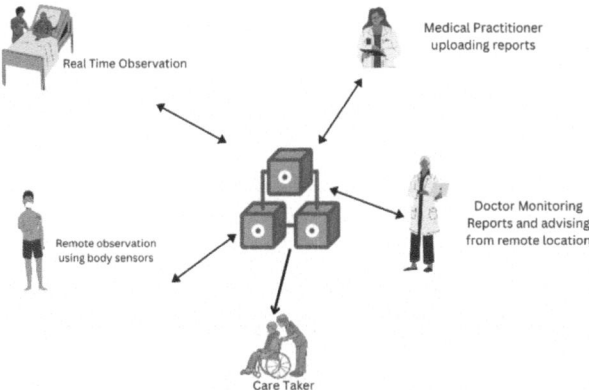

Fig. 2. Bloackchain based IoMT architecture

Blockchain's decentralized architecture employs cryptographic techniques, consensus algorithms, and cryptographic hashing to enhance data security. This ensures that data stored in the immutable ledger remains tamper-proof and resistant to unauthorized alterations, thereby upholding patient privacy and confidentiality. The use of smart contracts within blockchain further enhances security in IoMT, enabling automated and tamper-proof agreements for secure data sharing and access control among authorized entities [65]. Overall, the incorporation of decentralized mechanisms in IoMT architectures not only strengthens the security of medical data but also instills greater trust and reliability in healthcare systems, fostering innovation and efficiency in delivering patient-centric care.

4 AI-Driven Healthcare Solutions

4.1 Role of Machine Learning and AI Algorithms in Diagnosing Diseases

Machine Learning (ML) and AI algorithms have become crucial in disease diagnosis within the healthcare sector, especially in areas like paediatric care where clinical studies are limited [66]. The COVID-19 pandemic has further prompted organizations to utilize ML for streamlining healthcare operations and research [67]. AI and ML technologies are revolutionizing healthcare by extracting insights from challenging unstructured data, influencing patient care and operational strategies at scale [68]. These technologies, including machine learning and deep learning, play pivotal roles in developing clinical systems, patient record management, and disease diagnosis [69]. In disease identification, AI algorithms demonstrate efficiency in recognizing and diagnosing various ailments, contributing to early detection [70].

Despite challenges in integrating and harmonizing healthcare data, frameworks are being developed to aggregate and process large datasets for AI applications [71]. Researchers employ various AI techniques, such as machine and deep learning models, to detect and diagnose diseases like skin disorders, liver diseases, and heart conditions, achieving high accuracy rates [73]. For example, recurrent neural networks and supervised AI methods show high accuracies in diagnosing liver diseases and detecting skin disorders [74]. Moreover, the integration of IoT-based healthcare systems and AI algorithms enables real-time monitoring and early prediction of diseases like diabetes, hypertension, and tuberculosis, enhancing patient care [75, 76]. The combination of ML and AI algorithms transforms disease diagnosis, providing efficient and accurate tools for early detection and improved patient outcomes across various healthcare domains [77]. The continued application of these technologies holds promise for addressing existing challenges in disease diagnosis and patient care.

4.2   Enhancing Treatment Efficiency with AI-Enabled Decision Support Systems

Researchers extensively leverage artificial intelligence (AI) to enhance treatment efficiency by implementing advanced decision support systems in various medical fields [78, 79]. AI-based techniques, including machine learning and deep learning models like Boltzmann machines, K nearest neighbor (kNN), support vector machines (SVM), decision trees, logistic regression, fuzzy logic, and artificial neural networks, demonstrate high accuracies in early disease detection and diagnosis [78, 79]. For instance, backpropagation neural networks effectively diagnose skin diseases [78], while recurrent neural networks (RNNs) and residual neural networks successfully diagnose liver diseases and gastrointestinal disorders [79].

Innovative approaches explored in multiple studies incorporate machine learning algorithms, IoT-based healthcare systems, image-based retrieval systems, and AI-enhanced diagnosis systems for disease prediction, diagnosis, and treatment monitoring [80, 81]. These approaches highlight the potential of AI-enabled decision support systems to revolutionize healthcare, enabling early diagnosis, efficient treatment, and improved patient care with advanced predictive capabilities. AI-driven decision support systems contribute to a new era in healthcare, diagnosing diseases with greater accuracy and at earlier stages, facilitating prompt intervention and personalized treatment strategies [82, 83]. These systems extend beyond diagnosis to monitoring chronic conditions, predicting disease progression, and optimizing treatment plans.

These AI-based models address diverse healthcare challenges, predicting coronary heart disease [96], detecting tuberculosis [85], assessing diabetic risk [86], and developing early-stage cervical cancer prediction models. These efforts underscore the transformative impact of AI-driven decision support systems in healthcare, offering accurate diagnoses and paving the way for proactive healthcare management and disease prevention strategies.

5   Synergies Between Blockchain, AI, and IoMT in Healthcare

The convergence of Blockchain, Artificial Intelligence (AI), and Internet of Medical Things (IoMT) in healthcare is forging new frontiers in data management, security, and patient-centric accessibility. With the exponential growth in electronic health data and evolving patient data privacy regulations, the landscape for managing health data and facilitating patient access has evolved. Blockchain technology is emerging as a robust solution capable of addressing critical issues concerning data security, storage, transactions, and seamless integration, offering profound value to data-driven organizations, particularly in healthcare [87]. The concept of global scientific data sharing in healthcare is pivotal for enhancing healthcare quality and fostering smarter healthcare systems. However, existing operational mechanisms often limit patient access to their health records and transparency in data sharing. Blockchain technology stands out as a crucial enabler in revolutionizing health data sharing mechanisms, ensuring security, and facilitating convenient electronic health data exchange [88].

Blockchain technologies have also been instrumental in reinforcing data management strategies in healthcare settings. Systems and models proposed by some studies [89, 90] emphasize decentralized EMR management, controlled blockchain data management, and reshaped consent management, respectively. These models integrate blockchain to ensure secure data access and prevent unauthorized actions, thereby bolstering data security and user control in healthcare environments. In the realm of data storage, the integration of blockchain in cloud-based applications has demonstrated immense promise. The study [91] proposed a patient-centric healthcare data management system, utilizing blockchain in cloud environments to ensure data privacy and integrity.

Electronic Health Records (EHRs) are undergoing a transformative shift from paper-based to digital systems, significantly improving healthcare quality and disease management. Blockchain-enabled solutions such as secure frameworks for medical data sharing, attribute-based signature schemes for validating EHRs, and secure cloud-based EHR systems ensure encrypted data storage, integrity, and access control, strengthening the security and interoperability of healthcare records. Innovations like OmniPHR propose distributed models for integrating personal health records (PHRs) using parallel databases, facilitating seamless data integration across diverse formats. Additionally, frameworks employing genetic algorithms and cryptographic hash generators, as demonstrated by Hussein et al., further fortify medical record security, enhancing system immunity against potential attacks [92].

These advancements underscore the transformative potential of integrating Blockchain, AI, and IoMT in healthcare, ushering in an era of secure, interoperable, and patient-centric healthcare systems, poised to revolutionize data management, access, and security in the healthcare domain.

6 Challenges and Considerations

The convergence of Blockchain and the Internet of Things (IoT) presents significant potential but introduces challenges that require attention for effective integration [92]. Combining these technologies brings about complexities due to their inherent limitations and constraints. Scalability and storage pose significant hurdles when integrating Blockchain with IoT networks. Blockchain's scalability limitations are magnified within IoT setups due to the numerous interconnected devices, straining the storage capacities of these devices with the expanding memory requirements of a distributed ledger system.

A scarcity of expertise is a considerable challenge for successful IoT-Blockchain integration. Despite widespread use of IoT, limited understanding and expertise in Blockchain, often associated solely with cryptocurrencies like Bitcoin, restrict the available pool of skilled individuals for integration efforts. Legal and compliance challenges, especially regarding international collaboration, are inherent in intertwining Blockchain with IoT. Operating across borders without established legal precedents or compliance standards, Blockchain faces hurdles in navigating the regulatory landscape, hindering seamless integration within IoT frameworks.

Addressing these challenges requires collaborative efforts across technology development, skill enhancement, regulatory frameworks, and infrastructural adaptations. Innovative solutions are needed to optimize data management and processing within IoT-Blockchain systems, overcoming scalability and storage limitations. Bridging the knowledge gap through educational initiatives is crucial, and fostering multidisciplinary expertise is necessary for leveraging the potential of both technologies. Establishing legal frameworks and compliance standards tailored for Blockchain-enabled IoT environments is imperative to foster secure and globally harmonized integration efforts. These collective efforts will play a pivotal role in unlocking the full potential of Blockchain-IoT integration while mitigating associated challenges.

7 Conclusion

The integration of Blockchain, Artificial Intelligence (AI), and Internet of Medical Things (IoMT) has ushered in a new era of innovation and transformation within the healthcare industry. The amalgamation of these technologies holds great promise in addressing critical challenges, enhancing patient care, securing medical data, and revolutionizing healthcare delivery. The role of Blockchain in securing medical data within IoMT architectures is particularly noteworthy, offering robust solutions for privacy, data management, and seamless collaboration among stakeholders.

However, the integration of Blockchain, AI, and IoMT also presents formidable challenges that warrant attention for future advancements. The scalability and storage constraints of Blockchain, exacerbated within the expansive IoMT networks, necessitate innovative solutions to manage burgeoning data volumes. Additionally, addressing the dearth of expertise in Blockchain technology, navigating legal complexities, and establishing robust regulatory frameworks for global compliance remain imperative for seamless integration.

The future scope lies in concerted efforts towards overcoming these challenges and further harnessing the synergies between Blockchain, AI, and IoMT. Advancements in Blockchain scalability, coupled with AI-driven optimization techniques, can revolutionize data management within IoMT architectureres. Bridging the knowledge gap through education and skill enhancement initiatives will facilitate a more widespread understanding and application of Blockchain technology.

Furthermore, the development of tailored regulatory frameworks and legal standards will be pivotal in fostering secure and compliant Blockchain-IoMT ecosystems on a global scale. Collaborative endeavors between technology developers, healthcare professionals, regulatory bodies, and policymakers will drive the evolution of a more efficient, secure, and patient-centric healthcare landscape.

References

1. Joyia, G.J., Liaqat, R.M., Farooq, A., Rehman, S.: Internet of medical things (IoMT): applications, benefits and future challenges in healthcare domain. J. Commun. **12**(4), 240–247 (2017). https://doi.org/10.12720/jcm.12.4.240-247

2. Quwaider, M., Biswas, S.: On-body packet routing algorithms for body sensor networks. In: Proceedings of the 2009 First International Conference on Networks & Communications; December 2009; Chennai, India, pp. 171–177. IEEE (2009)

3. Rghioui, A., Lloret, J., Harane, M., Oumnad, A.: A smart glucose monitoring system for diabetic patient. Electronics **9**(4), 678 (2020). https://doi.org/10.3390/electronics9040678

4. Javaid, A., Niyaz, Q., Sun, W., Alam, M.: A deep learning approach for network intrusion detection system. Eai Endorsed Trans. Secur. Safety. **3**(9), e2 (2016). https://doi.org/10.4108/eai.3-12-2015.2262516

5. Mohamed Shakeel, P., Baskar, S., Sarma Dhulipala, V.R., Mishra, S., Jaber, M.M.: Maintaining security and privacy in health care system using learning based deep-q-networks. J. Med. Syst. **42**(10), 186 (2018). https://doi.org/10.1007/s10916-018-1045-z

6. Alsubaei, F., Abuhussein, A., Shandilya, V., Shiva, S.: IoMT-SAF: internet of medical things security assessment framework. Internet of Things (2019). https://doi.org/10.1016/j.iot.2019.100123.100123

7. Kumar, R., Rajasekaran, M.P.: An IoT based patient monitoring system using raspberry Pi. In: Proceedings of the 2016 International Conference on Computing Technologies and Intelligent Data Engineering (ICCTIDE'16); January 2016; Kovilpatti, India, pp. 1–4. IEEE (2016)

8. Jain, S., Nehra, M., Kumar, R., et al.: Internet of medical things (IoMT)-integrated biosensors for point-of-care testing of infectious diseases. Biosensors Bioelectron. (2021). https://doi.org/10.1016/j.bios.2021.113074.113074

9. Ahmed, I., Jeon, G., Piccialli, F.: A deep-learning-based smart healthcare system for patient's discomfort detection at the edge of internet of things. IEEE Internet of Things J. (2021). https://doi.org/10.1109/jiot.2021.3052067.10318

10. Ahmad, S., Khan, S., Fahad, M., et al.: Deep learning enabled disease diagnosis for secure internet of medical things. Comput. Mater. Continua. **73**(1), 965–979 (2022). https://doi.org/10.32604/cmc.2022.025760

11. Srivastava, J., Routray, S., Ahmad, S., Waris, M.M.: Internet of medical things (IoMT)-based smart healthcare system: trends and progress. Comput. Intell. Neurosci. **16**(2022), 7218113 (2022). https://doi.org/10.1155/2022/7218113.PMID:35880061;PMCID:PMC9308524

12. Alotaibi, Y.K., Federico, F.: The impact of health information technology on patient safety. Saudi Med. J. **38**, 1173 (2017). https://doi.org/10.15537/smj.2017.12.20631

13. Bragazzi, N.L., Dai, H., Damiani, G., Behzadifar, M., Martini, M., Wu, J.: How big data and artificial intelligence can help better manage the covid-19 pandemic. Int. J. Environ. Res. Public Health **17**(9), 3176 (2020). https://doi.org/10.3390/ijerph17093176

14. Sahoo, M.S., Baruah, P.K.: HBasechainDB – a scalable blockchain framework on Hadoop Ecosystem. In: Yokota, R., Wu, W. (eds.) Supercomputing frontiers, pp. 18–29. Springer International Publishing, Cham (2018). https://doi.org/10.1007/978-3-319-69953-0_2

15. Paul, D., Sanap, G., Shenoy, S., Kalyane, D., Kalia, K., Tekade, R.K.: Artificial intelligence in drug discovery and development. Drug Discov. Today **26**, 80–93 (2021). https://doi.org/10.1016/j.drudis.2020.10.010

16. Cha, Y., et al.: Drug repurposing from the perspective of pharmaceutical companies. Br. J. Pharmacol. **175**(2), 168–180 (2018). https://doi.org/10.1111/bph.13798

17. Siyal, A.A., Junejo, A.Z., Zawish, M., Ahmed, K., Khalil, A., Soursou, G.: Applications of blockchain technology in medicine and healthcare: challenges and future perspectives. Cryptography. **3**(1), 3 (2019). https://doi.org/10.3390/cryptography3010003

18. Feng, Q., He, D., Zeadally, S., Khan, M.K., Kumar, N.: A survey on privacy protection in blockchain system. J. Netw. Comput. Appl. **126**, 45–58 (2019). https://doi.org/10.1016/j.jnca. 2018.10.020

19. Lin, C., He, D., Huang, X., Khan, M.K., Choo, K.K.R.: DCAP: a secure and efficient decentralized conditional anonymous payment system based on blockchain. IEEE Trans. Inf. Forensics Secur. **15**, 2440–2452 (2020). https://doi.org/10.1109/TIFS.2020.2969565

20. Hang, L., Choi, E., Kim, D.H.: A novel EMR integrity management based on a medical blockchain platform in hospital. Electronics **8**, 467 (2019). https://doi.org/10.3390/electroni cs8040467

21. Khurshid, A.: Applying blockchain technology to address the crisis of trust during the COVID-19 pandemic. JMIR Med Informatics. **8**(9), e20477 (2020). https://doi.org/10.2196/20477

22. Andoni, M., et al.: Blockchain technology in the energy sector: a systematic review of challenges and opportunities. Renew Sust Energ Rev. **21**(100), 143–174 (2019). https://doi.org/ 10.1016/j.rser.2018.10.014

23. Hylock, R.H., Zeng, X.: A blockchain framework for patient-centered health records and exchange (healthChain): evaluation and proof-of-concept study. J. Med. Internet Res. **21** (2019)

24. Tagde, P., et al.: Blockchain and artificial intelligence technology in e-Health. Environ. Sci. Pollut. Res. **28**(38), 52810–52831 (2021). https://doi.org/10.1007/s11356-021-16223-0

25. Khan, H.A., Abdulla, R., Selvaperumal, S.K., Bathich, A.: IoT based on secure personal healthcare using RFID technology and steganography. Int. J. Electr. Comput. Eng. **11**(4), 3300 (2021). https://doi.org/10.11591/ijece.v11i4.pp3300-3309

26. Abdulmohsin, H.D., Rahim, H.A., Alkhayyat, A., Ahmad, R.B.: Body-to-body cooperation in internet of medical things: toward energy efficiency improvement. Future Internet. **11**(11), 239 (2019). https://doi.org/10.3390/fi11110239

27. Huang, C., Wang, J., Wang, S., Zhang, Y.: Internet of medical things: a systematic review. Neurocomputing **557**, 126719 (2023). https://doi.org/10.1016/j.neucom.2023.126719

28. Bibi, N., Sikandar, M., Ud Din, I., Almogren, A., Ali, S.: IoMT-based automated detection and classification of leukemia using deep learning. J. Healthcare Eng. **2020**, 1–12 (2020). https://doi.org/10.1155/2020/6648574

29. Ahammed, K., Satu, M., Abedin, M.Z., Rahaman, M., Islam, S.M.S.: Early detection of coronavirus cases using chest X-ray images. Employing Mach. Learn. Deep Learn. Approaches (2020). https://doi.org/10.1101/2020.06.07.20124594

30. Yang, H.S., Hou, Y., Vasovic, L.V., et al.: Routine laboratory blood tests predict SARS-CoV-2 infection using machine learning. Clin. Chem. **66**(11), 1396–1404 (2020). https://doi.org/10. 1093/clinchem/hvaa200

31. Iskanderani, A.I., et al.: Artificial intelligence and medical internet of things framework for diagnosis of coronavirus suspected cases. J. Healthc. Eng. **2021**, 1–7 (2021). https://doi.org/ 10.1155/2021/3277988

32. Kakria, P., Tripathi, N.K., Kitipawang, P.: A real-time health monitoring system for remote cardiac patients using smartphone and wearable sensors. In. J. Telemed. Appl. **2015**, 1–11 (2015). https://doi.org/10.1155/2015/373474

33. Sijobert, B., Azevedo, C., Pontier, J., Graf, S., Fattal, C.: A sensor-based Multichannel FES system to control knee joint and reduce stance phase Asymmetry in post-stroke gait. Sensors (Basel). **21**(6), 2134 (2021). https://doi.org/10.3390/s21062134

34. Sakamoto, K., Tsujioka, C., Sasaki, M., Miyashita, T., Kitano, M., Kudo, S.: Validity and reproducibility of foot motion analysis using a stretch strain sensor. Gait Posture **86**, 180–185 (2021). https://doi.org/10.1016/j.gaitpost.2021.03.007

35. Merchant, R., Szefler, S.J., Bender, B.G., et al.: Impact of a digital health intervention on asthma resource utilization. World Allergy Organ. J. **11**(1), 28 (2018). https://doi.org/10. 1186/s40413-018-0209-0

36. Cobelli, C., Renard, E., Kovatchev, B.: The artificial pancreas: a digital-age treatment for diabetes. Lancet Diab. Endocrinol. **2**, 679–681 (2014). https://doi.org/10.1016/S2213-858 7(14)70126-3

37. Sangave, N.A., Aungst, T.D., Patel, D.K.: Smart connected insulin pens, caps, and attachments: a review of the future of diabetes technology. Diabetes Spectr. **32**(4), 378–384 (2019). https://doi.org/10.2337/ds18-0069

38. Okamoto, J., Masamune, K., Iseki, H., Muragaki, Y.: Development concepts of a smart cyber operating theater (SCOT) using ORiN technology. Biomed. Tech. (Berl) **63**(1), 31–37 (2018). https://doi.org/10.1515/bmt-2017-0006

39. Joshi, A., Kim, K.H.: Recent advances in nanomaterial-based electrochemical detection of antibioics: challenges and future perspectives. Biosens. Bioelectron. **153**, 112046 (2020)

40. Plowman, R., Peters-Strickland, T., Savage, G.: Digital medicines clinical review on the safety of tablets with sensors. Expert Opinion on Drug Safet (2018). https://doi.org/10.1080/14740338.2018.150844717(9):849-852

41. Wessels, F., Schmitt, M., Krieghoff-Henning, E., et al.: Deep learning approach to predict lymph node metastasis directly from primary tumour histology in prostate cancer. BJU Int. (2021). https://doi.org/10.1111/bju.15386

42. Simoens, P., Dragone, M., Saffiotti, A.: The Internet of Robotic Things: a review of concept, added value and applications. Int. J. Adv. Robot. Syst. (2018). https://doi.org/10.1177/172 9881418759424

43. Pradhan, B., Bharti, D., Chakravarty, S., et al.: Internet of things and robotics in transforming current-day healthcare services. J Healthc Eng. (2021). https://doi.org/10.1155/2021/9999504

44. Mohammed, M., Hazairin, N.A., Al-Zubaidi, S., Ak, S., Mustapha, S., Yusuf, E.: Toward a novel design for coronavirus detection and diagnosis system using IoMT based drone technology. Int. J. Psychosoc. Rehabil. **24**(7), 2287–2295 (2020). https://doi.org/10.37200/IJPR/V24I7/PR270220

45. Zhang, T., Liu, M., Yuan, T., Al-Nabhan, N.: Emotion-aware and intelligent internet of medical things towards emotion recognition during COVID-19 pandemic. IEEE Internet Things J. (2020). https://doi.org/10.1109/JIOMT.2020.3038631

46. Lee, S.M., Lee, D.: Opportunities and challenges for contactless healthcare services in the post-COVID-19 Era. Technol Forecast Soc Change. **167**, 120712 (2021)

47. Jampani, N.D., Nutalapati, R., Dontula, B.S.K., Boyapati, R.: Applications of teledentistry: a literature review and update. J. Int. Soc. Prev. Community Dent. **1**(2), 37–44 (2011)

48. Sezgin, E., Huang, Y., Ramtekkar, U., Lin, S.: Readiness for voice assistants to support healthcare delivery during a health crisis and pandemic. npj Digital Medicine **3**(1), 122 (2020). https://doi.org/10.1038/s41746-020-00332-0

49. Amazon . Amazon. Alexa and Amazon Devices COVID-19 Resources (2020). https://blog.aboutamazon.com/devices/alexa-and-amazon-devices-covid-19-resources

50. Cohen, R., Fernie, G., Roshan, F.A.: Fluid intake monitoring systems for the elderly: a review of the literature. Nutrients **13**(6), 2092 (2021). https://doi.org/10.3390/nu13062092

51. Dwivedi, R., Mehrotra, D., Chandra, S.: Potential of internet of medical things (IoMT) applications in building a smart healthcare system: a systematic review. J. Oral Biol. Craniofacial Res. **12**(2), 302–318 (2022). https://doi.org/10.1016/j.jobcr.2021.11.010

52. Liaqat, S., Akhunzada, A., Shaikh, F.S., Giannetsos, A., Jan, M.A.: SDN orchestration to combat evolving cyber threats in Internet of Medical Things (IoMT). Comput. Commun. **160**, 697–705 (2020)

53. Huang, Q., Zhou, Y., Tao, L., et al.: A chan-vese model based on the Markov chain for unsupervised medical image segmentation. Tsinghua Sci. Technol. **26**(6), 833–844 (2021)

54. Sollins, K.R.: Iot big data security and privacy versus innovation. IEEE Internet Things J. **6**(2), 1628–1635 (2019)

55. Ma, M., He, D., Wang, H., Kumar, N., Choo, K.R.: An efficient and provably secure authenticated key agreement protocol for fog-based vehicular ad-hoc networks. IEEE Internet Things J. **6**(5), 8065–8075 (2019)

56. Wang, Y., Cai, Z., Zhan, Z.-H., Zhao, B., Tong, X., Qi, L.: Walrasian equilibrium-based multiobjective optimization for task allocation in mobile crowdsourcing. IEEE Trans. Comput. Soc. Syst. **7**(4), 1033–1046 (2020)

57. Moin, S., Karim, A., Safdar, Z., Safdar, K., Ahmed, E., Imran, M.: Securing iots in distributed blockchain: analysis, requirements and open issues. Futur. Gener. Comput. Syst. **100**, 325–343 (2019)

58. Pan, J., Wang, J., Hester, A., Alqerm, I., Liu, Y., Zhao, Y.: Edgechain: an edge-iot framework and prototype based on blockchain and smart contracts. IEEE Internet Things J. **6**(3), 4719–4732 (2019)

59. Karmakar, K.K., Varadharajan, V., Tupakula, U., Nepal, S., Thapa, C.: Towards a security enhanced virtualised network infrastructure for Internet of medical things (iomt). In: 2020 6th IEEE Conference on Network Softwarization (NetSoft), pp. 257–261. Ghent, Belgium (2020)

60. Din, S., Paul, A., Rehman, A.: 5g-enabled hierarchical architecture for software-defined intelligent transportation system. Comput. Netw. **150**, 81–89 (2019)

61. Garg, N., Wazid, M., Das, A.K., Singh, D.P., Rodrigues, J.J., Park, Y.: Bakmp-iomt: design of blockchain enabled authenticated key management protocol for Internet of medical things deployment. IEEE Access **8**, 95956–95977 (2020)

62. Kumar, R., Tripathi, R.: Towards design and implementation of security and privacy framework for internet of medical things (iomt) by leveraging blockchain and ipfs technology. J. Supercomput. **77**(8), 7916–7955 (2021)

63. Danzi, P., Kalør, A.E., Stefanović, Č, Popovski, P.: Delay and communication tradeoffs for blockchain systems with lightweight IoT clients. IEEE Internet Things J. **6**(2), 2354–2365 (2019)

64. Xu, Y., Ren, J., Wang, G., Zhang, C., Yang, J., Zhang, Y.: A blockchain-based nonrepudiation network computing service scheme for industrial IoT. IEEE Trans. Industr. Inf. **15**(6), 3632–3641 (2019)

65. Ali, I., Gervais, M., Ahene, E., Li, F.: A blockchain-based certificateless public key signature scheme for vehicle-to-infrastructure communication in VANETs. J. Syst. Archit. **99**, 101636 (2019). https://doi.org/10.1016/j.sysarc.2019.101636

66. Rafiq, W., Khan, M., Khan, S., Ally, J.: SecureMed: a blockchain-based privacy-preserving framework for internet of medical things. Wirel. Commun. Mob. Comput. **2023**, 1–14 (2023). https://doi.org/10.1155/2023/2558469

67. Saleem, T.J., Chishti, M.A.: Exploring the applications of machine learning in healthcare. Int. J. Sensor. Wireless Commun. Control **10**(4), 458–472 (2020)

68. Vyas, S., Gupta, M., Yadav, R.: Converging blockchain and machine learning for healthcare 2019 Amity International Conference on Artificial Intelligence (AICAI), IEEE, pp. 709–711 (2019)

69. Li, J.P., Haq, A.U., Din, S.U., Khan, J., Khan, A., Saboor, A.: Heart disease identification method using machine learning classification in e-healthcare. IEEE Access **8**, 107562–107582 (2020)

70. A. Dhillon, A. Singh: Machine learning in healthcare data analysis: a survey. J. Biol. Today's World, 8 (6) (2019), pp. 1–10

71. Chen, P.C., Liu, Y., Peng, L.: How to develop machine learning models for healthcare. Nat. Mater. **18**(5), 410–414 (2019). https://doi.org/10.1038/s41563-019-0345-0. PMID: 31000806

72. Shailaja, K., Banoth, S., Jabbar, M.: Machine learning in healthcare: a review, pp. 910–914 (2018). https://doi.org/10.1109/ICECA.2018.8474918

73. Chen, M., Hao, Y., Hwang, K., Wang, Lu., Wang, Lin: Disease prediction by machine learning over big data from healthcare communities. IEEE Access **5**, 8869–8879 (2017). https://doi.org/10.1109/ACCESS.2017.2694446

74. Qayyum, A., Qadir, J., Bilal, M., Al-Fuqaha, A.: Secure and robust machine learning for healthcare: a survey. IEEE Rev. Biomed. Eng. **14**, 156–180 (2021). https://doi.org/10.1109/RBME.2020.3013489

75. Prosperi, M., et al.: Causal inference and counterfactual prediction in machine learning for actionable healthcare. Nat. Mach. Intell. **2**, 1–7 (2020). https://doi.org/10.1038/s42256-020-0197-y

76. Rajendran, S., et al.: Emphasizing privacy and security of edge intelligence with machine learning for healthcare. Int. J. Intell. Comput. Cybern. (2021). https://doi.org/10.1108/IJICC-05-2021-0099

77. Han, T., Stone-Weiss, N., Huang, J., Goel, A., Kumar, A.: Machine learning as a tool to design glasses with controlled dissolution for healthcare applications. Acta Biomater. **15**(107), 286–298 (2020). https://doi.org/10.1016/j.actbio.2020.02.037. Epub 2020 Feb 28 PMID: 32114183

78. Javaid, M., Haleem, A., Singh, R., Suman, R., Rab, S.: Significance of machine learning in healthcare: features, pillars and applications. Int. J. Intell. Netw. (2022). https://doi.org/10.1016/j.ijin.2022.05.002

79. Dabowsa, N., Amaitik, N., Maatuk, A., Shadi, A.: A hybrid intelligent system for skin disease diagnosis. In: Conference on engineering and technology, pp. 1–6 (2017). https://doi.org/10.1109/ICEngTechnol.2017.8308157

80. Owasis, M., Arsalan, M., Choi, J., Mahmood, T., Park, K.: Artificial intelligence based classification of multiple gastrointestinal diseases using endoscopy videos for clinical diagnosis. J. Clin. Med. **8**, 786 (2019). https://doi.org/10.3390/jcm8070986

81. Ijaz, M.F., Attique, M., Son, Y.: Data-driven cervical cancer prediction model with outlier detection and over-sampling methods. Sensors **20**(10), 2809 (2020)

82. Srinivasu, P.N., SivaSai, J.G., Ijaz, M.F., Bhoi, A.K., Kim, W., Kang, J.J.: Classification of skin disease using deep learning neural networks with MobileNet V2 and LSTM. Sensors **21**(8), 2852 (2021)

83. Alfian, G., Syafrudin, M., Ijaz, M.F., Syaekhoni, M.A., Fitriyani, N.L., Rhee, J.: A personalized healthcare monitoring system for diabetic patients by utilizing BLE-based sensors and real-time data processing. Sensors **18**(7), 2183 (2018)

84. Katharine, E., Oikonomou, E., Williams, M., Desai, M.: A novel machine learning derived radiotranscriptomic signature of perivascular fat improves cardiac risk prediction using coronary CT angiography. Eur. Heart J. **40**, 3529–3543 (2019). https://doi.org/10.1093/eurheartj/ehz592

85. Gonsalves, A.H., Singh, G., Thabtah, F., Mohammad, R.: Prediction of coronary heart disease using machine learning: an experimental analysis. ACM Digit. Libr. (2019). https://doi.org/10.1145/3342999.3343015

86. Romero, M.P., et al.: Decision tree machine learning applied to bovine Alzheimer risk factors to aid disease control decision making. Prev. Vet. Med. **175**, 104860 (2020). https://doi.org/10.1016/j.prevetmed.2019.104860

87. Sarao, V., Veritti, D., Paolo, L.: Automated diabetic retinopathy detection with two different retinal imaging devices using artificial intelligence. Graefe's Arch. Clin. Exp. Opthamol. (2020). https://doi.org/10.1007/s00417-020-04853-y

88. Dimitrov, D.V.: Blockchain applications for healthcare data management. Healthc. Inform. Res. **25**, 51–56 (2019)

89. Frost, J.H., Massagli, M.P.: Social uses of personal health information within PatientsLikeMe, an online patient community: What can happen when patients have access to one another's data. J. Med. Internet Res. **10**, e15 (2008)

90. Zhu, L., Wu, Y., Gai, K., Choo, K.K.R.: Controllable and trustworthy blockchain-based cloud data management. Future Gen. Comput. Syst. **91**, 527–535 (2019)
91. Genestier, P., et al.: Blockchain for consent management in the ehealth environment: a nugget for privacy and security challenges. J. Int. Soc. Telemed. eHealth **5**, GKR-e24 (2017)
92. Khezr, S., Moniruzzaman, M., Yassine, A., Benlamri, R.: Blockchain technology in healthcare: a comprehensive review and directions for future research. Appl. Sci. **9**, 1736 (2019). https://doi.org/10.3390/app9091736

The Evolution and Potential of Conversational Agents in Healthcare

Mily Lal[1,2](✉) and S. Neduncheliyan[2]

[1] Dr. D. Y. Patil School of Science and Technology, Dr. D. Y. Patil Vidyapeeth, Sant Tukaram Nagar, Pimpri, Pune, India
milylike@gmail.com

[2] Department of CSE, School of Computing, Bharath Institute of Higher Education and Research (BIHER), Tamilnadu, Chennai, India

Abstract. Artificial intelligence (AI) tools known as conversational agents have shown great promise for managing chronic diseases in healthcare systems. This study explores the state of the art, recent developments, and possible future directions in the field of conversational agents for healthcare-assisted chronic illness management. This study provides a thorough analysis by looking at current innovations, trends, and implementation-related practical considerations. Important discoveries highlight conversational agents' acceptability, effectiveness, and usability and highlight how they can help with chronic illness self-management. The incorporation of conversational agents into healthcare practices is recommended as a practical measure to enhance patient engagement, self-care, and disease monitoring. In order to improve the efficacy and reliability of these agents, future research directions emphasize the necessity of thorough clinical trials, strict technical reporting, and personalized guidelines. The present study enhances our comprehension of the significance of conversational agents in transforming approaches to managing chronic illnesses.

Keywords: Conversational agents · Chronic disease management · Healthcare · Technology · Patient engagement · Self-care · Personalization

1 Introduction

1.1 Background

Globally, chronic illnesses present serious obstacles to both individual health and healthcare systems. Individuals are burdened with managing these conditions, which call for rigorous treatment adherence, ongoing monitoring, and major lifestyle changes [3, 4]. About 60% of Americans suffer from at least one chronic illness, and 40% manage several conditions [1, 2]. In order to monitor symptoms, adhere to treatments, and modify lifestyles, self-management has become essential [3, 5]. On the other hand, managing chronic illnesses presents a number of difficulties, including reliable health monitoring, regular self-care, and efficient communication with medical professionals [6, 7].

© The Author(s), under exclusive license to Springer Nature Switzerland AG 2025
M. Khurana et al. (Eds.): ICMLA 2024, CCIS 2238, pp. 209–220, 2025.
https://doi.org/10.1007/978-3-031-75861-4_18

Conversational agents, or chatbots, have gained popularity as potential tools to assist people in managing chronic conditions as a result of recent advances in Natural Language Processing (NLP) and Artificial Intelligence (AI) [8, 9]. These AI-driven chatbots converse with users while offering tailored advice and observation akin to therapy [10, 11]. They help with goal-setting, self-monitoring, and feedback by facilitating interactions, with the goal of averting emergencies and lowering associated costs [10, 11]. Although these agents have great potential, there are still issues with their design, user interaction, and customization for different user groups [12, 13]. Therefore, it is essential to comprehend the landscape—its advantages, difficulties, and potential future directions in order to advance healthcare interventions.

1.2 Overview of Conversational Agents

Dialogue systems, a synonym for conversational agents, converse with users in natural language. By understanding and producing content in natural language and going beyond predefined words or sentence structures, they enable human-computer interaction [9, 14]. Although they include chatbots, conversational agents use cutting-edge technologies such as natural language processing and deep learning, which allows for more complex and context-aware dialogues [15].

Text, voice, or multi-modal communication is used to categorize these agents [16]. Voice-based agents, such as Apple Siri or Amazon Alexa, comprehend and react to voice commands, while text-based agents, such as ELIZA, mimic text-based interactions [17]. Text, voice, gestures, and facial expressions are just a few of the communication modalities that multimodal agents combine [21].

Conversational agents can be further classified based on their capabilities as communication-focused (like ELIZA) or action-performing (like Alexa) [22, 23]. Some have no representation at all and only communicate through language interfaces [10], while others have physical embodiments like social or mobile robots and embody human-like traits in verbal and nonverbal communication.

Response generation techniques separate conversational agents into two categories: generative and rule-based systems. Rule-based systems use machine learning to generate and improve responses, while generative-based systems adhere to predefined rules and databases. Additionally, they differ in the domains they focus on: general agents serve a wide range of subjects, whereas domain-specific agents cater to niche markets like healthcare or education [24].

Conversational agents have their roots in Alan Turing's exploration of machine intelligence, which progressed from rule-based models to AI-driven systems. Personal assistants such as Siri and Alexa use deep learning algorithms that combine natural language processing, speech recognition, and contextual understanding to effectively carry out user tasks [18]18.

Social chatbots, such as Microsoft's XiaoIce, combine information retrieval and generative capabilities to create engaging conversations, exhibiting enhanced dialogue and emotional intelligence [20]. Conversational agents have become more flexible and intelligent as a result of their transition from rule-based to AI-powered models, which has changed user interactions and promoted personalized experiences.

1.3 Problem Statement

People with chronic illnesses bear an immense burden, as do healthcare systems and society at large. The use of technology, particularly conversational agents, has the potential to completely change how chronic illness is managed. These agents could improve healthcare delivery, increase patient engagement, and offer tailored support. The need for novel approaches that enhance patient outcomes, lower healthcare costs, and improve the general quality of life for people with chronic conditions is what spurs research into the use of conversational agents in the management of chronic diseases.

Although the use of conversational agents in the management of chronic diseases is gaining popularity and may have advantages, there are a number of obstacles and moral issues that must be resolved. Ensuring data privacy and security, preserving medical information accuracy, addressing potential biases in algorithms, and comprehending the limitations of conversational agent capabilities are just a few of the challenges that this paper attempts to investigate. Furthermore, it is important to carefully consider the ethical ramifications of using these agents in immersive contexts like virtual healthcare settings. In order to provide insights for future research and practical implementation in the healthcare domain, this study aims to provide a thorough review of the current landscape, highlighting the potential benefits, challenges, and ethical considerations surrounding the utilization of conversational agents in the management of chronic diseases.

The rest of the article is organized as follows: The benefits of using conversational agents in the management of chronic diseases are outlined in Sect. 2 and include improving patient engagement, encouraging self-care, and providing individualized healthcare. The difficulties in incorporating conversational agents into healthcare are covered in Sect. 3. The ethical ramifications of incorporating conversational agents into virtual healthcare environments are the main topic of Sect. 4. In Sect. 5, the state of conversational agent technologies is evaluated, along with possible paths for future advancement and application. Section 6 offers developers and healthcare professional's advice on how to use conversational agents to manage chronic illnesses. Finally, the main conclusions and ramifications of this investigation are outlined in Sect. 7.

2   Conversational Agents in Healthcare

In the healthcare industry, conversational agents have become essential, revolutionizing patient empowerment, accessibility, monitoring, and engagement. These agents enable patients to actively engage in their healthcare journey and provide personalized assistance by leveraging machine learning and natural language processing [10]. By acting as virtual companions and teaching patients about their conditions, medications, and lifestyle modifications, they improve health literacy and self-management skills. Constant communication empowers people to take charge of their health outcomes by promoting better behaviors, treatment compliance, and efficient chronic illness management [24].

These agents serve as watchful companions, facilitating ongoing health monitoring and immediate symptom management, especially in the context of remote monitoring chronic illnesses. Their AI-driven capabilities improve disease management and give patients continuous support by analyzing symptoms, predicting disease patterns, and making proactive recommendations [25]. Health chatbots implemented through mobile apps, particularly during the COVID-19 pandemic, became essential for remote triage and risk assessment, providing instant access to medical information and lessening the strain on overloaded systems. By providing services to underserved or remote areas and guaranteeing equitable access to healthcare resources, they close accessibility gaps in the healthcare system [24] (Fig. 1).

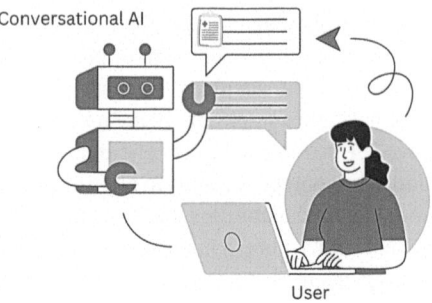

Fig. 1. Conversational Agent in Healthcare

By facilitating the management of chronic diseases through tailored interventions and increasing general healthcare accessibility, these agents greatly improve patient outcomes. Case studies demonstrate their effectiveness in providing mental health support [26], diagnosing illnesses during emergencies [33], managing chronic illnesses [34], and even educating medical professionals [27].

3   Challenges of Conversational Agents

Implementing conversational agents in various domains poses multifaceted challenges that span technical, ethical, and practical dimensions. These challenges are pivotal considerations for developers, healthcare professionals, and other stakeholders aiming to integrate conversational agents effectively.

- *Technical Challenges:* One of the biggest agent development is creating agents that can comprehend user queries with great accuracy and subtlety. These agents' Natural Language Processing (NLP) algorithms frequently have trouble understanding context, which causes them to misinterpret user intentions. Robust training datasets and continuously improving algorithms are necessary to ensure accuracy when understanding a variety of linguistic nuances, accents, and slang. Furthermore, complex user queries or unclear inputs pose ongoing technical challenges to the development of agents that can react appropriately in a variety of scenarios.
- *User Acceptance and Engagement:* It's still difficult to ensure user acceptance and continuous interaction with conversational agents. Adoption depends on creating agents that people find logical, beneficial, and reliable. It's difficult to strike a balance between the conversational tone, personality, and formality level to accommodate different user preferences. There are also difficulties in sustaining user interest over time and giving users ongoing value, avoiding engagement declines brought on by repetitive interactions or insufficient support from the agents.

- *Integration with Existing Systems:* There are practical challenges in smoothly integrating conversational agents into current technological infrastructures. Smooth integration of conversational agents is hampered by compatibility problems with legacy systems, interoperability issues, and aligning with diverse platforms or applications. Adaptable architectures and strategies are required to overcome these obstacles and integrate conversational agents with the various technological ecosystems that are used in various industries.

- *Evaluation and Performance Metrics:* It can be difficult to determine the right evaluation metrics for conversational agents when assessing their performance and efficacy. Robust evaluation frameworks are necessary for evaluating user satisfaction, response accuracy, and the quality of conversational experiences. One of the ongoing challenges is to develop standardized metrics that accurately assess the agent's performance across various domains and user interactions.

- *Clinical Accuracy and Liability:* It is critical to guarantee clinical accuracy during dialogue. Neglecting to read medical questions correctly or giving bad medical advice can have serious consequences. For conversational agents to provide trustworthy information without taking the place of expert medical advice, they must demonstrate a high degree of accuracy and possess current, evidence-based medical knowledge. Determining responsibility for mistakes or misunderstandings also poses a difficult problem.

- *Interoperability and Data Integration:* Healthcare is a multisystem industry, with patient data dispersed among wearable devices, various hospital systems, and Electronic Health Records (EHRs). A major challenge is ensuring that conversational agents can access, interpret, and update patient data in real-time across these disparate systems with seamless integration and interoperability. One of the challenges still standing is creating a coherent data ecosystem for precise diagnosis and individualized care.

- *Continual Adaptation and Learning:* Conversational agents must always change and adapt to the ever-evolving field of medicine, shifting guidelines for patient care, and shifting treatment paradigms. Strong learning mechanisms, frequent updates, and methods to incorporate the most recent medical developments are needed for this, which presents a big challenge in healthcare settings where knowledge is continuously changing.

- *Considerations for Biases and Limitations:* Conversational agents' algorithms may contain flaws that hinder their efficacy, equity, and precision. As a result, a deeper understanding is required to improve AI systems' performance for a wider range of users [28].

  - Training Data Biases: One major challenge is dealing with biases in training data that result from historical disparities or underrepresentation of particular demographics. Agents trained on data that primarily represents particular groups may find it difficult to comprehend or react to people from different backgrounds, which could lead to under-represented groups receiving subpar services.

- Nuances in Language and Culture: Due to inherent design limitations, agents may be unable to understand slang, regional dialects, or expressions unique to a community, which could result in misinterpretations of questions or the delivery of unnecessary information.
- Inherited Biases: It is possible for agents to inherit biases from training datasets, which could lead to responses or recommendations that reinforce societal prejudices or stereotypes. Biases present in historical data, like those based on gender or race, may unintentionally spread through user interactions.
- Limitations in Handling Complexity: Particularly in healthcare contexts where understanding complex human emotions is essential, agents may find it difficult to respond empathetically or to comprehend complex or sensitive interactions like emotional cues (Fig. 2).

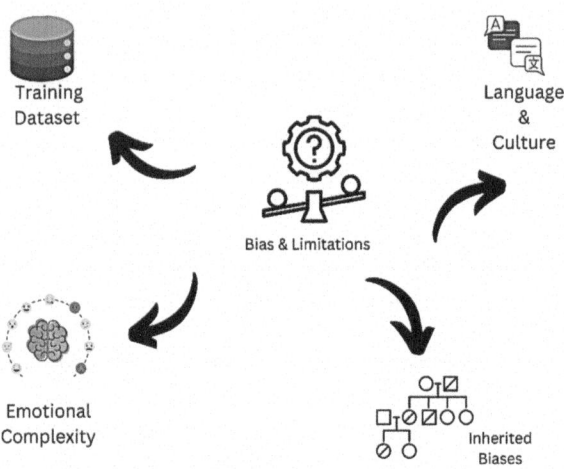

Fig. 2. Bias induced in Conversational Agents

4   Ethical Implications Within Immersive Contexts

More specifically in underserved areas and during emergencies, conversational agents—such as chatbots and embodied virtual agents—offer promising solutions for closing the gaps in health services. These agents provide a range of health-related services, such as treatment, mental health support, and health education. They are attractive because of their scalability, accessibility, and perceived impartiality, especially in environments with limited resources. But their implementation, design, and possible risks present ethical issues that need to be carefully considered [29], 30.

There are serious ethical issues with conversational agents' implicit, explicit, and underlying algorithmic biases. These prejudices, which originate from training data sets or programmers' values, have the potential to reinforce inequalities and give preference to some groups over others. If diverse end users are not taken into consideration during the design process, cross-cultural deployment may make these biases worse, necessitating ongoing research and inclusive testing procedures.

The potential harm that conversational agents might cause is another ethical concern. This risk is particularly present in situations where individuals are at risk, such as when they are expressing suicide thoughts. To reduce these risks, it is essential to ensure proper disclosure, conduct user screening, keep an eye on risks, and provide quick access to resources. Conversational agents collect sensitive user data, which raises privacy concerns and calls for compliance with a number of international privacy regulations. Furthermore, unequal access to technology and low levels of education may prevent some populations from taking advantage of these agents, which would increase healthcare disparities [29, 30].

The incorporation of conversational agents into virtual healthcare environments gives rise to ethical concerns concerning the interactions between patients and providers. Although these agents provide accessibility and assistance, the patient's relationship, confidence, and satisfaction with healthcare providers may be impacted by their integration. The use of conversational agents could change the traditional dynamics of the relationship between the patient and the provider, which could result in less empathy and individualized care—two things that are critical to the patient's well-being. The moral conundrum is how to employ these agents to improve healthcare accessibility without sacrificing the vital human touch and knowledge that medical professionals offer. Maintaining patient trust and ensuring ethical use of conversational agents in healthcare interactions requires open and honest communication about their role and limitations [31].

In order to avoid escalating healthcare disparities, diverse populations must have equitable access to conversational agents and their services. When creating and implementing these agents, careful consideration of technological, socioeconomic, linguistic, and cultural aspects is necessary. Ignoring these factors could make inequality worse since underprivileged communities might not have access because of poor technology infrastructure, inadequate educational opportunities, or communication difficulties. Proactive steps like creating agents that accommodate a range of cultural backgrounds, offering multilingual support, and guaranteeing that all populations have access to technology are necessary to address these issues. In order to prevent further marginalization of underprivileged groups, it is ethically required that efforts be made to close the digital divide and guarantee inclusivity in access to these healthcare technologies [31].

5   Current Landscape and Future Trajectories

Conversational agent technologies are still in their infancy as a field, but they have the potential to address a wide range of health goals, including self-care, monitoring, education, and therapy [32]. Nonetheless, the majority of conversational agents are still in the prototype phase and are not widely accessible. Table 1 provides an overview of the state of conversational agents as of right now.

Conversational agents have the potential to revolutionize human-machine interactions in a number of industries as they continue to evolve toward more intelligent, personalized, and ethical AI-driven interactions. Nonetheless, advancements must tackle

Table 1. Summary of the current landscape of conversational agents.

Aspect	Summary
Technological Approaches	Agents interact through multiple channels and varied input/output formats. Most remain in prototype stages. Studies focus on user experience rather than technical performance or health outcomes [34], 35
Geographical Focus	Concentrated research in North America and Europe, with minimal representation from Asia, Australia, and Africa. [33]
AI Techniques	Natural Language Processing (NLP) primarily used; less frequent application of deep learning and neural networks. Lack of technical details raises concerns about replicability and trust. [35]
Interoperability and Standards	Focus on specific chronic diseases lacks standardization in evaluation measures, hindering comparison between studies. Need for standardized definitions and terminology [33]

technical constraints, broaden the geographic scope of research, and prioritize uniform assessment metrics to ensure dependability and relevance in various demographics and chronic illness environments.

6 Practical Considerations for Implementation

Implementing conversational agents in healthcare requires careful consideration of various practical aspects to ensure their successful integration and utilization in improving patient care, engagement, and health outcomes.

- *Technical Infrastructure and Integration:*, data processingSecure communication, and storage require a strong technical infrastructure, which is required for the deployment of conversational agents. In order to guarantee data interoperability and maintain strict security and privacy standards, healthcare systems must make sure that these agents are compatible with and integrate seamlessly with the current Electronic Health Record (EHR) systems.
- *User-Centric Design and Usability:* Designing conversational agents from a user-centric perspective, with an emphasis on usability and user experience, is an essential component. These agents ought to be intuitive, simple to use, and able to comprehend a wide range of patient demographics, including varying literacy levels, languages spoken, and accessibility needs. For interactions to go smoothly, natural language processing and speech recognition technologies must be included.
- *Clinical Validation and Evidence-Based Practice:* Before deployment, conversational agents in healthcare must undergo rigorous clinical validation and testing to ensure their accuracy, reliability, and adherence to evidence-based medical practices. Collaborations between developers, healthcare professionals, and researchers are crucial to validate these agents' efficacy in various clinical scenarios, ensuring they provide accurate information and recommendations.

- *Continuous Monitoring and Improvement:* Conversational agents in healthcare must pass stringent clinical validation and testing procedures before being put into use in order to guarantee their accuracy, dependability, and adherence to best practices in medicine. Developers, medical professionals, and researchers must work together to validate the effectiveness of these agents in a range of clinical scenarios and to ensure that the information and recommendations they provide are accurate.
- *Training and Support for Healthcare Professionals:* For healthcare personnel to integrate these conversational agents into their workflow efficiently, they need to be properly trained and supported. The functionalities of the agents, their integration into clinical practice, the interpretation of information generated by the agents, and the handling of patient inquiries or concerns resulting from agent interactions should all be covered in training programs.
- *Cost-Benefit Analysis and Scalability:* It is imperative to evaluate the financial viability of deploying conversational agents in the healthcare industry. It entails assessing the outlay of funds, ongoing expenses, and possible savings in terms of patient outcomes, operational effectiveness, and healthcare delivery. In order to guarantee the agents' broad adoption without sacrificing performance or raising operating expenses, scalability considerations are crucial.

By addressing these practical considerations, healthcare organizations can harness the full potential of conversational agents to enhance patient care, improve healthcare delivery, and foster better patient-provider interactions.

7 Conclusion

The evaluation of conversational agents in the treatment of chronic illnesses brought to light both ongoing difficulties and encouraging developments. These agents showed promise in assisting with self-management and education for chronic conditions by demonstrating acceptance and usability among patients. There is conflicting data regarding the efficacy of AI-enabled conversational agents in treating chronic health conditions, though, due to ongoing concerns about technical implementation, particularly with regard to personalization and ontology.

These findings imply important ramifications for medical procedures. The inclusion of conversational agents in care plans has the potential to improve patient education, self-care, and engagement. It's imperative to address ontology and personalization-related technical issues in order to accommodate a range of patient needs. Developers, medical professionals, and researchers must work together to validate clinical efficacy and meet regulatory requirements. Agent development should be guided by user-centric design, taking accessibility, language, and patient preferences into account.

Further investigation is needed to advance conversational agents in the management of chronic illnesses. In-depth analyses should assess technical implementation, including AI techniques, to ascertain the actual influence of agents on health outcomes. To improve usability and effectiveness, research into personalization, ontology, and natural language processing can be conducted. Standardized protocols or frameworks for development, validation, and implementation could guarantee uniformity, openness, and integration with healthcare systems.

References

1. Centers for Disease Control and Prevention. About chronic diseases. https://www.cdc.gov/chronicdisease/about/index.html. Last accessed 27 Mar 2023
2. National Center for Chronic Disease Prevention and Health Promotion: 'National diabetes statistics report, 2017,' CS279910-A (2017). Last accessed 27 Mar 2023
3. Barlow, J., Wright, C., Sheasby, J., Turner, A., Hainsworth, J.: Self-management approaches for people with chronic conditions: a review. Patient Educ. Counsel. (Patient ed) **48**(2), 177–187 (2002). https://doi.org/10.1016/s0738-3991(02)00032-0
4. Rijken, M., Jones, M., Heijmans, M., Dixon, A.: Supporting self-management. In: Nolte, E., McKee, M. (eds.) Caring for People with Chronic Conditions. A Health System Perspective, pp. 116–142. Open University Press (2008)
5. Fort, M.P., Alvarado-Molina, N., Peña, L., Montano, C.M., Murrillo, S., Martínez, H.: Barriers and facilitating factors for disease self-management: a qualitative analysis of perceptions of patients receiving care for type 2 diabetes and/or hypertension in San José, Costa Rica and Tuxtla Gutiérrez, Mexico. BMC Fam. Pract. **14**, 1–9 (2013)
6. Wilde, M.H., Garvin, S.: A concept analysis of self-monitoring. J. Adv. Nurs. **57**(3), 339–350 (2007). https://doi.org/10.1111/j.1365-2648.2006.04089.x
7. Jurafsky, D., Martin, J.H., Speech and Language: Processing: An Introduction to Natural Language Processing, Computational Linguistics, and Speech Recognition, 2nd ed. Prentice Hall (2009)
8. Guy, I.: Searching by talking: Analysis of voice queries on mobile web search. In: the 39th International ACM SIGIR conference on Research and Development in Information Retrieval, Pisa, Italy, pp. 35–44 (2016). https://doi.org/10.1145/2911451.2911525
9. Montenegro, J.L.Z., da Costa, C.A., da Rosa Righi, R.: Survey of conversational agents in health. Expert Syst. Appl. **129**, 56–67 (2019). https://doi.org/10.1016/j.eswa.2019.03.054
10. Laranjo, L., et al.: Conversational agents in healthcare: a systematic review. J. Am. Med. Inform. Assoc. **25**(9), 1248–1258 (2018). https://doi.org/10.1093/jamia/ocy072
11. Agnihothri, S., Cui, L., Delasay, M., Rajan, B.: The value of mhealth for managing chronic conditions. Health Care Manag. Sci. **23**(2), 185–202 (2020). https://doi.org/10.1007/s10729-018-9458-2
12. Christensen, S., Johnsrud, S., Ruocco, M., Ramampiaro, H.: Context-aware sequence-to-sequence models for conversational
13. Fernandes, A.: NLP, NLU, NLG and how Chatbots work (2018). https://chatbotslife.com/nlp-nlu-nlg-and-how-chatbots-work-dd7861dfc9df
14. Vishnoi, L.: Conversational agent: A more assertive form of chatbots. Towards Data Science. https://towardsdatascience.com/conversational-agent-a-more-assertive-form-of-chatbots-de6f1c8da8dd Retrieved 21 March 2023
15. Allouch, M., Azaria, A., Azoulay, R.: Conversational agents: Goals, technologies, vision and challenges. Sensors **21**(24), 8448 (2021). https://doi.org/10.3390/s21248448
16. Gudivada, V.N., Rao, D., Raghavan, V.V.: Big data driven natural language processing research and applications. In: Big Data Analytics, pp. 203–238. Elsevier (2015). https://doi.org/10.1016/B978-0-444-63492-4.00009-5
17. Alesanco, Á., Sancho, J., Gilaberte, Y., Abarca, E., García, J.: Bots in messaging platforms, a new paradigm in healthcare delivery: Application to custom prescription in dermatology," presented at the EMBEC & NBC: Joint Conference of the European Medical and Biological Engineering Conference (EMBEC) and the Nordic-Baltic Conference on Biomedical Engineering and Medical Physics (NBC), Tampere, Finland, June 2017, 2018, pp. 185–188 (2017)

18. Garg, R., SenGupta, S.: Conversational technologies for in-home learning: Using co-design to understand children's and parents' perspectives," presented at the Proceedings of the 2020 CHI conference on human factors in computing systems. Honolulu. HI, pp. 1–13 (2020)

19. Storer, K.M., Judge, T.K., Branham, S.M.: 'All in the same boat': Tradeoffs of voice assistant ownership for mixed-visual-ability families, presented at the Proceedings of the 2020 CHI Conference on Human Factors in Computing Systems. Honolulu. HI, pp. 1–14 (2020)

20. Garg, R., et al.: The last decade of HCI research on children and voice-based conversational agents, pp. 1–19. New Orleans, LA (2022). https://doi.org/10.1145/3491102.3502016

21. Gehl, R.W.: Teaching to the Turing test with Cleverbot. Transform. J. Incl. Scholarsh. Pedagogy **24**(1–2), 56–66 (2013)

22. Maroengsit, W., et al.: A survey on evaluation methods for chatbots. Presented at the 7th International conference on information and education technology, Aizu, Japan, Mar 2019, pp. 111–119 (2019). https://doi.org/10.1145/3323771.3323824

23. Diederich, S., Brendel, A.B., Kolbe, L.M.: On conversational agents in information systems research: Analyzing the past to guide future work. Presented at the 14th International Conference on Wirtschaftsinformatik, pp. 1550–1564. Siegen, Germany (2019)

24. Pereira, J., Díaz, Ó.: Using health chatbots for behavior change: a mapping study. J. Med. Syst. **43**(5), 135 (2019). https://doi.org/10.1007/s10916-019-1237-1

25. Fadhil, A.: Beyond patient monitoring: Conversational agents role in telemedicine and healthcare support for home-living elderly individuals (2018)

26. Provoost, S., Lau, H.M., Ruwaard, J., Riper, H.: Embodied conversational agents in clinical psychology: A scoping review. J. Med. Internet Res. **19**(5), e151 (2017). https://doi.org/10.2196/jmir.6553

27. Kenny, P., Parsons, T., Gratch, J., Rizzo, A.: In: Proceedings of the 1st International Conference on Pervasive Technologies Related to Assistive Environments, pp. 1–4. Jonathan and Rizzo. https://doi.org/10.1145/1389586.1389594

28. Ueda, D., et al.: Fairness of artificial intelligence in healthcare: review and recommendations. Jpn. J. Radiol. (2023). https://doi.org/10.1007/s11604-023-01474-3

29. Luxton, D.D.: Ethical implications of conversational agents in global public health. Bull. World Health Organ. **98**(4), 285–287 (2020). https://doi.org/10.2471/BLT.19.237636. Epub January 27, 2020. PubMed: 32284654, PubMed Central: PMC7133471

30. Ruane, E., Birhane, A., Ventresque, A.: Conversational AI: Social and ethical considerations. In: Conference: AICS – 27th AIAI Irish Conference on Artificial Intelligence and Cognitive ScienceAt: Galway. Ireland (2019)

31. Conversational AI: What are the potential ethical implications and how can these be addressed? https://ceur-ws.org/Vol-2563/aics_12.pdf. Retrieved 20 Nov 2023

32. Roca, S., Sancho, J., García, J., Alesanco, Á.: Microservice chatbot architecture for chronic patient support. J. Biomed. Inform. **102**, 103305 (2020). https://doi.org/10.1016/j.jbi.2019.103305

33. Schachner, T., Keller, R., Wangenheim, F.V.: Artificial intelligence-based conversational agents for chronic conditions: Systematic literature review. J. Med. Internet Res. **22**(9), e20701 (2020). https://doi.org/10.2196/20701

34. Martinengo, L., et al.: Conversational agents in health care: expert interviews to inform the definition, classification, and conceptual framework. J. Med. Internet Res. **25**, e50767 (2023). https://doi.org/10.2196/50767

35. de Cock, C., Milne-Ives, M., van Velthoven, M.H., Alturkistani, A., Lam, C., Meinert, E.: Effectiveness of conversational agents (virtual assistants) in health care: Protocol for a systematic review. JMIR Res. Protoc. **9**(3), e16934 (2020). https://doi.org/10.2196/16934

A Computer Aided Detection System for Lung Nodules Classification

Maruf-uz-zaman$^{(\boxtimes)}$ and Rajeev Ranjan

Department of Electronics and Communication Engineering, Chandigarh University,
Mohali, India
marufshah95@gmail.com

Abstract. Lung cancer is the most popular type of cancer and a significant disease worldwide. Radiologists mostly prefer computer-aided detection (CAD) systems, and it's the most essential and popular method. The CAD systems have the necessary steps to acquire data to show results. The segmentation is an integral part of the CAD system, and region growth, Threshold, and active shape models are standard methods of lung segmentation. Accuracy is the close of an actual measurement percentage value. Alcohol and smoking is more common reason cause of lung cancer. In this work, we describe lung nodules and the comparison accuracy rate with Age vs alcohol is 84.6%, and Age vs. smoking is 86.7%. Age vs Alcohol positive predictive value of 23.1%, 87.6%, false discovery rates of 76.9%, 12.4%. In this work, Age vs Alcohol's true positive rates were 8.3% and 96.9%, and false negative rates (fnr) were 91.7% and 4.1%. Again, the Age vs smoking true positives rate (tpr) is 0% and 99.3%, and the false negatives rate is 100% and 0.7%.

Keywords: Segmentation · Nodules detection · Computed tomography (CT) · Computed Aided detection (CAD)

1 Introduction

Cancer is one of the most significant diseases in the world. Among all cancers, lung cancer is the deadliest. Cancer deaths are on the rise both in Asia and the United States. China and the Democratic Republic of South Korea have Asia's highest rates of cancer incidence (Ning et al., 2019). IARC (The international Agency for Cancer Research) states that tobacco smoking accounts for 75% of lung cancer cases. Day by day, the number of deaths is increasing. According to a British study, the danger of being a running smoker is 15 times higher than the chance of never smoking (El-Regaily et al., 2017). Cancer has a five-year survival rate between 10–16%, and approximately patients are diagnosed with advanced cancer without adequate treatment. However, if cancer is detected early enough, the five-year survival rate can rise to 52% (Henschke et al., 1999). The uncontrolled and uneven development of lung tissue is the first stage of lung cancer. Lung tissue is affected in many ways, but the most compelling reason is tobacco and alcohol. Radiologists recommended different imaging techniques for detecting pulmonary nodules, such as Computed tomography (CT), Positron Emission Tomography

© The Author(s), under exclusive license to Springer Nature Switzerland AG 2025
M. Khurana et al. (Eds.): ICMLA 2024, CCIS 2238, pp. 221–228, 2025.
https://doi.org/10.1007/978-3-031-75861-4_19

(PET), and Magnetic Resonance Images (MRI). To find these lung nodules, radiologists currently use CT city scans and chest radiographers (Ning et al., 2019). The CT is the most popular and most accurate technique. This work discusses the CT technique with some result analysis. Several detector rows With CT scanners, the entire chest may be scanned in a few seconds (El-Regaily et al., 2017). The CT has some challenges; significant challenges are slice thickness measurement and radiologist analysis of CT Images. Lung cancer is detected in two significant ways: analog and CAD systems. The CAD system follows some steps to recognize data and verify the result. This work discusses CAD systems briefly. The paper's outline is arranged as follows: Sect. 1 is the introduction of the work, Sect. 2 is the CAD system working principle, Sect. 3 is the Computed Aided Detection (CAD) steps and Sect. 7 is the conclusion.

2   Computed Aided Detection Systems

The CAD systems have classified by (CADe) and (CADx). It is anticipated that CADx will effectively minimized false positive in less dose Computed tomography lung cancer covering. A radiologist needs careful observation of the density of pulmonary nodules because, at an early stage, this density may resemble the densities of other lung parts (Wang et al., 2010). The CAD systems follow some steps. These are acquisition, pre-possessing, segmentation, nodule detection, false position reductions. Segmentation and nodule analysis are the major part of the CAD system step. Acquisition is played as an input.

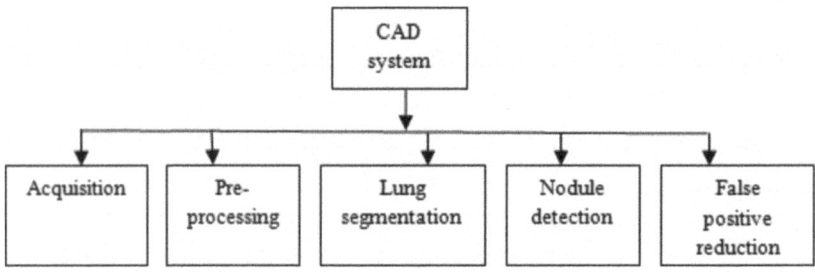

Fig. 1. Five important step of CAD system

2.1   Acquisition

Acquisition is played as an input. Acquisition is observes medical images using imaging modalities is known as image acquisition. There are several lung imaging methods, but CT has the most benefits since it allows for a more thorough and detailed examination of the whole lung parenchyma. Database units include acquisition. There is some popular pubic data set for lung cancer. Early Lung Cancer Action Program (ELCAP) and Lung Image Dataset Consortium (LIDC) are popular data sets. LIDC data set is a huge data set, and it is provided by the National Cancer Institute (Ning et al., 2019). Nederland-leavens lonkanker onderzoek (NELSON) (van Ginneken et al., 2010), Automatic Nodule detection 2009 (ANODE09) (Fernández-Carrobles et al., 2015), Lung Nodule Analysis 2016 (LUNA16) (Setio et al., 2017), Database of Japanese Society of Radio-logical Technology (JSRT) (Abhishek et al., 2023) are the more popular datasets. The acquisition has two sub-processes: selection and extraction.

2.2 Pre-processing

Removing unessential actuality (noise, artifacts, etc.) and recovering important information during data gathering are the two main goals of preprocessing raw CT images. The Preprocessing step is optional but essential in the CAD system. The per-processing step reduces noise, and data starts to be filtered in this step. It could be challenging to divide the lungs precisely and find low-contrast nodules with ground-glass opacity or these attached to the lungs (Chaudhary & Singh, 2012). The median filter is used with smooth algorithms in the preprocessing algorithms (Choi & Choi, 2013).

2.3 Lung Segmentation

Lung segmentation plays significant importance in nodule detection and analysis. In the CAD systems, lung segmentation reduces scanning time and minimizes search space. Segmentation has many difficulties and challenges. Because of lung structure similarities, nodules are attached to the lung wall. Simple algorithms avoid this kind of problem.

3   Segmentation Techniques

Lung segmentation has different techniques and methods. Standard lung segmentation methods include threshold, Region growth, Connected Component Analysis (CCA) (McNitt-Gray et al., 2007), and robust active shape model (Fig. 2).

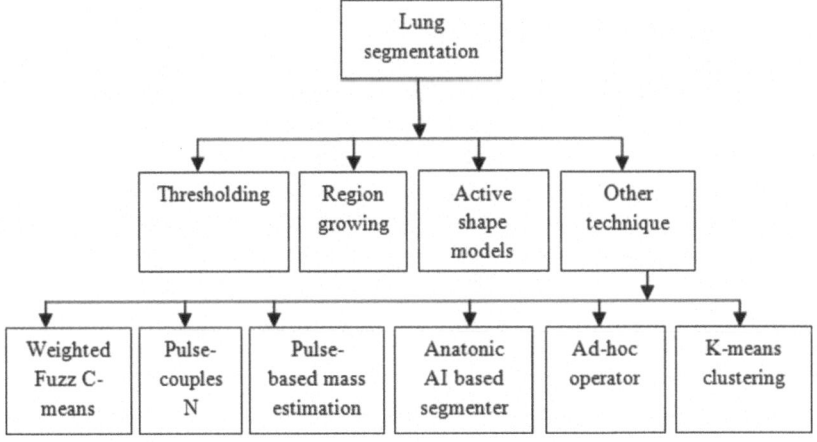

Fig. 2. Lung segmentation in different technique

3.1 Threshold

Threshold is one of the simplest ways to address typical segmentation issues, resulting in a two-level binary output picture with pixels over the defined threshold allocated to one and the remainder of pixels assigned to zero. In threshold, density advantage is the extra advantage, and it depends two ways: high density and low density. Threshold processing is usually followed by morphological processing and fills the gaps.

3.2 Region Growing

The region growing method is a popular segmentation technique that starts with a seed of a point. Seed points are grown from joining pixels. Voxels in 3D planes are also included in region-growing segmentation. Region-growing segmentation is included in some algorithms. The algorithm's performance can be improved by selecting the appropriate inclusion rule with the optimized threshold and initial seed points.

3.3 Active Shape Models

The shaped model's technique is used for the interpretation of images. This technique also works as adjacent structures. Active shape models have some algorithms, but Minimum description length (MDL) and Machine cube algorithms are helpful and popular.

3.4 Other Technique

Lung segmentation has some techniques that are used in different areas. Different techniques have different algorithms. Different algorithms work in different fields. K-means clustering, Ad-hoc operator, Anatomic AI-based segment, pulse-coupled NN, Region-based mass estimation, and Fuzz C-means are popular lung segmentation techniques.

4 Classification of Lung Segmentation

Segmentation is divided into parts, which are definable, available, and trainable. Lung segmentation is divided into some parts. Lung segmentation is classified into different types. Cluster-based lung segmentation, semantic-based lung segmentation, Edge segmentation and neural network-based lung segmentation are popular. Cluster-based lung segmentation is performed pixel-wise. Two approaches perform cluster-based lung segmentation: clustering by merging and clustering by diversity. Semantic lung segmentation is used in deep learning algorithms. Semantic lung segmentation is also used to recognize a collection of pixels. Edge-based lung segmentation worked as lung boundaries. Lung images are different from each other. Edge-based lung segmentation is essential to observe the different boundaries in different ways. Neural network-based lung segmentation plays a vital role in lung segmentation. Neural network based architecture used for segmentation. This segmentation consists of some layers. Neural networks also work on data sets.

5   Lung Segmentation Image Analysis

This section discusses lung segmentation and some parameters. These parameters discuss the Excel data set value. This paper discusses age vs smoking and age vs alcohol consumers' lung data sets. Table 1 describes about age vs alcohol and age vs smoking accuracy, total cost, prediction speed, training time and model size. In this Table 1 age vs alcohol accuracy much higher than age vs alcohol accuracy but age vs alcohol model size higher than age vs smoking model size.

Table 1. Age vs Alcohol and Age vs smoking parameters

Parameter	Age vs alcohol	Age vs Smoking
Accuracy	84.6%	86.7%
Total cost	43	41
Prediction speed	3300obs/sec	340obs/sec
Training time	3.6637 s	10.593 s
Model size (Compact)	8kB	4kB

Figure 3 described about age vs alcohol and age vs smoking roc curve. Figure 3 (a) roc curve is better than Fig. 3 (b) roc curve, because Fig. 3 (a) roc curve value is 0.707 and Fig. 3 (b) roc curve value is 0.523. ROC maximum value is 1 and minimum value is 0. If the vale is more than 0.5, it's called this roc curve is better.

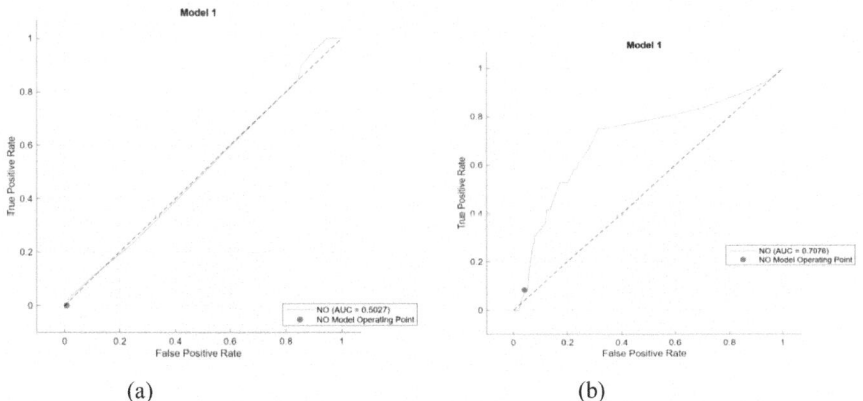

(a)   (b)

Fig. 3. Validation roc curve (a) Age vs alcohol (b) Age vs smoking

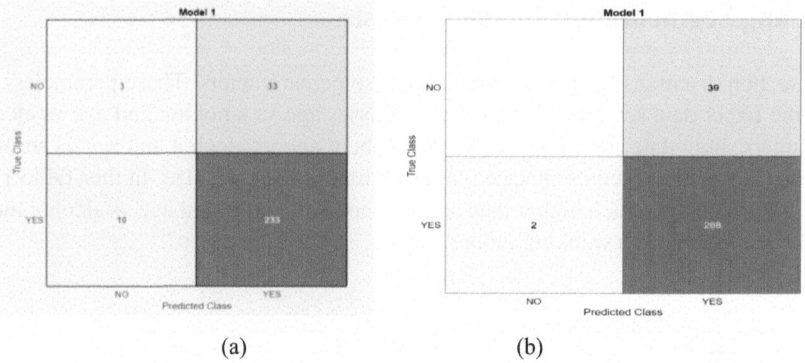

Fig. 4. Number of observation (a) Age vs alcohol (b) Age vs smoking

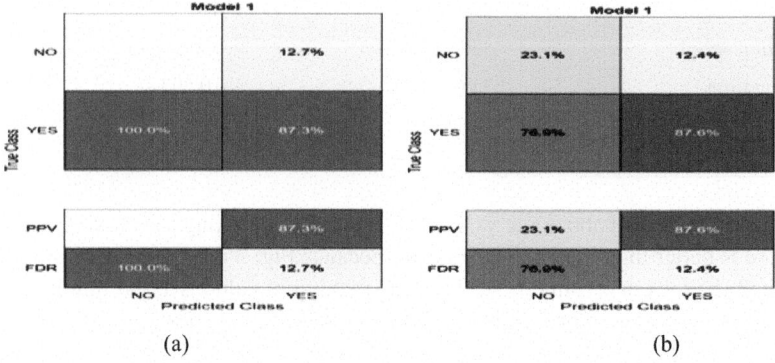

Fig. 5. (a) Positive predictive values for Age vs alcohol (b) Positive predictive values for age vs Smoking

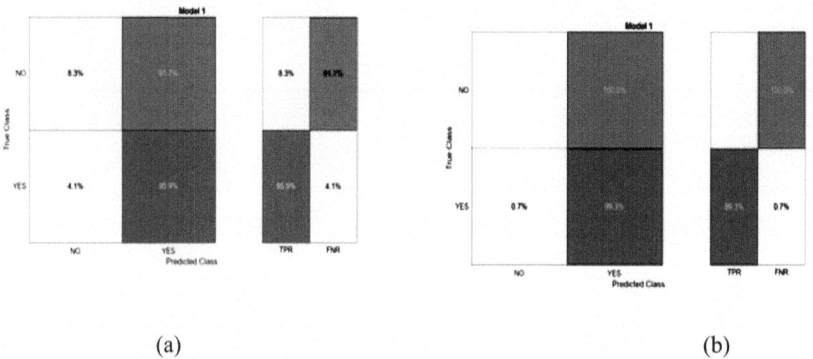

Fig. 6. (a) True positive rates and false negative rates for Age vs alcohol, (b) True positive rates and false negative rates for Age vs smoking

6 Result Analysis

In this simulation result, accuracy depends on some parameters. In this table, Age with alcohol and smoking accuracy are defined. Alcohol with smoking accuracy is better than Age vs alcohol accuracy. This table also indicates that Age vs smoking total cost, model size and prediction speed is less than other. Figure 3 compares the validation receiver operation characteristics (ROC) curve, which is the false positive rate and true positive rate ratio. When the area under a ROC curve (AUC) is 0.5 or more than 0.5, its means ratio is much better. In Fig. 1, both validation ROC curve AUC value is 0.7027 and 0.5027. In Fig. 4 shown the counting of observations and it's depend the data set's mean and defines true and false classes. The true class has two conditions, and the false class has two conditions, as shown in Fig. 4 and 5 define positive predictive value and false discovery rates. It is the percentage of true class and predicted class. Figure 5 (a) Age vs Alcohol positive predictive value 23.1% and 87.6%, false discovery rates 76.9% and 12.4%. Figure 5 (b) Age vs smoking positive predictive value 0% and 87.3%, false discovery rates 100% and 12.7%. Figure 6 defines true positive rates and false negative rates. Figure 6 (a) defines Age vs Alcohol true positive rates of 8.3% and 96.9% and false negative rates of 91.7% and 4.1%. Again Fig. 6 (b) defines Age vs smoking, with true positive rates of 0% and 99.3% and false negative rates of 100% and 0.7%.

6.1 Nodule Detection

The main problem is solved in this step. The nodule detection step separates the suspicious pulmonary nodule from the pulmonary nodule and highlights the true nodule location. In this step, collect true positive data and many false positives. Nodule detection is classified into different categories and different techniques. These techniques are intensity-based schemes, machine-learning-based schemes, and shape-based schemes. Threshold and region growth include intensity-based schemes and graph cuts methods, 3D detection boxes, spherical shape enhancement filters include shape-based schemes, and Convolutional neural networks (CNN) include machine learning schemes. In nodule detection intensity-based algorithms are used primarily. There are many nodule detection processes, and different nodules work differently. For example, the juxta vascular nodule works on the vessel, the pleural tail works on the pleural wall, and the juxta-pleural nodule works on the nodule wall.

6.2 False Positive Reduction

False positive reduction is the last step of segmentation. While the nodule detection step counts suspicious nodule candidates, it consists of true and false nodules. False nodule rate increases the true nodule rate. Sensitivity, accuracy rate, and maximum positive rate are defined in this step. The false positive reduction step was defined in two ways: the false nodule was removed for sensitivity and accuracy, and the true nodule rate increased. Four possible results appear in this step. When the pulmonary nodule is classified correctly, it is defined as true positive (TP) or false negative (FN). When the pulmonary nodule is rejected, it is defined as true negative (TN) or false positive (FP).

$$\text{Sensitivity rate} = [\text{TP}/(\text{TP} + \text{FN})] \tag{1}$$

$$\text{Specificity} = [TN/(TN + FP)] \tag{2}$$

7 Conclusion

This paper discusses the CAD system step and lung nodules classification with simulation results. The simulation result: Accuracy rate, Number of observations, Receiver Operating characteristics Curve, Positive predictive value, and false discover rate. This type of parameter comparison with age vs alcohol and age vs smoking data sets.

References

Abhishek, Ranjan, R., Sahni, R.: X-ray image classification for COVID-19 using transfer learning. https://doi.org/10.3233/atde221307(2023)

Chaudhary, A., Singh, S. S.: Lung cancer detection on CT images by using image processing. In: Proceedings Turing 100 - International Conference on Computing Sciences, ICCS 2012, pp. 142–146 (2012)

Choi, W.J., Choi, T.S.: Automated pulmonary nodule detection system in computed tomography images: a hierarchical block classification approach. Entropy 15(2), 507–523 (2013). https://doi.org/10.3390/e15020507

El-Regaily, S.A., Salem, M.A., Abdel Aziz, M.H., Roushdy, M.I.: Survey of computer aided detection systems for lung cancer in computed tomography. Curr. Med. Imaging Rev. 14(1), 3–18 (2017)

Fernández-Carrobles, M.M., et al.: A CAD system for the acquisition and classification of breast TMA in pathology. Stud. Health Technol. Inf. 210, 756–760 (2015). https://doi.org/10.3233/978-1-61499-512-8-756

Henschke, C.I., et al.: Early lung cancer action project: overall design and findings from baseline screening. Lancet 354(9173), 99–105 (1999)

McNitt-Gray, M.F., et al.: The lung image database consortium (LIDC) data collection process for nodule detection and annotation. Acad. Radiol. 14(12), 1464–1474 (2007). https://doi.org/10.1016/j.acra.2007.07.021

Ning, J., Zhao, H., Lan, L., Sun, P., Feng, Y.: A computer-aided detection system for the detection of lung nodules based on 3D-ResNet. Appl. Sci. (2019). https://doi.org/10.3390/app9245544

Setio, A.A.A., et al.: Validation, comparison, and combination of algorithms for automatic detection of pulmonary nodules in computed tomography images: the LUNA16 challenge. Med. Image Anal. 42, 1–13 (2017)

van Ginneken, B., et al.: Comparing and combining algorithms for computer-aided detection of pulmonary nodules in computed tomography scans: the ANODE09 study. Med. Image Anal. 14(6), 707–722 (2010)

Wang, H., et al.: Multilevel binomial logistic prediction model for malignant pulmonary nodules based on texture features of CT image. Eur. J. Radiol. 74(1), 124–129 (2010)

A Note on Interpolation of Hermitian Signal: Based on Centrosymmetric Property

Manpreet Kaur[1] and Vineet Bhatt[2]([✉])

[1] Department of Mathematics, Chandigarh University, Mohali, Punjab, India
[2] Department of Mathematics, Graphic Era Hill University, Dehradun, India
vineet.bhatt58@gmail.com

Abstract. The most significant use of a centrosymmetric matrix in signal processing is the subject of this article. A discrete-time decimation formula is the foundation of the derivation. Considered is a specific case of a signal that contains known samples and satisfies the Hermitian property. The two interpolation schemes are as follows. In an effort to find a solution to a square system of linear equations, the first strategy disregards the middle sample. The second technique uses an estimated least squares solution to an over determined system of equations and includes a middle sample. In the end, our solution has the benefit of reducing computing effort when compared to others.

Keywords: Centro-symmetric · Hermitian Signal · Fourier transform · Function

1 Introduction

In different areas of linear algebra, such as engineering and the sciences, matrices are taking on an important role. The symmetric matrix, which is symmetric about its diagonal, is among the most well-liked matrices (inclining). The Centrosymmetric matrix, which is symmetric about its geometric centre and can be described as the matrix $A \in R^{m \times n}$ satisfying $A = J_m A J_n$, where $J_k \in R^{k \times k}$ is a reverse identity matrix having antidiagonal entries are 1 and 0 elsewhere, has been the subject of extensive study since the 19th century [1]. The Centrosymmetric matrix has several significant properties [2, 3]. The properties of centrosymmetric matrices have been studied by many writers in a variety of domains, including signal processing, image recognition, differential quadrature, molecular recognition, etc. [6–9]. One of the properties is the Hermitian property, which is used in the unique use of Centrosymmetric matrices, namely, Hermitian signal interpolation. If the complex conjugate of a complex function, $f^*(x) = f(-x)$, is the same as the original function with a variable change of sign, then the complex function is said to be a Hermitian function. A method of finding new data points based on the range of a discrete set of existing data points is called interpolation [4, 5]. Electrical engineering's field of signal processing allows us to examine altering signals like images, sounds, and other types of signals [12]. In this paper, we discuss the Hermitian signal, a specific example of signal. The Hermitian property occurs in signal sampling. For the analysis of these samples, we created a method that was built on the discrete-time decimation algorithm.

M. Khurana et al. (Eds.): ICMLA 2024, CCIS 2238, pp. 229–238, 2025.
https://doi.org/10.1007/978-3-031-75861-4_20

As a result, this study is divided into two schemes. The first scheme yields solutions in the square system of equations for unknown interpolated locations of the signal but does not use the known samples at the origin. The second technique makes use of the origin's known samples, but it produces an over determined system of equations that must be solved using the least squares method. There are four sections in this work. The introduction is included in the first section. Section 2 is divided into subsections, such as Sect. 2.1, which discusses the definition and characteristics of centrosymmetric, and Sects. 2.2, 2.3, and 2.4, which define the Hermitian signal, discrete-time Fourier transform formula, and Sinc function, respectively. We covered the use of centrosymmetric matrices in signal processing in Sect. 3.

2 Notation and Some Important Definitions

2.1 Definition of a Contra-Identity Matrix

A matrix J_n is said to be a contra-identity matrix of order n (it is also known as reverse identity matrix), if entries of J_n satisfies:

$$[J]_{i,j} = \begin{cases} 1, i+j = n+1 \\ 0, elsewhere \end{cases}. \tag{1}$$

2.2 Definition of Centrosymmetric Matrix and Its Properties

A $n \times n$ matrix A is said to be a centrosymmetric matrix if entries of A satisfy the following condition:

$$[A_{ij}] = [A_{n-i+1,n-j+1}], 1 \leq i, j \leq n. \tag{2}$$

Let $CSR^{n \times n}$ be set of all $n \times n$ centrosymmetric matrices. According to [2], $CSR^{n \times n}$ is an algebra over the felid F. Further important properties of $CSR^{n \times n}$ are: if $X \in CSR^{n \times n}$ with an $m-$ dimensional eigenbasis, then its m eigenvectors can be chosen so that they satisfy either $x = Jx$ or $x = -Jx$, where J is the reverse identity matrix. If $X \in CSR^{n \times n}$ having distinct eigenvalues, then any matrix that commutes with X must be centrosymmetric. If $X \in CSR^{n \times n}$ is a non-singular matrix, then $X^{-1}, X^T \in CSR^{n \times n}$.

2.3 Introduction of Hermitian Function

A Hermitian function is a complex function that has a property that its complex conjugate is equal to the original function with the sign of variable change i.e., for all $x \in$ domain of f,

$$f^*(x) = f(-x). \tag{3}$$

2.4 Discrete-Time Fourier Transform Formula

The discrete-time Fourier transform of a discrete sequence of real or complex number $f[n], \forall n \in Z$, is a Fourier series, which produces a periodic function of a frequency variable. The frequency variable ω, has normalized units of a sample, the periodicity is 2π, and the Fourier series is

$$X_{2\pi}(\omega) = \sum_{n=-\infty}^{\infty} f[n] e^{-i\omega m}. \tag{4}$$

2.5 Sinc Function

Sinc function is a short form of Sine cardinal and also called the sampling function that arises in signal processing and the theory of Fourier transforms. According to [10, 11], the Sinc function is defined as:

$$sinc(x) = \begin{cases} 1, \, for \, x = 0 \\ \frac{\sin(\pi x)}{\pi x}, \, otherwise \end{cases}. \tag{5}$$

3 Application of Centrosymmetric in the Interpolation of Hermitian Signal

We go through the centrosymmetric matrix's significant use in signal processing in this section. Here, we will demonstrate the elegant centrosymmetric matrix used by the suggested approach, which proves that if the given samples possess the Hermitian property, the interpolated samples also have. So, we use a bandlimited signal $f(t)$ with known samples that correspond to sampling time period T. Mathematically, the given sequence is:

$$f[n] = f(nT). \tag{6}$$

The following sequence is to identify interpolated samples that correspond to a high sampling rate:

$$f_m[n] = f\left(n\frac{T}{a_0^m}\right), \tag{7}$$

where a_0 and m are positive integers.

$$\text{Form (6) and (7)}, f[n] = f_m[a_0^m n]. \tag{8}$$

With a decimation ratio a_0^m, the given sequence f [n], may be seen as a truncated version of the necessary sequence. Using definition Sect. 2.3 now,

$$f[n] = \frac{1}{a_0^m} \frac{1}{2\pi} \int_{-\pi}^{\pi} e^{j\omega n} \sum_{k \in Z} f_m[k] e^{-j\frac{\omega}{a_0^m}k} d\omega. \tag{9}$$

The result of reversing the sequence of integration and summation in the equation above is

$$f[n] = \frac{1}{a_0^m} \sum_{k \in Z} f_m[k] \frac{1}{2\pi} \int_{-\pi}^{\pi} e^{j\omega n} e^{-j\frac{\omega}{a_0^m} k} d\omega. \tag{10}$$

On performing (10) and using Sinc function, we get the Sinc interpolation form:

$$f[n] = \frac{1}{a_0^m} \sum_{k \in Z} f_m[k] sinc\left(n - \frac{k}{a_0^m}\right). \tag{11}$$

Using definition of Sinc function (5), then Eq. (11) reduces to the following form

$$f[n] = \frac{1}{a_0^m} \left\{ f_m[a_0^m n] + \sum_{k \neq a_0^m n} f_m[k] sinc\left(n - \frac{k}{a_0^m}\right) \right\}.$$

Putting (8) in the above equation, we get

$$(a_0^m - 1)f[n] = \sum_{k \neq a_0^m n} f_m[k] sinc\left(n - \frac{k}{a_0^m}\right). \tag{12}$$

Now, using Eqs. (6) and (7), the (12) becomes

$$(a_0^m - 1)f(nT) = \sum_{k \neq a_0^m n} f\left(n\frac{T}{a_0^m}\right) sinc\left(n - \frac{k}{a_0^m}\right). \tag{13}$$

This is the important formula for finding $f(t)$ at $t = \frac{kT}{a_0^m}$, $k \neq pa_0^m$, (p integer) given $f(t)$ at $t = nT$. Expressing the integer k as $k = k_1 a_0^m + k_2$, where $k \neq pa_0^m$, which give $k_2 \neq 0$, then (13) can be written as

$$(a_0^m - 1)f(nT) = \sum_{k_1=k_{1min}}^{k_{1max}-1} \sum_{k_2=1}^{a_0^m-1} f\left(\left(k_1 + \frac{k_2}{a_0^m}\right)T\right) sinc\left(n - k_1 - \frac{k_2}{a_0^m}\right). \tag{14}$$

Here, we assume that the unknown $f(t)$ values outside of the range $k_{1min}T < t < k_{1max}T$ on the right-hand side of (14) are negligible. The equation with the same number of n values should be written in order to evaluate the unknown $(a_0^m - 1)(k_{1max} - k_{1min})$ on the right side of (14) and produce a square linear system of algebraic equations that can be solved for $f\left(\left(k_1 + \frac{k_2}{a_0^m}\right)T\right)$, $k_1 = k_{1min}, \ldots, k_{1max} - 1$ and $k_2 = 1, \ldots, (a_0^m - 1)$.

Now, let's assume that special signals have Hermitian properties and that we are taking them rather than general signals [13, 14]. The (14) may be expressed as using (2).

$$(a_0^m - 1)f(nT) = \sum_{k_1=-K}^{K-1} \sum_{k_2=1}^{a_0^m-1} f\left(\left(k_1 + \frac{k_2}{a_0^m}\right)T\right) sinc\left(n - k_1 - \frac{k_2}{a_0^m}\right). \tag{15}$$

The number of interpolated points $f(t)$ to be evaluated is $2N$, where $N = LK$, where $L = (a_0^m - 1)$. To find the $2N$ interpolated points of $f(t)$ with the help of the following two schemes:

First scheme: In this initial scheme, we substituted the $2N$ values $n = -N, ..., 1, 1, ..., N$ (it should be noted that $n = 0$ has been eliminated) in (15). The resultant matrix is provided as follows:

$$Ax = y, \tag{16}$$

where $y = \left(a_0^m - 1\right)[f(-NT)...f(-T)f(T)...f(NT)]^T, \tag{17}$

and $x = [x_{-K} ... x_{-1} x_0 x_1 ... x_{K-1}]^T$ be $2K-$ partitioned vector with the $k_1{}^{th}$ partition being the $\left(a_0^m - 1\right)-$ dimensional vector is given below:

$$x_{k_1} = \left[f\left(\left(k_1 + \frac{1}{a_0^m}\right)T\right) f\left(\left(k_1 + \frac{2}{a_0^m}\right)T\right) ... f\left(\left(k_1 + \frac{a_0^m - 1}{a_0^m}\right)T\right) \right]^T. \tag{18}$$

The matrix A given in (21) can be written as a partition matrix

$$A = \begin{pmatrix} P & Q \\ R & S \end{pmatrix}, \tag{19}$$

where P, Q, R, S of order N are the $N \times K$ partitioned matrix as follows:

$$P = \begin{bmatrix} G_{-N,-K} & G_{-N,-(K-1)} & \cdots & G_{-N,-1} \\ G_{-(N-1),-K} & G_{-(N-1),-(K-1)} & \cdots & G_{-(N-1),-1} \\ \vdots & \vdots & \ddots & \vdots \\ G_{-1,-K} & G_{-1,-(K-1)} & \cdots & G_{-1,-1} \end{bmatrix}, \tag{20}$$

$$Q = \begin{bmatrix} G_{1,0} & G_{1,1} & \cdots & G_{1,(K-1)} \\ G_{2,0} & G_{2,1} & \cdots & G_{2,(K-1)} \\ \vdots & \vdots & \ddots & \vdots \\ G_{N,0} & G_{N,1} & \cdots & G_{N,(K-1)} \end{bmatrix}, \tag{21}$$

$$R = \begin{bmatrix} G_{-N,0} & G_{-N,1} & \cdots & G_{-N,(K-1)} \\ G_{-(N-1),0} & G_{-(N-1),1} & \cdots & G_{-(N-1),(K-1)} \\ \vdots & \vdots & \ddots & \vdots \\ G_{-1,0} & G_{-1,1} & \cdots & G_{-1,(K-1)} \end{bmatrix}, \tag{22}$$

$$S = \begin{bmatrix} G_{1,-K} & G_{1,-(K-1)} & \cdots & G_{1,-1} \\ G_{2,-K} & G_{2,-(K-1)} & \cdots & G_{2,-1} \\ \vdots & \vdots & \ddots & \vdots \\ G_{-N,-K} & G_{-N,-(K-1)} & \cdots & G_{-N,-1} \end{bmatrix}, \tag{23}$$

where term of $\left(a_0^m - 1\right)-$ dimensional row vector is defined by

$$G_{p,q} = \left[sinc\left(p - q - \frac{1}{a_0^m}\right) sinc\left(p - q - \frac{2}{a_0^m}\right) ... sinc\left(p - q - \frac{a_0^m - 1}{a_0^m}\right) \right]. \tag{24}$$

Now, using the following lemma, we demonstrate that A is a centrosymmetric matrix in Theorem 1.

Lemma 1. [13] If A is a centrosymmetric matrix if and only if $J_n A J_n = A$.

Theorem 1. Matrix A defined by (19) is a centrosymmetric matrix.
    Proof: Post multiplying by J_L to (24), we obtain

$$G_{p,q} J_L = \left[sinc\left(q+1-p-\frac{1}{a_0^m}\right) sinc\left(q+1-p-\frac{2}{a_0^m}\right) \dots sinc\left(q+1-p-\frac{a_0^m-1}{a_0^m}\right) \right], \tag{25}$$

where $L = \left(a_0^m - 1\right)$.

By comparing (24) and (25), we get $G_{p,q} J_L = G_{q-1,p} = G_{q,p-1}$. $\tag{26}$

Now, from (19), we get $J_{2N} A J_{2N} = \begin{pmatrix} J_N S J_N & J_N R J_N \\ J_N Q J_N & J_N P J_N \end{pmatrix}$, $\tag{27}$

where $J_{2N} = \begin{pmatrix} 0 & J_N \\ J_N & 0 \end{pmatrix}$ and $J_N = \begin{pmatrix} & & J_L \\ & \cdot^{\cdot^{\cdot}} & \\ J_L & & \end{pmatrix}$.

Now, from (27), we obtain (i, j) elements of $J_N S J_N$ is given as

$$[J_N S J_N]_{i,j} = G_{N-i+1,K-j} J_L, i = 1, \dots, N, j = 1, \dots, K. \tag{28}$$

Using (26), (28) reduces to

$$[J_N S J_N]_{i,j} = G_{K-j,N-i}. \tag{29}$$

From (20), the (i, j) elements of P is given as

$$[P]_{i,j} = G_{-N-i-1,-K+j-1}. \tag{30}$$

Now, vector $G_{p,q}$ satisfy the following property $G_{p,q} = G_{-q,-p}$, and (30) can be rewritten as

$$[P]_{i,j} = G_{-N+i,-K+j} = G_{K-j,N-i}. \tag{31}$$

Hence, equating (29) and (31), we get

$$J_N S J_N = P. \tag{32}$$

Here, using the property of J is $JJ = I$, I is an identity matrix, (32) become

$$J_N P J_N = S. \tag{33}$$

In similar way, we can show that

$$J_N R J_N = Q. \tag{34}$$

$$\text{and } J_N Q J_N = R. \tag{35}$$

Putting (32)–(35) into (27), we get

$$J_{2N}AJ_{2N} = \begin{pmatrix} P & Q \\ R & S \end{pmatrix} = A. \tag{36}$$

By Lemma 1, A is centrosymmetric matrix.

Due to the fact that the known samples of $f(t)$ satisfy the Hermitian condition. As a consequence, the vector y in form (17) is a Hermitian vector.

$$Jy^* = y. \tag{37}$$

Now, pre-multiplying by J and taking conjugate of both side of (16), we get

$$(JAJ)(Jx^*) = Jy^*.$$

Here, it is noted that A is real matrix and using the property of J is $JJ = I$ in above. Theabove equation become

$$A(Jx^*) = Jy^*. \tag{38}$$

Taking inverse of A on both side of (38), we get

$$Jx^* = x. \tag{39}$$

Therefore, it follows from the above equation that x is also a Hermitian vector, indicating that the interpolated sample of $f(t)$ satisfies the Hermitian property. Second scheme: Taking the $2N$ values $= -N, \ldots, -1, 0, 1, \ldots, N$, and putting in (15), we get resulting $2N$ constraints in following array,

$$Bx = z, \tag{40}$$

where x is defined in (18) and z is $(2N + 1)$ dimensional vector

$$z = (a_0^m - 1)[f(-NT) \ldots f(-T) f(0) f(T) \ldots f(NT)]^T. \tag{41}$$

Now, matrix B of order $(2N + 1) \times 2N$ in (40) given as

$$B = \begin{bmatrix} P & Q \\ g & h \\ R & S \end{bmatrix}, \tag{42}$$

where P, Q, R, S matrices are given as form as (20)–(23) and the $N-$ dimensional vectors g and h are defined by

$$g = \begin{bmatrix} G_{0,-K} G_{0,-(K-1)} \ldots G_{0,-1} \end{bmatrix}, \tag{43}$$

$$h = \begin{bmatrix} G_{0,0} G_{0,1} \ldots G_{0,(K-1)} \end{bmatrix}, \tag{44}$$

where vector $G_{p,q}$ is defined in (24).

Theorem 2: Matrix B defined in (42) is centrosymmetric matrix.

Proof: Pre- and post-multiplying by J_{2N+1} and J_{2N}, then (42) become

$$J_{2N+1}BJ_{2N} = \begin{bmatrix} J_N PJ_N & J_N QJ_N \\ hJ_N & gJ_N \\ J_N RJ_N & J_N SJ_N \end{bmatrix},$$

(45)

where $J_{2N+1} = \begin{pmatrix} & & J_N \\ & 1 & \\ J_N & & \end{pmatrix}$, $J_{2N} = \begin{pmatrix} 0 & J_N \\ J_N & 0 \end{pmatrix}$.

From (44), using partition matrix $J_N = \begin{pmatrix} & & J_L \\ & \cdots & \\ J_L & & \end{pmatrix}$ and (45), and applying (26),

we get

$$hJ_N = \begin{bmatrix} G_{K,0} \ldots G_{2,0} G_{1,0} \end{bmatrix}.$$

(46)

Using the property of vector G i.e., $G_{p,q} = G_{-q,-p}$ and above vector reduces to

$$hJ_N = \begin{bmatrix} G_{0,-K} \ldots G_{0,-2} G_{0,-1} \end{bmatrix}.$$

(47)

Equating (43) and (47), we get

$$hJ_N = g.$$

(48)

Similarly, we obtain

$$gJ_N = f.$$

(49)

Putting (32)–(35), (48), (49) in (45) and using lemma 1, we get

$$J_{2N+1}BJ_{2N} = B.$$

(50)

Therefore, B is centrosymmetric matrix.

Now, the least square solution vector x_{LS} to (42) can be evaluated from the normal Eq. [15],

$$B^T B x_{LS} = B^T z.$$

(51)

According to (50) and the property of J is $JJ = I$, we get

$$J_{2N+1}B^T BJ_{2N} = B^T B.$$

(52)

The coefficient matrix of the normal Eq. (51) is centrosymmetric, as shown by (52). Assuming the B matrix is real, we pre-multiply the complex conjugate of (51) by J_{2N} and then use $JJ = I$ to get

$$\left(J_{2N}B^T BJ_{2N} \right) J_{2N}x_{LS}^* = \left(J_{2N}B^T J_{2N+1} \right) J_{2N+1}z^*.$$

(53)

By definition in (41) and the Hermitian property (2), we get

$$J_{2N+1}z^* = z. \tag{54}$$

Putting (50), (52) and (54) in (53), we obtain

$$B^T B\left(J_{2N}x_{LS}^*\right) = B^T z. \tag{55}$$

The matrix $B^T B$ will be a nonsingular matrix if rank of matrix B is full, and the least squares solution vector x_{LS} in (51) will be unique. The result is that (51) and (55) become

$$J_{2N}x_{LS}^* = x_{LS}. \tag{56}$$

From (56), it is clear that the interpolated sample of $f(t)$ will have Hermitian property.

4   Conclusion

This paper covers the crucial function that centrosymmetric matrices play in signal processing. The interpolation technique used in this work was inspired by the wavelet theory's scaling operation, which obtains $f(t/a_0^m)$ is from the mother wavelet $f(t)$ via scaling. The current method has the advantage of requiring less computational work compared to the widely used spline interpolation technique in [16, 17] because the values of the intermediate points do not need to be substituted after determining the coefficients of each spline segment connecting every pair of subsequent points.

Acknowledgments. The author(s) express their hearty thanks and gratefulness to all those scientists whose masterpieces have been consulted during the preparation of the present research article.

Disclosure of Interests. The author(s) declare(s) that there is no conflict of interest regarding the publication of this paper.

References

1. Aitken, A. C.: Determinants and matrices. Read Books Ltd. (2017)
2. Weaver, J.R.: Centrosymmetric (Cross-Symmetric) matrices, their basic properties, eigenvalues, and eigenvectors. Am. Math. Mon. **92**(10), 711–717 (1985). https://doi.org/10.1080/000 29890.1985.11971719
3. Datta, L., Morgera, S.D.: On the reducibility of centrosymmetric matrices—applications in engineering problems. Circuits Syst. Signal Process **8**(1), 71–96 (1989)
4. Sheppard, W.F.: Interpolation. In: Chisholm, H. (ed.) Encyclopædia Britannica, vol. 14, pp. 706–710. Cambridge University Press, Cambridge (1911)
5. Crochiere, R.E., Rabiner, L.R.: Interpolation and decimation of digital signals—a tutorial review. Proc. IEEE **69**(3), 300–331 (1981)

6. Xiong, R.G., Liu, C.M., Wang, H., You, X.Z.: Molecular recognition of organic non-linear optical material by bis (perchlorate) tris (phenanthroline) cadmium (II). Polyhedron **16**(7), 1263–1265 (1997)
7. Stanhill, D., Zeevi, Y.Y.: Two-dimensional orthogonal filter banks and wavelets with linear phase. IEEE Trans. Signal Process. **46**(1), 183–190 (1998)
8. Noor, F.: Inverse and Eigenspace decomposition algorithms for statistical signal processing (1993)
9. Wang, X.: New developments in differential quadrature analysis of structural components. In: Computational Mechanics '95, pp. 457–462. Springer, Berlin, Heidelberg (1995)
10. Woodward, P. M.: (2014) Probability and Information Theory, with Applications to Radar: International Series of Monographs on Electronics and Instrumentation (Vol. 3). Elsevier
11. Bracewell, R.: (1999) The Filtering or Interpolating Function, sincx. The Fourier Transform and its Applications, 62–64
12. Sengupta, N., Sahidullah, M., Saha, G.: Lung sound classification using cepstral-based statistical features. Comput. Biol. Med. **75**(1), 118–129 (2016). https://doi.org/10.1016/j.compbiomed.2016.05.013.PMID27286184
13. Chan, Y.T.: Multiresolution analysis, wavelets and digital filters. In: Wavelet Basics, pp. 53–110. Springer, Boston, MA (1995)
14. Vetterli, M., Kovacevic, J.: (1995) Wavelets and subband coding (No. BOOK). Prentice-hall
15. Stewart, G.W.: Introduction to Matrix Computations. Elsevier (1973)
16. Chapra, S.C., Canale, R.P.: Numerical Methods for Engineers, vol. 1221. Mcgraw-Hill, New York (2011)
17. James, M. L., Smith, G. M., Wolford, J. C.: Applied Numerical Methods for Digital Computation with FORTRAN and CSMP. Addison-Wesley Longman Publishing Co., Inc. (1977)

An Efficient Algorithm for Hadamard Product of Centrosymmetric Matrices

Vineet Bhatt[1($\boxtimes$)], Sunil Kumar[1], and Seema Saini[2]

[1] Department of Mathematics, Graphic Era Hill University, Dehradun 248002, India
vineet.bhatt58@gmail.com
[2] Department of Mathematics, Graphic Era (deemed to be) University, Dehradun 248002, India

Abstract. This paper introduces Hadamard product of centrosymmetric matrices. Such matrices are symmetric about its centre and Hadamard product is an element wise product, which is applicable in various fields of applications. This work provides an algorithm for conducting Hadamard product for centrosymmetric matrices. By using this algorithm, we able to calculate Hadamard product for centrosymmetric matrices and analysis some important fact of this product. The algorithm illustrates how the output of the Hadamard product of centrosymmetric matrices is influenced by its rank characteristics. A comparison with general product of matrices is also given in this study.

Keywords: Hadamard Product · Centrosymmetric · Singular-Nonsingular Matrices

1 Introduction

The matrix product holds great significance in the fields of science and technology. It serves not only as a tool for solving systems of linear equations but also finds applications in various engineering domains, including network technology, digital image compression technology, and coordinate system transformations, among others. If we discuss on some special type of matrix products like Hadamard product, for which some properties are different than general matrix multiplication [3]. Hadamard product is named on French mathematics 'Jacques Hadamard'. It is also known as 'Schur product' named on German mathematics 'Issai Schur'. In an article [1], author represented a theorem named Schur's Product Theorem which states that Hadamard Product of two positive semi-definite matrices is always positive semi-definite. If we discuss about centrosymmetric matrix which is symmetric about its centre, always gives interesting results when different kinds of operations are used on it [2, 4, 5]. Some authors worked on centrosymmetric matrices, which are useful to apply in different-different fields of mathematics, engineering and technology [7]. This article discusses unique variations of matrix multiplication, 'Hadamard product' for centrosymmetric matrices. In this article, we observe the properties of the Hadamard products of special types of centrosymmetric matrices. The primary emphasis lies on the outcome of the Hadamard product when applied to various types of centrosymmetric matrices. We have seen the result of Hadamard product

© The Author(s), under exclusive license to Springer Nature Switzerland AG 2025
M. Khurana et al. (Eds.): ICMLA 2024, CCIS 2238, pp. 239–250, 2025.
https://doi.org/10.1007/978-3-031-75861-4_21

on taking two centrosymmetric matrices, where one is singular and other is non-singular or if both are non-singular or are singular. The research point struck on our mind when we observe that for all $n \leq 2$, the Hadamard product of non-singular and singular centrosymmetric matrices is non-singular if singular matrix is other than null matrix of order two. If it then, the product of non-singular and singular centrosymmetric is singular but Hadamard product of tow singular centrosymmetric matrix is always singular for all $n \leq 2$, i.e., $\begin{bmatrix} \alpha & \beta \\ \beta & \alpha \end{bmatrix}$ is centrosymmetric non-singular, if $\alpha \neq \beta$ and singular only, if $\alpha = \beta$ or $\alpha = \beta = 0$. However, if we observe Hadamard product of centrosymmetric matrix of order $n > 2$, the result is different, for instance, the Hadamard product of two singular centrosymmetric matrix is non-singular, i.e.,

$$\begin{bmatrix} 7 & 3 & 7 \\ 4 & 0 & 4 \\ 7 & 3 & 7 \end{bmatrix} \odot \begin{bmatrix} -8.544 & 3 & 7 \\ 4 & -15.544 & 4 \\ 7 & 3 & -8.544 \end{bmatrix} = \begin{bmatrix} -59.8080 & 9 & 49.000 \\ 16 & 0 & 16 \\ 49.000 & 9 & -59.8080 \end{bmatrix}.$$

Now the question arise, how can we observe such type of product for any order of matrix and what conditions on a Hadamard product for two matrices to find singular or non-singular matrices? Can we generate a proper algorithm for calculating such type's products? Our main focus is to generate an algorithm for Hadamard product of centrosymmetric matrices of any order which is applicable in different fields of applications.

We gather all the results/data on applying the Hadamard product on mentioned matrices. This data is thus used to generate a theorem as well as a formula for Hadamard product on special kinds of centrosymmetric matrices. This paper is divided into sections parts. First section is introduction of article. Second contain some important results related to Hadamard Product, which are very useful for the next sections. This section is also associated with the discussion about a special type of matrix known as centrosymmetric matrix with its some significant conditions like positive definite, positive semi definite, negative definite and negative semi definite cases. Section three is related to the main result of this article. In this section every mathematical calculation and computational analysis has been performed. Section four has extensively covered the significant applications aligned with our vision for future endeavors. Section five contains conclusion based on this study. References and appendix are also mentioned in the last.

2   Some Important Definition and Operations

2.1   Hadamard Product

For two matrices of same dimensions, Hadamard Product is the element-wise product. For instance, let we have two matrices A and B of order $m \times n$, then Hadamard Product $(A \odot B)$ is $(A \odot B)_{ij} = (A_{ij})(B_{ij})$ [6]. Hadamard Product satisfies the commutatively, associativity and distributive properties under addition operation. If we want to know about the Hadamard Product of identity matrix with any matrix, we always find the resultant is a diagonal matrix. If we take two vectors $u \& v$ then their corresponding diagonal matrices will be $D_u \& D_v$ and their main diagonals will be as above-mentioned vectors respectively. The identity which holds good for such cases is

$$u^*(A \odot B)v = \text{tr}\left(D_u^* A D_v B^T\right),$$

where u^* is the conjugate transpose of u. In a particular case, if we use vector of ones then the resultant is; the trace of AB^T is the sum of all the elements in the Hadamard Product. The related result for squares of $A \& B$ is; the diagonal elements of AB^T is the row-sums of their Hadamard Product, i.e.,

$$\sum_i (A \odot B)_{ij} = \left(B^T A\right)_{jj} = \left(AB^T\right)_{ii}$$

Similarly, $(vu^*) \odot B = D_v A D_u^*$.

The Hadamard Product also satisfy the rank inequality, i.e.,

$rank(A \odot B) \leq rank(A) rank(B)$. Also, if $A \& B$ are the positive definite matrices, the Hadamard Product is valid for the following inequality: $\Pi_{i=k}^n \lambda_i(A \odot B) \geq \Pi_{i=k}^n \lambda_i(AB)$, $k = 1, 2, \ldots, n$, where $\lambda_i(A)$ is the i_{th} largest eigenvalue of A. Also, let $E \& F$ be the diagonal matrices, then

$$E(A \odot B)F = (EAF) \odot B = (EA) \odot (BF) = (AF) \odot (EB) = A \odot (EBF);$$

and if $x \& y$ are two vectors, then Hadamard Product of these vectors is same as the matrix multiplication of one vector by corresponding diagonal matrix of the other vector $x \odot y = E_x y$. Now, the Hadamard product of two singulars (i.e., determinant of the matrix is 0)$n \times n$ matrices, where $n \leq 2$, is always singular.

Example: Consider $A = \begin{pmatrix} 1 & 2 \\ 1 & 2 \end{pmatrix}$ and $B = \begin{pmatrix} 3 & 12 \\ 2 & 8 \end{pmatrix}$, such that both A and B are singular

$as |A| = 0$ and $|B| = 0$, and their Hadamard product $A \odot B = \begin{pmatrix} 3 & 24 \\ 2 & 16 \end{pmatrix}$, whose determinant is also zero, therefore, it is a singular matrix. Similarly, the Hadamard Product of two non-singular matrices of order 2 is always non-singular and the Hadamard product of a singular and a non-singular matrix of order 2 is again non-singular. Similar outcomes are achieved for the Hadamard product of two singular lower triangular matrices and two upper triangular matrices for order 2. Now, a question arises here; Do we have two singular matrices whose order is 2 but their Hadamard product is non-singular? Similarly, if we talk about two singular $n \times n$ matrices, where $n = 3$ and after applying the Hadamard product, two cases are formed. One, when the Hadamard product is singular and second, when the Hadamard product is non-singular. When we talk about the Hadamard Product of two non-singular matrices of order 3, the result is always non-singular. Now another question arises that in this case, does it hold true for all values of n? i.e., the Hadamard product of any two non-singular matrices of any order will always be non-singular? Now one main topic which we want to discuss is that for the above-mentioned analysis if we replace the ordinary singular and non-singular matrices with the special types of centrosymmetric matrices, then what will our result look like?

2.2 Centrosymmetric Matrix

Matrices that exhibit symmetry around their center are referred to as centrosymmetric matrices [2]. A square matrix $A = (a_{ij})(n \times n)$ is centrosymmetric, when its (i, j)th elements satisfy the conditions:

$$aij = a(n - i + 1, n - j + 1) \text{ for } i \geq 1, n \geq j.$$

or we can say that a matrix A is centrosymmetric if and only if $A \times J_n = J_n \times A$, where Jn is the reverse identity matrix of an identity matrix In denoted by:

$$\begin{pmatrix} 000\ldots001 \\ 000\ldots010 \\ 000\ldots100 \\ \ldots \\ 100\ldots000 \end{pmatrix}_{n \times n}.$$

Lemma 2.1: Let $A = \left[a_{ij}\right]_{n \times n}$ be a centrosymmetric matrix in the following form: $A = \begin{pmatrix} A_1 & J_mA_2J_m \\ A_2 & J_mA_1J_m \end{pmatrix}$. Such that, for $B, C \in R^{m \times m}$, $a_1, a_2 \in R^{m \times 1}$, we have

$$Q_1^T AQ_1 = \begin{pmatrix} A_1 - J_mA_2 & 0 \\ 0 & A_1 + J_mA_2 \end{pmatrix}, \text{ where } Q_1 =$$

$$\frac{1}{\sqrt{2}} \begin{pmatrix} I_m & I_m \\ -J_m & J_m \end{pmatrix}.$$

Theorem 2.1: Let there be two centrosymmetric matrices $A = \left[a_{ij}\right]_{n \times n}$ and $B = \left[b_{ij}\right]_{n \times n}$ where, $A = \begin{pmatrix} A_1 & J_mA_2J_m \\ A_2 & J_mA_1J_m \end{pmatrix}$ and $B = \begin{pmatrix} B_1 & J_mB_2J_m \\ B_2 & J_mB_1J_m \end{pmatrix}$, then $A \odot B$ is a centrosymmetric matrix.

Proposition 2.1: Let there be two positive centrosymmetric matrices $A = \left[a_{ij}\right]_{n \times n}$ and $B = \left[b_{ij}\right]_{n \times n}$ then,

$$\rho(A \odot B) \leq \rho\left(A_1^T B_1\right) + m^{\frac{1}{p}} \rho\left(A_2^T B_2\right) \text{ for } n = 2m, \text{ and}$$

$$\rho(A \odot B) \leq \max\{\|K\|_\rho . \|A_1 \odot B_1 - J_m(A_2 \odot B_2)\|_\rho\} \text{ for } n = 2m + 1.$$

Theorem 2.2: Let A and B be two centrosymmetric matrices such as

$$A = \begin{pmatrix} P - J_mS & 0 \\ 0 & P + J_mS \end{pmatrix} \text{ for } n = 2m \text{ and } B = \begin{pmatrix} P - J_mS & 0 \\ 0 & K \end{pmatrix} \text{ for } n = 2m + 1.$$

and let $2 \leq p < \infty$, then for $n = 2m$ and $n = 2m + 1$ respectively, we have

$$\rho(A \odot B) \leq 2^{\frac{1}{p}}(\rho - 1)\left(1 + m^2\right)\left(\|P\|_\rho^2 + \|S\|_\rho^2\right)^{\frac{1}{2}}$$

$$\rho(A \odot B) \le ||P - J_m S||_\rho + 2^{\frac{1}{p}} \left(\frac{(\rho - 1)}{2} \left(\alpha^2 \beta^2 + 2M^2 + ||P + J_m S||_\rho^2 \right) + \frac{2 - \rho}{4} \left(\alpha \beta + ||P + J_m S||_\rho \right)^2 \right)^{\frac{1}{2}}$$

where $P = A_1 \odot B_1$, $S = A_2 \odot B_2$, and

$$M = \{ \frac{1}{2} \left(\left\| \sqrt{2} J_m (a_2 \odot b_2) \right\|_p^p + \left\| \sqrt{2} \left(a_1^T \odot b_1^T \right) \right\|_p^p \right\}^{\frac{1}{p}}.$$

2.3 Special Conditions Applied on Centrosymmetric Matrix

The Hadamard Product has been utilized on the aforementioned unique centrosymmetric matrices, and an analysis of their outcomes will be conducted. Presently, it is observed that:

a) The question arises here is that, do we have two singular matrices whose order $(1,2, 3, \dots)$ but their Hadamard product is non-singular?

b) Does the Hadamard product of two singular centrosymmetric matrices result in a singular centrosymmetric matrix?

Let $A = \begin{pmatrix} \beta & \varphi & \gamma \\ \delta & \alpha & \delta \\ \gamma & \varphi & \beta \end{pmatrix}$, where $n = 3$; Matrix A will be singular if $|A| = 0$;

i.e., $\beta = \varphi = \gamma = 0 \, or \, \alpha = \beta = \gamma = \delta = \varphi \, at least \, one \, \lambda_i = 0 (i.e \, at least \, one \, eigen \, value \, of \, A \, is \, zero)$

Example: $\begin{pmatrix} 0 & 0 & 0 \\ 0 & 5 & 0 \\ 0 & 0 & 0 \end{pmatrix} \odot \begin{pmatrix} 1 & 0 & 0 \\ 0 & 1 & 0 \\ 0 & 0 & 1 \end{pmatrix} = \begin{pmatrix} 0 & 0 & 0 \\ 0 & 5 & 0 \\ 0 & 0 & 0 \end{pmatrix}$. Here the resultant matrix is singular

centrosymmetric matrix and its rank is 1. Similarly, if we take the Hadamard Product of any two singular centrosymmetric matrices of order $n = 3,4, 5, \dots, and \, so \, on$, then the resultant matrix will always be singular centrosymmetric matrix. In this case, the rank of the resultant matrix depends on the behavior of the input matrices. In every input matrix there will be rows of zeros i.e., in the above example the first input matrix has 2 rows of zeroes. So, the rank of the resultant matrix will be the $n - (No. \, of \, zero \, rows)$. The rank can also be zero if the rows of zeros of one input matrix are interchanged with the non-zero rows and are formed as the other input matrix, the rank of the resultant matrix will be,

$$r = \begin{cases} n - (No. \, of \, zero \, rows), \\ 0. \end{cases}$$

c) Is the Hadamard product of two non-singular centro-symmetric matrices is singular or non-singular centro-symmetric?

Let $A = \begin{pmatrix} x & y & z \\ q & p & q \\ z & y & x \end{pmatrix}$ be a 3×3 centro-symmetric matrix.

The condition to form a non-singular centro-symmetric matrix is that the $|A| \ne 0$;

i.e.

$$|A| = x(px - qy) - y(qx - qz) + z(qy - pz)$$

$$= x^2 p - xyq - yqx + yqz + zqy - pz^2$$

$$= x^2 p - pz^2 - 2xyq + 2yzq$$

$$= p\left(x^2 - z^2\right) - 2yq(x - z)$$

$$= (x - z)\{p(x + z) - 2yq\}$$

$$= (x - z)\{px + pz - 2yq\},$$

Therefore,

- If $x \neq z$ then $|A| \neq 0$

  or

- If $p \neq 0$ and y or $q \neq 0$ then $|A| \neq 0$

  or

- If $x \neq -z$ and y or $q \neq 0$ then $|A| \neq 0$

We also want to know about the conditions on a pair of centro-symmetric positive semi-definite matrices are sufficient to ensure their Hadamard product is non-singular.

3   The Hadamard Product of Centrosymmetric Matrix

Theorem 3.1: If there are two centro-symmetric matrices of order $n \times n$ such that

a) First matrix is singular and second is non-singular then their Hadamard product is always non-singular, when the elements of the above matrices are generated by the rule $(a_{ij} \pm \alpha)$, where a_{ij} are the elements of input centrosymmetric matrix.

Above theorem is true for matrices of order $n = 4, 5$ and6. Our analysis strikes on the point that, will the above statement satisfy for every value of n. If this is true then what happens when the matrix is positive definite, positive semi-definite and negative definite type centro-symmetric matrix.

3.1   Rank of Special Types Matrices Using Hadamard Product

Theorem 3.2: Let $A, B \in M_n$ be positive semidefinite such that A and B have no zero main diagonal entries and $\max\{k_A + r_B, r_A + k_B\} > n$, then $A \odot B$ is positive definite, where k_A and k_B are the Kruskal rank of matrices A and B respectively and r_A and r_B are rank of matrices A and B respectively [8].

Corollary: Suppose that $A, B \in M_n$ are positive semidefinite and have no zero main diagonal entries, then

$$r_{A \odot B} \geq \min\{k_A + r_B - 1, n\}$$

and

$r_{A \odot B} \geq \min\{r_A + k_B - 1, n\}$. In particular, if $k_A \geq 2$ and B is singular, then $r_{A \odot B} > r_B$.

3.2 Algorithm for Hadamard Product for Centrosymmetric Matrices

Analysis for the outcomes of an algorithm: In this algorithm, we have denoted matrix order by n. Always generated matrices will be centrosymmetric, whenever, one fixes the value of n, the generated matrix will be centro-symmetric $n \times n$. If we come to some important analyses: for $n = 2$, the generated matrix order is 2×2 and the generated Hadamard product is singular only for the cases when both are singular or one of them is non-singular. For instance, we have two examples, $\begin{bmatrix} 0 & 0 \\ 0 & 0 \end{bmatrix} \odot \begin{bmatrix} 6 & 0 \\ 0 & 6 \end{bmatrix} = \begin{bmatrix} 0 & 0 \\ 0 & 0 \end{bmatrix}$ and

$\begin{bmatrix} 7 & 7 \\ 7 & 7 \end{bmatrix} \odot \begin{bmatrix} 7 & 7 \\ 7 & 7 \end{bmatrix} = \begin{bmatrix} 0 & 0 \\ 0 & 0 \end{bmatrix}$. There are some examples in which Hadamard product of two singular matrix is non-singular,

$$\begin{bmatrix} 6 & 2 & 6 \\ 6 & 0 & 6 \\ 6 & 2 & 6 \end{bmatrix} \odot \begin{bmatrix} -7.7460 & 2 & 6 \\ 6 & -13.7460 & 6 \\ 6 & 2 & -7.7460 \end{bmatrix} = \begin{bmatrix} -46.4758 & 4 & 36.000 \\ 36 & 0 & 36 \\ 36 & 4 & -46.4758 \end{bmatrix}.$$

Similarly, we can observe such types another cases for $n = 4$.

$$\begin{bmatrix} 4 & 7 & 4 & 9 \\ 9 & 8 & 3 & 4 \\ 4 & 3 & 8 & 9 \\ 9 & 4 & 7 & 4 \end{bmatrix} \odot \begin{bmatrix} -20 & 7 & 4 & 9 \\ 9 & -16 & 3 & 4 \\ 4 & 3 & -16 & 9 \\ 9 & 4 & 7 & -20 \end{bmatrix} = \begin{bmatrix} -80 & 49 & 16 & 81 \\ 81 & -128 & 9 & 16 \\ 16 & 9 & -128 & 81 \\ 81 & 16 & 49 & -80 \end{bmatrix}.$$

According to our analysis we can observe that for $n > 5$, we can't arrange a singular Hadamard product of any two centro-symmetric matrices. Also, some orders matrices in which we can't generated singular matrices such as $n = 11$.

Example: According the given MATLAB code, for $n = 6$, matrix

$$B = \begin{pmatrix} 0 & 0 & 0 & 0 & 0 & 0 \\ 0 & 0 & 0 & 0 & 0 & 0 \\ 0 & 0 & 0 & 0 & 0 & 0 \\ 0 & 0 & 0 & 0 & 0 & 0 \\ 0 & 0 & 0 & 0 & 0 & 0 \\ 0 & 0 & 0 & 0 & 0 & 0 \end{pmatrix}, \text{ if } n = 6 \text{ the value of } r = 0 \text{ and } q = 3.$$

$$A = \begin{pmatrix} 8 & 9 & 1 & 9 & 6 & 0 \end{pmatrix} \text{ and } C = \begin{pmatrix} 0 & 6 & 9 & 1 & 9 & 8 \end{pmatrix}$$

Generated non-singular centrosymmetric matrix $B = \begin{pmatrix} 8 & 9 & 1 & 9 & 6 & 0 \\ 2 & 5 & 9 & 9 & 1 & 9 \\ 9 & 4 & 8 & 1 & 4 & 9 \\ 9 & 4 & 1 & 8 & 4 & 9 \\ 9 & 1 & 9 & 9 & 5 & 2 \\ 0 & 6 & 9 & 1 & 9 & 8 \end{pmatrix}$, $|B| =$

910805

Set of eigenvalues of B is

$E = [4.2787 + 0.0000i - 5.6393 + 6.3100i - 5.6393 - 6.3100i 6.0000 + 4.1231i 6.0000 - 4.1231i 7.0000 + 0.0000i]$

$$G = \begin{matrix} 34.2787 & 0 & 0 & 0 & 0 & 0 \\ 0 & 34.2787 & 0 & 0 & 0 & 0 \\ 0 & 0 & 34.2787 & 0 & 0 & 0 \\ 0 & 0 & 0 & 34.2787 & 0 & 0 \\ 0 & 0 & 0 & 0 & 34.2787 & 0 \\ 0 & 0 & 0 & 0 & 0 & 34.2787 \end{matrix}$$

$$H = \begin{matrix} -26.2787 & 9.0000 & 1.0000 & 9.0000 & 6.0000 & 0 \\ 0.2.0000 & -29.2787 & 9.0000 & 9.0000 & 1.0000 & 9.0000 \\ 9.0000 & 4.0000 & -26.2787 & 1.0000 & 4.0000 & 9.0000 \\ 9.0000 & 4.0000 & 1.0000 & -26.2787 & 4.0000 & 9.0000 \\ 9.0000 & 1.0000 & 9.0000 & 9.0000 & -29.2787 & 2.0000 \\ 0.0000 & 6.0000 & 9.0000 & 1.0000 & 9.0000 & -26.2787 \end{matrix}$$

H is a singular matrix. Hadamard product of B and H is non-singular, i.e.,

$$\begin{pmatrix} -210.2295 & 81.0000 & 1.0000 & 81.0000 & 36.0000 & 0 \\ 4.0000 & -146.3935 & 81.0000 & 81.0000 & 1.0000 & 81.0000 \\ 81.0000 & 16.0000 & -210.2295 & 1.0000 & 16.0000 & 81.0000 \\ 81.0000 & 16.0000 & 1.0000 & -210.2295 & 16.0000 & 81.0000 \\ 81.0000 & 1.0000 & 81.0000 & 81.0000 & -146.3935 & 4.0000 \\ 0 & 36.0000 & 81.0000 & 1.0000 & 81.0000 & -210.2295 \end{pmatrix}$$

Hadamard Product of Singular matrix H and nonsingular matrix B is non-singular.

3.3 Comparison

Simple Product	Hadamard Product
Number of columns of the first matrix should be equal to the number of rows of the second matrix	Dimensions of the two matrices should be same
Not Commutative	Commutative
The matrix which has diagonal elements are 1 and all other are 0 is identity matrix	The matrix which has all the elements as 1 is the Identity matrix
Inverse under Simple product is possible only when the determinant of the matrix is non-zero	Inverse under Hadamard product is only possible if and only if none of the elements are equal to 0
Simple Product of an upper triangular matrix and lower triangular matrix does not give a particular kind of matrix	Hadamard Product of an upper triangular matrix and lower triangular matrix yields a diagonal matrix
Simple product of two symmetric matrices is not always symmetric	Hadamard product of two symmetric matrices is always symmetric
Simple product of two centro-symmetric matrices is not always symmetric	Hadamard product of two centro-symmetric matrices is always centro-symmetric

4 Application

There are many applications of Hadamard Product and centro-symmetric matrices. The unique properties of these two make them very special. A centro-symmetric matrix is applicable in different varieties of work, like speech analysis, electric and mechanical systems, quantum physics, pattern recognitions, etc. [9, 10, 14, 15]. It is also used in complex algorithms to find their computational effective outcomes. Similarly, Hadamard Product is very useful in different fields. It is used to calculate the generating functions of random sequences. It is also applied in image compression techniques such as JPEG and LSTM (Long Short-Term Memory) cells of Recurrent Neural Networks (RNNs). Thus, is not only used in problems of enumerative combinatory and discrete probability theory, but also in problems of complex analysis, linear programming, and mathematical physics. The author [9] seeks to establish a structured approach for restoring different types of image degradation using the Hadamard product. This method aims to produce a clear image by considering the degradation factors involved. Moreover, studies have revealed the implicit possibilities within latent domains by examining datasets [9]. Hadamard product is also applicable in weighted Bergman spaces [10]. Author [11] also set an application of hadamard powers of line, which is applicable in star configurations. In [12] author worked on centrosymmetric matrices for left-right eigenvalue problem under a centrosymmetric sub-matrix. [13] worked the convex set of nonnegative centrosymmetric matrices, in which given a method for finding extreme point of polytope of all $n \times n$ centro-symmetric doubly sub-strochastic matrices.

5 Conclusion

The present study investigates the Hadamard product of various kinds of centro-symmetric matrices and their characteristics. Additionally, it demonstrates the contrast between the standard product and the Hadamard product of centro-symmetric matrices. Moreover, a MATLAB program has been created to compute the Hadamard product of centro-symmetric matrices and derive certain theorems related to them. With the aid of the MATLAB code developed, it is feasible to generate centro-symmetric matrices of any order and compute the Hadamard product for matrices of any order effortlessly. The provided MATLAB script can be utilized to identify unique Hadamard products where the outcome is non-singular despite having both singular and non-singular matrices as inputs. By examining the Hadamard product of various matrix types, specific criteria can be established to ascertain whether the resulting matrices are singular or non-singular. Additionally, the rank characteristics of the product of two singular centro-symmetric matrices are dependent on the properties of the input matrices.

Acknowledgments. The authors express their heartfelt thanks and appreciation to all the researchers whose outstanding contributions have been cited in the making of this research paper.

Disclosure of Interests. No conflicts of interest associated with this paper.

Appendix

```
Appendix File:
clear all; close all;
clc;
for j=1:100
n = 10;
B = zeros(n,n);
r = rem( n , 2 );
q = fix(n/2);
if r~=0
    for i = 1:q+1
        if i<=q
            A = randi([0 9],1,n);
            C = fliplr(A);
            B(i,:)=A;
            B(n+1-i,:)=C;
        end
        if i==q
            A = randi([0 9],1,q);
            C = fliplr(A);
            D = randi([0 9],1,1);
            B(q+1,:)=[A D C];
        end
    end
end
if r==0
    for i = 1:q
        A = randi([0 9],1,n);
        C = fliplr(A);
        B(i,:)=A;
        B(n+1-i,:)=C;
    end
end
if det(B)==0
    fprintf('Generated matrix is singular\t\t\n');
    disp(B);
    sprintf('Determinat is = %16.f',det(B))
        else
            fprintf('Generated matrix is non-singular\t\n');
            disp(B);
            sprintf('Determinat is = %16.f',det(B))
end

E(1,:)=eig(B);
G=zeros(n,n);
for i=1:n
    if isreal(E(1,i))==1
            G=(E(1,i)).*eye(n);
            break
    end
end
    %disp(G);
    %disp('Corresponding singular matrix is');
    H=B-G;
    sprintf('Determinat is = %16.f',det(H))
        O = B.*H;
        sprintf('Hadmard Product is :')
        disp(O);
        if abs(det(O))<=0.00001
            fprintf('Hadmard Product is singular\n');
        else
            fprintf('Hadmard Product is non-singular\n');
        end
end
```

References

1. Horn, R.A., Johnson, C.R.: Matrix Analysis, 2nd edn. Cambridge University Press (2013)
2. Bhatt, V., et al.: A note on factorization for centrosymmetric real matrix that preserves centro-symmetry. Electron. J. Math. Anal. Appl. **8**(2), 272–290 (2020)
3. Weaver, J.R.: Centrosymmetric (cross-symmetric) matrices, their basic properties, eigenvalues, and eigenvectors. Am. Math. Mon. **92**(10), 711–717 (1985)
4. Liu, Z.Y.: Some properties of centrosymmetric matrices. Appl. Math. Comput. **141**(2–3), 297–306 (2003)
5. Zhao, L.: Submatrix constrained left and right inverse eigenvalue problem for centrosymmetric matrices. Inverse Prob. Sci. Eng. **20**(10), 1–17 (2020)
6. Sun, L., Zheng, B., Zhou, J., Yan, H.: Some inequalities for the Hadamard product of tensors. Linear Multilinear Algebra **66**(6), 1199–1214 (2018)
7. Datta, L., Morgera, S.D.: On the reducibility of centrosymmetric matrices—applications in engineering problems. Circuits Syst. Signal Process. **8**(1), 71–96 (1989)

8. Horn, R.A., Yang, Z.: Rank of a Hadamard product. Linear Algebra Appl. **591**, 87–98 (2020)
9. Wang, Y., Ma, L., Liu, R.: A convergent framework with learnable feasibility for Hadamard-based image recovery. Comput. Vision Image Understanding. **202**, 103095 (2021)
10. Karapetrović, B., Mashreghi, J.: Hadamard products in weighted Bergman spaces. J. Math. Anal. Appl. **494**(2), 124617 (2021)
11. Bocci, C., et al.: Hadamard products of linear spaces. J. Algebra. **448**, 595–617 (2016)
12. Zhao, L.: Submatrix constrained left and right inverse eigenvalue problem for centrosymmetric matrices. Inverse Probl. Sci. Eng. **29**(10), 1412–1428 (2021)
13. Chen, Z., Cao, L., Koyuncu, S.: The extreme points of centrosymmetric transportation polytopes. Linear Algebra Appl. **608**, 214–235 (2021)
14. Atar, B., et al.: Hadamard products and binomial ideals. J. Pure Appl. Algebra **228**(6), 107568 (2024)
15. Guo, C., Chen, C.H., Hwang, F.J., Chang, C.C., Chang, C.C.: Multi-view spatiotemporal learning for traffic forecasting. Inf. Sci. **657**, 119868 (2024)

Investigation on Combined Impacts of Different Clustering Techniques and Enhanced K-means Algorithm

Neeshu Sharma[(✉)] and Rohit Katyal

University Institute of Engineering and Technology, Chandigarh University, Mohali 140413,
India
neeshusharma.cse@cumail.in

Abstract. In the Proposed work, firstly we are discussing the various techniques of clustering and most widely used k-means clustering Algorithm and then design a hybrid technique from various enhancements done on k-means Algorithm that improves the performance of the system. Clustering techniques are used to implement unsupervised learning. In clustering K-means algorithm with various advancements is used for making clusters in large datasets. There are some limitations in basic k-means Algorithm like choosing of number of clusters initially before making the clusters and finding the initial centroids. In this work elbow technique using sum of square errors (SSE) is used for calculating number of clusters and Back Propagation Algorithm is used to do the iterations to check the Euclidian distance of different observations with centroids. By using this strategy, performance of the system is increased in terms of reducing the number of iterations for large datasets and all the data sets are comprised properly in different clusters. After studying all enhancements of k-means Algorithm, a hybrid Algorithm is designed that works efficiently for large datasets and scaling is used so that the unclustered datasets should also be included in clusters.

Keywords: Arithmetic Mean · Back Propagation · Clustering · K-Mean

1 Introduction

Data mining is mechanism of uncovering inquisitive arrangement from a large dataset which can be reside in databases, information distribution centers and storehouses of data [1]. There are different key features of data mining like patter recognition, predicting the behavior of data, extracting information from data, making clusters of datasets etc. Data mining incorporates as an absorption machines through various strategies, for instance, insights, database innovation, machine learning, neural network systems, immense accomplishment make sense of and pattern recognition, data perception, recuperation and so on. This zone clarifies the different strides in KDD plan. The term KDD, implies expansive system of predicting patterns, uncover useful information from data, and emphasize different data mining. Methods to extract knowledge from large datasets. Knowledge data discovery (KDD) is very important and crucial step in number of different applications like Customer Segmentation, Recommender Systems, Decision Support Systems and classification [2]. It is nontrivial technique to search the relationship between different data items, to identify useful information from dataset. Nontrivial means knowledge data discovery (KDD) is an iterative process to analyze the hidden relationships in different attributes of data. It involves different steps like data selection, data pre-processing, data transformation, data mining, pattern evaluation, knowledge representation. The techniques of knowledge data discovery are most widely work with unstructured data. All missing values are filled and irrelevant entries are deleted from the dataset in data cleaning process. Data cleaning process is very important in knowledge data discovery (KDD) because false interpretation gives inaccurate results. If there are more attributes in the dataset then it's very complex to analyze the data and plot that data to extract some useful information. So scaling is performed on the dataset to align all the points in a particular range.

In data transformation dataset is transformed into required form for mining process by which relevant data is collected which is used to uncover the relationship of attributes of data. Different models and historical records with clustering and classification are used to represent the patterns in dataset. Finally patterns are recognized; trends are identified and converted into the human understandable form [3].

KDD has been extraordinarily fascinating theme for the researchers as it become the important stride in many fields like AI, machine learning, pattern recognition, customer segmentation etc. Traditionally, manual analysis is used to convert data into knowledge but nowdays digital data is growing rapidly and require a new tool or computatitonal theory to extract some useful information from large data sets.

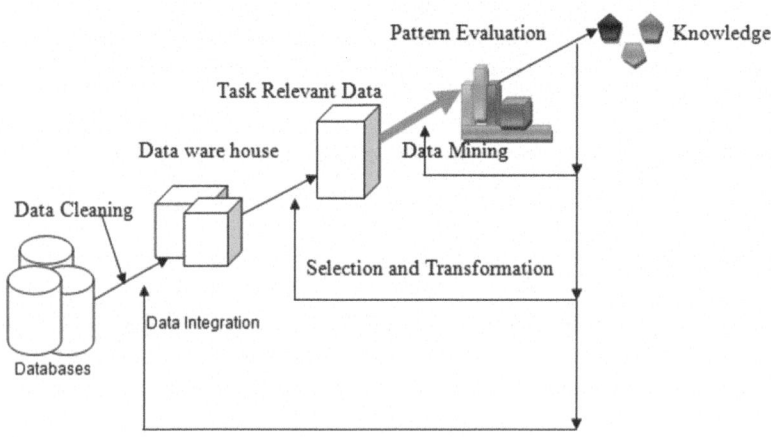

Fig. 1.1 KDD process

1.1 Clustering in Data Mining

Clustering is a hard problem for high dimensional space datasets. It looks very easy for two dimensional space datasets and where the attributes in the data sets are less.

Clustering is a methodology by which different groups are made based on some attributes of data sets, the objects which are similar they are contained in same cluster and the object of one cluster is different in behavior with the object of another cluster. In business, in view of acquiring patterns, clustering can assist advertisers with discovering interests of their clients and describe them into groups of clients.

Clustering is an unsupervised learning technique that functions on unlabeled and unstructured datasets. Based on the attributes of dataset different groups are made. Scaling technique is used to integrate all data points in a particular range. Clustering is a strategy which is utilized to categorize different music files. Search engines also used clustering techniques to categorize documents based on set of keywords. In real word applications points are in high dimensional space and grouping of points are defined using a distance measure like Jaccard distance, Euclidean distance, Cosine distance [4]. Sometimes clusters are overlap with each other and mix into each other then it is very hard to find the boundary of clusters.

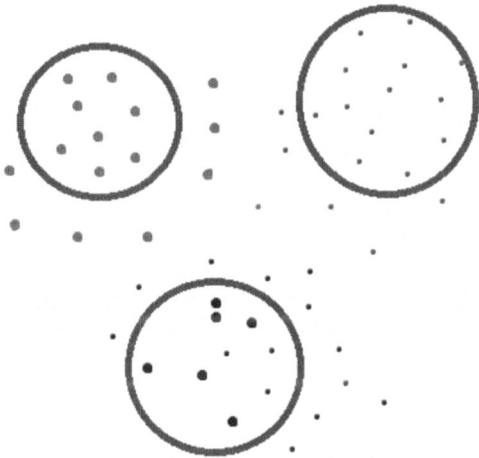

Fig. 1.2 Output of clustering

2   Clustering Methods

Partitioning Methods: The essential idea of partitioning includes the gathering of test samples in clusters. These clusters might have substantial measure of divergence among definite clusters. The greater part of the partitioning strategies depends on distance based techniques. For instance, suppose k to be the amount of sections which are utilized to assemble the clusters. An underlying advance for partitioning is first taken by this strategy and further iterative reallocation framework is utilized which enhances the partitioning. The change or upgrade is finished by moving the items from one cluster and after that from another. In this partitioning, the items that are available in a comparable cluster are identified with each other. Nonetheless, the items that are available in various clusters are assorted. Methodologies like greedy algorithm which include k-means and k-mediod algorithm is utilized as a part of different applications. For this, a piece of data set is taken which contains n items in an arrangement of k clusters. This is done to limit the Θ criterion. Here, finding a fragment of k clusters which enhance the picked partitioning measure is the principle point. The input parameter is k and the cases are k-means and k-centroid [5].

Fig. 2.1 Partitioning Clustering [5]

Hierarchical Methods: The hierarchical decomposition should be possible by both of two methodologies i.e. by agglomerative approach or divisive based approach. The agglomerative approach is a base strategy which begins by uncovering each item making a different group and after that combining the groups which are nearer to each other. This is preceded until the point when every one of the groups is combined and one cluster is shaped. The divisive approach is a kind of top down approach. It begins when the clusters that is like to each other. These clusters are isolated or split into littler bunches in every emphasis step. This is done until the point when each item is shaped in one cluster. Dendrogram, a tree like structure is utilized for speaking to the hierarchical decomposition. The clusters which are utilized as data sources are not to be included by any stretch of the imagination. Partitions are seen at various levels of granularities in this sort of clustering. This is finished utilizing different sorts of K and the cases are Level Clustering [6].

Density Based Methods: The vast majority of the clustering objects partitioning procedures are rely on the distance between the articles. These methods can be utilized to find the spherical formed clusters. The finding clusters of self-assertive shapes are a touch of intense undertaking to perform. Another technique known as density based strategy which is utilized for the self-assertive shapes. The cluster continues developing with the length of density in these techniques. Alongside this the edge of the framework is likewise expanded. The idea of density is the essential thing of this strategy. The primary idea along these lines depends in the way that each point display in the cluster has a radius which needs to have in any event risen to base number of focuses. The subjective molded clusters are found through this method. This system likewise handles the noise display in the data. The examination of this strategy is done once a while. The density parameters are likewise required also [6].

Grid Based Methods: A grid structure is framed utilizing a limited number of cells introduce in the applications. The space is quantized into grid structure by the framework based techniques. This technique is quick. This system just relies on number of cells which are available in each measurement of quantized space. The objects assemble to

Fig. 2.2 Density based clustering

frame a grid and further the cells acquired shape a structure known as grid. The density of the entire zone is figured and the grid cells are doled out to objects. The object which has the density underneath the threshold value is dispensed with. The cluster of density groups is shaped here and furthermore because of the evasion of distance calculations, the procedure is speedier. The cluster which is available in the neighbor is anything but easy to be resolved and the shapes are association. The complexity of the clustering depends totally on the grouping if the gathering of cells [7]. The space is register into limited number of grids in the grid based approach and the activities are performed here. These methodologies have preference of quick handling time.

K-Means Clustering Algorithm: The division technique is used by the k-mean clustering algorithm. This technique is useful for clustering various work and best for the ones with low measurement datasets. The k is utilized as a parameter which is utilized to confine n objects into k clusters [8]. These articles are then moved to comparative clusters where the items introduce are comparative in each cluster, in any case, disparate in various clusters. The cluster focuses (C1… Ck) are discovered until the point when the sum of the squared distances between each datum point xi, $1 \leq i \leq N$, and the nearest cluster focus Cj, $1 \leq j \leq k$, is reduced. As a matter of first importance the k objects are chosen haphazardly and every one of them speak to a cluster mean or focus. Furthermore, each item xi in the data collection is approved to the cluster focus that is nearest to it [9]. This cluster focus is the most near focus accessible. The most recent mean of each cluster is figured and each object is reassigned to the nearest most recent focus. The emphasis proceeds till each difference in the task of articles has been finished. The sum of squares error is the sum of the squared distances from each observation to its cluster focus. **Algorithm:** The k-means algorithm is a partitioning method in which the mean value of the items in a cluster is used to describe the center of each cluster [10].

Input: 1: **convert the large data set into a set of attributes and perform scaling on the observations of all objects to set in particular range.**

2: Choose number of clusters k from data objects using elbow technique with the help of finding sum of square errors.

Output:
All data points are grouped into k number of clusters exclusively.

Procedure:

(1) Initially, select k observations as centroid from whole dataset randomly.
(2) For every object find the Euclidean distance with all centroids.
(3) Assign all the observations to particular cluster with minimum Euclidean distance with centroids.

(4) Calculate the centroids of freshly created clusters once again.
(5) Repeat step 3 and 4 until no conflictions;

The k-means method has significant disadvantages, despite its widespread use in many applications:

1. The k-means method assumed that the number of clusters k in the database was known beforehand. In real- world applications, this is not possible.
2. Because it is an iterative approach, the k-means algorithm is particularly well suited to the selection of starting centers [11].
3. The k-means method has the potential to reach local minima.

3   Proposed Work

K-means Clustering is used in different fields like data mining, big data and machine learning to extract useful information from unstructured data. This technique is used to categorize large data set into k number of clusters. Though, there are many applications of k-means clustering algorithm, still there are some flaws existed in this method. In real applications it's very difficult to predict the number of cluster in the first instance of algorithm, selection of centroid is random it play crucial role in creating clusters [12]. The shape of cluster depends upon the number of clusters and initial centroid selection. Normalization technique is also required to scale up the data sets, so that every data set categorized in cluster.

In the proposed work a new hybrid technique is design that will resolve the issue of choosing number of clusters and of number of iteration to recompute the centroid. It will also improve the performance of clustering algorithm.

3.1   Objectives:

a) To investigate the issues in K-means clustering Principle.
b) To investigate various augmentation already done on K-means technique.
c) To accomplish a new hybrid K-means clustering algorithm that will improves the performance of clustering.
d) To do comparative analysis of proposed work with traditional methods

3.2   Basic Design of Proposed Work

1. Create training datasets for clustering.
2. Using Principal Component Analysis technique converts the high dimension dataset into key features.
3. Use Percentile Method to split the dataset into different partition based upon number of clusters [13].
4. Choose central point's randomly from partitioned trained datasets.
5. Determine the Euclidian distance for all observation with the center locations [14].
6. Find the closest centroid according to minimum Euclidian distance and simply used an array to memorize the distance between every dataset to its nearest centroid.

7. Find the distance of dataset to the new cluster allocated in its previous iteration if the calculated distance is less than or equal to the old centroid then there is no need to calculate distance to other centroids. Because that dataset stay in same cluster [15].
8. If the calculated distance is larger than to the older centroid then calculate distance with all centroids to find the local minimum.
9. Allocate the dataset to nearest cluster and update the count of allocated cluster and store the cluster id and its distance to the array to use for next iteration [16].
10. Steps 6 to 9 are repeated until to get same cluster from last two iterations. This will reduce the time complexity of algorithm [17].

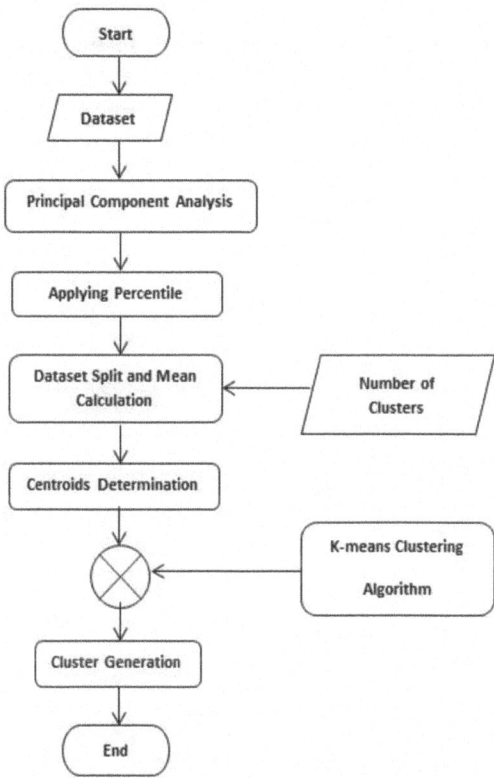

Fig. 5. Basic design of the system

The following figures describe the proposed idea. Here in Fig 6 the objects of data set are plotted with 3 centroids

Here Fig. 7 shows the shifting of initial centroid to the new centroids based upon the calculated average of the data set.

Here Fig. 8 shows the output of final clustering

We have applied the proposed algorithm on the following dataset [18]. By using proposed methodology the complexity of the process to find the cluster is reduced. In

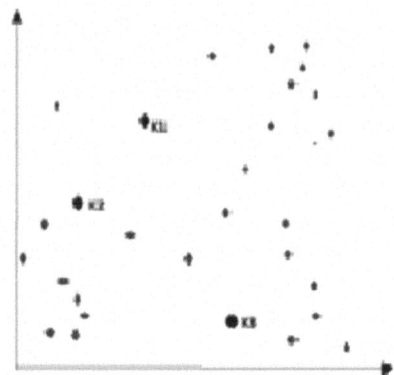

Fig. 6 Selection of initial centroid from data set

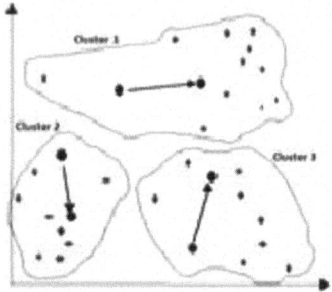

Fig. 7 Identifying the location of new centroids

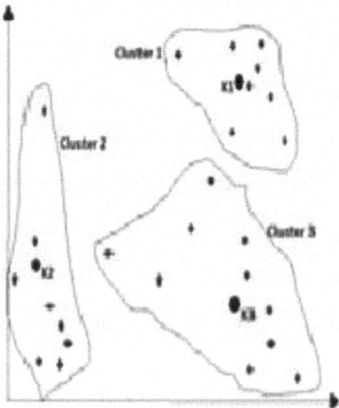

Fig. 8 Final location of centroids

the basic K-Mean algorithm there are large numbers of iterations. In enhanced K-mean algorithm the Euclidian distance of previous iteration is stored in array data structure. If the Euclidian distance of object to the new centroid is less than the stored Euclidian distance of previous iteration then data set stay in the same cluster and there is no need to calculate Euclidian distance with all centroids [19]. The almost half of the iterations are reduced by using the proposed methodology.

Here in this figure comparison of traditional K mean and enhanced k mean algorithm is represented.

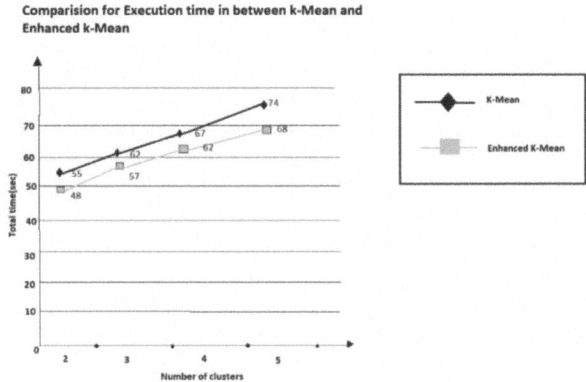

Fig. 9 Final location of centroids

4   Conclusion and Future Scope

In this proposed work we have discussed enhanced K mean algorithm with an appropriate example. Here also we indicated the issues of k-mean algorithm. The principle component analysis method is used to pre-process the data and percentile method is used to split the data into partitions [20]. In this paper the comparative analysis of traditional k-mean and enhanced k-mean algorithm is implemented on same data set and found that enhanced k- mean algorithm is more efficient than basic k-mean algorithm.

References

1. Yu, D., Deng, L., He, X., Acero, A.: Large-margin minimum classification error training: a theoretical risk minimizationperspective. Comput. Speech Lang. (2008). https://doi.org/10.1016/j.csl.2008.03.002
2. He, X., Deng, L., Chou, W.: Discriminative learning in sequential pattern recognition: a unifying review for optimization-oriented speech recognition. IEEE Signal Process. Mag. (2008). https://doi.org/10.1109/MSP.2008.926652
3. Renals, S., Morgan, N., Bourlard, H., Cohen, M., Franco, H.: Connectionist probability estimators in hmm speech recognition. IEEE Trans Speech Audio Process. (1994). https://doi.org/10.1109/89.260359
4. Hinton, G.E., Osindero, S., Teh, Y.-W.: 2006 Dbn. Neural Comput. **18**, 1527–1554 (2006)
5. Mohamed, R., Yu, D., Deng, L.: Investigation of full-sequence training of deep belief networks for speech recognition. In: Proc. 11th Annu. Conf. Int. Speech Commun. Assoc. INTERSPEECH 2010, 2846–2849 September 2010

6. Collobert, R., Weston, J.: A unified architecture for natural language processing: Deep neural networks with multitask learning. In: Proc. 25th Int. Conf. Mach. Learn., pp. 160–167, 2008

7. Narayan, Y., Mathew, L., Chatterji, S.: SEMG signal classification with novel feature extraction using different machine learning approaches. J. Intell. Fuzzy Syst. (2018). https://doi.org/10.3233/JIFS-169794

8. Jiang, W., Wang, Z., Jin, J.S., Han, X., Li, C.: Speech emotion recognition with heterogeneous feature unification of deep neural network. Sensors (Switzerland) (2019). https://doi.org/10.3390/s19122730

9. Wang, K., An, N., Li, B.N., Zhang, Y., Li, L.: Speech emotion recognition using Fourier parameters. IEEE Trans. Affect. Comput. (2015). https://doi.org/10.1109/TAFFC.2015.2392101

10. AbdelWahab, M., Busso, C.: Domain adversarial for acoustic emotion recognition. IEEE/ACM Trans. Audio Speech Lang. Process. (2018). https://doi.org/10.1109/TASLP.2018.2867099

11. Nicholson, J., Takahashi, K., Nakatsu, R.: Emotion recognition in speech using neural networks. In: ICONIP 1999, 6th International Conference on Neural Information Processing - Proceedings, 2000, https://doi.org/10.1109/ICONIP.1999.845644

12. Park, C.H., Sim, K.B.: Emotion recognition and acoustic analysis from speech signal. Proc. Int. Jt. Conf. Neural Networks (2003). https://doi.org/10.1109/ijcnn.2003.1223975

13. Dang, N.C., Moreno-García, M.N., De la Prieta, F.: Sentiment analysis based on deep learning: a comparative study. Electron. (2020). https://doi.org/10.3390/electronics9030483

14. Agarwalla, N., Panda, D., Modi, M.K.: Deep learning using restricted Boltzmann machines. Int. J. Comput. Sci. Inf. Secur. 7(3), 1552–1556 (2016)

15. Zheng, W. Q., Yu, J. S., Zou, Y. X.: An experimental study of speech emotion recognition based on deep convolutional neural networks. In Proc. Int. Conf. Affect. Comput. Intell. Interact. (ACII), 827–831 Sep. 2015

16. Kim, Y.: Convolutional neural networks for sentence classification, arXiv:1408.5882. https://arxiv.org/abs/1408.5882 (2014)

17. Zhao, J., Mao, X., Chen, L.: Learning deep features to recognize speech emotion using merged deep CNN. IET Signal Process. 12(6), 713–721 (2018)

18. Cen, L., Ser, W., Yu, Z. L.: Speech emotion recognition using canonical correlation analysis and probabilistic neural network. In: Proc. 7th Int. Conf. Mach. Learn. Appl. (ICMLA), 859–862, Dec. 2008

19. Huang, K.-C., Kuo, Y.-H.: A novel objective function to optimize neural networks for emotion recognition from speech patterns. In: Proc. IEEE 2nd World Congr. Nature Biolog. Inspired Comput. (NaBIC), 413–417 Dec. 2010

20. Fayek, H.M., Lech, M., Cavedon, L.: Evaluating deep learning architectures for speech emotion recognition. Neural Netw. 92, 60–68 (2017)

Comprehensive Exploration of IoT Communication Protocol: CoAP, MQTT, HTTP, LoRaWAN and AMQP

Lalhriatpuii[1]([✉]), Ruchi[1], and Vikas Wasson[2]

[1] ECE Department, Chandigarh University, Mohali, Punjab, India
hriatpuii4133@gmail.com
[2] Computer Science Engineering Department, Chandigarh University, Mohali, Punjab, India

Abstract. The world of the Internet of Things (IoT) is offering significant business prospects and paving the way for exciting innovations. Smart devices utilize a variety of IoT application protocols to ensure compatibility among the different nodes within the IoT network. As the number of connected devices keeps on increasing, it is highly belief that the Internet will generate around five quintillion bytes of data on a daily basis. It's important to recognize that not all these connections and ecosystems are safe, and security vulnerabilities are on the rise. This research aims to explore and analyze the features and abilities of six protocols: Constrained Application Protocol (CoAP), Message Queue Telemetry Transport (MQTT), HTTP, LoRaWAN, AMQP, and WebSocket. It delves into their potential to operate efficiently on constrained devices, considering factors such as security measures and adaptability to various network conditions.

Keywords: CoAP · MQTT · LoRaWAN · HTTP · AMQP · WebSocket · IoT Protocol

1 Introduction

In recent times, the propagation of interconnected devices within the Internet of Things (IoT) ecosystem has triggered the requirement for effective communication protocols designed for restricted environments. The seamless exchange of data among these devices, often characterized by limited resources such as constrained bandwidth, processing power, and energy, requires protocols capable of accom-modating these limitations without compromising functionality. The objective of these studies is to explore and understand the characteristics and capabilities of six protocols: the Constrained Application Protocol (CoAP), Message Queue Teleme-try Transport (MQTT), Hypertext Transfer Protocol (HTTP), Wide Area Network (LoRaWAN), Advanced Message Queuing Protocol (AMQP), and WebSocket.

To facilitate communication, within a system there are two used methods today. These protocols, MQTT and CoAP are designed to facilitate communication, at the application layer. MQTT, specifically designed to ease machine-to-machine (M2M) communication,

M. Khurana et al. (Eds.): ICMLA 2024, CCIS 2238, pp. 261–274, 2025.
https://doi.org/10.1007/978-3-031-75861-4_23

has gained acknowledgment as an ISO standard (ISO/IEC 20922;2016) [1]. On the flip side, CoAP, developed by the Internet Engineering Task Force (IETF), utilizes a specific set of HTTP methods to assist devices with limited resources that work over UDP [2]. The ThingsBoard platform provides a CoAP API, a representation of HTTP RESTful API, facilitating connectivity with CoAP-based IoT device application protocols. Additionally, ThingsBoard offers an MQTT API, another HTTP RESTful interface serving the same connectivity purposes but tailored for MQTT-based devices [3]. Several studies have integrated AMQP with QUIC and evaluated its performance. The results show that using QUIC at the transport level improves communication time, startup latency, and total communication time compared to TCP. Additionally, AMQP over QUIC outperforms AMQP over TCP in terms of packet loss and throughput [4]. The LoRaWAN standard, which is founded by LoRa Alliance [5], represents in LPWAN (Low-Power Wide-Area Network) technology and is rooted in the foun-dational framework of the LoRa physical layer, enjoying extensive acceptance and utilization. LoRaWAN is experiencing growing adoption because it operates with-in the ISM unlicensed bands, enabling adaptable adjustments in transmission rates and coverage. The evaluation aims to assess the feasibility of these protocols to effectively operate within the limitations posed by devices functioning with lim-ited resources. Such investigation involves a comprehensive analysis of how these protocols address challenges related to bandwidth optimization, energy efficiency, scalability, and interoperability while ensuring reliable and secure communication in resource-restricted environments.

2   Description of the Protocols

2.1   CoAP

The Constrained Application Protocol (CoAP) was crafted by the IETF's CoRE (Constrained RESTful Environments) [6, 7] working group with a specific goal: to facilitate data sharing, applications, and services among low-power constrained nodes operating within an unreliable and resource-constrained radio frequency network.

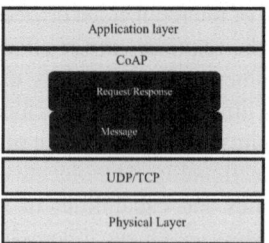

Fig. 1. Layering of CoAP

CoAP clients are designed to operate on extremely resource-constrained devices, such as 8, 16, or 32-bit microcontrollers, which typically have limited RAM and ROM. Despite these limitations, even in environments prone to packet loss, CoAP can attain throughput rates reaching up to 0.1 mbps. Figure 1 demonstrates the CoAP protocol stack. CoAP predominantly functions via UDP, but it also offers the possibility of utilizing CoAP over DTLS over UDP. This variation incorporates provisions for dependable delivery, uncomplicated congestion control, and flow control to facilitate communication between endpoints. Although feasible, employing CoAP over TCP leads to heightened overhead, additional roundtrips, elevated RAM usage, and increased resource demands on devices, culminating in larger packet sizes. As a result, CoAP over UDP remains the preferred choice for most IoT scenarios.

Implementation of CoAP. Multiple CoAP implementations are available, encompassing both open-source and proprietary iterations. The used example are as follows:

- *CoAP implemented on Contiki OS*
- *CoAP-Based Semantic Resources Directory in Californium*
- *CoAP-based IoT Framework Smart Contract Security Guarantee*

[8] The use of Contiki OS allows for the creation of secure connections for small IoT devices that may not have the capability to establish secure connections on their own. Experimental results of [9] affirm its feasibility and alignment with the CoRE WG's RD specification, showing satisfactory performance in semantic applications. IoT Framework was developed using IoTtalk as a base [10]. This framework integrates the CoAP-based IoT Proxy and smart contract-based IoT Chain. Its goal is to expand the practical use of IoTtalk within IoT networks by offering identity verification and ensuring data confidentiality. Table 1 provides a breakdown of the various sizes of CoAP messages specifically at the application layer. To accurately calculate the total size, it's essential to account for not just these message lengths, but also incorporate the additional data from the UDP header (8 bytes), and further, the IP header (20 bytes for IPv4 and 40 bytes for IPv6).

Table 1. Variations in the Lengths of CoAP Messages

Message	App Layer (bytes)	On wire (bytes)
ACK	4	64
ACK + answer (ans)	28	72
CON (req)	9	62
CON (ans)	28	72
NON (req)	9	62
NON (ans)	28	72

2.2 MQTT

MQTT is positively not a communication protocol specifically tailored for nodes with limited resources in the area IoT [11]. It functions as a client-server publish-subscribe messaging TP, operating over TCP/IP as illustrated in Fig. 2. MQTT is renowned for its simplicity, energy efficiency, and small memory usage, rendering it suitable for networks with resource-constrained nodes The primary focus of this protocol lies in ensuring reliable message delivery, which is achieved through the incorporation of Quality of Service (QoS) levels [12]. In spite of the absence of inherent security measures, MQTT possesses the capability to enforce security measures at different levels such as authentication, authorization, and data confidentiality. The MQTT specification provides a comprehensive list of security concerns that implementers should take into account [13]. MQTT-SN is a customized version of MQTT designed specifically for wireless sensor networks [14].

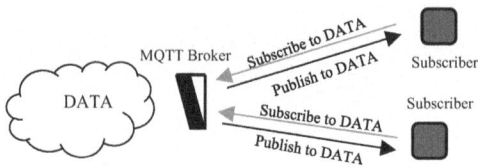

Fig. 2. MQTT Protocol Architecture

The primary focus of this protocol lies in ensuring reliable message delivery, which is achieved through the incorporation of Quality of Service (QoS) levels [12]. In spite of the absence of inherent security measures, MQTT possesses the capability to enforce security measures at different levels such as authentication, authorization, and data confidentiality. The MQTT specification provides a comprehensive list of security concerns that implementers should take into account [13]. MQTT-SN is a customized version of MQTT designed specifically for wireless sensor networks [14].

Implementation of MQTT: In different situations and uses, people have put the MQTT protocol to work. take Mangushev's case, for instance [15]. He talks about this device that sends data and handles commands using MQTT online. This gadget runs on an STM32 microcontroller paired with a GSM/GPRS SIM800 module, it can link up with lots of different add-ons. Pramono at al [16] came up with a system to track the chiller's temperature. They used MQTT and a PT-100 sensor to do this. Their setup nailed an accuracy rate of 99.32% and keeps an eye on the temperature right as it happens. Moreover, among the open-source versions of MQTT, there's Paho MQTT, crafted by Eclipse, offering its source code accessible in various programming languages like Java, C, Python, and more. Moquette stands as the C implementation, while Moquette serves as its Java counterpart for the MQTT broker. Figure 3 illustrates the fixed header format used in MQTT

Bit	7	6	5	4	3	2	1	0
Byte 1	MQTT Control Packet type				MQTT Control Packet contains specific flags.			
Byte 2	The length remaining							

Fig. 3. MQTT fixed header format

2.3   LoRaWAN

LoRaWAN (Long Range wide Area Network) is a widely used way of connecting devices in the Internet of Things (IoT) world its purpose is to facilitate communication among IoT devices across extended distances while minimizing power consumption. LoRaWAN is part of the LoRa (Long Range) technology stack and defines the network layer and some aspects of the data link layer for IoT communication. By employing the grant-free approach, LoRaWAN can potentially find applications in the context of mIoT (massive IoT). However, the increased volume of IoT access attempts leads to a multitude of concurrent transmissions within LoRa networks. These concurrent transmissions can result in co-SF interference, occurring among end-devices employing the same Spreading Factor (SF), and inter-SF interference, which takes place among end-devices using different SFs.

Table 2. LoRaWAN Protocols important details

Protocol	Message type	Constrained network GPRS, 2G, 3G, 4G	Security	Usage	Transport	Architecture
LoRaWAN	Various (e.g., Join request, Data ACK)	Compatible with LPWAN technologies	End-to-End Encryption	IoT, Low-Power Applications	Over LoRa Physical Layer	Star Topology, Gateways

Table 2 gives the important details of LoRaWAN protocols. When there are more devices and increased data moving through the system, LoRaWAN coverage and how much data it can handle tend to decrease noticeably [17, 18]. Fortunately, LoRa benefits from a unique modulation scheme known as the "capture effect" [19, 20]. This situation enables successful receiving of signals even when there's interference within the same spreading factor (SF) or across different spreading factors (inter-SF). It works as long as the signal-to-interference-plus-noise ratio (SINR) exceeds the required threshold. In a previous work, a heuristic SF-allocation strategy, combined with a power control algorithm, was introduced. In [21], they used a simple method to assign end-devices with similar path losses to the same channel but with different spreading factors (SFs), depending how close they were to the gateway. While they acknowledged the problem of interference between different spreading factors, their proposed solution didn't address this issue. Furthermore, their solution demanded that all end-devices possess the capability to effectively receive all Spreading Factors (SFs).

Implementation of LoRaWAN Protocol. In addition to server implementations, the implementation of LoRaWAN also involves the design and configuration of base stations and gateway nodes. an instance of this is the creation of a low-power LoRaWAN autonomous base station. this station was built using a RAK7246 gateway along with arduino MKR WAN 1310 slave gateway nodes [22]. It's designed to both send and receive messages, and they've also enhanced the antenna design to improve its coverage and performance [23].

2.4   HTTP

HTTP and web sockets are common standardized ways used to send and receive data. They're often utilized to transmit information, usually in formats such as XML or JavaScript Object Notation (JSON). JSON acts as abstraction layer, allowing web developers to create strong web applications that stay connected to a web server without interruptions, as shown in Fig. 4. Meanwhile, HTTP forms the essential foundation of the client-server model that's used widely throughout the World Wide Web.

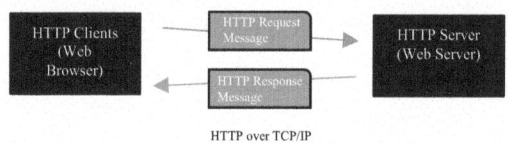

HTTP over TCP/IP

Fig. 4. HTTP Protocol

A more secure implementation of HTTP involves configuring it for IoT devices to function exclusively as clients rather than servers. HTTP provides different operations or methods like GET, POST, PUT, DELETE. The reliable transmission of data is guaranteed by its functionality relying on the TCP protocol [24]. Moreover, HTTP breaks down large data requests and responses into smaller, more manageable pieces for smoother transmission over networks. HTTP is designed to facilitate communication strictly between two systems at a time, which might not align with the needs of many IoT applications. In these cases, HTTP falls short in enabling the essential one-to-many communication necessary between sensors and the server [25]. Table 3 provides significant performance insights into HTTP.

Implementation of HTTP. Http can be put into action in different setups. For instance, one way is creating an http proxy server that has a user interface through an embedded www server. This setup needs sockets and works best with a multithreaded architecture [26]. Another example involves integrating http/1.1 into a portable device, like a Wireless PDA, enabling applications to access the web [27].

Furthermore, HTTP's implementation is vital in adaptive and dynamic streaming tech. This technology helps design a system for smooth video playback in various network conditions, ensuring continuous viewing pleasure [28].

2.5   AMQP

The AMQP protocol, functioning at the application level, has undergone thorough testing and development specifically for IoT (Internet of Things) applications. It prioritizes environments geared towards seamless message exchange [29]. The system takes

Table 3. LoRaWAN Protocols important details

Protocol	Message type	Constrained network GPRS, 2G, 3G, 4G	Security	Usage	Transport	Architecture
HTTP	Request/ Response	Excellent	Low	Gateways, home services	TCP/IP	Client Server

great pride in its diverse range of messaging-related features, which include reliable queuing, topic-driven publish-and-subscribe messaging, flexible routing, and the facilitation of transactions [30]. AMQP is constructed based on the publish/subscribe model, harnessing a messaging queue that is both dependa-ble and efficient.

Specifically, within AMQP communication, exchanges act as a conduit for publishers and consumers to locate each other, while message queues serve as destina-tions to which consumers establish bindings to receive messages relevant to their interests as shown in Fig. 5. Within this framework, exchanges are responsible for routing messages to the appropriate queues, where these messages can be stored.

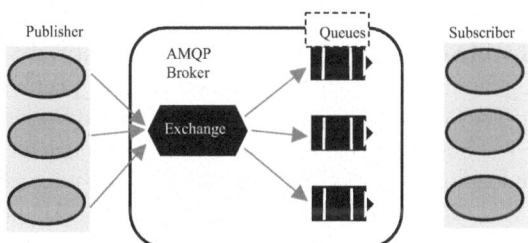

Fig. 5. AMQP Architecture

One notable AMQP limitation aspect is its deficiency in automated discovery mechanisms, which renders it less suitable for resource-constrained environments and real-time applications [31]. These limitations are significant considerations, particularly in the context of designing efficient solutions for smart grids, where automation, constrained environments, and real-time capabilities are integral components of a robust IoT system. Performance table is given in Table 4.

Implementation of AMQP. Santoso employed the AMQP protocol within a system to monitor the indoor air quality by utilizing intelligent IoT technology [32]. Kostromina et al. relied on AMQP to enhance the reliability of a weather monitoring system situated in Thailand [33].

Llamuca and her team implemented AMQP within a hypothetical factory in order to efficiently manage operations and facilitate communication between different sections of the factory [34].

Table 4. Performance of AMQP

Protocol	Security	Architecture	Transport	Bandwidth	Power Consumption	Packet Received ratio
AMQP	SSL/TLS	Publish/ Subscribe	TCP	Low	High	Low

2.6 WebSocket

WebSocket is a communication protocol that facilitates two-way, simultaneous, and immediate communication via a solitary TCP connection. It has been specifically developed to surmount the limitations inherent in customary request-response protocols, such as HTTP, which are well-suited for numerous web applications but not optimally equipped for situations necessitating uninterrupted updates and instantaneous interactions. [35] WebSocket is extensively employed in a wide array of web applications and serves as a valuable instrument for constructing interactive and dynamic functionalities, including chat applications, online gaming, collaborative editing, and real-time dashboards. WebSocket, a communication protocol defined by RFC 6455 and RFC 7936, offers a versatile and efficient means of enabling real-time, bidirectional interactions on the web [36].

Implementation of WebSocket: At SWCU STARS, they put WebSockets into action using the Erlang programming language [37]. In another setup, people used WebSockets for location sharing. They made this happen using NodeJS and Socket.IO [38]. There's talk about using the WebSockets method to control devices through the web. this method seems to lighten the load on networks and helps transfer data more efficiently than the usual HTTP-based way. Also, there's this neat thing - a WebSocket proxy based on the Californium open-source framework. It's designed to work better than an HTTP/CoAP proxy, especially when handling lots of requests at the same time. This means faster responses when dealing with multiple requests [39]. The performance of WebSocket is shown in Table 5

Table 5. Performance of WebSocket

Protocol	Message type	Constrained Network (GPRS, 2G, 3G, 4G)	Security	Usage	Transport	Architecture
WebSocket	Text/ Binary	Works with modern network technologies	Secure (TLS/SSL)	Real-time Communication, Web applications, Gaming, chat application	Over TCP	Client-Server Model, Event-Driven Approach

WebSocket is further reinforced by client libraries available in multiple programming languages, such as JavaScript, Python, and Java, while server implementations like Node.js with the "ws" library, Python's "websockets," and Java's "javax websocket" provide robust support. However, while WebSocket connections are secure within the same-origin policy, for sensitive data, additional security measures like TLS/SSL are highly recommended. Moreover, ensuring data integrity and mitigating security vulnerabilities, like Cross-Site Scripting (XSS) attacks, demands adequate data validation and sanitization procedures.

3   The Protocols in Different Aspect of Quality

In a security aspect of IoT Protocol they differ in several ways, including authentication, encryption, access control and key management. Affirmation and validation are important facets of ensuring the protection of communication protocols within the domain of Internet of Things (IoT) and data exchange we can see it from Table 6. MQTT offers various options for authentication such as username/password or client certificates, although the methods for authorization may differ. Conversely, CoAP supports authentication and authorization through widely recognized standards like OAuth 2.0, as well as customized security solutions. HTTP, being a versatile protocol, offers various ways to confirm identity, like Basic Authentication, OAuth, and JWT. It also helps manage precise access rights through role-based access control (RBAC) methods.

Table 6. Specification of MQTT, CoAP and LoRaWAN

Specification	MQTT	CoAP	LoRaWAN
Architecture	Client/Broker	Client/Server or Client/Broker	Client/ Server
Abstraction	Publish/ subscribe	Request/ Response or Publish/ Subscribe	uplink (device-to-gateway) and downlink (gateway-to-device) data transfer
Header Size	2 Byte	4 Byte	13 Byte
Message Size	Variable (256 MB max size)	Small and Undefined (usually fits a single IP Datagram)	Varies by region; typically, 51 bytes
Transport protocol	TCP, can use UDP for MQTT-SN	UDP, DTLS (Datagram Transport Layer Security)	LoRa physical layer

WebSocket depends on transport layer security (TLS/SSL) to confirm identities and introduces special methods to improve overall security measures. LoRaWAN incorporates essential security measures, including device authentication and session keys for guarantee secure communication. On the other hand, AMQP grants robust authentication and authorization mechanisms that include role-based access control and customizable security policies, thereby establishing itself as a trustworthy preference for secure data exchange in IoT and other applications. Table 7 shows the protocols in terms of their security.

3.1   In Terms of Security

Table 7. MQTT, CoAP, HTTP and WebSocket in terms of Security

Security	MQTT	CoAP	HTTP	WebSocket
Authentication and Authorization	Username/Password Client certificates	Authentication mechanisms (eg. OAuth)	– Multiple authentica-tion methods (eg. Basic Auth, OAuth) – Fine-grained access control (RBAC)	– Typically relies on transport layer security (TLS/SSL) – Custom security mechanisms
Encryption (TLS/SSL)	Contained	Present	Contained	Present
Data Integrity	Encryption - Message integrity checks (e.g., HMAC, digital signatures)	Encryption - Message integrity checks	Encryption - Message integrity checks	Encryption - Message integrity checks
Device Identity Management	Client certificates Username/Password	Custom security solutions OAuth	Client certificates- OAuth, JWT	transport layer security (TLS/SSL) - Custom security mechanisms
Access Control Lists	Varies	Access Control Lists (ACLs)	Fine-grained access control (RBAC)	Varies
Certificate Based Security	Client certificates	Certificate based security	Client certificates	TLS/SSL
Security Bootstrapping	Varies	Secure bootstrapping process	Varies	Varies

3.2 In Terms of Strengths

IoT protocols vary in their strengths and performance characteristics. MQTT and CoAP are lightweight and efficient for constrained devices and low bandwidth networks, with MQTT using a publish/subscribe model and CoAP designed for resource-constrained devices. HTTP/HTTPS is widely adopted and secure but may have higher overhead. AMQP supports various communication patterns but may have higher overhead, while Zigbee offers low-power wireless mesh networks. LoRaWAN is for long-range, low-power IoT, and Bluetooth is suitable for short-range applications. Selecting the appropriate protocol relies on factors such as the purpose of use, limitations of the device, prevailing network conditions, and security needs. Each protocol comes with its unique advantages and compromises, making the selection a critical aspect for the success of an IoT project. IoT Protocols in terms of their strength are summarized it in Table 8.

Table 8. IoT Protocols in terms of their Strength

IoT Protocols					
WebSocket	MQTT	CoAP	LoRaWAN	HTTP	AMQP
Full-Duplex Communication Bi-Directional	Low Bandwidth Usage Efficiency Reliability (QoS)	Suitable for Constrained Devices RESTful Interface Efficiency	Long-Range Communication Low Power Consumption Scalability	Widely supported Security Features (TLS/SSL)	Robust Messaging Interoperability

4 Conclusion

In our study, we've looked into how well-known IoT protocols like MQTT, CoAP, HTTP, AMQP, WebSocket, and LoRaWAN perform, understanding these protocols is crucial for engineers and developers making informed choices. Each protocol has its strengths, weaknesses, structure, usage, security features, and power consumption parameters, essential factors for decision-making. The choice of protocol hinges on specific IoT application requirements, including data volume, power constraints, communication range, network topology, and data transmission reliability. Often, a combination of these protocols addresses diverse communication needs within a single IoT ecosystem. This paper can be valuable for researchers aiming to explore various protocols within an IoT network.

References

1. Prayogo, S. S., Mukhlis, Y., & Yakti, B. K.: The use and performance of MQTT and CoAP as internet of things application protocol using NodeMCU ESP8266. In: 2019 Fourth International Conference on Informatics and Computing (ICIC), pp. 1–5. IEEE (2019)

2. Keoh, S.L., Kumar, S.S., Tschofenig, H.: Securing the internet of things: a standardization perspective. IEEE Internet Things J. **1**(3), 265–275 (2014)
3. Jang, S. I., Kim, J. Y., Iskakov, A., Fatih Demirci, M., Wong, K. S., Kim, Y. J., Kim, M. H.: Blockchain Based Authentication Method for ThingsBoard. In: Advances in Computer Science and Ubiquitous Computing: CSA-CUTE 2019, pp. 471–479. Springer, Singapore (2019)
4. Zhang, Z., Li, S., Ge, Y., Xiong, G., Zhang, Y., Xiong, K.: PBQ-enhanced QUIC: QUIC with deep reinforcement learning congestion control mechanism. Entropy **25**(2), 294 (2023)
5. Cheikh, I., Aouami, R., Sabir, E., Sadik, M., Roy, S.: Multi-layered energy efficiency in LoRa-WAN networks: a tutorial. IEEE Access **10**, 9198–9231 (2022)
6. Espes, D., Lagrange, X., Suárez, L.: A cross-layer MAC and routing protocol based on slotted aloha for wireless sensor networks. Annals of Telecommunications-annales des télécommunications **70**, 159–169 (2015)
7. Villaverde, B. C., Pesch, D., Alberola, R. D. P., Fedor, S., Boubekeur, M.: Constrained application protocol for low power embedded networks: A survey. In: 2012 sixth international conference on innovative mobile and internet services in ubiquitous computing, pp. 702–707. IEEE (2017)
8. Colitti, W., Steenhaut, K., De Caro, N., Buta, B., Dobrota, V.: Evaluation of constrained application protocol for wireless sensor networks. In: 2011 18th IEEE Workshop on Local & Metropolitan Area Networks (LANMAN) pp. 1–6. IEEE (2011)
9. Behal, A., Sandhu, J. K., Gupta, G.: Constrained Application Protocol (COAP) Implementation on Contiki OS For Anatomization of Low-Power and Lossy Networks In IOT. In: 2022 International Conference on Futuristic Technologies (INCOFT) pp. 1–4. IEEE (2022)
10. Wang, Y., Wei, G.: An Implementation of CoAP-Based Semantic Resource Directory in Californium. In: Sun, X., Pan, Z., Bertino, E. (eds) Artificial Intelligence and Security. ICAIS 2019. Lecture Notes in Computer Science, vol 11634. Springer, Cham (2019)
11. S. -P. Hsu, Y. -M. Chang, S. -R. Yang P. -H. Hsieh.: The Design and Implementation of a Lightweight CoAP-based IoT Framework with Smart Contract Security Guarantee. In: 2020 12th International Symposium on Communication Systems, Networks and Digital Signal Processing (CSNDSP), pp. 1–6. Springer, IEEE, Porto, Portugal (2020)
12. Silva, D., Carvalho, L.I., Soares, J., Sofia, R.C.: A performance analysis of internet of things networking protocols: evaluating MQTT CoAP OPC UA. Appl. Sci. **11**(11), 4879 (2021)
13. Reddy, T.B., Karthigeyan, I., Manoj, B.S., Murthy, C.S.R.: Quality of service provisioning in ad hoc wireless networks: a survey of issues and solutions. Ad Hoc Netw. **4**(1), 83–124 (2006)
14. Sadio, O., Ngom, I., Lishou, C.: Lightweight security scheme for mqtt/mqtt-sn protocol. In: 2019 sixth international conference on internet of things: systems, management and security (IOTSMS), pp. 119–123. IEEE (2019)
15. da Rocha, H., Monteiro, T.L., Pellenz, M.E., Penna, M.C., Alves Junior, J.: An MQTT-SN-Based QoS Dynamic Adaptation Method for Wireless Sensor Networks. In: Barolli, L., Takizawa, M., Xhafa, F., Enokido, T. (eds.) Advanced Information Networking and Applications. AINA 2019. Advances in Intelligent Systems and Computing, vol 926. Springer, Cham (2020)
16. Pham, C., Ehsan, M.: Dense deployment of LoRa networks: expectations and limits of channel activity detection and capture effect for radio channel access. Sensors **21**(3), 825 (2021)
17. Gkamas, A.: LoRa Technology Benefits in Educational Institutes. In: Internet of Things, Infrastructures and Mobile Applications: Proceedings of the 13th IMCL Conference 13, pp. 413–424. Springer, International Publishing (2021)

18. Cuomo, F., Campo, M., Caponi, A., Bianchi, G., Rossini, G., Pisani, P.: EXPLoRa: Extending the performance of LoRa by suitable spreading factor allocations. In: 2017 IEEE 13th International Conference on Wireless and Mobile Computing, Networking and Communications (WiMob), pp. 1–8. IEEE (2017)

19. Mangushev, A.A.V.: Mqtt client implementation based on a single-chipmicrocontroller for remote equipment management tasks. Izvestiâ ÛFU. Engineering Science, pp. 75–84 (2022)

20. Pramono, B. S., Slamet, I., Wahyu, J.: The Imple-mentation of MQTT Protocol using PT-100 for Monitoring the Vaccine Temperature. Jurnal RESTI (Rekayasa Sistem dan Teknologi Informasi), 6(2), 346–351 (2022)

21. Xiao, W., Kaneko, M., El Rachkidy, N., Guitton, A.: Integrating LoRa collision decoding and MAC protocols for enabling IoT massive connectivity. IEEE Internet Things Mag. 5(3), 166–173 (2022)

22. Mohammadi, S., Farahani, G.: Scalability Analysis of a LoRa Network Under Co-SF and Inter-SF Interference in Large-scale IoT Applications. In: 2021 5th International Conference on Internet of Things and Applications (IoT), pp. 1–6. IEEE (2021)

23. Carbonel, J. B. M., Pestaño, A. C. S., Roque, G. M., Tan, M. A. C., Villa-rubin, R. T. A., Arada, G. P.: Design and Implementation of LoRaWAN autonomous base station as a communication network for rural areas. In: 2022 IEEE 14th International Conference on Humanoid, Nanotechnology, Information Technology, Communication and Control, En-vironment, and Management (HNICEM), pp. 1–6. IEEE (2022)

24. Dumitru, M. C., Pietraru, R. N., Moisescu, M. A.: Lo-RaWAN as Open Scalable IoT Ecosystem. In: 2023 13th International Sym-posium on Advanced Topics in Electrical Engineering (ATEE), pp. 1–6. IEEE (2023)

25. Seufert, M., Egger, S., Slanina, M., Zinner, T., Hoßfeld, T., Tran-Gia, P.: A survey on quality of experience of HTTP adaptive streaming. IEEE Commun. Surv. Tutorials 17(1), 469–549 (2014)

26. Nielsen, H. F., Gettys, J., Baird-Smith, A., Prud'hommeaux, E., Lie, H. W., Lilley, C.: Network performance effects of HTTP/1.1, CSS1, and PNG. In: Proceedings of the ACM SIGCOMM'97 conference on Applications, technologies, architectures, and protocols for computer communication, pp. 155–166 (1997)

27. Begović, T., Kukrić, N., Lubura, S.: HTTP web server implementation in air parameters monitoring system. In: 2023 22nd Interna-tional Symposium INFOTEH-JAHORINA (INFOTEH), pp. 1–6. IEEE (2023)

28. Riihijarvi, J., Mahonen, P., Saaranen, M.J., Roivainen, J., Soininen, J.P.: Providing network connectivity for small appliances: a functionally minimized embedded web server. IEEE Commun. Mag. 39(10), 74–79 (2001)

29. Sun-Chul, J., Tae-Hak, B., Hoe-Kyung, J.: An implementation of dynamic and adaptive streaming system over HTTP. J. Korean Inst. Inf. Commun. Eng. 16(3), 476–481 (2012)

30. Prajapati, A.: AMQP and beyond. In: 2021 International Conference on Smart Applications, Communications and Networking (SmartNets), pp. 1–6. IEEE (2021)

31. Vinoski, S.: Advanced message queuing protocol. IEEE Internet Comput. 10(6), 87–89 (2006)

32. Fernandes, J. L., Lopes, I. C., Rodrigues, J. J., Ullah, S.: Performance evaluation of RESTful web services and AMQP protocol. In: 2013 Fifth international conference on ubiquitous and future networks (ICUFN), pp. 810–815. IEEE (2023)

33. Kostromina, A., Siemens, E., Babich, Y.: A concept for a high-reliability meteorological monitoring system using amqp. In: Proceedings of the 6th International Conference on Applied Innovations in IT, pp. 109–115 (2018)

34. Llamuca, E. S., Garcia, C. A., Naranjo, J. E., Garcia, M. V.: Cyber-physical production systems for industrial shop-floor inte-gration based on AMQP. In: 2019 International Conference on Information Systems and Software Technologies (ICI2ST), pp. 48–54. IEEE (2019)

35. Caiza, G., Llamuca, E. S., Garcia, C. A., Gallardo-Cardenas, F., Lanas, D., Garcia, M. V.: Industrial shop-floor integration based on AMQP protocol in an IoT environment. In: 2019 IEEE Fourth Ecuador Technical Chapters Meeting (ETCM), pp. 1–6. IEEE (2019)
36. Pimentel, V., Nickerson, B.G.: Communicating and displaying real-time data with websocket. IEEE Internet Comput. **16**(4), 45–53 (2012)
37. Huang, M., Wu, X., Dai, W.: Using Dynamic Interface Function Block in IEC 61499-based Industrial Edge Applications. In: 2022 IEEE 17th Conference on Industrial Electronics and Applications (ICIEA), pp. 845–850. IEEE (2022)
38. Tanaem, P. F., Manuputty, A. D., Wijaya, A. F.: STARS: Websocket Design and Implementation. In: 2022 International Seminar on Application for Technology of Information and Communication (iSemantic), pp. 167–171. IEEE (2022)
39. Maia, A. D. O., Silva, D. A.: Proposal to use of the websocket protocol for web device control. In: Proceedings of the 23rd Brazillian Symposium on Multimedia and the Web, pp. 253–260 (2017)

A Systematic Review on Blockchain Technology

Kapil Madan[1]([✉]) [iD] and Dinesh Vij[2] [iD]

[1] Jaypee Institute of Information Technology, Noida, India
dr.kapilmadan@gmail.com
[2] Institute of Engineering and Technology, Chitkara University, Rajpura, Punjab, India

Abstract. Despite being a technology that possesses attractive characteristics such as decentralization, transparency, auditability, anonymity, integrity, and verification, Blockchain is still in its infancy. The initial focus of this paper is to conduct an extensive analysis of blockchain technology, with particular emphasis on its evolution, Zero-Knowledge Proofs, intricacies of cryptography encompassing public key cryptography, hash functions utilized in blockchain, quantitative comparisons of consensus algorithms, and an inclusive inventory of blockchain applications. This paper also concentrates on the security of blockchain technology. This paper presents the challenges and current research trends that need to be addressed to increase the scalability and attain secure blockchain systems that can support vast deployments. This paper explores various intriguing use cases of blockchain technology that are unrelated to cryptocurrencies. In fact, beyond its initial application in the financial sector and cryptocurrencies, blockchain technology is increasingly being utilized in numerous diverse applications where its distinct characteristics have enabled the development of innovative and occasionally groundbreaking solutions. This paper specifically considers the following application scenarios: supply chain management, management of healthcare records, access control systems, systems for identity management, and end-to-end verifiable electronic voting. In each of these situations, we begin by analyzing the issue, the related needs, and the possible advantages that adopting blockchain technology could provide. Subsequently, we discuss various related solutions that have been proposed in the academic and corporate spheres.

Keywords: Blockchain · Cryptocurrency · Smart Contract · Consensus · Voting · Healthcare

CCS CONCEPTS: Distributed Systems · Networks P2P · Blockchain · Distributed Ledger Technology

1 Introduction

The adoption of blockchain, a growing technology, is continuously expanding in both academic and corporate settings. Blockchain technology was initially presented to aid cryptocurrencies such as Bitcoin, and as a result, blockchain-based systems and applications focused on cryptocurrencies are commonly referred to as Blockchain 1.0. While

M. Khurana et al. (Eds.): ICMLA 2024, CCIS 2238, pp. 275–286, 2025.
https://doi.org/10.1007/978-3-031-75861-4_24

the primary accomplishment of cryptocurrency is the decentralization of the transfer of value between different participants, this groundbreaking innovation can serve as the foundation for numerous other complex applications. The incorporation of smart contracts to enable Decentralized applications, smart property, financial data recording, supply chain management, smart tokens, and other similar applications have facilitated the development of automated financial applications. They operate using cryptocurrencies. The integration of smart contracts with digital currencies has enabled the development of various innovative financial applications, which are collectively referred to as Blockchain 2.0. Cryptocurrencies are merely one aspect of Distributed Ledger Technology (DLT). In fact, distributed ledgers can store any type of information, regardless of whether it pertains to financial transactions. Applications of blockchain technology are based on the non-cryptocurrency category and distributed ledgers. This category is generally classified as Blockchain 3.0. It is worth noting that, although these applications are different from cryptocurrencies, yet these could gain benefits from integrating with cryptocurrencies. Practically, these applications are commonly implemented on a blockchain that utilizes cryptocurrencies, such as Bitcoin, Tether, or Dogecoin. In the past years, the popularity of blockchain technology has led to a plethora of proposals for Blockchain applications, with new ideas emerging almost daily as companies seek to implement blockchain-based solutions.

The main focus of this paper is on the different blockchain applications such as supply chains, cryptocurrency, healthcare, and government services. The review methodology is based on framing the research questions, after a detailed analysis of the previous research works. The framed research questions are as follows:

Q1 How does blockchain concept help to resolve challenges of different applications?
Q2 What are various domains where blockchain is useful for government services?
Q3 What are the research trends and future scopes related to different areas such as scalability, privacy issues, and secure software codes?

This follows a discussion on the challenges of developing standards and regulations of different systems such as cryptocurrencies, and verification of COVID-19 vaccine certificates. This paper is organized as follows: A background concerning blockchain and decentralized Ledger is presented in Sect. 2, followed by Sect. 3, which presents different applications such as cryptocurrencies, supply chain, healthcare records management scenarios, and smart India Center of Excellence in Blockchain. Section 4 discusses the various challenges and research trends of blockchain technology. Section 5 presents a detailed discussion on Blockchain trends. Finally, Sect. 6 concludes the paper along with future directions for the research community.

2 Background and Related Work

2.1 Distributed Ledger

A distributed ledger is a data repository that is maintained and managed by numerous candidates without requiring trust to be established between them [1]. In a distributed ledger, typically all participants possess comparable rights and authority over the repository. They communicate with one another directly by making use of a peer-to-peer approach to propose and inform updates to the repository. The updates typically adhere to an append-only policy to ensure that the data remains immutable. In a distributed ledger, the need for intermediaries and centralized controllers is eliminated, because the participants use a distributed consensus algorithm to collectively agree on any repository update [2]. All participants have equal authority and control, and they communicate with one another directly in a peer-to-peer way. In computer science, distributed typically implies a network of independent entities that communicate with each other to achieve a shared objective, such as a distributed ledger. However, in general, there is no assumed coherence among the participants. A distributed system can involve nodes that execute completely distinct tasks while exchanging their results as data to other nodes. An instance of the use case of such a system is to orchestrate multiple threads executing different algorithms. Each node in a distributed ledger is required to adhere to the same protocol to ensure that a consistent outcome is obtained each time consensus is achieved. This implies that every truthful participant should obtain an indistinguishable copy of the ledger. Moreover, there is no single agreed-upon ledger. Each node, instead of relying on a central authority, stores and manages a copy of the same ledger. The correct version of ledger is determined by the majority of the network. To be more precise, it is appropriate to state that the ledger is replicated, as each participant stores a copy of it and performs the same management operations locally.

2.2 Blockchain Technology

A distributed ledger can be implemented using various technologies. It can be implemented in a blockchain by aggregating records into blocks. These blocks are secured against tampering by incorporating a cryptographic signature to the block data. The block header typically includes the entire cryptographic hash of the block content to secure it against tampering. The block headers are connected in a tamper-resistant way by linking each block to its previous one. The usual approach to achieve tamper resistance in the blocks' chaining is through the use of cryptographic hash functions [3]. In this, we add the hash of the previous block in the header of the next block. Ensuring that every block in the chain depends on its own content as well as the preceding block in the chain. A blockchain ensures that every block is dependent on all the content of the previous blocks leading up to the first block created, known as the genesis block [4]. Figure 1 shows the basic structure of a blockchain. This recursive dependency of blocks on all previous blocks makes it impossible to change any data within a block without rendering all upcoming blocks invalid.

Due to historical reasons, the data stored in each block are often referred to as transactions [6]. Blockchain technology was initially developed to facilitate cryptocurrencies as demonstrated by the Bitcoin cryptocurrencies protocol [7]. The primary challenges associated with implementing distributed ledgers using blockchain technology are related to scalability and efficiency [5]. If the system's trust assumptions are relaxed, it becomes possible to design more efficient and simplified consensus algorithms. While pure blockchains often use expensive distributed consensus algorithms, there are alternative approaches that can relax trust assumptions and use simple and efficient consensus algorithms. Generally, as our trust in entities increases, the system becomes more efficient, but it may also become more centralized. A distinction between different types of blockchains can be made based on the level of trust required for read-and-write operations. To clarify, a write operation refers to the capability to modify or update the ledger by adding new content, whereas a read operation refers to the capability to access and view the existing content of the blockchain. A blockchain has the capability to store any form of data, including code. However, there is a distinction between storing and executing code. Integrating blockchain technology into executable code has led to the development of smart contracts.

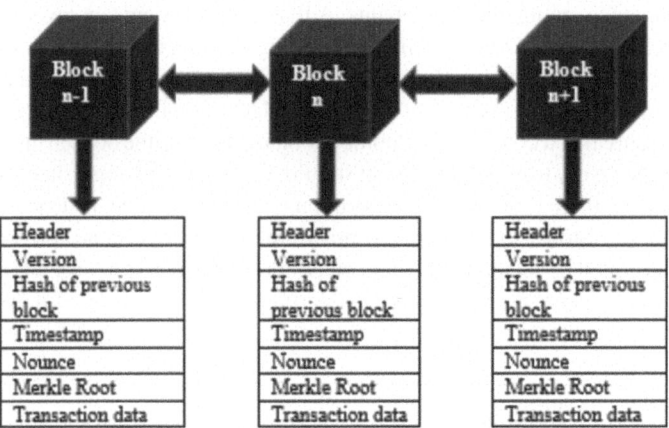

Fig. 1. Basic Structure of a Blockchain

2.3 Decentralized Consensus Protocol

To keep the blockchain network operational, its peers must reach a consensus on the decentralized ledger's recent state and on the method of grouping data into blocks. A decentralized consensus protocol is used to establish an agreement among peers in a blockchain network regarding the state of decentralized ledger and the method of packing data into blocks [2]. This agreement ensures the validation of the sequential order of transactions that are generated [9]. It confirms that a sufficient number of peers in the blockchain network comply with the exact state of the shared ledger, which in turn determines the order of appending the new blocks to the ledger [10].

Several distributed consensus protocols are utilized in blockchain networks, such as Proof-of-Activity, Proof-of-Deposit (PoD), Proof-of-Burn (PoB), Proof-of-Importance, Delegated-Proof-of-Stake, Proof-of-Stake (PoS), and Proof-of-Work (PoW). A decentralized consensus protocol specifies the process by which a network selects peer responsible for creating and validating the most recent block with unconfirmed and unformatted data. Choosing a peer randomly to make and finalize the latest block with non-formatted and unconfirmed data is a simple approach, but it may not be effective in ensuring the network's span and could be unsafe for the network as some peers might choose to attack it [11]. The Bitcoin network uses the PoW consensus protocol, which leverages computational power to choose designated peer [7, 9]. In Bitcoin, PoW consensus protocol is utilized, where peers compete by hashing unconfirmed transactions. When a peer is chosen as winner, it receives a reward, currently set at 6.25 BTC new generated bitcoins that are then credited to account of peer [12] (Table 1).

Table 1. Comparative analysis of different techniques based on the different protocols.

Sr. No.	Author (Year)	Protocol	Description	Limitations	Future Work
1	H. Tian et al. (2021) [37]	PoW, PoD	Cryptocurrency for trading between different cryptocurrencies	The smart contract is not tested on actual Etherum network	Scalability issues can be addressed
2	K. Karantias et al. (2020) [38]	PoB	The proposed technique is compatible with popular cryptocurrencies	Proposed technique is not tested for majority of dishonest federation	Privacy concerns can be improved in the future
3	P. Gazi et al. (2019) [28]	Proof of Stake	Technique for cross-chain certification	Ambiguous stake parties are the main bottleneck	Work can be done on the assumption of uncorrupted parties
4	M. Borge et al. (2017) [13]	Proof-of-personhood	Designed a cryptocurrency that prevents Sybil attacks	This model is based on at least one trusty authority	Technique for Public and Private Key Generation
5	S. Nakamoto (2008) [7]	Proof of work	Created a trustworthy system for electronic transactions based on PoW	Privacy concern are there When owner key is revealed	Incentives regulation can be formed further for optimized results

3 Blockchain Applications

Blockchain technology finds applications in various fields, such as property title registries, digital ownership management, digital records, law and enforcement, cryptocurrency, voting, education, identity management, finance (crowdfunding, stock exchange, etc.), reputation systems, security and privacy (privacy protection, security enhancement, etc.) [15], agricultural sector [16], healthcare, insurance, copyright protection, energy, defense, automotive [17], advertising [18]. Some researchers have proposed efficient data storage mechanisms using blockchain for wireless sensor networks [39, 40]. It is anticipated that increased use cases for blockchain systems will emerge in the future. The subsequent sections will delve into further details about the first application, i.e., cryptocurrency, followed by supply chain as a commonly used scenario, and the Centre of Excellence in Blockchain Technology has been proposed by the Indian government where they are working on document chain, property chain, drug logistic chain [19].

3.1 Cryptocurrency

Bitcoin, the first cryptocurrency, was stated in 2008 and introduced in 2009. At maximum, 21 million Bitcoins can exist in total. When a mining node (miner) discovers a nonce value which satisfies the complexity level and successfully has accepted a block, it receives a transaction fee (currently valued at 6.696 USD/tx for May 04, 2023) [20]. After every 210,000 blocks (approx. after four years), the mining reward is halved with a mining reward of 3.125 BTC in the year 2024 [21]. Presently, nearly 90% of all BTC has been mined. Following Bitcoin, Ethereum (ETH) is the second-largest cryptocurrency and has a market cap that is approximately 19% of Bitcoin's size [22].

3.2 Supply Chains

Through the usage of blockchain technology, decentralized ledgers can be created that offer a shared and permanent record of all transactions. The participants that are authorized have visibility into these recorded transactions, and the ledger itself is both traceable and immutable, meaning it cannot be altered or revoked. Consequently, blockchain technology is now being utilized more and more frequently for sharing of data within supply chains. IBM, for instance, has developed data-sharing solutions for supply chains that are based on permissioned blockchains, with an emphasis on logistics [7]. VeChain has created a cold-chain solution that leverages this technology to oversee and trace logistics information, thus promoting secure, regulated, transparent, and dependable sharing of data [24].

3.3 The Need for Blockchain in Healthcare

Healthcare is an area where blockchain is believed to hold enormous potential [25]. To revolutionize healthcare, emphasis should be placed on managing data that can leverage the ability to combine dissimilar systems and enhance the precision of electronic health records. This technology has the potential to aid drug prescription, and data management of pregnancies and related risks, in addition to facilitating data sharing, access control, and audit trail management of medical activities. The healthcare industry is undergoing a transformation to adopt a patient-centric approach. By giving patients greater control over their healthcare records, implementing healthcare systems based on blockchain could improve the dependability and security of patient data. Protecting patient data is crucial as it is sensitive information and is often targeted by cyber-attacks. Therefore, securing all sensitive data is of utmost importance in the healthcare industry. Patient control over their data is a key element of healthcare. Blockchain technology is highly resilient to failures and attacks, and provides various access control methods. The private blockchain is the most suitable for personal medical data. The model proposed by Würst and Gervais suggests that a blockchain can be employed in situations wherein multiple untrusting parties need to communicate and share data, instead of depending on a Trusted Third Party [26]. The blockchain, depends on consensus protocols, thereby not requiring a central authority, provides a viable resolution to this issue. Figure 2 shows the predictions of global blockchain growth in the healthcare domain from 0.76 billion (2022) to 14.25 billion US Dollar (2032) which is quite significant [23].

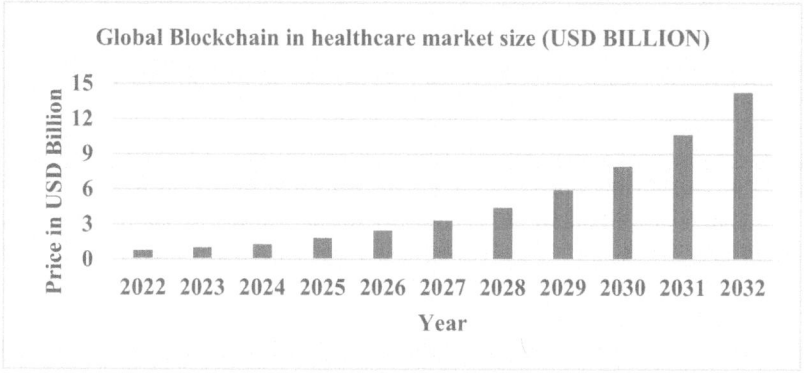

Fig. 2. Growth of global blockchain in healthcare domain

3.4   Smart India Centre of Excellence in Blockchain

According to a report from the New York Times, blockchain helps to improve accountability and improve the transparency in government services [14]. India is making significant investments in the Centre of Excellence to leverage blockchain technology for transforming the government's role [19]. The city is allocating funds for implementing blockchain technology at multiple levels, including:

- To implement government services such as drug logistics & warehousing society, death and birth certificate.
- CBSE academic documents are stored in blockchain from 2019 onwards.
- The Government of India is working on the remote voting system with help of proof of concept as per direction from the Election Commission of India.
- Promoting growth of the blockchain industry among startups and businesses.
- Establishing a groundbreaking model for government services.

4   Research Trends and Challenges

Several existing works have identified the research trends and future scopes for blockchain technology. For instance, Refs. [15, 23] discuss the smart contracts, blockchain testing, preventing centralization, and blockchain applications. Ref. [27] presents a hybrid consensus mechanism, application hardening, and trusted computing for privacy protection, more efficient consensus, and code obfuscation. Ref. [16] promotes awareness of blockchain technology and explores several use cases and applications suggesting to resolve bugs in blockchain technology. Apart from these, our paper emphasizes the following research trends and challenges:

4.1   Scalability

The issue of transaction scalability is significant for blockchain technology. The number of transactions that can be processed per second (TPS) varies widely between different blockchain platforms, from a maximum of 27 for Bitcoin to 3,996 for Electro Optical System (EOS). The consensus algorithm used by a blockchain also affects its TPS capabilities; for instance, PoW blockchains such as Bitcoin can process anywhere from 10 to 27 TPS worldwide. Furthermore, a sidechain architecture has been proposed for sidechain systems based on PoS to enhance their scalability [28]. Currently, Ethereum and Bitcoin are only capable of processing an average of approximately 5 KB or 10 TPS. On the other hand, HIE can achieve a significantly greater throughput of 8,000–20,000 TPS. It should be noted, though, that Visa's payment network was reported to be able to process more than 65,000 TPS as of August 2017 [29]. The fact remains that achieving scalability with blockchain technology, particularly in a truly distributed environment, continues to be a major challenge when it comes to TPS. Both IBM [30] and VeChain [24] provide supply chain solutions based on blockchain technology, but their scalability is limited by storing shared data directly on blockchain. This issue is compounded by the fact that there are typically a large number of stakeholders involved and a substantial amount of data that needs to be shared among them, which may not be limited solely to logistics data. To address this challenge and benefit from blockchain technology, it is possible to share data through an off-chain dedicated channel, while still recording data sharing proof on blockchain for purpose of tracking.

4.2   Securer Software Codes

Every year, there are occurrences of attacks on smart contracts and software code, underscoring the significance of security in any asset-related software. The smart contracts' security is of paramount importance as they manage valuable information like tokens, cryptocurrencies, and other digital assets. Smart contracts are designed to be immutable, meaning that transactions made through them cannot be reversed, and any bugs in their software code can be difficult to fix [31]. To ensure blockchain environment's security, there are certain limitations imposed on smart contracts.

4.3 Anomaly Detection, Zero Trust, and Audit

The process of auditing smart contracts is crucial before their deployment. A tool called Erays was introduced for converting the smart contract into high-level pseudocode, thereby analyzing it to evaluate various contract properties [33]. One potential avenue of research is to enhance the audit tools for smart contracts, with the goal of automatically auditing more, if not all, of the contract properties. Another research trend pertains to zero trust network access, which aims to provide continuous authentication of endpoints for improved endpoint security.

4.4 Monitor and Anomaly Detection

Securing blockchain systems requires continuous network monitoring and detecting attacks or anomalies. One research trend in this area is to leverage machine learning and deep learning techniques to analyze data, behaviors, logs, and transactions, in addition to the current approach based on parsing [33], to improve network security. One instance of using machine learning for attack detection is the detection of eclipse attacks by ETH-EDS, which employed random forest classification [34].

4.5 Privacy Preserving

As the quantity of data stored on the blockchain increases, organizations and individuals are becoming increasingly concerned about privacy leaks. To address this concern, various techniques such as homomorphic encryption, trusted execution platforms, code obfuscation, and smart contracts for preserving the privacy are promising directions to explore. American Association of Retired Persons (AARP) reported that in 2022, due to large-scale frauds and investment scams, Americans lost around $8.8 billion [32]. Various Blockchain-based enterprises such as Civic, Evernym Inc., Ligero and Ocular are working to prevent the identity using blockchain.

4.6 Impact of Quantum Computing

Elliptic Curve Digital Signature Algorithm is a commonly employed cryptographic algorithm for transaction signing. The algorithm computes a public key from a private key using an efficient one-way function based on elliptic curve multiplication. However, it is not feasible to compute the private key from the public key because of the hardness of the mathematical discrete logarithm problem. As a result, it is virtually impractical to compute the private key from the public key through reverse engineering. As such, users in blockchain can prove their ownership by signing a digital signature with their private key. Several major players such as Microsoft, IonQ, D-Wave, IBM, Rigetti, Google, and Intel are currently focusing on the research and development of quantum computing. Shor et al. introduced a quantum algorithm that has potential to undermine the security of commonly used public key cryptography algorithms [35].

4.7 IOTA Security

The issues of scalability and transaction fees faced by Bitcoin and Ethereum cryptocurrencies have led some to consider Internet of Things Application (IOTA) as a potential alternative. IOTA is distributed ledger for Internet of Things. Its use of Directed Acyclic Graphs instead of blocks and chains provides a different structure of vertices and edges. Through its Tangle technology, IOTA aims to achieve unlimited scalability and zero transaction fees. Curl raises concerns about its ability to store transactions' orders correctly and vulnerabilities in its proprietary hash function [36]. These issues must be addressed for IOTA to gain wider adoption. Once the technology is mature, it has a huge potential for adoption on a wide scale in the rapidly growing IoT industry [8].

5   Discussions

Three trends can be observed in the current development of blockchain technology. The first trend is the growing popularity of cryptocurrencies, which offer convenience and cost-saving benefits for fund transactions. However, this also poses a challenge to national financial policies and control. The second trend is the emergence of more international blockchain applications, such as the use of blockchain systems to confirm COVID-19 vaccine injection certificates. Therefore, it is essential to establish agreements and regulations between countries to unanimously recognize vaccination certificates kept on blockchain systems. The emergence of international blockchain applications, such as the blockchain-based verification of COVID-19 vaccine certificates, highlights the significance of this development. Thirdly, agreeing to use blockchain as a shared infrastructure within a country can be a considerable challenge, and establishing a common or international standard adds another layer of complexity. Thus, developing standards and regulations can be a significant hurdle for the widespread adoption of blockchain systems. To sum up, the range of blockchain components employed in the analyzed studies is wide and includes network structures with varying participant limitations, such as consortium and private networks, which offer greater control over record creation and access. Among the publications reviewed, Ethereum and Hyperledger Fabric are the most commonly utilized platforms, both supporting smart contracts, thus offering a wide range of features. To make the efficient smart contracts, various companies such as Chainlink Labs, DFINITY, and BurstIQ are using blockchain [32]. However, smart contracts are not as frequently used as one might expect. Hence, we believe that further exploration of smart contract capabilities and the use of less restrictive consensus algorithms can be some of the future directions to boost the integration of blockchain technology in the healthcare field.

6   Conclusion and Future Work

The security benefits and decentralized structure of blockchain have garnered significant interest in the realm of decentralized information systems. It offers a unique way of storing, sharing, and updating data, which would be important for future interactive systems like the supply chain systems and internet of things. However, the growing privacy concerns might present an obstacle to the extensive use of blockchain technology. This survey explores various challenges and presents a comprehensive analysis of different blockchain applications. By reviewing and discussing these mechanisms, we identify potential research directions for diverse application domains. We aim to present the current state of research in blockchain and its applications in different domains like healthcare, supply chain, and cryptocurrency domain. Owing to the sensitive nature of healthcare data, blockchain technology is considered to have substantial prospects for use in the industry. Moreover, since blockchain is a comparatively new technology in the field of healthcare, there is still room for discovering and researching new ways to utilize it. In conclusion, it is advisable to use blockchain in situations where it is necessary and justifiable.

References

1. Underwood, S.: Blockchain beyond bitcoin. Commun. ACM **59**(11), 15–17 (2016)
2. Xiao, Y., Zhang, N., Li, J., Lou, W., Hou, Y.T.: Distributed consensus protocols and algorithms. Blockchain Distrib. Syst. Sec. **25**(40), 1–31 (2019)
3. Kuznetsov, A., Oleshko, I., Tymchenko, V., Lisitsky, K., Rodinko, M., Kolhatin, A.: Performance analysis of cryptographic hash functions suitable for use in blockchain. Int. J. Comput. Netw. Inf. Sec. **13**(2), 1–15 (2021)
4. Bhadoria, R.S., Arora, Y., Gautam, K.: Blockchain hands on for developing genesis block. In: Advanced Applications of Blockchain Technology, pp. 269–278 (2020)
5. Gao, W., Hatcher, W.G., Yu, W.: A survey of blockchain: techniques, applications, and challenges. In: 27th International Conference on Computer Communication and Networks (ICCCN), pp. 1–11. IEEE (2018)
6. Gupta, S., Sadoghi, M.: Blockchain Transaction Processing, pp. 1–21 (2021). arXiv preprint arXiv:2107.11592
7. Nakamoto, S.: Bitcoin: A peer-to-peer electronic cash system. Decentralized Bus. Rev., 1–9 (2008)
8. Panarello, A., Tapas, N., Merlino, G., Longo, F., Puliafito, A.: Blockchain and iot integration: a systematic survey. Sensors **18**(8), 2575(1)-2575(37) (2018)
9. Zheng, Z., Xie, S., Dai, H., Chen, X. and Wang, H.: An overview of blockchain technology: architecture, consensus, and future trends. In: IEEE international congress on big data (BigData congress), pp. 557–564. IEEE (2017)
10. Aste, T., Tasca, P., Di Matteo, T.: Blockchain technologies: the foreseeable impact on society and industry. Computer **50**(9), 18–28 (2017)
11. Vokerla, R.R., et al.: An overview of blockchain applications and attacks. In: International Conference on Vision Towards Emerging Trends in Communication and Networking (ViTECoN), pp. 1–6. IEEE (2019)
12. Rosenfeld, M.: Analysis of Bitcoin Pooled Mining Reward Systems, pp. 1–50 (2011). arXiv preprint arXiv:1112.4980.
13. Borge, M., Kokoris-Kogias, E., Jovanovic, P., Gasser, L., Gailly, N., Ford, B.: Proof-of-personhood: redemocratizing permissionless cryptocurrencies. In: IEEE European Symposium on Security and Privacy Workshops (EuroS&PW), pp. 23–26. IEEE (2017)
14. Governments Explore Using Blockchains to Improve Service. https://www.nytimes.com/2018/06/27/business/dealbook/governments-blockchains-servivces.html. Accessed 30 Dec 2023
15. Zheng, Z., Xie, S., Dai, H.-N., Chen, X., Wang, H.: Blockchain challenges and opportunities: a survey. Int. J. Web Grid Serv. **14**(4), 352–375 (2018)

16. Dave, D., Parikh, S., Patel, R., et al.: A survey on blockchain technology and its proposed solutions. Procedia Comput. Sci. **160**, 740–745 (2019)
17. Joshi, A.P., Han, M., Wang, Y.: A survey on security and privacy issues of blockchain technology. Math. Found. Comput. **1**(2), 121–147 (2018)
18. Chen, W., Xu, Z., Shi, S., Zhao, Y., Zhao, J.: A survey of blockchain applications in different domains. In: International Conference on Blockchain Technology and Applications (ICBTA), pp. 17–21. ACM (2018)
19. Centre of Excellence - Blockchain Technology. https://blockchain.gov.in/. Accessed 27 Dec 2023
20. Bitcoin Average Transaction Fee. https://ycharts.com/indicators/bitcoin_average_transacti on_fee. Accessed 27 Dec 2023
21. Binance News. https://www.binance.com/en-IN/feed/post/473567. Accessed 27 Dec 2023
22. Cryptoslate, Coin rankings. https://cryptoslate.com/coins/. Accessed 27 Dec 2023
23. Blockchain in Healthcare Market Size. https://www.precedenceresearch.com/blockchain-in-healthcare-market. Accessed 30 Dec 2023
24. VeChain Solution Overview. https://vechain.com/solution/logistics. Accessed 27 Dec 2023
25. Agbo, C.C., Mahmoud, Q.H., Eklund, J.M.: Blockchain technology in healthcare: a systematic review. Healthcare **7**(2), 56(1)-56(30) (2019)
26. Wüst, K., Gervais, A.: Do you need a blockchain? In: Crypto Valley Conference on Blockchain Technology (CVCBT), pp. 45–54. IEEE (2018)
27. Li, X., Jiang, P., Chen, T., Luo, X., Wen, Q.: A survey on the security of blockchain systems. Future Gener. Comput. Syst. **107**, 841–853 (2020)
28. Gazi, P., Kiayias, A., Zindros, D.: 40th IEEE Symposium on Security and Privacy, pp. 139–156. IEEE (2019)
29. Visa fact sheet. https://usa.visa.com/dam/VCOM/download/corporate/media/visanet-techno logy/aboutvisafactsheet.pdf. Accessed 27 Dec 2023
30. IBM Blockchain Supply Chain Solutions. https://www.ibm.com/uk-en/blockchain/industries/ supply-chain. Accessed 27 Dec 2023
31. Zou, W., Lo, D., Kochhar, P.S., et al.: Smart contract development: challenges and opportunities. IEEE Trans. Softw. Eng. **47**(10), 2084–2106 (2019)
32. Blockchain Applications and Real-World Use Cases. https://builtin.com/blockchain/blockc hain-applications. Accessed 29 Dec 2023
33. Kalodner, H., Moser, M., Lee, K., et al.: BlockSci: design and applications of a blockchain analysis platform. In: 29th USENIX Security Symposium, pp. 2721–2738. USENIX Association, Berkeley, CA, USA (2020)
34. Xu, G., et al.: Am I eclipsed? A smart detector of eclipse attacks for ethereum. Comput. Secur. **88**(101604), 1–10 (2020)
35. Shor, P.W.: Algorithms for quantum computation: discrete logarithms and factoring. In: 35th Annual Symposium on Foundations of Computer Science, pp. 124–134. IEEE (1994)
36. IOTA Price Prediction 2021 and Beyond: What to Expect?. https://www.bitdegree.org/cry pto/tutorials/iota-price-prediction. Accessed 31 Dec 2023
37. Tian, H., Xue, K., et al.: Enabling cross-chain transactions: a decentralized cryptocurrency exchange protocol. IEEE Trans. Inf. Forensics Secur. **16**, 3928–3941 (2021)
38. Karantias, K., Kiayias, A., Zindros, D.: Proof-of-burn. In: 24th International Conference In Financial Cryptography and Data Security, pp. 523–540. Springer International Publishing, Kota Kinabalu, Malaysia (2020)
39. Bhola, J., Mohammad, S., et al.: Performance evaluation of multilayer clustering network using distributed energy efficient clustering with enhanced threshold protocol. Wirel. Pers. Commun. **126**(3), 2175–2189 (2022)
40. Rani, S., Deepika, K., et al.: An optimized framework for WSN routing in the context of industry 4.0. Sensors **21**(19), 6474(1)-6474(15) (2021)

A Survey on Datasets, Feature Extraction and Classification Techniques Used in Personality Classification from Handwriting

Parul Garg[✉] and Naresh Kumar Garg

Department of Computer Science and Engineering, GZSCCET,
MRSPTU, Bathinda, Punjab, India
parul2707garg@gmail.com, naresh_cse@mrsptu.ac.in

Abstract. Prediction of demographic characteristics of a person from handwriting is an upcoming new area of research as previous research mainly focused on the recognition of characters, words or paragraph. Predicting demographic characteristics like age, gender, handedness, nationality and many more can aid in applications like bank cheque processing, document forgery systems etc. The research on prediction of mood from handwriting lacks somewhere and its research can help in medical line to check the stress level of the person. Since there is no detailed survey on this so, this paper presents a detailed literature on the datasets, feature extraction and selection, and classification techniques. It is observed that CNN, hybrid CNN-RNN, SVM, Random Forest and many more gave good results for the classification. The future challenges in this research area are proposed.

Keywords: SVM · Classification · Handwriting recognition · personality traits · Random Forest

1 Introduction

Handwriting is considered to be an important factor of communication. The written communication encompasses number of hidden demographic stats like gender [6], age [5], nationality [5], handedness [21], and personality traits like stress, mood and anxiety [16]. The machine learning has wide number of applications in areas like Writer Identification, Radio Wireless Networks [8]. There are many research developments like prediction of gender [32], age and handedness [35] and prediction of stress, anxiety [2] etc. from handwriting [1] have been carried out by using online and offline handwriting datasets. The data is to be pre-processed for the applicability of machine learning algorithms. The basic process of machine learning includes the following steps:

a) Data Collection
b) Data Pre-processing
c) Feature selection and extraction
d) Classification

M. Khurana et al. (Eds.): ICMLA 2024, CCIS 2238, pp. 287–296, 2025.
https://doi.org/10.1007/978-3-031-75861-4_25

1.1 Basic Process of Machine Learning

Data Collection

The process of collecting data samples either in online or offline mode is known as data collection. The online handwriting dataset is made by using e-beam interface which tracks the pen movements of the person and generates their corresponding X, Y coordinates [3]. The offline dataset is made with the help of OCR, their image dataset is created [3]. Many researchers have built their own datasets in different languages like English, Telugu, Urdu, Arabic that can be used for the task of predicting the personality traits. The Table 1 shows the datasets mostly used in the previous studies along with their applications.

Table 1. Datasets used in Literature

Dataset Name	Dataset Language and number of participants	Online/Offline	Dataset publication year	Papers that refer the dataset	Advantages
IAM	English and 657 for Offline dataset 221 for Online dataset	Both	1999	[3, 21]	Word recognition, mental condition prediction
QUWI	Arabic and English 1017	Offline	Unknown	[4, 22]	Gender Classification
KHATT	Arabic and 1000	Offline	2015	[21]	Age, Gender and handedness prediction
MSHD	French and Arabic and 100	Offline	Unknown	[4, 23]	Signature Verification, writer identification
IIIT-HW-Telugu	Telugu and 11	Offline	2015	[25]	Word recognition
CEDAR	English and 55	Both	1994	[3]	signature verification
ADAB	Arabic and 165	Online	2009	[24]	Handwriting recognition

Data Pre-processing

The data from data collection is the raw form which cannot be used directly for further processing. Firstly, the collected data has to be pre-processed. The process of pre-processing include data cleaning, data integration, data transformation and data reduction. Data Cleaning refers to the process of cleaning the data by removal of noisy data and filling the missing values. Data Integration is the process of merging data from multiple sources into a single data source. After the process of data cleaning, cleaned data is represented in some other form by either changing the value, structure or format as per requirement of algorithm of machine learning. The data transformation is mainly done with the help of generalization and normalization. Data warehouse contains a large amount of data which is difficult to process so data is reduced to some extent. The dimensionality reduction method is most commonly used for the same.

Feature Extraction and Selection

Features refer to the data values which can project the useful information about the data. The dataset contains large number of features. Some features are selected and extracted using some algorithms. Most of the researchers have used Convolutional Neural Networks [11] for the same. The geometric features [3], textural features [6], gradient features [3], local features and global features [3] can be used for this. The detail of each feature is given in this section.

Geometric Features

The features which are built from some arithmetic operations are known as geometric features. The arithmetic features include curves, lines, and surfaces. Many researchers [20] used this feature for the application of writer identification. The four types of geometric features are- Tortuosity feature, Curvature feature, Chain code features and edge based directional features. The identification of fast or slow writer [20] is done with the help of tortuosity.

Gradient Features

The features that calculate the gradient value for each pixel of an image is known as gradient features. The examples are Histogram of Oriented Gradient (HOG) and Global Local Binary Pattern (GLBP). The Table 2 depicts the different features used in previous studies.

Table 2. List of features extracted in literature

Author Name	Year	Dataset Used	Features Extracted	Single feature or ensemble features
Manogna Pallapothu et.al [2]	2021	IAM	Baseline Angle, letter size, line spacing, word spacing, top margin features are extracted	All are single features

(continued)

Table 2. (*continued*)

Author Name	Year	Dataset Used	Features Extracted	Single feature or ensemble features
Maadeed And Hassaine [5]	2014	QUWI	direction, curvature, tortuosity, chain code, edge-based directional Feature	All are single features
Mahreen Ahmeda et.al [6]	2017	QUWI	Word Error Rate and Character Error Rate are calculated	Ensemble Classifiers
Nesrine Bouadjenek et. al [26]	2014	IAM	Histogram of gradients, Local binary patterns	Both are single features
Hussaine et al. [27]	2013	QUWI	Chain code features and edge based features	Both are single features
Chitlangia et al. [31]	2019	Own dataset	Histogram of gradients	Single features
Srihari et al. [36]	2001	Own Dataset	Line separation, character shapes, slant	Single features
Al-Maadeed et al. [20]	2014	IAM and QUWI	direction, curvature, and tortuosity	Single features

Classification

Classification is the process of categorizing the data elements into different categories or classes on the basis of some properties. There are many algorithms available for the classification like K Nearest Neighbor, Support Vector Machine [34], Decision Tree, and Random Forest etc. Among these, SVM [13, 37] can be used with SVMDT multi class classifier [14]. The Table 3 depicts the classification techniques used in previous studies.

Table 3. Classification Techniques used with results.

Author Name	Year	Dataset Used	Classification technique used	Research findings
Kartik Dutta, Praveen Krishnan, Minesh Mathew and C.V. Jawahar [28]	2018	IIIT-HW-Dev, IAM and RoyDB	CNN RNN Hybrid architecture	Calculated word error rate and character error rate

(*continued*)

Table 3. (*continued*)

Author Name	Year	Dataset Used	Classification technique used	Research findings
Ali Mirza, Momina Moetesum, Imran Siddiqi and Chawki Djeddi [29]	2016	QUWI and MSHD	Feed forward neural network	68% classification rate
Manogna Pallapothu, Pragati Shinde, and Vinish Marito [2]	2021	IAM dataset	Support Vector Machine (SVM)	Anxiety prediction accuracy of 98.58% is achieved
Mina Rahmanian, Mohammad Amin Shayegan [7]	2021	KHATT and IAM	CNN	Gender and handedness classification accuracy of 84% for IAM and 84% accuracy for KHATT
Shaveta Dargan and Munish Kumar [30]	2021	own dataset from 200 writers	SVM, MLP, KNN, Random forest	Gender classification accuracy of 90.57% and writer identification accuracy of 87.76% is reported
Bi et al. [12]	2019	Own Dataset	SVM	slant, curvature, line separation, chain code, character shapes
Payal Maken, and Abhishek Gupta [15]	2021	Own Dataset	SVM	Features like slanteness, area and perimeter are considered
Morera et al. [19]	2018	IAM and KHATT	CNN	Handedness and gender prediction accuracies of 83.19% and 68.90%

2   Literature Survey

Joshi et al. [1] used 100 handwriting samples for prediction of personality trait. Manogna Pallapothu et al. [2] classified the IAM dataset on the basis of stress, anxiety and depression with an accuracy of 98.58% with SVM classifier. Maken Payal et al. [3] presented survey on different datasets, techniques used for feature extraction especially gradient features like histogram of gradient, local binary pattern and classification. Siddiqi et al. [4] achieved accuracy of 68.75% is achieved for QUWI database and 73.02% accuracy is achieved for MHSD database using ANN and SVM. Somaya Al Maadeed and Abdelaali Hassaine [5] predicted the personality traits from QUWI dataset using ensemble features and achieved 74.05% accuracy for gender prediction, 55.76% accuracy for age range prediction and 53.66% for nationality prediction. Ahmed et al. [6] identified gender by using textural features. Classification is done using artificial neural networks (ANN), support vector machine (SVM), nearest neighbor classifier (NN), decision trees (DT) and random forests (RF). Mina Rahmanian et al. [7] used Convolutional Neural Network to classify handwritings on the basis of gender and handedness on two databases named IAM and KHATT. The accuracy achieve in gender classification is 84% and in handedness classification is 99.14%. Dargan et al. [8] presented a survey on deep learning techniques and its applications. U. Pal and B.B. Chaudhuri [9] surveyed the work done for handwriting recognition in different scripts like Bangla, Gujarati, kannada, Kashmiri, Devnagri, Oriya etc. Mahmoud et al. [10] presented an offline Handwritten Text database in Arabic language, KHATT. The writers developed tools to extract sentences, paragraphs from text documents or XML documents. Hidden Markov Model (HMM) and syntactic classifier are used for classification. Weixin Yang, Lianwen Jin, and Manfei Liu [11] introduced a system called DeepWriterID for writer identification. They achieved an identification rates of 95.72% for Chinese text and 98.51% for English text. Bi et al. [12] determined gender of a person from their handwriting samples by proposing KMI. The classification was done by SVM. Batta Mahesh [13] presented a review of machine learning techniques used for number of applications like forensics department, medical department, food science department, robotics and many more. Zhang et al. [14] developed Triboelectric Nanogenerator. The accuracies achieved for three languages were: 99.66%, 91.36% and 93.63% respectively. Payal Maken and Abhishek Gupta [15] identified gender from handwriting using SVM. Nishigandha Vyawahare et al. [16] considered pen pressure as a feature to evaluate the stress from handwriting. The handwriting samples from 100 persons are taken. Monica S et al. [17] extracted number of features from handwriting of a person to know his/her mental state. The Classification is done using Convolutional Neural Networks. Shaveta Dargan and Munish Kumar [18] determined the writer or author of the text from their handwriting. Their study is related to the writing styles, feelings, perception and behavior. Morera et al. [19] predicted gender and handedness on IAM and KHATT database and achieved 83.19% accuracy for handedness prediction and 68.90% accuracy for gender prediction. Al-Maadeed et al. [20] achieved accuracy of 82% and 87% on IAM and QUWI datasets. Nesrine Bouadjenek, Hassiba Nemmour and Youcef Chibani [21] predicted age, gender and handedness from HOG and GLBP features using SVM on IAM and KHATT datasets and achieved accuracy of 70%. Mirza et al. [22] classified gender from QUWI dataset using feed forward neural network with an accuracy of 68%. Djeddi Chawki et al. [23] introduced a new multi-script offline handwriting database named LAMIS-MSHD. Azeem et al. [24] Presented handwriting recognition database in Arabic language. They manually segmented the Arabic Characters database from existing database ADAB. Dutta Kartik et al. [25] proposed a new dataset named IIIT-HW-Telugu. The dataset is interpreted using UTF-8. Their dataset contains 12945 words. Nesrine Bouadjenek, Hassiba Nemmour and Youcef Chibani [26] proposed the use of local descriptors like Histogram of Gradient and Global Local Binary Predictor on IAM-OnDB dataset. The accuracy rate achieved was 70%. The task is done using SVM classifier. Al Maadeed et al. [27] presented a new dataset named Qatar University Writer Identification dataset (QUWI). This dataset is combination of both Arabic and English Handwriting samples from 1017 writers. Kartik Dutta et al. [28] released a new hand-written word dataset for Devanagari named IIIT-HW-Dev. They have used CNN-RNN hybrid architecture. Ali Mirza et al. [29] presented classification of writer demographics from offline handwritten documents using QUWI dataset Shaveta Dargan and Munish Kumar [30] classify the Gurmukhi Handwriting scripts on the basis of gender and writer identification. Number of classifiers namely SVM, NN, RF and MLP are used to get the classification results. An accuracy of 90.57% is achieved for gender classification and 87.67% is achieved for writer identification. Aditya Chitlangia, G.Malathi [31] catego-rized authors into five personality traits- Introvert, Extrovert, Sloppy, Optimistic, and Energetic. HOG feature was extracted using SVM and achieved an accuracy rate of 80% by using polynomial kernel. Cordasco et al. [32] performed the gender identification from online handwriting and drawing analysis. They took the samples from 126 males and 114 females. Maadeed et al. [33] proposed a model for the detection of handedness feature from handwriting. They stated performance of 67.06% for gender classification and 84.66% for handedness classification. Yousaf et al. [34] provided details about a new dataset known as Handwritten Characters Dataset (HCD). Erika Griechisch and Erika Bencsik [35] predicted handedness from online handwriting samples. They achieved an accuracy of 83.55% with Left-Condition (LC) method and 85.97% with Majority Voting (MV) method. Sargur N. Srihari, Sung-Hyuk Cha and Sangjik Lee [36] achieved accuracy of 98% for writer identification. Fahimeh Alaei & Alireza Alaei [37] presented a comparative study for demographic attributes detection from handwriting based on machine learning and deep learning approach.

3   Gaps in Research

It has been observed that most of the work regarding prediction of personality traits is done on Online Available databases. The amount of work on offline databases is very less as compare to the online databases. In addition to this, mostly research is carried out on the prediction of gender, age, handedness and nationality but somehow, very less attention is paid to feelings, emotions, state of mind of the writer during writing the document. The previous study focuses on the English, Arabic, Gurmukhi scripts. The consideration of Devanagari script is lacking behind somewhere. The classification is carried out using SVM, CNN algorithms. There could be more algorithms which can yield good consequences on ensemble the different features.

4   Methodology

This section presents the basic process of classification of different demographic stats followed by most of the research developments (Fig. 1).

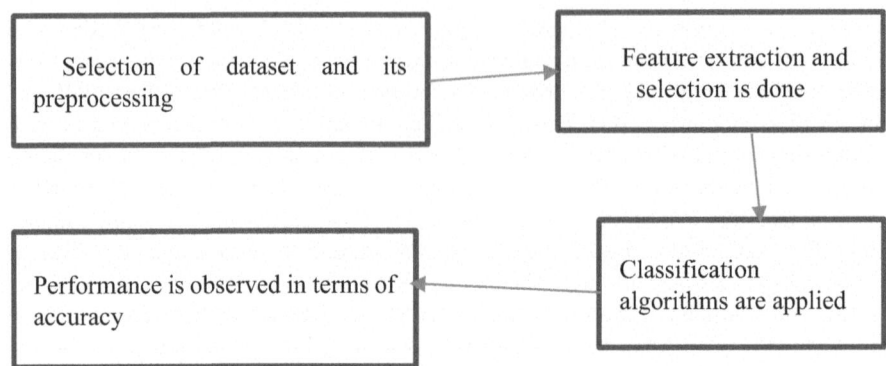

Fig. 1. Methodology used for demographic stats classification

The dataset is selected from number of available datasets and preprocessing is done using different techniques. After that, features are extracted using different feature extraction techniques and required features are selected. The extracted features are fed to classification algorithms and their performance is observed in terms of accuracy.

5 Conclusion

This paper presents a comprehensive study of the previous work done for classification of demographic stats from handwriting that has great impact in fields like forensic investigations, document examination. It is observed that most of the research developments are done in different languages like English, Urdu, Punjabi, Telugu and many more but there is a need to pay attention on Hindi language due to lack of availability of standard Hindi dataset. There is a need to pay focus on standard Hindi dataset. This paper proposes a new area for researchers to explore the advanced machine learning, deep learning algorithms for Hindi language since most of the research findings are based on usage of traditional machine learning approaches.

References

1. Joshi, P., Agarwal, A., Dhavale, A., Suryavanshi, R., Kodolikar, S.: Handwriting analysis for detection of personality traits using machine learning approach. Int. J. Comput. Appl. **130**(15), 40 (2015)
2. Manogna, P., Pragati, S., Vinish, M.: Mental health analysis using handwriting by generating writing prompts. Int. J. Creat. Res. Thoughts (IJCRT) **9**, 2320–2882 (2021)
3. Maken, P., Gupta, A., Gupta, M.K.: A study on various techniques involved in gender prediction system: a comprehensive review. Cybernet. Inf. Technol. **19**(2), 51–73 (2019)
4. Siddiqi, I., Djeddi, C., Raza, A., Souici-Meslati, L.: Automatic analysis of handwriting for gender classification. Pattern Anal. Appl. **18**(4), 887–899 (2015)
5. Al Maadeed, S., Hassaine, A.: Automatic prediction of age, gender, and nationality in offline handwriting. EURASIP J. Image Video Process. **2014**(1), 1–10 (2014)
6. Ahmed, M., Rasool, A.G., Afzal, H., Siddiqi, I.: Improving handwriting based gender classification using ensemble classifiers. Expert Syst. Appl. **85**, 158–168 (2017)

7. Rahmanian, M., Shayegan, M.A.: Handwriting-based gender and handedness classification using convolutional neural networks. Multimed. Tools Appl. **80**, 1–24 (2021)
8. Dargan, S., Kumar, M., Ayyagari, M.R., Kumar, G.: A survey of deep learning and its applications: a new paradigm to machine learning. Arch. Comput. Methods Eng. **27**, 1–22 (2019)
9. Pal, U., Chaudhuri, B.B.: Indian script character recognition: a survey. Pattern Recog. **37**(9), 1887–1899 (2004)
10. Mahmoud, S.A., et al.: KHATT: an open Arabic offline handwritten text database. Pattern Recogn. **47**(3), 1096–1112 (2014)
11. Yang, W., Lianwen, J., Manfei, L.: Deepwriterid: an end-to-end online text-independent writer identification system. IEEE Intell. Syst. **31**(2), 45–53 (2016)
12. Bi, N., Suen, C.Y., Nobile, N., Tan, J.: A multi-feature selection approach for gender identification of handwriting based on kernel mutual information. Pattern Recogn. Lett. **121**, 123–132 (2019)
13. Mahesh, B.: Machine learning algorithms-a review. Int. J. Sci. Res. (IJSR) **9**, 381–386 (2020)
14. Zhang, W., et al.: Multilanguage-handwriting self-powered recognition based on triboelectric nanogenerator enabled machine learning. Nano Energy **77**, 105–174 (2020)
15. Maken, P., Gupta, A.: A method for automatic classification of gender based on text-independent handwriting. Multimed. Tools Appl. **80**, 1–30 (2021)
16. Vyawahare, N., Ashtaputre-Sisode, A.: Relation between stress, anxiety and handwriting. Journal of the Maharaja Sayajirao University of Baroda ISSN 25: 0422
17. Monica, S., Kumar, K., Anaga, A., Anagha, S.: Mental health status detection through handwriting analysis. Int. J. Res. Appl. Sci. Eng. Technol. IJRASET **10**, 981 (2022)
18. Dargan, S., Kumar, M.: Writer identification system for indic and non-indic scripts: state-of-the-art survey. Arch. Comput. Methods Eng. **26**(4), 1283–1311 (2019)
19. Morera, Á., Sánchez, Á., Vélez, J.F., Moreno, A.B.: Gender and handedness prediction from offline handwriting using convolutional neural networks. Complexity (2018)
20. Al-Maadeed, S., Hassaine, A., Bouridane, A., Tahir, M.A.: Novel geometric features for off-line writer identification. Pattern Analy. Appl. **19**(3), 699–708 (2016)
21. Bouadjenek, N., Nemmour, H., Chibani, Y.: Age, gender and handedness prediction from handwriting using gradient features. In: 2015 13th International Conference on Document Analysis and Recognition (ICDAR), pp. 1116–1120. IEEE (2015)
22. Mirza, A., Moetesum, M., Siddiqi, I., Djeddi,C.: Gender classification from offline handwriting images using textural features. In: 15th International Conference on Frontiers in Handwriting Recognition (ICFHR), pp. 395–398. IEEE (2016)
23. Djeddi, C., Gattal, A., Souici-Meslati, L., Siddiqi, I., Chibani, Y., El Abed, H.: LAMIS-MSHD: a multi-script offline handwriting database. In: 14th International Conference on Frontiers in Handwriting Recognition, pp. 93–97. IEEE (2014)
24. Azeem, S.A., Ahmed, H.: Recognition of segmented online Arabic handwritten characters of the ADAB database. In: 10th International Conference on Machine Learning and Applications and Workshops, vol. 1, pp. 204–207. IEEE (2011)
25. Dutta, K., Krishnan, P., Mathew, M., Jawahar, C.V.: Towards spotting and recognition of handwritten words in indic scripts. In: 16th International Conference on Frontiers in Handwriting Recognition (ICFHR), pp. 32–37. IEEE (2018)
26. Bouadjenek, N., Nemmour, H., Chibani, Y.: Local descriptors to improve off-line handwriting-based gender prediction. In: 6th International Conference of Soft Computing and Pattern Recognition (SoCPaR), pp. 43–47. IEEE(2014)
27. Al Maadeed, S., Ayouby, W., Hassaine, A., Aljaam, J.M.: QUWI: an Arabic and English handwriting dataset for offline writer identification. In: 2012 International Conference on Frontiers in Handwriting Recognition, pp. 746–751. IEEE (2012)

28. Dutta, K., Krishnan, P., Mathew, M., Jawahar, C.V.: Offline handwriting recognition on devanagari using a new benchmark dataset. In: 2018 13th IAPR International Workshop on Document Analysis Systems (DAS), pp. 25–30. IEEE (2018)
29. Mirza, A., Moetesum, M., Siddiqi, I., Djeddi, C.: Gender classification from offline handwriting images using textural features. In: 2016 15th International Conference on Frontiers in Handwriting Recognition (ICFHR), pp. 395–398. IEEE (2016)
30. Dargan, S., Kumar, M.: Gender classification and writer identification system based on handwriting in gurumukhi script. In: 2021 International Conference on Computing, Communication, and Intelligent Systems (ICCCIS), pp. 388–393. IEEE(2021)
31. Chitlangia, A., Malathi, G.: Handwriting analysis based on histogram of oriented gradient for predicting personality traits using SVM. Procedia Comput. Sci. **165**, 384–390 (2019)
32. Cordasco, G., et al.: Gender Identification through Handwriting: an Online Approach. In: 2020 11th IEEE International Conference on Cognitive Info communications (CogInfoCom), pp. 197–202. IEEE (2020)
33. Al-Maadeed, Somaya, Fethi Ferjani, Samir Elloumi, Abdelaali Hassaine, and Ali Jaoua. "Automatic handedness detection from off-line handwriting." In 2013 7th IEEE GCC Conference and Exhibition (GCC), pp. 119–124. IEEE(2013)
34. Yousaf, A., Jaleed Khan, M., Imran, M., Khurshid, K.: Benchmark dataset for offline handwritten character recognition. In: 2017 13th international conference on emerging technologies (ICET), pp. 1–5. IEEE (2017)
35. Griechisch, E., Bencsik, E.: Handedness detection of online handwriting based on horizontal strokes. In: 2015 13th International Conference on Document Analysis and Recognition (ICDAR), pp. 1272–1277. IEEE (2015)
36. Srihari, S.N., Cha, S.-H., Lee,S.: Establishing handwriting individuality using pattern recognition techniques. In: Proceedings of Sixth International Conference on Document Analysis and Recognition, pp. 1195–1204. IEEE (2001)
37. Alaei, F., Alaei, A.: A comparison of demographic attributes detection from handwriting based on traditional and deep learning methods. In: International Conference on Document Analysis and Recognition, pp. 167–179. Springer Nature Switzerland, Cham (2023). https://doi.org/10.1007/978-3-031-41501-2_12

Image Processing

A Generic Approach for Detection of Copy-Move Forgery Detection Scheme from Digital Images

R. Sudha[1], V. Akilandeswari[2], B. Durgalakshmi[3], K. Palaniammal[4], and Mahesh K. Singh[5($\boxtimes$)]

[1] Department of Computer Science and Engineering, Sathyabama Institute of Science and Technology, Chennai, India
[2] Department of Computer Science and Engineering, Velammal College of Engineering and Technology, Madurai, India
akilandeswari.v2024@gmail.com
[3] Tagore Institute of Engineering and Technology, Deviyakurichi, Attur, India
durgalakshmi.cse@tagoreiet.ac.in
[4] Department of Computer Science, Government Arts and Science College for Women, Nilakottai, Tamilnadu, India
drkpalaniammal@gmail.com
[5] Department of ECE, Aditya University, Surampalem, India
mahesh.092002.ece@gmail.com

Abstract. Digital forensics is a research area designed by retrieving historical records to verify the originality of digital images. Digital photos are rendered easily using various techniques that change the images' meaning and authenticity. The structures and algorithms for the recognition and classification of digital images into genuine and classified images within a category are included in this article. There was a review and comparison of optimization-based methods for detecting and categorising forgeries. The work proposes the identification of forgery using copy-moving pictures to illustrate how you can apply the forgery detection classification. This paper concerned the extraction features based on "Bi-orthogonal Wavelet Transform with Single Value Decomposition" (BWT-SVD) based transform transforms. New approaches and extensions to detect moving copies with an effective classification using hybrid feature extraction are described in this paper. Both methods address the issue of classifying genuine and false images in a noisy environment, and split imaging is one of them as well.

Keywords: Image forgery · BWT · Digital forensics · Image Identification · Classification

1 Introduction

The use of digital images in our daily lives has increased every day due to massive technological development. The digital picture has proven simpler and less comprehensible because of this classification. Today's technology has started to disintegrate picture authenticity, to falsify and to falsify by switching to Mega Pixels, thereby giving a new

M. Khurana et al. (Eds.): ICMLA 2024, CCIS 2238, pp. 299–309, 2025.
https://doi.org/10.1007/978-3-031-75861-4_26

route for falsehood with the aid of digital technology. Imitation is not new to mankind, its past-generation issues [1]. The manufacture and composition in the past had little influence on the whole population. Image-making is practically as ancient as painting. Photography has been the traditional and fascinating art for creating portraits for two decades. And photographers earn income by making classification possible by replacing their photographs. Picture classification has gained more acknowledgment and unbeliev-able study in various fields, such as computer display, imaging processing, biomedical equipment, unethical analysis, forensics, etc. [2]. It was difficult to verify if an image was affected by the naked eyes by advanced software. It received more publicity and demanded [3].

When we come to the sentence "Pictures speak louder than words", this means that the point is with a picture [4]. You can convey much more information than you can with words. A picture may be placed in the form of a difficult-to-understand definition. This makes it easy to use fake image manipulation to trick others. Manipulation of pictures nothing but fake news has become a most noticeable problem in the present day of the situation [1, 5, 18]. Fake news output in recent years has been active for two key reasons: firstly, the cost of decreasing image-generation technologies. Software for photo editing is now readily accessible and image editing is too straightforward [6].

The development of a fake picture is based on a new software computer graphics version, frequent in the photoshop, GIMP, and Corel paint store. The edited area is sometimes very hard to identify from the original image. Two modes of framing are realistic (watermarking, splicing) and passive (copy-moving attack). Copy-move forgery is a degree of classification when part of an image is copied and pasted into the same picture at a different location. For creating a false image a new computer graphics editing software is simple and is available mostly in photoshop and BWT-SVD paint shop [2, 7]. It is also very difficult to discern the edited region of the original image. Classification of the image plays a significant role to render the image real in forensics. Figure 1 divides the methods of picture falsehood into two methods.

Perhaps a copy transfer fake detection is the most vigorously investigated topic. Nowadays, It has become the most common and important digital forensic technology. The copy-moved pieces from the source image are difficult to detect [8]. Thus, features like noise and color match the rest of the images [9]. Thus, few techniques for forgery detection based on image characteristics are not applicable. Copy-move classification can be detected in two ways:

(1) Block-based copy-move forgery detection
(2) Copy-move key point-based detection of classification.

Both methods are used to define the BWT-SVD when determining the reconstruction parts of a single image. The forgery method based on a block divides the original images into superimposed and regular image blocks and matches the various bits [9, 10]. The key points are taken by identifying the reconstruction regions and matched with the whole image in the key points-based process.

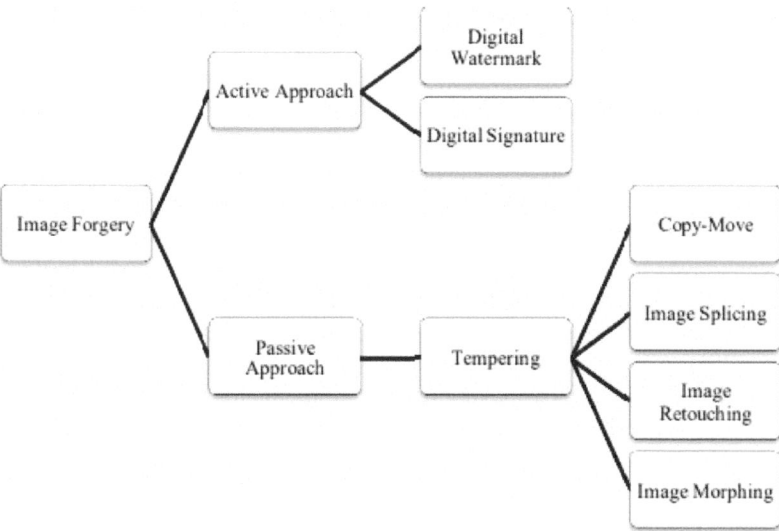

Fig. 1. Classification of forgery image

2 Related Work

Digital picture plays a vital role in real-time technology and computation. Modern technology in the field of digital cameras, computers, and software tools has become easy to modify and manipulate images [1, 12]. We must carry out further studies with modern image processing and algorithms to enhance the detection of falsifying. Few scholars have in recent times worked and researched issues concerning the recognition of the forged image and cannot detect the difference between simple image handling and the original. The learning and optimization of the machines is used to achieve optimal results. This thesis explores digital forensics and forms and focuses on copy-move classification [13].

This is described different forgery detection strategies in the movement of a copy. The survey is presented about optimization techniques for better classifications based on copy-move forgery detection. Based on optimization technology for the current identification and classification of forged movements, comparative research and analysis were conducted [2]. The topic also covers the classification of vector machines so that a hybrid mix of extreme learning machines with Fuzzy can be detected effectively. For successful picture feature training, several neural networks based on copy-move forgery detection were also discussed.. In the final section, the limitations of current techniques and the survey are defined [14].

For the setup of a test image, Functional vector elements and the general law are restricted using DCT. The method employs equity between elements one by one. Functional vectors, instead of the Euclidean distance or transverse relation, so that the image testing is used to determine threshold values [15]. The results show that copied and pasted areas can in various circumstances be detected and that higher precision ratios with a minimal incorrect rate compared with existing algorithms can be achieved. The demo function invariant to flutter or noise is discussed. These extracted properties are invariant to few geometric changes such as scaling and rotation that could be linked to the copied region before insertion [3]. As an indicator and bloom filter counter for the improvement of time complexity (Fourier Mellon Transform (FMT)). The results show that in this FMT function block JPEG compression is more energetic [4].

There is a robust method to detect copy-move. The DCT and quantitative division coefficients are quantified. This approach was applied to any image further divided into sub-blocks that do not overlap [16]. By utilising the greatest unique value of each block, SVD is employed to decrease the block size of the extracted attribute. A histogram is used to sort the vectors, and picture blocks are recognised based on how similar they are to each other and the shift frequency [17]. The copy forgery method successfully prevents JPEG, AWGN, and Gaussian bubbles, according to the experimental results. It is analyzed that video copy-moving forgery differentiates the area of video smithy. To eliminate noise and identify the forged field, this method used several preprocessing steps. You can edit your pictures and double-motion using the video editing application. Video frame from another position is used for a special segment to recognize the copied and pasted frame. In the extraction process, the SIFT algorithm is used to form the corners. The experimental results indicate the qualified processes for noise reduction to enhance image quality [18]. This method works with an effective rating accuracy of time pictures.

The proposed detection of classification the procedure of reducing columns and rows is the foundation of adopting DCT and IDC technologies. The new approach simplifies picture computation while cutting costs and saving time. The image was originally divided into columns and rows into grids. DCT is associated with each row and column using lines and segments and has changed with different estimates in different sections [6]. Finally, the copied image is processed from the containment point of view. Here is constructed (Polar Harmonic Transform (PHT) that matches the block for copy-move classification to be detected. The features extracted from the circular block by PHT include endless image features. These blocks are compared with PHT features. MATLAB generates the input images from data sets that are open to the public. The output results show the robust, compressive JPEG and object-rotating system proposed [7]. A method for copying the detection phase of image fraud by detecting the irregularity of the noise variation in HSV space shading has been proposed here [8].

The picture is initially moved from RGB to HSV color space and is then divided into different block sizes. The forged images were haphazardly decorated in various places with white Gaussian noise. The results of the assessments indicate the best results from the sound assessment for 32/32 panels [9]. It is suggested that an invariant rotation be selected that shares the benefits of vectors shift at an exclusively very extensive machine cost (SATS). Thus, the technique simply recoups the comparatively changing copied field parameters. This is the process. The results show that, independently of the image size, SATS exceeds rotating vectors when the region is copied [10].

Digital image forgery has attracted significant attention and is a field of a recent study. Researchers are founded on the truths of images and recordings [11]. The forensic image detected such counterfeiting to avoid illegal problems due to rapidly evolving images. Many forged images are detected using techniques, but accuracy and time complexity must also be emphasized. In some cases, few policies are ideal for broken, noisy, or cutting the copied element. Few methods are nice because they are rotated and scaled efficiently with fewer computational complications [12]. Further analysis needs to be done with image processing techniques and other algorithms for better detection of forgings. Recently, a few scholars participated in and analyzed the identification of the counterfeit image and could not detect the contrast of basic knowledge regarding manipulation with the original image. The learning and optimization of the machines are used to achieve optimal efficiency. The main subject of this survey is copy-moving classification with digital forensics and their forms [8, 9].

3   Methodology for Digital Image Forgery Detection

The work proposes a method for the identification of forgery using copy-moving pictures to illustrate how you can apply the forgery detection classification. Image processors such as preprocessing, feature extraction, match similarity, and classification are used as image authentication or forgery. There is a lot of procedure in digital image forensics. A variety of methodologies and detection techniques are applied to analyze the image. It was necessary to use two images to identify the forgery, one of which was the original and one forgery. The service offered is very creative nowadays, but there are some issues with picture forgery methods. For each of the two groups, you need to use class-based methods for image sampling and classification. Picture classification employs a variety of different methods such as painting, image enhancement, duplication, and photo manipulation. It is shown in Fig. 2.

A fabricated image can be difficult to understand from the eyes of the viewer. He has copied the region from an image and pasted it in another identical region in another image. It is imperative to ensure that such artifacts are checked in this manner. The availability of state-of-of-the-the-the-the-the-art software currently boosts the integrity and reliability of the digital picture In the 'real world', such as newspapers and photography, people have to deal with deceptive photo processing. The focus of this investigation has been on copying issues such as they are a big concern in digital forensics. There are many ways to manage these issues. Geometric transformations in particular, such as rotation, scaling, and particularly time complexity, are problematic with current strategies. Dates in this section: preprocessing, extractions of functions, and classification of the copy-moved images different algorithms are applied to each task.

It uses several well-known benchmarks, such as the confusion matrix, to test the copy-move method Determine which proposals areas equivalent to which already exist in practice. The parameter for example G-mean, F-measure, Recall, Precision, Accuracy, Specificity, and Sensitivity are utilized. These parameters are evaluated separately for each image and the average image value for the following parameters is calculated.

True Positive **(Tp):** constitutes a positive outcome of forgery image classification

True Negative **(Tn):** is the negative outcome of the forgery image classification

False Positive **(Fp):** refers to a good outcome for negative image forgery classification

False Negative **(Fn):** shows a negative outcome for a positive forgery picture classification

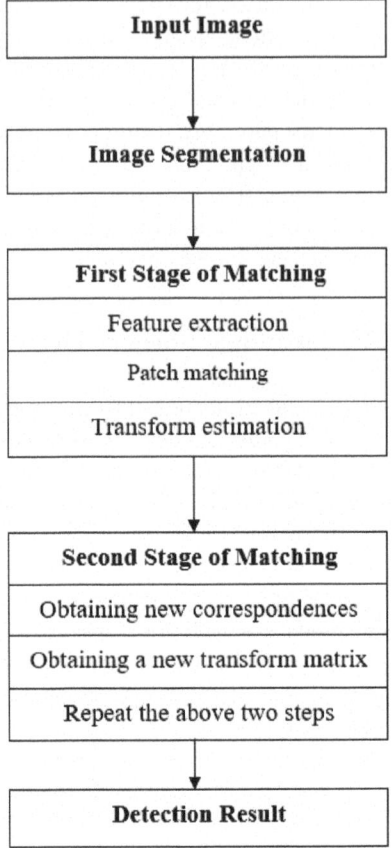

Fig. 2. Flowchart of digital image forgery detection

False Negative Rate **(FNR):** It is the percentage of positive results that the classification produces

$$\text{False Negative Rate}(FPR) = 1 - sensitivity \tag{1}$$

False Positive Rate **(FPR):** It is the percentage of positive results that the classification produces.

$$\text{False Positive Rate}(FPR) = 1 - specificity \tag{2}$$

Negative Predictive Value **(NPV):** It is the chance of a negative test that shows the lack of a classification picture.

$$\text{Negative Predictive Value (NPV)} = \frac{(T_n)}{(T_n + F_p)} \tag{3}$$

Positive Predictive Value (**PPV**): It shows that there is a chance of a positive test showing a fake picture.

$$Positive\ Predictive\ Value\ (PPV) = \frac{(T_p)}{(T_p + F_p)} \tag{4}$$

F-Measure: This is the test accuracy measurement and it is measured as:

$$F - Measure = 2\frac{(p.r)}{(p + r)} \tag{5}$$

Recall (**r**): The unity of the process is a fraction of the relevant picture of forgery that is found successfully.

$$Recall(r) = \frac{(T_n)}{(T_n + F_p)} \tag{6}$$

Precision (**p**): The exact and correct classification detection is evaluated as a classified image. It shows the strategy's accuracy.

$$Precision(p) = \frac{(T_p)}{(T_p + F_p)} \tag{7}$$

Specificity (**Spc**): The fact that the actual Condition is absent does not prevent the fake photo from being classed, nevertheless.

$$Specificity(Spc) = \frac{(T_n)}{(T_n + F_p)} \tag{8}$$

Accuracy (**Acc**): In order to generate an appropriate classification from the erroneous picture, the classifier takes measurements.

$$Accuracy(Acc) = \frac{(T_p + T_n)}{(T_p + F_n + T_n + F_p)} \tag{9}$$

Sensitivity (**Sen.**): It represents the hopeful classification of a fabricated image while the actual situation remains unresolved.

$$Sensitivity\ (Sen.) := \frac{(T_p)}{(T_p + F_n)} \tag{10}$$

A public domain database is a dataset for this assessment. For the identification of a copy-move classification, this database consists of 260 classified images in two groups, small (512×512) and big (3000×2000). Images are divided into 5 classes by application of handlings, such as translation, rotation, scale, combination, and distortion. The analysis of the proposed method involves (512×512) images.

In this method, to determine the previous usage of the ICs we use the FFs.

4   Detection Rate Based on Training and Testing Process

The database is used, for checking. The image datasets contain both initial and classifying the images in the training phase. Due to the input image, the function vectors are pre-processed and forwarded. In the classifier, classify inputs to determine the good weighted values by using the sigmoid kernel. The values of the preparation for an efficient classification are assessed and photographs are checked and differentiated as original and forged images according to the weighted value obtained. In this assessment, 500 photos are taken that are tested and trained. The test pictures as authenticating and forging pictures are seen here. In each parameter metrics including precision, reminder, specificity, sensitivity, F-measure, and G-mean, the output comparison of the proposed IRVMs is shown in Table 1. The current methods are assessed in each parameter. The database includes 500 pictures, which divide in each collection 25 pictures. To estimate the output of the proposed process, each set is evaluated by parameters. The series of images have been checked and educated and the average value of their parameters is forecast.

Table 1. Performance result for image detection for BWT-SVD technique (proposed)

Input sets	No. of real images	No. of false images	True Positive (Tp)	True Negative (Tn)	False Positive (Fp)	False Negative (Fn)	Accuracy (Acc)	Specificity	Sensitivity
Set_1	25	25	20	25	0	5	0.90	1	0.8
Set_2	25	25	20	25	0	5	0.90	1	0.8
Set_3	25	25	21	25	0	4	0.92	1	0.84
Set_4	25	25	24	21	4	1	0.90	0.84	0.96
Set_5	25	25	24	23	2	1	0.94	0.92	0.96
Total	125	125	109	119	6	16	0.91	0.952	0.872

Table 2. Performance result for image detection SVM technique (Existing)

Input sets	No. of real images	No. of false images	True Positive (Tp)	True Negative (Tn)	False Positive (Fp)	False Negative (Fn)	Accuracy	Specificity	Sensitivity
Set_1	25	25	16	21	4	9	0.74	0.84	0.64
Set_2	25	25	17	20	5	8	0.74	0.8	0.68
Set_3	25	25	18	18	7	7	0.72	0.72	0.72
Set_4	25	25	19	17	8	6	0.72	0.68	0.76
Set_5	25	25	16	19	6	9	0.70	0.76	0.64
Total	125	125	86	95	30	39	0.72	0.76	0.68

Each method performance evaluated for all parameters is shown in Tables 1 and 2 above. It shows that the proposed IRVM produces positive performance, in comparison to other methods.

Displays the performance compared to accuracy, sensitivity, and specificity forgery detection schemes proposed BWT-SVD and compared with SVM. The accuracy of the proposal is increased in Fig. 3, due to the excellent weight forecast and the SVD based stop working that decreases processing time and noise, increasing performance. The proposed BWT-SVD scheme is shown to be 91.12% accurate in all data sets and has reached 72.7% accuracy concerning the proposed method beyond the SVM schemes.

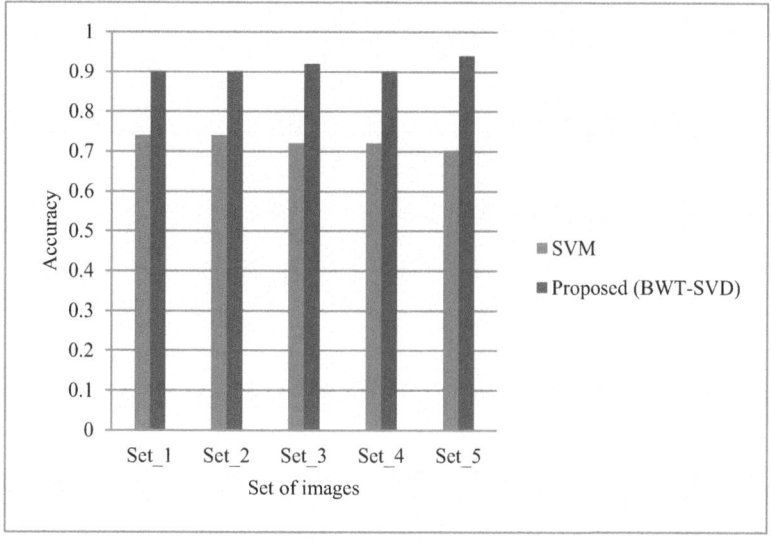

Fig. 3. Accuracy comparison result for a proposed and existing method

Figure 4 shows the results for the sensitivity comparison between the new SVD and the current SVM systems of five sets of cross-experimental installations. It shows that the sensitivity of the proposed SVD can be controlled by a 68.4% to 87.2% increase in the sensitivity rate.

The extracted features are fed to the images as an input for the classification process. The feature vectors of the segmented image are determined at different points as the form and texture image. These features should be trained in the classification systems to improve the classification rate. In general, classification is used for decision-making. The weight parameter of SVD is optimized by optimizing Glow-worm swarm optimization to improve accuracy. Therefore, this integrated technology was described as improved. The SVM method is common in many actual classification situations because of its simple implementation and consistently high precision of classification. The results (SVD-BWT) are based on a linear formulation of the model, leading to a satisfactory preliminary sparse presentation. As a consequence, it can be generalized and deduced at the low cost of measurement.

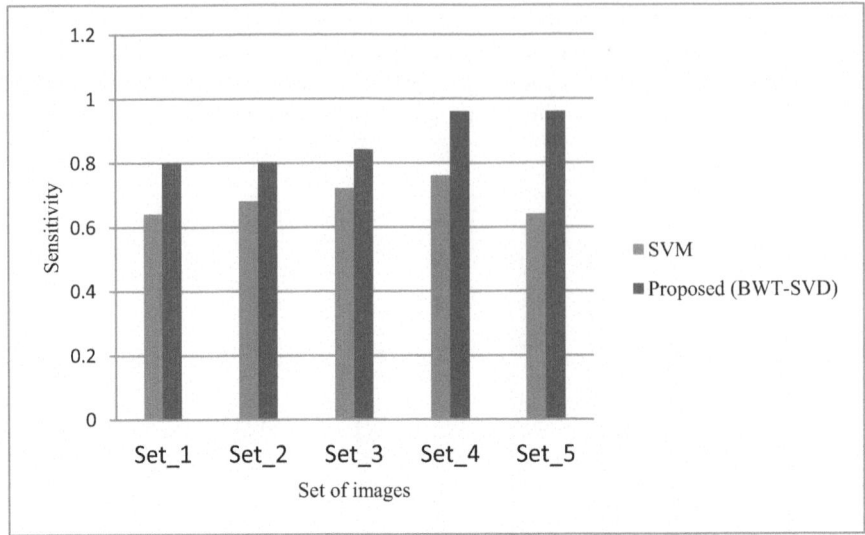

Fig. 4. Sensitivity comparison result for a proposed and existing method

5 Conclusion and Future Scope

This article addressed the proposed improved SVM based on image counterfeiting using the SVD technique. Certain techniques are used to detect copy-move classification on each picture block, which reduces the computational problem. The images are pre-processed to reduce noise, and BWT-SVD is used to separate and compute the complications. The extracted vectors are arranged graphically and identify the similarity between the two successive vectors. In the end, the SVD-BWT classifies the counterfeit image using an optimized BWT algorithm. The evaluation of output shows that the suggested solution effectively results in all parameters in comparison with the existing algorithms. This method will further enhance the exact classification of the counterfeit picture using different algorithms. It is clear that the methodology proposed generates productive results relative to other existing approaches. The SVD-BWT achieves an important sensitivity with its low specialty of high training time. The proposed BWT system achieved high performance with high accuracy, sensitivity, and specificity rates.

References

1. Lowe, D.G.: Distinctive image features from scale-invariant keypoints. Int. J. Comput. Vis. **60**(2), 91–110 (2004)
2. Huang, H., Guo, W., Zhang, Y.: Detection of copy-move forgery in digital images using SIFT algorithm. In: 2008 IEEE Pacific-Asia Workshop on Computational Intelligence and Industrial Application, vol. 2, pp. 272–276. IEEE (2008)
3. Sushma, K., Satyanarayana, V., Singh, M.K.: A copy and move image forged classification by using hybrid neural networks. In: International Conference on Artificial Intelligence and Data Science, pp. 101–111. Springer Nature Switzerland, Cham (2021). https://doi.org/10.1007/978-3-031-21385-4_9

4. Anushka, R.L., Jagadish, S., Satyanarayana, V., Singh, M.K.: Lens less cameras for face detection and verification. In: 2021 6th International Conference on Signal Processing, Computing and Control (ISPCC), pp. 242–246. IEEE (2021)
5. Christlein, V., Riess, C., Jordan, J., Riess, C., Angelopoulou, E.: An evaluation of popular copy-move forgery detection approaches. IEEE Trans. Inf. Forensics Secur. **7**(6), 1841–1854 (2012)
6. Li, J., Li, X., Yang, B., Sun, X.: Segmentation-based image copy-move forgery detection scheme. IEEE Trans. Inf. Forensics Secur. **10**(3), 507–518 (2014)
7. Pun, C.-M., Yuan, X.-C., Bi, X.-L.: Image forgery detection using adaptive over segmentation and feature point matching. IEEE Trans. Inf. Forensics Secur. **10**(8), 1705–1716 (2015)
8. Singh, M., Nandan, D., Kumar, S.: Statistical analysis of lower and raised pitch voice signal and its efficiency calculation. Traitement du Signal **36**(5), 455–461 (2019)
9. Cozzolino, D., Poggi, G., Verdoliva, L.: Efficient dense-field copy–move forgery detection. IEEE Trans. Inf. Forensics Secur. **10**(11), 2284–2297 (2015)
10. Chandana Sri, K., Deepika, Y., Radha, N., Singh, M.K.: Using convolution networks to remove stripes noise from infrared cloud images. In: International Conference on Artificial Intelligence and Data Science, pp. 530–539. Springer Nature Switzerland, Cham (2021)
11. Singh, M.K.: Feature extraction and classification efficiency analysis using machine learning approach for speech signal. Multimed. Tools Appl. **83**, 1–16 (2023)
12. Snigdha, K., Gurjar, A.A.: Image forgery types and their detection: a review. Int. J. Adv. Res. Comput. Sci. Softw. Eng. **5**(4), 174–178 (2015)
13. Kalyan, M.P., Kishore, D., Singh, M.K.: Local binary pattern symmetric centre feature extraction method for detection of image forgery. In: International Conference on Artificial Intelligence and Data Science, pp. 89–100. Springer Nature Switzerland, Cham (2021). https://doi.org/10.1007/978-3-031-21385-4_8
14. Tahaoglu, G., Ulutas, G., Ustubioglu, B., Ulutas, M., Nabiyev, V.V.: Ciratefi based copy move forgery detection on digital images. Multimed. Tools Appl. **81**(16), 22867–22902 (2022)
15. Veerendra, G., Swaroop, R., Dattu, D.S., Jyothi, C.A., Singh, M.K.: Detecting plant diseases, quantifying and classifying digital image processing techniques. Mater. Today Proc. **51**, 837 (2021)
16. Padma, U., Jagadish, S., Singh, M.K.: Recognition of plant's leaf infection by image processing approach. Mater. Today Proc. **51**, 914 (2021)
17. Satya, P.M., Jagadish, S., Satyanarayana, V., Singh, M.K.: Stripe noise removal from remote sensing images. In: 2021 6th International Conference on Signal Processing, Computing and Control (ISPCC), pp. 233–236. IEEE (2021)
18. Singh, M.K.: DWT and LBP hybrid feature based deep learning technique for image splicing forgery detection. Soft Comput. 1–9 (2024)

An Effective Biometric Medical Image Watermarking System Designed for e-Health Application

Jothi Prabha Appadurai[1], R. S. Nancy Noella[2], Minu Susan Jacob[2], K. Arunasakthi[3], B. S. Kiruthika Devi[4], and Mahesh K. Singh[5]([⊠])

[1] Department of Computer Science and Engineering (AI&ML), Kakatiya Institute of Technology and Science,
Warangal, India
ajp.csm@kitsw.ac.in

[2] Department of CSE, Sathyabama Institute of Science and Technology, Chennai, India
{nancynoella.cse,minususanjacob.cse}@sathyabama.ac.in

[3] Department of Computer Science and Engineering, Velammal College of Engineering and Technology, Madurai, India
aruna.sakthi10@gmail.com

[4] Department of CSE, School of Computing, Sathyabama Institute of Science and Technology, Chennai, India
kiruthikadevi.b.s.cse@sathyabama.ac.in

[5] Department of ECE, Aditya University, Surampalem, India

Abstract. In E-Health care, one of the primary requirements is the Electronic transmission of medical images. This transmission must and should not be liable to the hackers to manipulate the whole medical image by modifying only one part of it. For this purpose we use the watermarking techniques. In this manuscript represented two dissimilar watermark algorithms for medical images. Image hiding is designed for safety application particularly to defend an unauthorized person's secret message. Due to the immense growth and use of the Internet, the problem of Internet security is growing. Under this situation, it needs more safety to convert the data commencing the source to the recipient. Academic research has also suggested the use of a compression algorithm and cryptography in an effective medical images watermarking method inside E-healthcare applications. Only trust and reliability are guaranteed by the system. To solve this, the author recommended a biometrics system that produces an authentication, confidentiality and reliability systems base on an effective medical images watermark in the E-healthcare applications. The classification suggested includes the verification biometric fingerprint, privacy crypto to graphing and reversible integrity watermarking. The projected method consists essentially of two phases, for example (i) watermarking embedded procedure and (ii) watermarking extraction method. The test was conducted with electronic health records for the different medical pictures and the efficiency of the projected algorithms is analyze using a peak signal to noise ratio (PSNR) and standardized correlation.

Keywords: Electronic health records · Feature extraction · Health care · Image watermarking · Medical image processing

M. Khurana et al. (Eds.): ICMLA 2024, CCIS 2238, pp. 310–317, 2025.
https://doi.org/10.1007/978-3-031-75861-4_27

1   Introduction

Now-a-days the usage of internet plays a vital responsibility in every person daily existence. The area similar to electronic-healthcare (E-healthcare), electronic-banking (E-banking), and electronic-shopping (E-shop) are achieving new heights day by day. E-healthcare refers to an internet base method where patients can take treatments from professional doctors available at any corner of the world [1, 2]. Creation of high eminence health care accessible to popular individual beings around the world is a main disquiet to the administrator, government and politicians. Generally, In order to take treatment the doctor and patient should be physically nearby at the same place. But in the current advance in ICT have increased the amount of way in which healthcare be able to circulate to all inaccessible area across the globe [3–5].

Digital multimedia watermarking content has grown rapidly in recent years as the Internet technology progresses. In order to cover digital copyright and to differentiate contents with modified material on the Internet, this watermark aims to secure. Worldwide unprotected internet routing servers. Data on the Internet The watermarking system is also used to encrypt data sent to these unsecured servers. Typically normal communication is characterized as a source to destination information flow [6–8]. In this data transfer process, information may be changed or manipulated by various types of attacks. There are different four types of attacks discussed below: Fabrication: An attack on accuracy. An illegal party inserts the dummy things addicted to the organization. Interception: An attack on privacy. An illegal party gets admittance to the information. Disruption: An attack on accessibility. In order to destroyed or become occupied. Modification: An attack on reliability [9, 10]. An illegal party not merely gains admittance, but also change the information. These attacks can be separated additional into two categories, according to the character of the attack [11]. Active Attacks: These attack engage variation of the information flow or the formation of a false stream. Passive Attacks: These attacks engage eavesdrop, monitor or communication shown in Fig. 1.

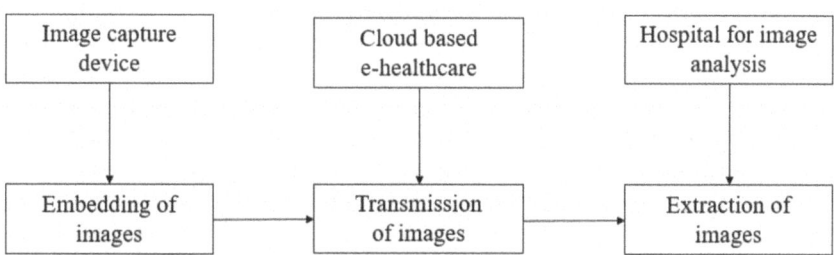

Fig. 1. Framework internet based E-healthcare platform.

Digital watermarking technique is used to offer privacy and honesty in the medical image. Further controlling and interfere recognition of digital information are the major aims of digital rights management (DRM).The digital information in the image cannot be distributed uniformly. A number of parts of images may contain more information than other [12–14]. There are various technologies that are used to separate different regions in images. The Region of Interest part contains more information and is used for diagnosis purpose so it has to be take care of healthcare application [15, 16].

2 Related Work

New issues have been solved by advancements in information technology in healthcare. It offers an alternative solution for the storage, stockpiling, retrieval and authentication of data in a range of medical data management and distribution issues. By incorporating four types of watermarks in the medical imaging wavelet coefficients, the proposed multiple watermarking systems solve above problems [1]. In 2008, a number of digital medical files are produced most frequently by different radiological methods, due to the advances in medical information technology. These files should be protected, in particular as they contain key medical information, against unauthorized materiel changes. They are proposing a new watermarking technique where medical images from MRIs are used to incorporate areas of concern in the area of non-interest. For integrity inspection, a simple segmentation algorithm is used [2, 17].

In 2009, in this dissertation discusses two methods for inserting watermarks: one is additive and the other replacement. Five reversible watermarking methods were developed and compared in this context. Each approach has different capabilities and efficiency of invisibility. Therefore, the following approaches take the specificity of the signal into account. The most vulnerable methods need high intensity. Several such methods have been formulated so far. This is one of the barriers to progress [3]. Here is an algorithm with solid watermarks DWT and DFT. A portion of the sign sequence of the DWT and DFT is used for enhancing the robustness against attacks such as rotation attacks, attacks in sizes, crossing and cutting attacks. The proposed algorithm with many watermarks can be easily incorporated with various keys, matching different watermarks. Furthermore, the watermark can be removed without the original medical image. The algorithm for the geometric attack is strong in nature from experimental proof [4, 18].

In 2013, The digital watermark picture is an integration process for digital data. The watermarking is considered invisible if an image does not appear in the cover, but is removable with a removal algorithm and key. A good watermark image from a sturdy watermark algorithm must achieve a good watermark performance. A new DCT domain watermarking algorithm is proposed in this paper. The algorithm is simple and benefits from the DCT inter-blocks correlation. The algorithm for watermark extraction is tested with and without image treatment [5]. In 2014, This paper proposes a new location and recovery strategy for medical image authentication. The EPR and reshaped region of interest (ROI) is sparsely coded into the Noninterest region's transforming domain. The first part is the storage of patient data with the picture, and the second part is for authentication. For various purposes EPR coverage, ROI authentication and the recovery of a manipulated region may be included in the proposed technique [6, 19].

In 2015, The dissemination of Digital Media was encouraged and facilitated by multimedia progress and Internet growth. It affects information security and improves the risk of attack. This paper presents the accurate and robust technology known as the intermediate bit plane incorporation (ISBPE). Two levels of integrated data security techniques are employed by ISBPE. Centered on the merits of the proposed communication conversion technology ISBPE [7, 10]. In 2016, Multimedia innovation and internet development promoted and promoted the diffusion of digital media. It affects protection of information and enhances the risk of attack. The precise and robust technology called interim bit plane integration is presented in this paper (ISBPE). ISBPE employs two types of advanced technology on data protection. Based on the merits of the ISBPE communication technology [8].

In 2018, The authors have proposed a biometric medical image aquamarking E-Healthcare application to create authentication, privacy and reliability. The proposed architecture uses fingerprint biometrics for automation, confidentiality encryption procedures, and reversible watermarking for integrity. In this study, we explained a MIW based on various algorithms. The fingerprint is biometric and uses and strengthens the security of the water signage system. Also, experimental results reveal that the proposed approach has a better watermark and better integrating performance [9]. The protection of data becomes very important to prevent the development of telemedicine applications. Protect data from manipulation and handling when transmitting the Internet. It presents a new blind image of marking medical water for the protection of personal information's security and confidentiality. The decomposition combination is useful to integrate watermark bits. With considerably greater imperceptibility and solidity against the most attacks, the results are very good [10].

3   Methodology for Watermarking in E-health

The process of embedding data in medicinal imagery is dissimilar commencing from that of the embedding in additional average images. As the E-health medical images contained responsive information and which is used for the purpose of e-diagnosis. As we know that E-health images are separated into two parts: one is region of non-Interest (RONI) and other is region of interest (ROI). This method called DCT, which one is used to convert the images into the frequency domains. The transformation equation which are used to change spatial sphere of influence into frequency domains by using the DCT as follows:

$$D_C[k] = \sum_{m-0}^{M-1} x[m] cos \frac{2\pi(2n+1)k}{4M} \text{ for k } = 0 : M-1 \tag{1}$$

The inverse DCT can be represented as follows:

$$x(m) = \frac{1}{M} X[0] + \frac{2}{M} \sum_{K=1}^{M-1} D[k] cos \frac{2\pi(2n+1)}{4N} \tag{2}$$

Generally in an image DCT is approved out in two behaviors. They are inclusive DCT and obstruct base DCT. Block dependent DCT used in preferred algorithm.

Structural similarity index metric (SSIM): SSIM is used to stego (S) and pristine (P) of e-healthcare image:

$$SSIM(P, S) = \frac{(2\alpha_P \alpha_S + P_1)(2\sigma_{P,S} + P_2)}{\left(\alpha_P^2 + \alpha_S^2 + P_1\right)\left(\sigma_P^2 + \alpha_S^2 + P_2\right)} \tag{3}$$

Universal image quality (UIQ): UIQ used to differentiate stego (S) and pristine (P) are shown below,

$$UIQ(P, S) = \frac{4\sigma_{P,S}\alpha_P\alpha_S}{(\alpha_P^2 + \alpha_S^2)(\sigma_P^2 + \alpha_S^2)} \tag{4}$$

Normalized Absolute Error (NAE): NAE is formulated as following equation:

$$NAE(P, S) = \frac{\sum_{a=1}^{L}\sum_{b=1}^{M}(P(a, b) - S(a, b))}{\sum_{a=1}^{L}\sum_{b=1}^{M}(C(a, b))} \tag{5}$$

Image fidelity (IF): IF mathematically calculated as:

$$NAE(P, S) = \frac{\sum_{a=1}^{L}\sum_{b=1}^{M}(P(a, b) - S(a, b))^2}{\sum_{a=1}^{L}\sum_{b=1}^{M}(C(a, b))} \tag{6}$$

Average difference (AD): It is measured as following equation:

$$AD(P, S) = \frac{\sum_{a=1}^{L}\sum_{b=1}^{M}(P(a, b) - S(a, b))}{h * w} \tag{7}$$

Normalized Cross Correlation (NCC): It is formulated by following equation:

$$NCC(P, S) = \frac{\sum_{a=1}^{L}\sum_{b=1}^{M}(P(a, b) - \alpha_P)(S(a, b) - \alpha_P)}{\sqrt{\sum_{a=1}^{L}\sum_{b=1}^{M}(P(a, b) - \alpha_P)^2}\sqrt{\sum_{a=1}^{L}\sum_{b=1}^{M}(P(a, b) - \alpha_S)^2}} \tag{8}$$

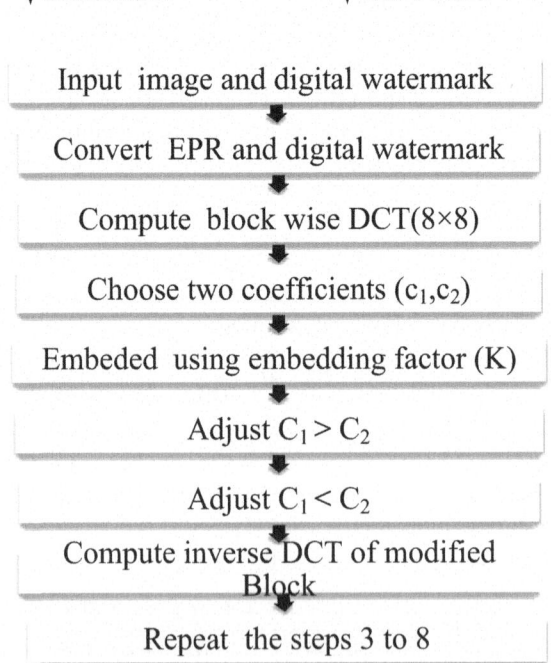

Fig. 2. DCT based watermarking system in E-Healthcare

There are two proposed algorithms for embedding data and its embedding procedure is given below Fig. 2:

In this method shown in Fig. 2, both EPR and the watermarking method are implanted in DCT, ROI and RONI of medical images. The results in embedding a watermarking in the RONI of a medical image by maintain for ROI. This method helps to separate ROI or not to manipulate the serious distance of the medical image that is used by the doctor for analysis purposes. Initially separating the ROI from the innovative medicals image that provide RONI for embedded watermarking. Embedding several watermarks will raise the protection level of a medical images as it carry prominent information.

4   Experimental Result and Discussion:

The experimentations are done using a platform called MATLAB R2014a software for different watermark medical images of patients. Further, EPR is also entrenched in those examination medical images. Separation of ROI is done using color thresh older relevance of MATLAB R2014a. The e-healthcare image quality detections are studied and calculated by PSNR, SSIM, UIQ, NCC, NAE, IF and AD detection approach by applying various images. The above values used for the detection are shown in Table 1.

Table 1. Performance analysis of e-healthcare image watermarking

Detection method	Image 4(a)	4(b)	4(c)	4(d)	4(e)
AD	−0.025	−0.038	0.001	0.008	0.001
NAE	0.008	0.011	0.006	0.018	0.124
NCC	0.895	0.997	0.978	0.988	0.983
UIQ	0.892	0.879	0.986	0.965	0.906
SSIM	0.963	0.989	0.985	0.969	0.896
PSNR	43.23	44.25	41.67	42.36	44.28

Inspired by the usefulness and deceptive protection provided by quantum computing, we have implemented a robust protocol on steganography based on a protocol of quantum walks to ensure images are transmitted on cloud E-Health platforms. In contrast to traditional QIP-based methods, the proposed methodology is based on conventional Quantico mechanics renderings that make it tenable for physical quantum hardware, subject to necessary refinements, to securing images from violence. Specifically, for the integration and extraction processes, the scheme employs the potential of our virtually bare bones classical transcription, whereas the programme is designed for medical photos. The virtual protocol determines that hidden bits are overlaid with the areas of the carrier picture. Our proposal does not require pre- or post encryption or removal, which means that the secret image is to be obtained only from the stego image and the primary state of the CAQWs. A data collection comprised of colour and grayscale images will be used to thoroughly evaluate our proposed method is discussed Table 2.

Table 2. Imperceptibility analysis

Images	Proposed method (2209 Bits)		
	NAE	SSIM	PSNR (dB)
Image 1	55.45	0.972	0.0023
Image 2	55.32	0.972	0.0021
Image 3	55.21	0.972	0.0031
Image 4	55.12	0.972	0.0036
Image 5	55.96	0.972	0.0040

The Imperceptibility of a projection is judged by comparing those images, through various random filters such as a salt and pepper noise. JPEG compression, Wiener filtering, and median filtering are used by varying processes such as tone translation, Gaussian noise, using a transformation, JPEG compression, and GND.

5   Conclusion and Future Scope

E-Health care plays an important role now days in medical system. It made that the distance is irrelevant in health care. E-Health care makes use of wide range of technologies which allows health users to communicate with health professionals all over the world. This manuscript presented two dissimilar techniques for watermark of medical images. In first method watermark is embedded in both ROI and RONI whereas in second method embedded tasks takes place only in RONI only. The results are very highly imperceptible and vigorous to a different in E-Healthcare. The method proposed is evaluated on multiple simulation-based studies on a color and gray-scale medical image data collection. Results confirm the effectiveness of the new system in terms of high visual quality, data loss tolerance, high embedding capability and robust safety. In addition, the study of performance indicates that, in comparison to other state-of- the art techniques, the proposal of a quantum-inspired programme suggests possible uses for the proposed scheme as an effective medical-image steganography strategy for future computer paradigms.

References

1. Parah, S.A., Sheikh, J.A., Ahad, F., Loan, N.A., Bhat, G.M.: Information hiding in medical images: a robust medical image watermarking system for E-healthcare. Multimed. Tools Appl. **76**(8), 10599–10633 (2017)
2. Aparna, P., Kishore, P.V.V.: Biometric-based efficient medical image watermarking in E-healthcare application. IET Image Proc. **13**(3), 421–428 (2018)
3. Aparna, P., Kishore, P.V.V.: A blind medical image watermarking for secure e-healthcare application using crypto-watermarking system. J. Intell. Syst. **29**(1), 1558–1575 (2020)
4. Aparna, P., Kishore, P.V.V.: An efficient medical image watermarking technique in E-healthcare application using hybridization of compression and cryptography algorithm. J. Intell. Syst. **27**(1), 115–133 (2018)

5. Gull, S., Loan, N.A., Parah, S.A., Sheikh, J.A., Bhat, G.M.: An efficient watermarking technique for tamper detection and localization of medical images. J. Ambient. Intell. Humaniz. Comput. **11**(5), 1799–1808 (2020)

6. Giri, K.J., Bashir, R., Bhat, J.I.: A discrete wavelet based watermarking scheme for authentication of medical images. Int. J. E-Health Med. Commun. (IJEHMC) **10**(4), 30–38 (2019)

7. Singh, M.K.: A text independent speaker identification system using ANN, RNN, and CNN classification technique. Multimed. Tools Appl. **83**, 1–13 (2023)

8. Vaidya, S.P., Kishore, V.R.: Adaptive medical image watermarking system for e-health care applications. SN Comput. Sci. **3**(2), 107 (2022)

9. Singh, M.K.: Feature extraction and classification efficiency analysis using machine learning approach for speech signal. Multimed. Tools Appl. **83**, 1–16 (2023)

10. Hassan, B., Ahmed, R., Li, B., Hassan, O.: An imperceptible medical image watermarking framework for automated diagnosis of retinal pathologies in an eHealth arrangement. IEEE Access **7**, 69758–69775 (2019)

11. Singh, M.K.: Speaker emotion recognition system using artificial neural network classification method for brain-inspired application. J. Circuits Syst. Comput. (2023)

12. Sushma, K., Satyanarayana, V., Singh, M.K.: A copy and move image forged classification by using hybrid neural networks. In: International Conference on Artificial Intelligence and Data Science, pp. 101–111. Springer Nature Switzerland, Cham (2021). https://doi.org/10.1007/978-3-031-21385-4_9

13. Chauhan, D.S., Singh, A.K., Kumar, B., Saini, J.P.: Quantization based multiple medical information watermarking for secure e-health. Multimed. Tools Appl. **78**, 3911–3923 (2019)

14. Kalyan, M.P., Kishore, D., Singh, M.K.: Local binary pattern symmetric centre feature extraction method for detection of image forgery. In: International Conference on Artificial Intelligence and Data Science, pp. 89–100. Springer Nature Switzerland, Cham (2021). https://doi.org/10.1007/978-3-031-21385-4_8

15. Soualmi, A., Alti, A., Laouamer, L.: Multiple blind watermarking framework for security and integrity of medical images in e-health applications. Int. J. Comput. Vis. Image Process. (IJCVIP) **11**(1), 1–16 (2021)

16. Chandana Sri, K., Deepika, Y., Radha, N., Singh, M.K.: Using convolution networks to remove stripes noise from infrared cloud images. In: International Conference on Artificial Intelligence and Data Science, pp. 530–539. Springer Nature Switzerland, Cham (2021). https://doi.org/10.1007/978-3-031-21385-4_43

17. Sivaprakash, A., Rajan, S.N., Selvaperumal, S.: Privacy protection of patient medical images using digital watermarking technique for e-healthcare system. Curr. Med. Imag. **15**(8), 802–809 (2019)

18. Abdi, H., Hacene, I.B.: An optimized medical image watermarking approach for e-health applications. Med. Technol. J. **5**(1), 594–603 (2023)

19. Choura, H., Chaabane, F., Frikha, T., Baklouti, M.: Robust and secure watermarking technique for E-health applications. In: International Conference on Soft Computing and Pattern Recognition, pp. 997–1008. Springer International Publishing, Cham (2020). https://doi.org/10.1007/978-3-030-73689-7_94

Exploring Diverse Techniques in Image and Video Forgery

Neha Dhiman[✉], Hakam Singh, and Abhishek Thakur

School of Engineering and Technology, Chitkara University, Baddi, Himachal Pradesh, India
{neha81.phd21,hakam.singh,abhishek}@chitkarauniversity.edu.in

Abstract. The versatility of leading editing tools, access to the internet and technological advancement have led to a sudden rise in visual forgery, resulting in manipulation, tampering and fabrication of content or information. The digital divide in affordable access and infrastructure led to prominent innovation and challenges. The visual forgery gives rise to challenges or threats to multimedia data's trustworthiness and integrity, which requires robust detection and prevention methods. This paper reviews existing approaches and techniques to authenticate and identify fabricated images and videos. In addition to this, technological advancements in deep learning and computer vision have led to the development of better algorithms. Mitigating visual forgery problems requires a multidisciplinary approach combining traditional forensics technologies, machine learning and ethical consideration.

Keywords: Temporal · Forgery · Splicing · Active · Passive · Copy-move · Spatial

List of Acronyms

AI	Artificial Intelligence
AUC-PR	Area Under the Precision-Recall Curve
AUC-RO	Area Under the Receiver Operating Characteristic Curve
CMF	Copy-move Forgery
CNN	Convolution Neural Network
DL	Deep Learning
FP	False Positives
FN	True Negatives
MSE	Mean Squared Error
ML	Machine Learning
TP	True Positives
TN	True Negatives
RNN	Recurrent Neural Network
SF	Splicing forgery

M. Khurana et al. (Eds.): ICMLA 2024, CCIS 2238, pp. 318–328, 2025.
https://doi.org/10.1007/978-3-031-75861-4_28

1 Introduction

Image processing is a method that performs certain operations on an image to get an enhanced image that extracts useful information from the image and implicates tasks like handling visual distortions, encoding and editing pictures, and identifying video frames. The process has been streamlined by combining image and video editing tools, enabling individuals with minimal computer proficiency to alter and override various image characteristics effortlessly. Due to this, digital forgeries have become prevalent, particularly on social media platforms. Visual data plays a crucial part in evidence in several frameworks, such as television news, where accuracy, credibility, and persuasive value are paramount [1, 2]. Significantly, occurrences of criminal activity recorded by CCTV cameras are often utilized as evidence in legal reports [3, 4]. The potential danger to our culture arises from the widespread existence of digital forgeries, particularly when these altered images infiltrate newspapers or courtrooms. Additionally, the accessibility of digital visual media brings along its own set of drawbacks.

2 Literature Survey

I-Cheng (2013) introduced an innovative algorithm for detecting forged areas in inpainted images commonly used for image manipulation [1]. Chi-Man (2015) have combined the block and keypoint-based methods to improve forgery detection in terms of accuracy [2]. Amneet (2018) presented a method that utilizes an image acquisition device matrix to extract features for identifying colour filter array artifacts in specific image regions [3]. Lichao (2017) reported the SWDT algorithm for identifying duplicated regions in videos using EFMs [4]. Qingzhong (2016) introduced an approach that commences with comprehensive feature extraction within the discrete transform domain [5]. Neenu and Jini (2014) presented an approach that combines the assessment of illumination discrepancies with the characteristics of resampling to identify manipulated photos [6]. Edoardo (2015) designed an advanced approach for image forgery, a common form of tampering [7]. T. Hoang (2015) introduced a feature-based segmentation method that accurately delineates the beard/mustache region within the identified facial hair area [8]. Jiangbin (2016) categorized the copy-move forged image into keypoint and block-based methods [9]. Maoguo (2017) presented a framework that integrates two essential components: neural networks and change feature extraction using superpixels and hierarchical difference representations [10]. Can (2017) design a new DL framework for detecting and localizing forged edges [11]. Haiqiang (2017) introduced an approach to developing a comprehensive network for human skin detection by incorporating RNN layers into FCNs [12]. Lichao (2019) proposed an algorithm to address various shortcomings of current forgery detection methods in video foreground removal, primarily focusing on enhancing efficiency and effectiveness [13]. Paul (2021) reviewed existing techniques for improving search warrant integrity, acknowledged their limitations and introduced the AFES, which offers various features to bolster integrity, including an unalterable record and biometric authentication of officers involved in search warrant execution [14]. Mubbashar (2020) introduced a method to distinguish between authentic and forged video clips [15]. Pawel and Nasir (2019) presented a novel approach in which neural imaging processes are trained to replace internal processes related to digital cameras [16]. Ana Lucila (2019) proposed techniques to counter forensic analysis for digital videos [17]. Yi-Xiang (2022) has designed the DAFDN that enables the CNN to capture distinctive indicators left by image manipulation effectively [18]. Sebastiano (2021) employed a custom-designed convolutional neural network to extract features and detect coding artifacts seamlessly [19]. Shuo (2021) proposed an analytical model that examines the variation in the alignment of flow positioning between neighbouring optical flow fields, a valuable indicator for detecting frame deletions [20]. Savita (2021) combined two distinct feature extraction approaches utilizing colour attributes and deep features extracted from the luminance component of the picture [21]. Van-Nhan (2022) reported a method known as MDD, which employs meta-learning to create a robust deepfake detection approach [22]. Dongping (2020) described a group of twelve sets of first and second-order differences in various orientations that were employed as image features to detect variations among pixels in an image [23]. In-Jae (2020) introduced a MCNet that leverages features from various domains, like spatial, spectral, and compressed domains [24]. Giulia (2020) introduced a method that operates on grayscale images and demonstrates resilience against various attacks, including image compression [25]. Esteban (2020) employed the colour filter array interpolation pattern for validating digital images [26]. Mohameed (2020) investigated using sequential analysis to identify video forgery [27]. Brian (2019) described the authentication of videos and the creation of the video-ACID [28]. Quist-Aphetsi (2019) successfully employed SHA-256 to perform hashing on the images used for digital forensic analysis [29]. Yijun (2020) developed the Warwick Image Forensics Dataset [30]. Naheed (2022) reported an exhaustive review of current passive techniques for detecting video tampering in a structured manner [31]. Ankit (2022) provided a depth study of image forgery methods in daily life [32]. Menglu (2022) developed a JPEG-resistant image forgery localization method using wavelet-based compression representation learning [33]. Abhishek and Neeru (2020), proposed a technique named as colour illumination, employing a DCNN transfer learning approach [34]. Harpreet and Neeru (2020) reported a practical approach for visual forensics, focusing on identifying manipulation through image splicing and duplicate object forgery [35]. Harpreet and Neeru (2020) highlight the contemporary challenge of preserving software authenticity in the context of images and videos [36]. Qian (2023) reported a method consisting of three key steps: first, the identification of moving objects within a video; second, the creation of a reference SPN from frames devoid of moving things; and third, the computation of the SPCE at the block level to distinguish manipulated from unaltered objects [37].

A1. Image Forgery

Image forgery refers to the alteration of the digital image to conceal information. A digital image can be manipulated or forged through various meanslike the addition of noise, rotation, resizing, splicing, etc.

A2. Forgery Detection Approaches

It is essential to comprehend the diverse methods employed in image manipulation thoroughly. Primarily these methods categorized into active and passive techniques Fig. 1. Each category can be further divided into specific subcategories [35].

a) **Active Approach** for detecting forgery includes preprocessing the image and incorporating a cipher key, which is subsequently utilized to authenticate the received image through digital watermarking or signatures.

b) **Passive Approach** doesn't necessitate image preprocessing; instead, it relies on recognizing irregular statistical characteristics or pixel intensities that arise from changes made to the original image. The objective is to detect and control tampered images. The advantage of not requiring prior image information makes this technique preferable over an active approach [31].

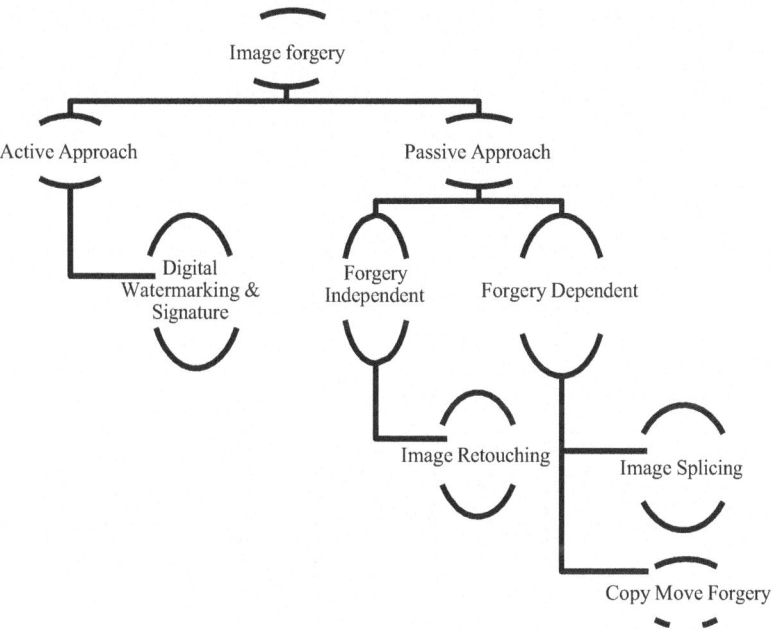

Fig. 1. Forgery techniques

A3. Forgery Methods in Digital Videos and Its Categories

Visual forgery refers to the creation or manipulation of visual content, such as images or videos, with the intent to deceive or mislead. There are various techniques used in visual forgery, and they can be broadly categorized into two types: spatial forgery and temporal forgery. The classification primarily revolves around digital video forgery techniques as spatial forgery (intra-frame), temporal forgery (intra-frame) and spatio-temporal forgery Fig. 2 [20, 27, 36].

a) **Spatial Tampering:** refers to the manipulation of visual elements within a video frame along the x-y axis. This type of tampering can manifest through the alteration of adjacent pixel bits in a video sequence or a single frame. It can be executed at different levels, such as pixel, block, or shot/scene levels. Techniques within this category of manipulation encompass crop and replace, morphing, as well as object addition and deletion [15, 18, 19, 37].

b) **Temporal Tampering** refers to manipulating the consecutive sequence of frames within a video. This type of editing predominantly affects how the device captures the chronological order of visual data over time. Generally, tampering occurs at the frame level, including adding, deleting, and rearranging frames [19].

c) **Spatio-temporal Tampering Tampering** refers to integrating or manipulating both spatial and temporal aspects within digital content. This manipulation encompasses modifying both time sequences and visual data, affecting both the concatenated frame sequence and the visual contents within the video frames [27]. In the present era, many users use social media platforms to share personal photos and valuable information, uploading billions of digital images.

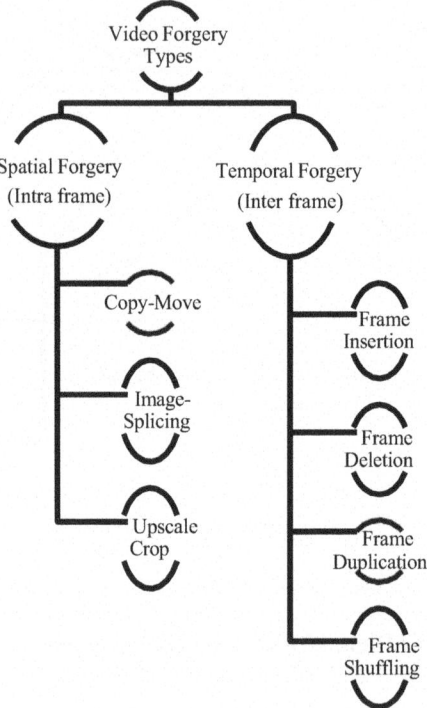

Fig. 2. Video forgery method

Unfortunately, these images are susceptible to being downloaded by anyone and could be exploited for illicit purposes, such as the generation of false propaganda. Editing software has made digital image manipulation a relatively uncomplicated procedure. Standard techniques involve manipulating JPEG compression, adjusting scale, and rotating images.

A4. Artificial Intelligence

Machine and Deep Learning are crucial components within the broader domain of Artificial Intelligence. Artificial Intelligence encompasses ML, which employs conventional techniques for feature extraction. Figure 2 illustrates the categories of artificial intelligence. Within ML, there exists a subset known as DL. Algorithms detect image forgery by inspecting image features or characteristics to identify potential alterations. Vast number of numerous pictures makes it practically impossible for a single individual to identify each one, given the constraints of limited capacity and computational power. Conversely, computers have the capability to operate continuously, exhibiting high computational power and efficiency (Fig. 3).

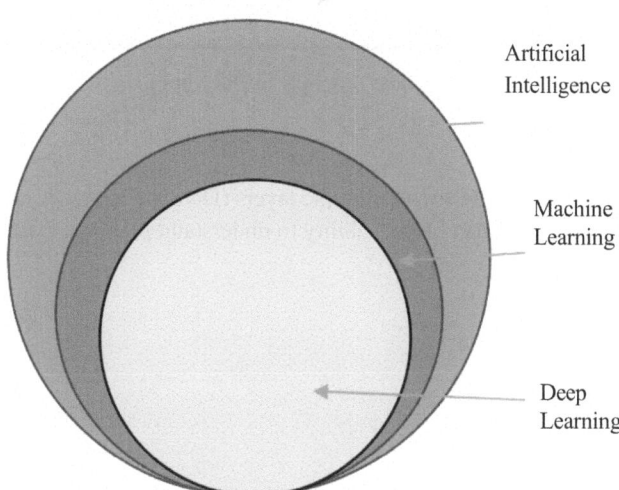

Fig. 3. Artificial Intelligence and its types

A5. Machine Learning

ML is a subset of AI that represents an approach to attaining intelligence. It assimilates knowledge from previous encounters and enhances the efficiency of intelligent programs. Further ML is categorized into three types (Fig. 4).

Supervised Learning: is a type of ML where an algorithm is trained using labeled datasets, the input data utilized for training is matched with corresponding output labels. This learning process enables the algorithm to make precise predictions or classifications.
Unsupervised Learning: is a type of ML where an algorithm is not provided with labeled training datasets.This learning process uncover the inherent structure or patterns within the data without explicit guidance or predefined output labels.
Reinforcement Learning: is a type of ML paradigm where an agent interacts with the environment and learns through rewards or penalties.

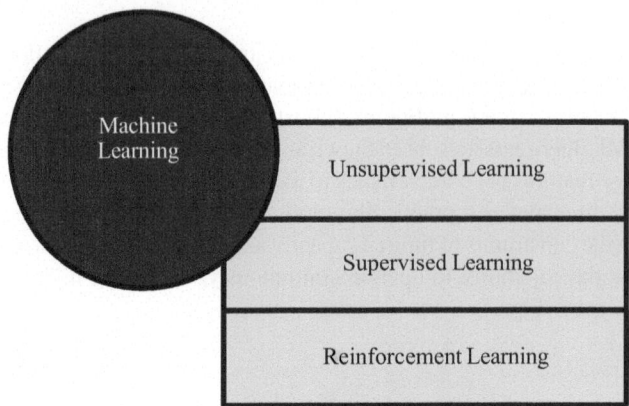

Fig. 4. Machine Learning and its types.

A6. Deep Learning

DL is a subset of ML that uses NN with multiple layers (DNN) to analyze and learn from data. It aims to mimic the human brain's ability to understand patterns, extract insights, and make decisions (Fig. 5).

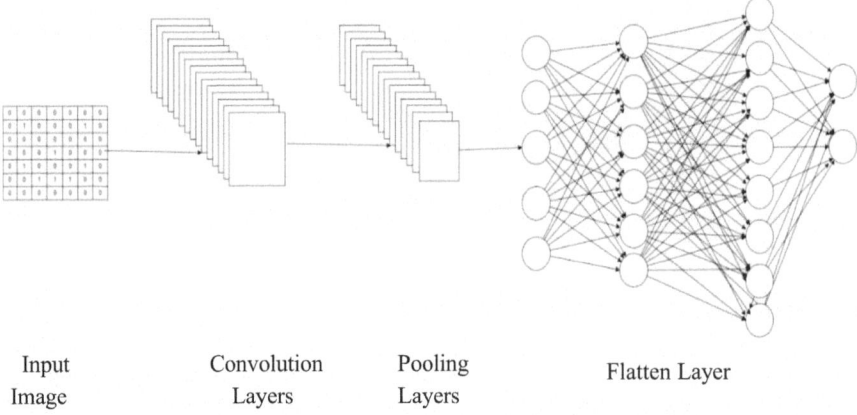

| Input | Convolution | Pooling | |
| Image | Layers | Layers | Flatten Layer |

Fig. 5. General framework of the deep convolution neural network [36]

A7. Impact of ML and DL Algorithms on Visual Forgery

ML and DL algorithms have significantly impacted the field of visual forgery detection. Visual forgery refers to creating or manipulating digital images or videos to deceive or mislead viewers. Applying ML and DL techniques in this context has enhanced the ability to detect various forms of forgery, including image manipulation, deepfake videos, and other types of visual misinformation.

A8. Tools and Datasets Utilized in Detecting Visual Forgery

Detecting visual forgery involves using various tools and datasets, often employing traditional image processing and ML methods. Here are some tools and datasets commonly utilized in the field:

Tools like Adobe Photoshop, image forensic software, ImageJ, MATLAB, Deep learning frameworks, and Camera identification tools are used, and datasets like UCID-uncompressed colour image database, CoMoFod -Columbia-MFO image forensics database, ImageNet are used.

A9. Various Performance Metrics Among Visual Forgery Detection Techniques

Visual forgery detection techniques are crucial in identifying manipulated or forged images, videos, or multimedia content. Here are some key performance metrics used in the assessment of visual forgery detection methods:

Accuracy: The ratio of correctly classified instances (authentic or forged) to the total number of instances.

Precision: is the ratio of TP to the sum of TP and FP.

Recall: is the ratio of TP to the sum of TP and FN.

F1Score: is the harmonic mean of precision and recall.

AUC-RO: A metric that assesses the model's ability to discriminate between authentic and forged instances across different threshold values. The ROC curve plots the TP rate against the FP rate.

AUC-PR: Is similar to AUC-ROC, but focuses on the trade-off between precision and recall. UC-PR measures the area under the precision-recall curve, providing insights into the model's performance across various decision thresholds.

Confusion Matrix: is a table that presents a summary of the model's performance, including TP, TN, FP, and FN.

MSE for Localization: MSE can measure accuracy by comparing predicted and ground truth bounding box coordinates.

A10. Effects and Future Outlook of Techniques for Detecting Visual Forgery

Visual forgery detection techniques play a crucial role in identifying manipulated or tampered images and ensuring the authenticity of visual content. The effects of visual forgery detection techniques are profound, contributing to trust, mitigating misinformation, and impacting legal outcomes. The future outlook involves technological advancements, increased integration with other technologies like blockchain, and a focus on real-time detection and interpretability. The field is expected to evolve to keep pace with emerging threats and challenges in the digital landscape.

3 Discussion on Machine and Deep Learning

See Table 1.

Table1. Comparative Study

References	Technique and Dataset	Accuracy
[3] Amneet Singh et al. (2018)	Markov Transition Probability Matrix (MTPM) technique is used	Accuracy = 90.58%
[12] Haiqiang Zuo et al. (2017)	RNN and CNN layers integration, COMPAQ and ECU skin datasets are used to validate the effectiveness	Integrated RNN layers in skin detection algorithm
[13] Lichaao Su, et al. 2019	AVIBE algorithm is used on SULFA and SYSU-OBJFORG Dataset	The average accuracy of 88.12% and 90.64%
[15] Mubbashar Saddique et al. (2020)	Inspection of tampered video clips, validation on six datasets	Accuracy achieved is 98.89%
[20] Shuoli et al. (2021)	Frame deletion detection technique and	TP rate can reach 90.12%, FP rate is 7.71%
[27] Mohammed Aloraini et al. (2020)	Sequential and patch analyses technique and SYSU-OBJFORG dataset is used	Robust results for trampled videos, low computational complexity
[34] Abhishek et al. (2020)	Semantic segmentation to localize forged pixel. GRIP, DVMM, CMFD and BSDS300 datasets are used	98% accuracy achieved
[37] Qian Li et al. (2023)	SPN (Sensor Pattern Noise) technique is used	Accuracy = 0.914, F1-score = 0.912

4 Conclusion

The extensiveness of digital image editing, fueled by advances in computing technology has fundamentally changed how information is shared and accessed. Internet as a platforms, social media and visual data continue to grow rapidly, requirement for authentic methods for identification of tampered as well as visually forged has become very crucial. Now machine and deep learning techniques have become powerful tools, which helps in detecting forged and tampered data. These machine and deep learning techniques excel at uncovering precisely altered features within images. As there is tremendous advancement in technology continues, the need of hour is strong collaboration between researcher, industry and policy makers will be required in staying ahead of emerging forgery techniques and mitigating their potential harm to credibility in the digital era.

References

1. Chang, I.C., Yu, J.C., Chang, C.C.: A forgery detection algorithm for exemplar-based inpainting images using multi-region relation. Image Vis. Comput. **31**(1), 57–71 (2013)

2. Pun, C.M., Yuan, X.C., Bi, X.L.: Image forgery detection using adaptive over segmentation and feature point matching. IEEE Trans. Inf. Forensics Secur. **10**(8), 1705–1716 (2015)
3. Singh, A., Singh, G., Singh, K.: A Markov based image forgery detection approach by analyzing CFA artifacts. Multimed. Tools Appl. **77**, 28949–28968 (2018)
4. Su, L., Li, C., Lai, Y., Yang, J.: A fast forgery detection algorithm based on exponential- Fourier moments for video region duplication. IEEE Trans. Multimed. **20**(4), 825–840 (2017)
5. Liu, Q., Sung, A. H., Zhou, B., Qiao, M.: Exposing inpainting forgery in jpeg images under recompression attacks. In: 2016 15th IEEE International Conference on Machine Learning and Applications (ICMLA), pp. 164–169. IEEE (2016)
6. Neenu, H.U., Cheriyan, J.: Image forgery detection based on illumination inconsistencies & intrinsic resampling properties. In: 2014 Annual International Conference on Emerging Research Areas: Magnetics, Machines and Drives (AICERA/iCMMD), pp. 1–6. IEEE (2014)
7. Ardizzone, E., Bruno, A., Mazzola, G.: Copy–move forgery detection by matching triangles of keypoints. IEEE Trans. Inf. Forensics Secur. **10**(10), 2084–2094 (2015)
8. Le, T.H.N., Luu, K., Savvides, M.: Fast and robust self-training beard/moustache detection and segmentation. In: 2015 International Conference on Biometrics (ICB), pp. 507–512. IEEE (2015)
9. Zheng, J., Liu, Y., Ren, J., Zhu, T., Yan, Y., Yang, H.: Fusion of block and keypoints based approaches for effective copy-move image forgery detection. Multidimension. Syst. Signal Process. **27**, 989–1005 (2016)
10. Gong, M., Zhan, T., Zhang, P., Miao, Q.: Superpixel-based difference representation learning for change detection in multispectral remote sensing images. IEEE Trans. Geosci. Remote Sens. **55**(5), 2658–2673 (2017)
11. Chen, C., McCloskey, S., Yu, J.: Image splicing detection via camera response function analysis. In: Proceedings of the IEEE Conference on Computer Vision and Pattern Recognition, pp. 5087–5096 =(2017)
12. Zuo, H., Fan, H., Blasch, E., Ling, H.: Combining convolutional and recurrent neural networks for human skin detection. IEEE Signal Process. Lett. **24**(3), 289–293 (2017)
13. Su, L., Luo, H., Wang, S.: A novel forgery detection algorithm for video foreground removal. IEEE access **7**, 109719–109728 (2019)
14. Black, P., Gondal, I., Brooks, R., Yu, L.: AFES: an advanced forensic evidence system. In: 2021 IEEE 25th International Enterprise Distributed Object//Computing Workshop (EDOCW), pp. 67–74. IEEE (2021)
15. Saddique, M., Asghar, K., Bajwa, U.I., Hussain, M., Aboalsamh, H.A., Habib, Z.: Classification of authentic and tampered video using motion residual and parasitic layers. IEEE Access **8**, 56782–56797 (2020)
16. Korus, P., Memon, N.: Content authentication for neural imaging pipelines: Endto-end optimization of photo provenance in complex distribution channels. In: Proceedings of the IEEE/CVF Conference on Computer Vision and Pattern Recognition, pp. 8621–8629 (2019)
17. Orozco, A.L.S., Huamán, C.Q., Quintero, J.A.C., Villalba, L.J.G.: Digital video source acquisition forgery technique based on pattern sensor noise extraction. IEEE Access **7**, 157363–157373 (2019)
18. Luo, Y.X., Chen, J.L.: Dual attention network approaches to face forgery video detection. IEEE Access **10**, 110754–110760 (2022)
19. Verde, S., Cannas, E.D., Bestagini, P., Milani, S., Calvagno, G., Tubaro, S.: Focal: a forgery localization framework based on video coding self-consistency. IEEE Open J. Signal Process. **2**, 217–229 (2021)
20. Li, S., Huo, H.: Frame deletion detection based on optical flow orientation variation. IEEE Access **9**, 37196–37209 (2021)
21. Walia, S., Kumar, K., Kumar, M., Gao, X.Z.: Fusion of handcrafted and deep features for forgery detection in digital images. IEEE Access **9**, 99742–99755 (2021)

22. Tran, V.N., Kwon, S.G., Lee, S.H., Le, H.S., Kwon, K.R.: Generalization of forgery detection with meta deepfake detection model. IEEE Access **11**, 535–546 (2022)
23. Wang, D., Gao, T., Zhang, Y.: Image sharpening detection based on difference sets. IEEE Access **8**, 51431–51445 (2020)
24. Yu, I.J., Nam, S.H., Ahn, W., Kwon, M.J., Lee, H.K.: Manipulation classification for jpeg images using multi-domain features. IEEE Access **8**, 210837–210854 (2020)
25. Boato, G., Dang-Nguyen, D.T., De Natale, F.G.: Morphological filter detector for image forensics applications. IEEE Access **8**, 13549–13560 (2020)
26. Vega, E.A.A., Fernández, E.G., Orozco, A.L.S., Villalba, L.J.G.: Passive image forgery detection based on the demosaicing algorithm and JPEG compression. IEEE Access **8**, 11815–11823 (2020)
27. Aloraini, M., Sharifzadeh, M., Schonfeld, D.: Sequential and patch analyses for object removal video forgery detection and localization. IEEE Trans. Circuits Syst. Video Technol. **31**(3), 917–930 (2020)
28. Hosler, B.C., Zhao, X., Mayer, O., Chen, C., Shackleford, J.A., Stamm, M.C.: The video authentication and camera identification database: a new database for video forensics. IEEE Access **7**, 76937–76948 (2019)
29. Quist-Aphetsi, K., Senkyire, I.B.: Validating of digital forensic images using SHA-256. In: 2019 International Conference on Cyber Security and Internet of Things (ICSIoT), pp. 118–121). IEEE (2019)
30. Quan, Y., Li, C.T., Zhou, Y., Li, L.: Warwick image forensics dataset for device fingerprinting in multimedia forensics. In: 2020 IEEE International Conference on Multimedia and Expo (ICME), pp. 1–6. IEEE (2020)
31. Akhtar, N., Saddique, M., Asghar, K., Bajwa, U.I., Hussain, M., Habib, Z.: Digital video tampering detection and localization: review, representations, challenges and algorithm. Mathematics **10**(2), 168 (2022)
32. Kumar, A., Singh, K.U., Swarup, C., Singh, T., Raja, L., Kumar, A.: Detection of copy-move forgery using euclidean distance and texture features. Traitement du Signal **39**(3), 781 (2022)
33. Wang, M., Fu, X., Liu, J., Zha, Z.J.: Jpeg compression-aware image forgery localization. In: Proceedings of the 30th ACM International Conference on Multimedia, pp. 5871–5879 (2022)
34. Abhishek, Jindal, N.: Copy move and splicing forgery detection using deep convolution neural network, and semantic segmentation. Multimed. Tools Appl. **80**, 3571–3599 (2020)
35. Kaur, H., Jindal, N.: Image and video forensics: a critical survey. Wirel. Pers. Commun. **112**, 1281–1302 (2020)
36. Kaur, H., Jindal, N.: Deep convolutional neural network for graphics forgery detection in video. Wirel. Pers. Commun. **112**, 1763–1781 (2020)
37. Li, Q., Wang, R., Xu, D.: A video splicing forgery detection and localization algorithm based on sensor pattern noise. Electronics **12**(6), 1362 (2023)
38. Nemade, V., Pathak, S., Dubey, A.K.: A systematic literature review of breast cancer diagnosis using machine intelligence techniques. Arch. Comput. Methods Eng. **29**(6), 4401–4430 (2022)
39. Barhate, D., Pathak, S., Dubey, A.K.: 'Hyperparameter-tuned batch-updated stochastic gradient descent', plant species identification by using hybrid deep learning. Eco. Inform. **75**, 102094 (2023)

Enhancing the Accuracy of Automatic Bone Age Estimation Using Optimized CNN Model on X-Ray Images

Nivedita$^{(\boxtimes)}$ and Shano Solanki

Department of Computer Science and Engineering, National Institute of Technical Teachers Training and Research, Chandigarh, India
nivedita341993@gmail.com

Abstract. Bone age assessment helps to detect and schedule treatment for several disorders. Estimating bone age differs from determining physical maturity based on the individual's date of birth. Assessing bone age determines development and progress, identifying and treating juvenile illnesses. Challenges in bone age assessment notably arise from poor-quality X-images, obscured bone structures, and the complexity of feature extraction due to degraded image quality, significantly impacting model performance. The proposed methodology involves utilizing pre-trained neural networks—InceptionV3, DenseNet201, XceptionNet, and MobileNetV2—finetuned by adding dense layers alongside dropout and kernel initializers adjustments. The hyperparameters for each pre-trained model are rigorously defined, and the performance evaluation encompasses a spectrum of optimizers such as Adam, Nadam, Adamax, RMSprop, and SGD. Notably, the implementation of the Adamax optimizer yields superior results, demonstrating exceptional accuracy in bone age assessment on the RSNA dataset, particularly with DenseNet201, InceptionV3, and XceptionNet models. This research presents a comprehensive comparative analysis showcasing the enhanced accuracy of bone age estimation.

Keywords: Bone Age Assessment · Tanner Whitehouse 3 · Pretrained CNN models · Optimizers · Hand X-ray images · BAA

1 Introduction

Bone age reflects an individual's skeletal and biological maturation, unlike chronological age derived from the individual birth date. Pediatricians and endocrinologists frequently seek bone age assessments (BAA) in parallel with chronological age to diagnose conditions that contribute to deviations in children's stature, be they excessive or restricted growth [1]. Evaluating bone age can be a supplementary approach to diagnosing various endocrine conditions, including precocious puberty and idiopathic dwarfism. That facilitates the timely and effective implementation of treatments for individuals experiencing abnormal growth. BAA often plays a crucial role in determining athletes' eligibility

M. Khurana et al. (Eds.): ICMLA 2024, CCIS 2238, pp. 329–340, 2025.
https://doi.org/10.1007/978-3-031-75861-4_29

and in legal forensics, ensuring the accuracy and reliability of these processes [2]. Earlier, the BAA used several manual techniques. The Greulich & Pyle Atlas (G&P) [3] method and the Tanner Whitehouse Methods (TW, TW1, TW2) [4] are well known traditional approaches. The rich abundance of the hand and wrist bones and less radiation are required to take X-rays of the hand, enhancing its suitability for BAA tasks. The one major issue with the traditional methods is the inconsistency between the same or different observer observations [5] and the need for a very high BAA expertise, which is very expensive. These issues emphasize the importance of facilitating the BAA process. The main goal of this process is to create an automated BAA that reduces the limitations of manual BAA.

Machine Learning (ML) has several advantages but also some associated disadvantages. ML model needs manual feature extraction, which is time-consuming. It may not take the appropriate information or only a smaller portion of the data. ML models take lots of time to train to achieve the desired level of performance. DL eliminates the problem of ML and can identify complex patterns from the raw data that eliminate the laborious process of feature engineering. DL can capture intricate patterns with tiny details of the radiographs, hence capable of improving the model's performance. A Convolutional Neural Network (CNN) or ConvNet is a regularized feed-forward neural network that can learn intricate features and optimize performance [6].

1.1 Pre-trained CNN Models

Pre-trained models are trained DL and ML models on the large dataset used for specific tasks like object detection, regression, natural language processing, and image classification. The pre-trained model is used as is or by using transfer learning.

InceptionV3. It is a 48-layered architecture developed by Google researchers with a lower error rate than Inception's previous models. It can classify images into 1000 categories. The main advantages are that it reduces the no of parameters to maintain the computational efficiency, reduces the grid's size, auxiliary classifiers used which act as a regularizer, asymmetric convolution, and replaces the bigger convolution with a smaller convolution [7].

XceptionNet. François Chollet introduces the 71-layered, depth-wise separable, and extreme version of the Inception neural network. XceptionNet is better than InceptionV3; the main difference is that InceptionNet adds the non-linearity by adding the RELU function after each operation, but Xception does not add non-linearity [8].

DenseNet201. It is a densely connected DL architecture with 201 layers and is linked from one layer to the other in a feed-forward manner. It eliminates the problem of the vanishing gradient. Its main advantages are feature reuse, improved feature propagation, and reduced number of layers [9].

MobileNetV2. It is tailored for mobile and embedded devices. It has 54 layered and inverted residual structures with an efficient and lightweight design. In the expansion layer, the features are filtered by lightweight convolutions; by doing this, the non-linearity of narrow layers is removed. The depth-wise separable model minimizes computational cost. It is widely applicable to computer vision tasks like image recognition and classification [10].

1.2 Optimizers

Optimizers in DL adjust model parameters to minimize errors during training, optimizing the learning process for improved accuracy and faster convergence. They are crucial in updating neural network weights based on calculated gradients, facilitating efficient model training in DL (Table 1).

Table 1. Overview of different optimizers

Optimizers	Formula	Description
Adam (Adaptive Moment Estimation) [11, 12]	$m_t = \beta_1 \cdot m_{t-1} + (1 - \beta_1) \cdot g_t$ $v_t = \beta_2 \cdot v_{t-1} + (1 - \beta_2) \cdot g_t^2$ $\widehat{m}_t = 1 - \frac{m_t}{1-\beta_1^t}$ $\hat{v}_t = 1 - \frac{v_t}{1-\beta_2^t}$ $\theta_{t+1} = \theta_t - \frac{\eta}{\sqrt{v_t}+\epsilon} \cdot m_t$	Combines momentum and adaptive learning rates for efficient optimization
Nadam (Nesterov-accelerated Adaptive Moment) [11, 13]	$m_t = \beta_1 \cdot m_{t-1} + (1 - \beta_1) \cdot g_t$ $v_t = \beta_2 \cdot v_{t-1} + (1 - \beta_2) \cdot g_t^2$ $\widehat{m}_t = 1 - \frac{m_t}{1-\beta_1^t}$ $\hat{v}_t = 1 - \frac{v_t}{1-\beta_2^t}$ $\theta_{t+1} = \theta_t - \frac{\eta}{\sqrt{v_t}+\epsilon} \cdot (\beta_1 \cdot \widehat{m}_t + \frac{(1-\beta_1) \cdot g_t}{1-\beta_1^t})$	Integrates Nesterov momentum into Adam for faster convergence and handling noisy gradients
RMSprop (Root Mean Square Propagation) [11, 13]	$v_t = \beta \cdot v_{t-1} + (1 - \beta) \cdot g_t^2$ $\theta_{t+1} = \theta_t - \frac{\eta}{\sqrt{v_t}+\epsilon} \cdot g_t$	Adapts learning rates based on recent gradient magnitudes to handle sparse gradients
Adamax (Adaptive Moment Estimation with Infinity Norm) [11, 13]	$m_t = \beta_1 \cdot m_{t-1} + (1 - \beta_1) \cdot g_t$ $u_t = \max(\beta_2 \cdot u_{t-1}, \|g_t\|)$ $\theta_{t+1} = \theta_t - \frac{\eta}{u_t+\epsilon} \cdot m_t$	Utilizes the L_∞ norm for more stable updates, suitable for large parameter spaces
SGD (Stochastic Gradient Descent) [11, 13]	$\theta_{t+1} = \theta_t - \alpha \cdot \nabla J(\theta_{ti})$	Updates parameters using random mini-batches for faster convergence

m_t: Represents the first moment estimate (mean) of gradients.

v_t: Denotes the second moment estimate (uncentered variance) of gradients.

u_t: Refers to the exponentially weighted infinity norm (max norm) of gradients.

$\widehat{m}_t$ and $\hat{v}_t$: Indicate bias-corrected moment estimates.

θ_t and θ_{t+1}: Signify parameters at times t and t + 1.

g_t: Represents the gradient at time t.

η: Stands for the learning rate, governing the step size for parameter updates.

β_1 and β_2: Define exponential decay rates for moment estimates, regulating their impact.

ϵ: Represents a small value added to the denominator for numerical stability, preventing division by zero.

$\nabla J(\theta_t)$: Denotes the gradient of the loss function J concerning parameters θ at time t. It signifies both the direction and magnitude of the steepest increase of the loss function at a specific point θ_t.

1.3 Evaluation Metrics

In this study, the performance metrics to evaluate BAA are MSE (Mean Squared Error), (RMSE) Root Mean Squared Error, MAE (Mean Absolute Error), and Accuracy.

MSE: It measures the average squared difference between the actual (true) and the predicted values in a dataset.

$$MSE = \frac{1}{n} \sum_{i=1}^{n} \left(y\_true - y\_pred \right)^2 \tag{1}$$

RMSE: It is the square root of the MSE, measuring the average magnitude of error between actual and predicted values.

$$RMSE = \sqrt{MSE} \tag{2}$$

MAE: It measures the average absolute differece between the actual(true) and the predicted values in a dataset.

$$MAE = \sum_{i=1}^{n} \frac{|y\_true - y\_pred|}{n} \tag{3}$$

The key contributions of research paper are:

- The research paper is based on proposed methodology which involves fine-tuning of the dense layers of the CNN models. By modifying these layers, model's performance for BAA is enhanced in terms of improved prediction accuracy.
- The study investigates the impact of Adam, Nadam, SGD, Adamax, and RMSprop optimizers on the last layer of pre-trained networks to determine best optimizer for BAA.

The paper is organized into various sections. Sect. 2 focuses on the latest advancements in the field, discussing ongoing research and current developments. Following this, Sect. 3 describes the proposed methodology for BAA. Sect. 4 presents the outcomes derived from the proposed research. Finally, Sect. 5 encapsulates the conclusions drawn from the study and Future scope.

2   Related Work

Earlier, BAA research predominantly centered on established methods such as the Greulich-Pyle [3, 14] and Tanner-Whitehouse [4] techniques. These approaches rely on Radiographic atlases and involve comparing X-rays to evaluate bone maturity. The Gilsanz-Ratib [15] digital atlas provides improved accuracy via categorized images tailored to specific age groups and genders. Within Computer-Aided Diagnosis (CAD), the initial focus was accurately segmenting X-ray images to isolate bone structures. This pursuit faced challenges distinguishing bone from soft tissue and backgrounds, prompting exploration into various techniques.

Lee et al. (2017) achieved BAA with a model showing an MAE result of 6.1 months using ImageNet with DICOM images via OsiriX software [6]. Spa et al. (2017) introduced BoNet using (Digital Hand Atlas) DHA data, achieving an MAE of 9.6 months, showcasing a groundbreaking 0.8 decades efficiency gap between traditional and automated skeletal bone age evaluation on a universal scale for all age groups, countries, and genders [16]. Liu et al. (2019) introduced a novel BAA method, merging NSCT with CNNs, enhancing BAA on DHA with VGGNet-16, achieving an MAE of 8.28 months by employing multi-scale data fusion, surpassing traditional spatial domain techniques [17]. Bui et al. (2019) integrated TW methodologies with fully CNN, employing Faster-RCNN and InceptionV4 networks for ROI detection in BAA, achieving a remarkable MAE of 7.08 months [18]. Liang et al. (2019) introduced a deep automated skeletal BAA model using Faster RCNN, achieving MAEs of 6.12 and 5.76 months on different datasets, surpassing limitations of CNN-based approaches by leveraging large-scale X-ray images for accurate ossification center identification and bone age prediction [2]. Lee et al. (2020) introduced RT-FuseNet, utilizing ResNet and RT-FuseNet on RSNA 2017 and a hospital dataset with 5286 images, employing attention modules and spatial pyramid pooling for improved learning from hand radiographs and text information, achieving MAEs of 6.91 months with ResNet and 6.11 months with RT-FuseNet, effectively identifying radiographs of varying ages by emphasizing pertinent qualities [19]. Lee et al. (2020) proposed a DL approach for bone age prediction using carefully chosen hand photos from the RSNA dataset, marking specific features for key ROI extraction, training various DL architectures, achieving a minimum MAD of 8.890 months with GoogleNet [20]. Gao et al. (2020) devloped a deep CNN-based technique utilizing U-Net for semantic segmentation to extract finger bones sections and VGGNet16 with an attention strategy for classification on RSNA, achieving an MAE of 9.97 months, surpassing standard methods for BAA [21]. Wibisono et al. (2020) proposed a decision support system using ML and DL, employing RB-FCL for specific hand image areas and DL models (DenseNet121, InceptionV3, InceptionResNetV2) to extract bone-related features, achieving an MAE of 6.97 months on RSNA, outperforming traditional DL models and improving upon a conventional DNN's score of 9.41 months in bone age forecasting from X-ray scans [22]. Li et al. (2021) introduced a DL-based computer-aided assessment for BAA, utilizing MobileNet and (Multi-Layered Perceptron) MLP with one hidden layer, incorporating unsupervised learning for informative area identification, achieving an MAE of 5.1 months on the Clinical dataset by including sex information in the prediction process for enhanced clinical research and 6.2 months on the RSNA dataset [23]. Xu et al. (2022) introduced a hierarchical CNN, YOLOv5, for

BAA, employing (Regions of Interest) ROI identification and bone score categorization on a dataset from Xuzhou Central Hospital (2158 X-ray images), achieving an MAE of 6.53 months on the public RSNA dataset and 7.68 months on the clinical dataset, demonstrating competitive performance and outperforming existing fine-grained image classification methods in BAA [24].

However, challenges persist in employing DL models for BAA. These include accurately capturing essential image features, establishing evaluation criteria for key points, handling image quality and variability, complex background management, detecting ROIs, and adapting to diverse image resolutions. Addressing these challenges entails selecting appropriate pre-trained models and employing various optimization strategies. The objective is to refine feature capture, establish robust evaluation criteria, enhance adaptability to image variability, and ensure resilience in complex background scenarios, ultimately improving the precision of BAA.

3   Proposed Methodology

The proposed methodology optimizes BAA accuracy by refining a pre-existing neural network through fine-tuning. Additionally, it seeks to determine how different optimizers impact the model's performance. The outlined steps are as follows:

3.1   X-Ray Image Acquisition

This study utilizes the RSNA Paediatric Bone Age Challenge (2017) dataset containing 12,611 X-ray images for BAA [25]. The dataset covers an age range from 0 to 217 months, with 6,833 male and 5,778 female records, featuring 0 months as a minimum age and 216 months as a maximum for estimation purposes (Fig. 1).

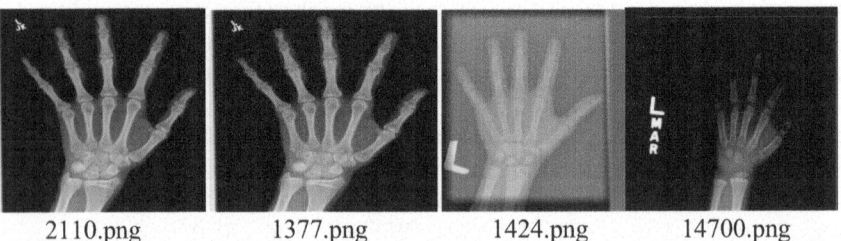

2110.png   1377.png   1424.png   14700.png

Fig. 1. Example of RSNA dataset

3.2   Data Split: Training and Testing

The RSNA dataset for BAA is divided into an 80:20 ratio, allocating 10,088 radiographs for training and 2,523 for testing.

3.3   Fine Tuning of the Last Layer of the Pre-trained Neural Network

The process involves modifying a pre-trained model (such as InceptionV3, XceptionNet, DenseNet201, or MobileNetV2) specifically for BAA. The model undergoes transfer learning, starting with the architecture of a pre-existing model while excluding its top classification layers. A CNN architecture is then created by integrating several dense layers with varying neuron counts (128, 64, and 32), culminating in a final output layer. This layer is designed for regression tasks and utilizes a linear activation function to predict continuous values effectively. All layers within the network are configured to be trainable, ensuring thorough fine-tuning and optimized results in BAA. The model incorporates a Dropout layer, typically with a rate around 0.3, to mitigate overfitting by deactivating some neurons during the training phase. Diverse optimizers, including Adam, Nadam, SGD, RMSprop, and Adamax, are employed, each maintaining a fixed learning rate of 0.001 to minimize the MSE. The fine-tuning strategy is designed for optimized results, leveraging diverse optimizers such as Adam, Nadam, SGD, RMSprop, and Adamax. Each optimizer's learning rate and specific parameters are carefully adjusted to augment model learning and convergence (Fig. 2).

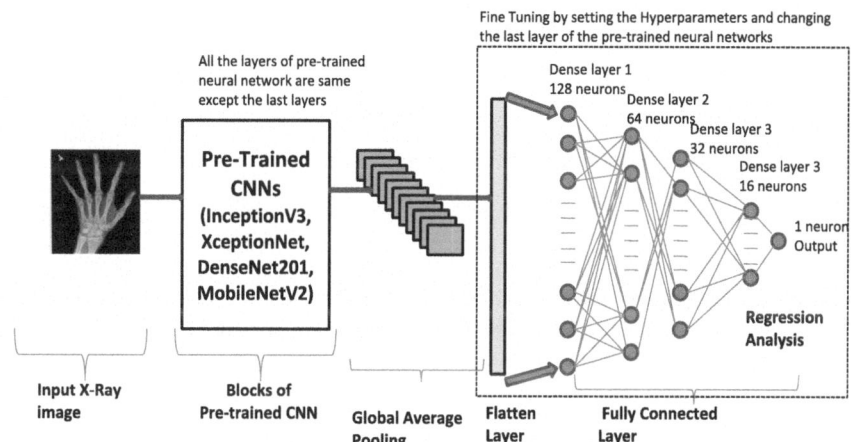

Fig. 2. Block diagram of the fined tuned CNN

Additionally, the dense layers used the RandomUniform kernel initializer. There are several callbacks which support training such as: ModelCheckpoint saves the optimal weights, ReduceLROnPlateau dynamically adjusts learning rates according to testing loss, EarlyStopping stops training if testing loss remains unchanged. The training protocol spans 50 epochs, although batch sizes are 32 for training and testing. These fine-tuning settings and optimizer variations aim to optimize model learning and convergence, leading to superior performance and enhanced BAA accuracy.

3.4   Comparative Analysis of the Different Pre-trained Model

The comparison involves studying optimizers' impacts on different pre-trained models like InceptionV3, XceptionNet, MobileNetV2, and DenseNet201 for BAA. This analysis explored how optimizers like Adam, Nadam, RMSProp, Adamax, and SGD affect the model's performance. Through hyperparameter tuning, adjustments to learning rates, batch sizes, and specific layer configurations to assess their influence on predicting accuracy. This investigation aims to optimize each model's potential, identifying the best pre-trained architectures and optimizer combinations. Ultimately, the goal is to enhance diagnostic precision, particularly in evaluating growth disorders in children, by leveraging the most effective model-optimizer pairings.

4   Result and Discussion

This section presents a comparative analysis of diverse pre-trained models with various optimizers. Furthermore, it assesses the performance of proposed model against the previous Automatic BAA models (Table 2).

Table 2. InceptionV3 model with different optimizers

Optimizer	MAE	RMSE	Accuracy %
Adam	6.72	8.91	97.05%
Nadam	7.76	11.04	96.59%
RMSprop	6.85	9.85	96.99%
SGD	8.58	11.88	96.23%
Adamax	5.85	8.70	97.55%

In BAA, InceptionV3 performs best with an Adamax optimizer, achieving a 5.85 months MAE, closely followed by 6.72 months with Adam. However, employing SGD leads to the least favorable results for this model (Table 3).

Table 3. DenseNet201 model with different optimizers

Optimizer	MAE	RMSE	Accuracy %
Adam	9.28	13.15	95.92%
Nadam	6.51	9.57	97.14%
RMSprop	10.19	16.28	95.53%
SGD	10.10	13.14	95.57%
Adamax	5.17	7.73	97.73%

The pre-trained DenseNet201 model, when fine-tuned with the Adamax optimizer, achieves the best performance with a 5.17 months MAE on the RSNA dataset for BAA, while the Nadam optimizer follows closely with a 6.51 months MAE. However, the RMSprop optimizer resulted in the least favorable outcomes for this task (Table 4).

Table 4. MobileNetV2 model with different optimizers

Optimizer	MAE	RMSE	Accuracy %
Adam	11.23	15.09	95.07%
Nadam	10.34	14.42	95.46%
RMSprop	9.90	13.62	95.65%
SGD	12.26	16.74	94.62%
Adamax	11.92	15.91	94.77%

When utilizing the MobileNetV2 architecture, the most favorable outcomes, with an MAE of 9.90, are achieved with the RMSprop optimizer. Similarly, the Nadam optimizer yields a competitive MAE of 10.34. SGD results in the least desirable performance for BAA using this architecture (Table 5).

Table 5. XceptionNet model with different optimizers

Optimizer	MAE	RMSE	Accuracy %
Adam	6.72	9.55	97.05%
Nadam	6.98	9.78	96.93%
RMSprop	7.90	10.72	96.53%
SGD	30.97	39.08	86.84%
Adamax	5.86	8.92	97.42%

When optimized with Adamax, the XceptionNet achieves the best MAE of 5.86 months, slightly increasing to 6.72 months with Adam, while SGD produces the least favorable MAE for BAA.

Compared to existing approaches for automatic BAA, our model demonstrates superior optimization, yielding the most refined results with the pre-trained model DenseNet201 and Adamax optimizer by setting the epsilon $1e-07$, the learning rate 0.001, beta_1 0.9, beta_2 0.999 (Table 6).

Table 6 The performance by different automated BAA models using public dataset RSNA

Authors	Models	MAE	Accuracy %
Hao et al. [19]	ResNet, RT FuseNet	6.29	97.08%
Lee et al. [20]	GoogleNet	8.890	96.10%
Gao et al. [21]	UNet, VGGNet16	9.997	95.61%
Wibisono et al. [22]	RB-FCL, InceptionV3	6.67	97.07%
Li et al. [23]	MLP, MobileNet	6.2	97.28%
Xu et al. [24]	YOLOV5	6.53	97.13%
Chen et al. [26]	Xception, InceptionNet	6.4	97.19%
Proposed Methodology	DenseNet201	**5.17**	**97.73%**

5 Conclusion and Future Scope

After conducting a comprehensive analysis for BAA, it was observed that the pre-trained models, specifically DenseNet201, InceptionV3, and XceptionNet, consistently delivered superior performance in terms of MAE, RMSE, and Accuracy metrics. Their robust architecture and feature extraction capabilities contributed significantly to BAA. Adamax optimizer performed well with all pre-trained CNN models except MobileNetV2. The RMSprop, Adam, and Nadam optimizers showed competitive performance across the evaluated pre-trained models. On the other hand, the Stochastic Gradient Descent (SGD) optimizer demonstrated the least favorable outcomes among the tested optimizers across the pre-trained models. Adamax is the primary choice for most CNN architectures. This study emphasizes the significance of model selection and optimization methods in BAA tasks, showcasing the nuanced impact these choices can have on the overall performance of deep learning models. Further research could delve deeper into fine-tuning strategies and architectural adjustments tailored to improve the performance of BAA. Expanding the dataset size and using high-quality images for the BAA can also improve the accuracy. Accurately finding the ROI can also minimize the error and enhance the performance of the pre-trained models.

References

1. Mughal, A.M., Hassan, N., Ahmed, A.: Bone age assessment methods: a critical review. Pak. J. Med. Sci. **30**, 211 (1969). https://doi.org/10.12669/pjms.301.4295
2. Liang, B., et al.: A deep automated skeletal bone age assessment model via region-based convolutional neural network. Futur. Gener. Comput. Syst. **98**, 54–59 (2019). https://doi.org/10.1016/j.future.2019.01.057
3. Alshamrani, K., Messina, F., Offiah, A.C.: Is the Greulich and Pyle atlas applicable to all ethnicities? A systematic review and meta-analysis. Eur. Radiol. **29**, 2910–2923 (2019). https://doi.org/10.1007/s00330-018-5792-5
4. Ostojic, S.M.: Prediction of adult height by Tanner-Whitehouse method in young Caucasian male athletes. QJM **106**, 341–345 (2013). https://doi.org/10.1093/qjmed/hcs230

5. Satoh, M.: Bone age: assessment methods and clinical applications. Clin. Pediatr. Endocrinol. **24**, 143–152 (2015). https://doi.org/10.1297/cpe.24.143
6. Lee, H., et al.: Fully automated deep learning system for bone age assessment. J. Digit. Imaging **30**, 427–441 (2017). https://doi.org/10.1007/s10278-017-9955-8
7. Szegedy, C., Vanhoucke, V., Ioffe, S., Shlens, J., Wojna, Z.: Rethinking the inception architecture for computer vision. In: 2016 IEEE Conference on Computer Vision and Pattern Recognition (CVPR), pp. 2818–2826. IEEE, Las Vegas, NV, USA (2016). https://doi.org/10.1109/CVPR.2016.308
8. Chollet, F.: Xception: deep learning with depthwise separable convolutions. In: 2017 IEEE Conference on Computer Vision and Pattern Recognition (CVPR), pp. 1800–1807. IEEE, Honolulu, HI (2017). https://doi.org/10.1109/CVPR.2017.195
9. Huang, G., Liu, Z., Van Der Maaten, L., Weinberger, K.Q.: Densely connected convolutional networks. In: 2017 IEEE Conference on Computer Vision and Pattern Recognition (CVPR), pp. 2261–2269. IEEE, Honolulu, HI (2017). https://doi.org/10.1109/CVPR.2017.243
10. Howard, A., et al. : MobileNets: Efficient Convolutional Neural Networks for Mobile Vision Applications (2017)
11. Bashetty, S., Raja, K., Adepu, S., Jain, A.: Optimizers in deep learning: a comparative study and analysis. IJRASET **10**, 1032–1039 (2022). https://doi.org/10.22214/ijraset.2022.48050
12. Kingma, D.P., Ba, J.: Adam: A Method for Stochastic Optimization (2017). http://arxiv.org/abs/1412.6980
13. Kandel, I., Castelli, M., Popovič, A.: Comparative study of first order optimizers for image classification using convolutional neural networks on histopathology images. J. Imaging **6**, 92 (2020). https://doi.org/10.3390/jimaging6090092
14. Maggio, A., Flavel, A., Hart, R., Franklin, D.: Assessment of the accuracy of the Greulich and Pyle hand-wrist atlas for age estimation in a contemporary Australian population. Aust. J. Forensic Sci. **50**, 385–395 (2018). https://doi.org/10.1080/00450618.2016.1251970
15. Adler, B.H.: Vicente Gilsanz, Osman Ratib: bone age atlas. Pediatr. Radiol. **35**, 1035 (2005). https://doi.org/10.1007/s00247-005-1527-2
16. Spampinato, C., Palazzo, S., Giordano, D., Aldinucci, M., Leonardi, R.: Deep learning for automated skeletal bone age assessment in X-ray images. Med. Image Anal. **36**, 41–51 (2017). https://doi.org/10.1016/j.media.2016.10.010
17. Liu, Y., Zhang, C., Cheng, J., Chen, X., Wang, Z.J.: A multi-scale data fusion framework for bone age assessment with convolutional neural networks. Comput. Biol. Med. **108**, 161–173 (2019). https://doi.org/10.1016/j.compbiomed.2019.03.015
18. Bui, T.D., Lee, J.-J., Shin, J.: Incorporated region detection and classification using deep convolutional networks for bone age assessment. Artif. Intell. Med. **97**, 1–8 (2019). https://doi.org/10.1016/j.artmed.2019.04.005
19. Hao, P., et al.. Radiographs and texts fusion learning based deep networks for skeletal bone age assessment. Multimed. Tools Appl. **80**, 16347–16366 (2021). https://doi.org/10.1007/s11042-020-08943-1
20. Lee, J.H., Kim, Y.J., Kim, K.G.: Bone age estimation using deep learning and hand X-ray images. Biomed. Eng. Lett. **10**, 323–331 (2020). https://doi.org/10.1007/s13534-020-00151-y
21. Gao, Y., Zhu, T., Xu, X.: Bone age assessment based on deep convolution neural network incorporated with segmentation. Int. J. CARS **15**, 1951–1962 (2020). https://doi.org/10.1007/s11548-020-02266-0
22. Wibisono, A., Mursanto, P.: Multi region-based feature connected layer (RB-FCL) of deep learning models for bone age assessment. J. Big Data **7**, 67 (2020). https://doi.org/10.1186/s40537-020-00347-0
23. Li, S., Liu, B., Li, S., Zhu, X., Yan, Y., Zhang, D.: A deep learning-based computer-aided diagnosis method of X-ray images for bone age assessment. Complex Intell. Syst. **8**, 1929–1939 (2022). https://doi.org/10.1007/s40747-021-00376-z

24. Xu, X., Xu, H., Li, Z.: Automated bone age assessment: a new three-stage assessment method from coarse to fine. Healthcare **10**, 2170 (2022). https://doi.org/10.3390/healthcare10112170

25. Halabi, S.S., et al.: The RSNA pediatric bone age machine learning challenge. Radiology **290**, 498–503 (2019). https://doi.org/10.1148/radiol.2018180736

26. Chen, C., Chen, Z., Jin, X., Li, L., Speier, W., Arnold, C.W.: Attention-guided discriminative region localization and label distribution learning for bone age assessment. IEEE J. Biomed. Health Inform. **26**, 1208–1218 (2022). https://doi.org/10.1109/JBHI.2021.3095128

Machine Learning-Based Detection of Forgery in Digital Images

Navneet Kaur$^{(\boxtimes)}$, Monika Parmar, Ramamani Tripathy, Hakam Singh, and Sandhya Sharma

Chitkara University School of Engineering and Technology, Chitkara University, Baddi, Himachal Pradesh 174103, India

`navneet.kaur_cse@chitkarauniversity.edu.in`

Abstract. The rapid progression of digital technology has led to an increasing number of advantages associated with acquiring voluminous information through internet access. The proliferation of media alteration software has facilitated the manipulation of multimedia data. As a result, the authentication and integrity of images are crucial in numerous disciplines. Image forensics is a burgeoning field utilized to assess the reliability of digital pictures. The purpose of image forgery is to produce obscure images while concealing vital and useful information. Copy-move forgery (CMF) is a form of forgery that exposes both the general public and image forensics specialists to grave danger. For effective forgery detection, the aim of this research is to hybridize discrete cosine transform (DCT) and accelerated KAZE (AKAZE). DCT is initially applied to each block subsequent to block division. In addition, segmentation is performed utilizing k-mean clustering, which is subsequently followed by feature extraction via AKAZE. Following this, the k-nearest neighbor algorithm is utilized to match the characteristics. Ultimately, morphological processes are employed to photographs to expose fabricated regions. The experimental outcomes are implemented on standard datasets for image manipulation. A range of performance parameters, including precision, recall, F_1 score, and F_2 score, are assessed and demonstrated to be superior in comparison to the currently employed methodologies.

Keywords: CMF · Recall · Segmentation · AKAZE · Precision

1 Introduction

Digital pictures are among the more prevalent forms of communication in the age of digital computing. The utilization of image editing software facilitates the modification of images [1]. The significance of photograph originality cannot be overstated, given their extensive use as evidence in a variety of contexts [2, 3]. Image forensics is primarily concerned with determining the legitimacy of digital photographs. Forgery detection procedures and image source device identification techniques have emerged as two main research directions. The primary focus of is on procedures for spotting forgery [4, 5].

The pixel-based CMF detection (CMFD) scheme is utilized because of its efficacy and simplicity. Typically, this is done with the intent of concealing specific information

© The Author(s), under exclusive license to Springer Nature Switzerland AG 2025
M. Khurana et al. (Eds.): ICMLA 2024, CCIS 2238, pp. 341–350, 2025.
https://doi.org/10.1007/978-3-031-75861-4_30

or duplicating particular image characteristics [6, 7]. Occasional use is made of image processing procedures including scaling, blurring, compression, and rotation to produce convincing forgeries. In contrast, the replicated regions are portion of the exact image and will retain their most notable attributes, including color quality, dynamic range, and noise component, while also being attuned with the remainder of the picture. In recent years, a substantial number of procedures have been developed to identify CMF [8, 9]. A description of CMF is provided in Fig. 1.

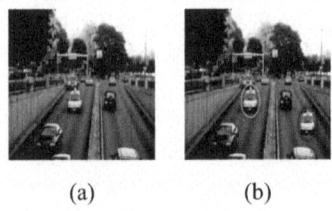

(a)   (b)

Fig. 1. Depiction of a CMF: a) Genuine image, and b) CMF representation where the yellow circle indicates the duplicated section while the red circle denotes the altered/falsified portion

The proposed scheme incorporates a methodology that combines block-based and keypoint-based approaches, namely DCT and AKAZE. In general, block-based procedures exhibit resilience to post-processing procedures such as compression in replicated regions. However, their effectiveness is limited when it comes to detecting geometric operations such as scale and rotation. Furthermore, these processes exhibit a significant computational complexity. On the contrary, keypoint-based techniques exhibit a minimal computational complexity; though, they are not proficient of handling compression efficiently [7, 10]. The aim of integrating both methodologies is to capitalize on their respective strengths and enhance the detection capabilities of CMF. Highlights of the suggested research are succinctly outlined as follows:

- Improving the forgery detection system's performance metrics including precision, F_1 score, recall, and F_2 score-is the primary objective of the proposed system. These metrics demonstrate an expansion over those of the current techniques.
- The methodology being evaluated exhibits promising results in detecting copy-move forgeries involving both a single and multiple copies.
- In the face of geometrical challenges, including compression and rotation, the proposed method exhibits remarkable resilience and achieves superior results in comparison to well-established approaches.

2   Related Work

Detection of CMF is accomplished via block-based or keypoint mechanisms [11–17]. The first method involves splitting the photograph into blocks or sections, and then extracting characteristics from each block. In addition, the similarity between duplicate regions is assessed through feature matching. Fridrich et al. [17] introduced the DCT method, which makes use of overlapping blocks. In order to correlate the units, lexicographic sorting was utilized. This method identifies fabricated components even if the replicated area has been enhanced. The block features were obtained through the implementation of Principal Component Analysis (PCA) by Popescu et al. [18]. While this scheme demonstrated resistance to noise and JPEG compression assaults, it was unable to detect multiple CMFs. Li et al. [19] utilized DWT and SVD to extract the characteristics of a photograph. The efficacy of this procedure extended to JPEG compression attacks. Using a one-dimensional (1-D) descriptor, Bravo et al. [20] have suggested a system that decreases the rate of false alarms. A scrutiny of various block-based CMF methods is presented in Table 1.

Table 1. Block based CMF techniques.

Author	Feature Extraction	Dataset	Observations
Huang [21]	DCT	DVMM	Less computational complexity and robust against JPEG compression, blurring, AWGN
Cao [22]	DCT	DVMM; Internet; Kodak	Robust to multiple CMFs with low computational complexity
Zhao [23]	DCT; SVD	USC-SIPI; Kodak; Internet	Works on compressed images, blurred and contaminated with noise
Li [24]	PST	UCID; Internet	Achieves higher detection accuracy
Cozzolino [25]	CHT	GRIP; Image manipulation	Provides novel detection with noteworthy enhancement in efficiency and swiftness

The operation of keypoint-based technique is predicated on keypoint extraction process. SIFT is an algorithm that was introduced by Huang et al. [26] to identify duplicated regions within a picture. The study by Pan et al. [27] presented the research on detecting CMF through keypoint matching. Although the fact that this methodology successfully handles geometric transformations, a more resilient method was still necessary. SIFT was implemented on the images that had been subjected to geometrical transformations by Amerini et al. [28]. In this method, the MICC-F600 and MICC-F220 datasets were utilized. Additional literature pertaining to keypoint-based techniques is given in Table 2.

Table 2. Keypoint based CMF techniques.

Author	Feature Extraction	Dataset	Observations
Amerini [29]	SIFT	MICC-F2000; SATA-130; MICC-F600	Introduced a clustering method that works in the domain of geometrical transformation also deal with multiple cloning
Jaberi [30]	MIFT	CASIA TIDE v2.0	Detect cloned areas with high accuracy, mainly with small size cloned region
Pandey [31]	SURF; SIFT	MICC-F220	Less processing time and invariable to different affine transformations like scaling and rotation
Yang [32]	KAZE; SIFT	Image manipulation	Effectively identifying duplicated image regions even amidst diverse geometric alterations
Prakash [33]	AKAZE; SIFT	Image manipulation	Rectify duplicated regions vulnerable to geometrical assaults

3   Proposed Method

Hybridizing block-based and keypoint-based techniques, specifically DCT and AKAZE, is an element of the proposed methodology. Overall, block-based techniques exhibit resilience to post-processing procedures such as compression in replicated regions. Instead, keypoint-based procedures possess the ability to handle geometrical alterations like scaling and rotation. Furthermore, their computational complexity is minimal. Thus, the purpose of integrating the two methods is to effectively increase the discovery of CMF by capitalizing on their respective benefits. Figure 3 illustrates the conceptual structure of the counterfeit detection procedure that has been proposed. The following is an explanation of the numerous stages that comprise the suggested system.

Initially, the image manipulation dataset (IMD) comprising 96 images: 48 original and 48 tampered image - created by Christlein et al. [34] is utilized. Images undergo various processing operations prior to progressing to the subsequent stage. The RGB image is converted to grayscale as:

$$Y = 0.299R + 0.587G + 0.114B \tag{1}$$

where B, R, and G are blue, red, and green modules of the image Y, correspondingly [35]. The picture is split into overlying sections of size $N \times N$ pixels, i.e., adjoining sections that only have one distinct column or row. Each section is represented as M_{cd} where

d and c designates the initial point of the section's column and row correspondingly. $(W - N + 1) \times (V - N + 1)$ show total sections for the image of $N \times N$ pixels. For each section, 2D DCT is performed in order to attain the corresponding DCT coefficient matrices. 2D-DCT for picture of size $W \times V$ is specified in equation beneath [23]:

$$c(p, q) = \alpha(p)\alpha(q) \sum_{x=0}^{W-1} \sum_{y=0}^{V-1} f(x, y) \cos\left(\frac{\pi(2x + 1)p}{2W}\right) \cos\left(\frac{\pi(2y + 1)q}{2V}\right) \quad (2)$$

where, $\alpha(p) = \begin{cases} \dfrac{1}{\sqrt{W}}, & p = o \\ \dfrac{\sqrt{2}}{W}, & 1 \leq p \leq W - 1 \end{cases}$ and $\alpha(q) = \begin{cases} \dfrac{1}{\sqrt{V}}, & q = o \\ \dfrac{\sqrt{2}}{V}, & 1 \leq q \leq V - 1 \end{cases}$

The value $c(p, q)$ is the image's DCT constants. To quantize the DCT coefficients, the quantization table is utilised. As a result, 64 coefficients are characteristics of each section in total. Following this, each section's total coefficients are utilised for further processing [23] (Fig. 2).

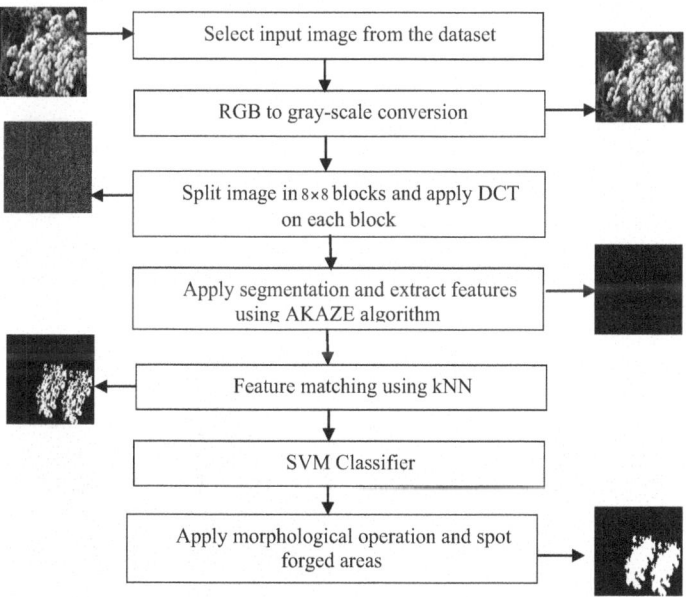

Fig. 2. The structure of the proposed forgery detection method

The AKAZE algorithm is particularly useful for identifying critical nodes within uniform regions during CMF detection. The diffusion process is regulated by the divergence of a flow function via a partial differential equation. The diffused radiance of a photo is caused by scale-space of the partial differential equation [36]. The AKAZE generates scale photos at every level through the utilization of fast explicit diffusion (FED). FED is an explicit technique that is straightforward and significantly quicker than conventional explicit techniques by utilizing varying time step sizes. Due to its invariance to rotation, and its enhanced individualism at different scales resulting from nonlinear scale spaces, AKAZE has been implemented [33, 36]. Features are utilized to delineate the various attributes of an image. The aim of feature vector extraction is to transform the input into a collection of features. This is carried out due to the substantial magnitude of the input required for processing, which is also anticipated to be redundant (i.e., it contains a great deal of data but little information).

A radial basis function (RBF) kernel-based SVM classifier was utilized to discriminate between genuine and forged images. SVM is selected for the proposed task because, among classification algorithms, it is most robust and accurate. After the SVM classifier classifies the images as fabricated, feature matching is executed in order to precisely identify the falsified regions. Utilizing a K-nearest neighbour (kNN) search, the region's corresponding critical points are identified. The kNN search algorithm is utilized for feature matching due to its efficiency, ease of implementation, and effectiveness. Additionally, numerous fabricated regions are identified in digital images. The identified regions that have been fabricated by copy-move are then produced using the morphological close operation. The morphological procedure avoids minor areas, diminishes isolated pixels, and replaces tiny holes in the detected forgery.

4   Experimental Results

A range of trials are conducted to assess the efficacy and resilience of the suggested scheme for detecting forgery. To evaluate the proposed method, three datasets were utilized. To begin, the IMD dataset comprising 48 images is utilized. The dimensions of the high-resolution images range between 3888×2592 and 800×533 pixels. The dataset comprises images that fall into various categories, including natural, artificial, organic, and blended. Illustration of two instances of CMF is offered in Fig. 3.

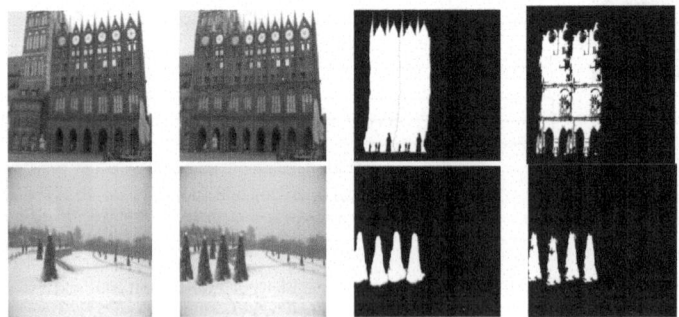

Fig. 3. Examples of test cases, first column: authentic images; second column: fabricated images; third column: ground truth; and fourth column: experimental outcomes.

Since Christlein et al. recommended each benchmark method in particular and utilised the same dataset they supplied, experimental results have been contrasted with copy-move evaluation results on IMD in this paper. Furthermore, a comparative analysis is conducted between current block-based, keypoint-based, and hybrid techniques. The comparison among the suggested methodology and existing methodologies, and evaluation of different performance parameters, is given in Table 1. Furthermore, Fig. 4 illustrates the graphical representation of the comparative analysis conducted with established techniques (Table 3).

Table 3. Assessing the performance metrics (%) of the proposed approach against existing methodologies

Methods	Precision	Recall	F_1 score	F_2 score
Sun [7]	90.91	83.33	86.96	84.74
Pun [12]	95.92	97.92	96.91	97.51
Yang [32]	90.27	78.61	84.04	80.69
Pan [27]	88.37	79.17	83.52	80.92
Bravo [20]	87.27	98.78	93.20	97.16
Prakash [33]	92.30	87.80	89.98	88.65
Cozzolino [25]	92.15	97.92	94.95	96.71
Proposed method	98.92	99.46	99.19	99.35

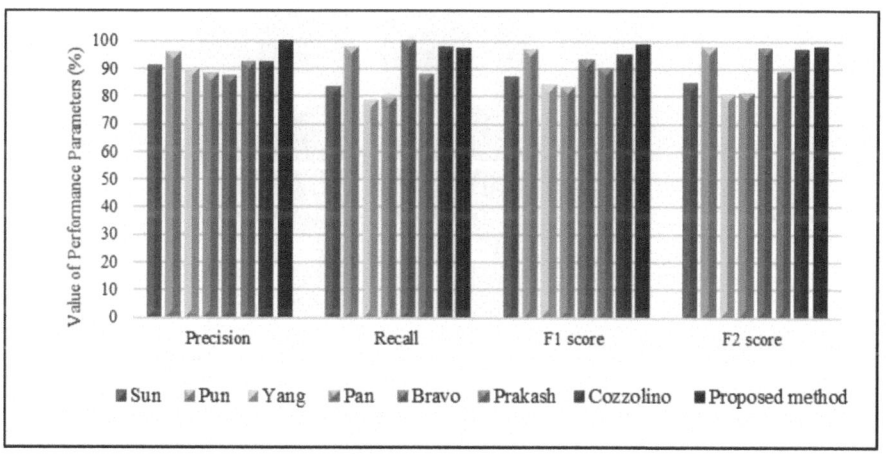

Fig. 4. Performance parameter comparison between the suggested and existing methods

4.1 Detection Results Under Different Attacks

Rotation: The computation of the scheme's robustness was performed in the presence of rotation attack alterations to copied areas. Alternating the replicated regions is a rotation angle ranging from 2° to 10°. Tests are conducted on a total of $48 \times 5 = 240$ images in

Fig. 5. Forged images with varying degrees of rotation in the first column; ground truth in the second column; detection results utilizing a rotation attack in the third column.

this instance. The detection outcomes for rotation attacks at rotation degrees of 2°, and 10° are demonstrated in Fig. 5.

JPEG Compression: Additionally, the resilience of the suggested methodology has been assessed in the face of a JPEG compression assault. The trial is conducted on a grand total of 432 images, or 48 × 9. The detection outcome of the suggested system for various quality factors of JPEG compression is illustrated in Fig. 6.

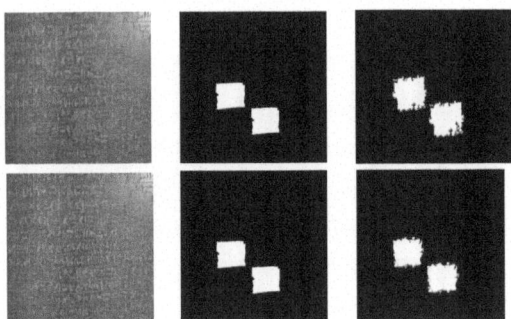

Fig. 6. Results of a detection attack utilizing JPEG compression.

5   Conclusion

Widespread use of copy-move operations generates digital forgery images, which makes their detection an extremely difficult task. An enhanced hybrid methodology has been integrated to efficiently identify both single and multiple CMFs through the utilization of keypoint-based and block-based techniques, namely DCT and AKAZE. Experiments specify that the suggested scheme achieves superior results to alternative approaches when using ordinary CMF, as dignified by precision, recall, F_1 score, and F_2 score. Furthermore, the suggested methodology possesses the capability to accurately identify duplicated areas within the provided tampered photo, regardless of the likelihood that they are subjected to geometrical assaults like rotation, and JPEG compression. The trial outcomes reveal that suggested structure detects image manipulation datasets with outstanding accuracy.

Acknowledgments. The authors declare that no funds, grants, or other support were received during the preparation of this manuscript.

Disclosure of Interests. The authors have no competing interests with anyone related to subject matter.

References

1. Redi, J.A., Taktak, W., Dugelay, J.L.: Digital image forensics: a booklet for beginners. Multimed. Tools Appl. **51**(1), 133–162 (2011)
2. Zanardelli, M., Guerrini, F., Leonardi, R.: Image forgery detection: a survey of recent deep-learning approaches. Multimed. Tools Appl. **82**, 17521–17566 (2023)
3. Singh, T.P., et al.: Visualization of customized convolutional neural network for natural language recognition. Sensors **22**(8), 2881 (2022)
4. Kaur, N., Jindal, N., Singh, K.: A deep learning framework for copy-move forgery detection in digital images. Multimed. Tools Appl. **82**, 17741–17768 (2022)
5. Sharma, S., Gupta, S., Kumar: A detailed study on the recognition of text using machine learning. In: AIP Conference Proceedings, vol. 2357, no. 1. AIP Publishing (2022)
6. Fan, J., Chen, T., Kot, A.C.: EXIF-white balance recognition for image forensic analysis. Multidimension. Syst. Signal Process. **28**(3), 795–815 (2017)
7. Sun, Y., Ni, R., Zhao, Y.: Nonoverlapping blocks based copy-move forgery detection. Secur. Commun. Netw. **2018**, 1–11 (2018)
8. Birajdar, G.K., Mankar, V.H.: Digital image forgery detection using passive techniques: a survey. Digit. Investig. **10**(3), 226–245 (2013)
9. Wang, H., Wang, H.X., Sun, X.M., Qian, Q.: A passive authentication scheme for copy-move forgery based on package clustering algorithm. Multimed. Tools Appl. **76**(10), 12627–12644 (2017)
10. Shahrokhi, M., Akoushideh, A., Shahbahrami, A.: Image copy-move forgery detection using combination of scale-invariant feature transform and local binary pattern features. Int. J. Image Graph. **22**(05), 2250048 (2022)
11. Kaur, N., Jindal, N., Singh, K.: An improved approach for single and multiple copy-move forgery detection and localization in digital images. Multimed. Tools Appl. **81**(27), 38817–38847 (2022)
12. Pun, C.M., Yuan, X.C., Bi, X.L.: Image forgery detection using adaptive over segmentation and feature point matching. IEEE Trans. Inf. Forensics Secur. **10**(8), 1705–1716 (2015)
13. Kaur, N., Jindal, N., Singh, K.: A passive approach for the detection of splicing forgery in digital images. Multimed. Tools Appl. **79**(43), 32037–32063 (2020)
14. Muniappan, T., Abd Warif, N.B., Ismail, A., Abir, N.A.M.: An evaluation of convolutional neural network (CNN) model for copy-move and splicing forgery detection. Int. J. Intell. Syst. Appl. Eng. **11**(2), 730–740 (2023)
15. Kaur, N.: AI-based COVID-19 disease detection in medical images: advancements and implications in healthcare. J. Auton. Intell. **6**(3), 1 (2023)

16. Lu, S., Hu, X., Wang, C., Chen, L., Han, S., Han, Y.: Copy-move image forgery detection based on evolving circular domains coverage. Multimed. Tools Appl. **81**(26), 37847–37872 (2022)

17. Fridrich, A.J., Soukal, B.D., Lukáš, A.J.: Detection of copy-move forgery in digital images. Proceed. Digit. Forensic Res. Workshop **3**(2), 652–663 (2003)

18. Popescu, A.C., Farid, H.: Exposing digital forgeries by detecting duplicated image regions. Dept. Comput. Sci., Dartmouth College, Tech. Rep. TR2004–515, pp. 1–11 (2004)

19. Li, G., Wu, Q., Tu, D., Sun, S.: A sorted neighborhood approach for detecting duplicated regions in image forgeries based on DWT and SVD. In: IEEE international conference on multimedia and expo, pp. 1750–1753. IEEE, Beijing (2007)

20. Bravo-Solorio, S., Nandi, A.K.: Exposing duplicated regions affected by reflection, rotation and scaling. In: IEEE International Conference on Acoustics, Speech and Signal Processing ICASSP, pp. 1880–1883. IEEE, Prague (2011)

21. Huang, Y., Lu, W., Sun, W., Long, D.: Improved DCT-based detection of copy-move forgery in images. Forensic Sci. Int. **206**(1–3), 178–184 (2011)

22. Cao, Y., Gao, T., Fan, L., Yang, Q.: A robust detection algorithm for copy-move forgery in digital images. Forensic Sci. Int. **214**(1–3), 33–43 (2012)

23. Zhao, J., Guo, J.: Passive forensics for copy-move image forgery using a method based on DCT and SVD. Forensic Sci. Int. **233**(1–3), 158–166 (2013)

24. Li, L., Li, S., Zhu, H., Wu, X.: Detecting copy-move forgery under affine transforms for image forensics. Comput. Electr. Eng. **40**(6), 1951–1962 (2014)

25. Cozzolino, D., Poggi, G., Verdoliva, L.: Efficient dense-field copy–move forgery detection. IEEE Trans. Inf. Forensics Secur. **10**(11), 2284–2297 (2015)

26. Huang, H., Guo, W., Zhang, Y.: Detection of copy-move forgery in digital images using SIFT algorithm. In: IEEE Pacific-Asia Workshop on Computational Intelligence and Industrial Application, vol. 2, pp. 272–276. IEEE, Wuhan (2008)

27. Pan, X., Lyu, S.: Region duplication detection using image feature matching. IEEE Trans. Inf. Forensics Secur. **5**(4), 857–867 (2010)

28. Amerini, I., Ballan, L., Caldelli, R., Del Bimbo, A., Serra, G.: A sift-based forensic method for copy–move attack detection and transformation recovery. IEEE Trans. Inf. Forensics Secur. **6**(3), 1099–1110 (2011)

29. Amerini, I., Ballan, L., Caldelli, R., Del Bimbo, A., Del Tongo, L., Serra, G.: Copy-move forgery detection and localization by means of robust clustering with J-Linkage. Signal Proc. Image Commun. **28**(6), 659–669 (2013)

30. Jaberi, M., Bebis, G., Hussain, M., Muhammad, G.: Accurate and robust localization of duplicated region in copy–move image forgery. Mach. Vis. Appl. **25**(2), 451–475 (2014)

31. Pandey, R.C., Singh, S.K., Shukla, K.K., Agrawal, R.: Fast and robust passive copy-move forgery detection using SURF and SIFT image features. In: 9th International Conference on Industrial and Information Systems, ICIIS, pp. 1–6. IEEE, Gwalior (2014)

32. Yang, F., Li, J., Lu, W., Weng, J.: Copy-move forgery detection based on hybrid features. Eng. Appl. Artif. Intell. **59**, 73–83 (2017)

33. Prakash, C.S., Panzade, P.P., Om, H., Maheshkar, S.: Detection of copy-move forgery using AKAZE and SIFT keypoint extraction. Multimed. Tools Appl. **78**(16), 23535–23558 (2019)

34. Christlein, V., Riess, C., Jordan, J., Riess, C., Angelopoulou, E.: An evaluation of popular copy-move forgery detection approaches. IEEE Trans. Inf. Forensics Secur. **7**(6), 1841–1854 (2012)

35. Emam, M., Han, Q., Niu, X.: PCET based copy-move forgery detection in images under geometric transforms. Multimed. Tools Appl. **75**(18), 11513–11527 (2016)

36. Alcantarilla, P.F., Nuevo, J., Bartoli, A.: Fast explicit diffusion for accelerated features in nonlinear scale spaces. In: British machine vision conference, BMVC, Bristol (2013)

FaceEvoke: Eliciting Emotions Through Facial Analysis

Aayushi Gupta[(⊠)], Ayushya Srivastava, and Manoj Kumar Shukla

Amity University, Noida, Uttar Pradesh, India
aayushigupta26802@gmail.com, mkshukla@amity.edu

Abstract. Eliciting Emotions through Facial Analysis delves into the interpretation of facial expressions in dogs. Understanding these expressions is crucial for strengthening the dog-human bond and ensuring canine well-being. This paper introduces FaceEvoke, a novel custom CNN model meticulously crafted for the precise detection of changes in eye shape, mouth position, and ear movement. The model seamlessly integrates multi-scale features and convolution, empowering observers to decipher a dog's emotional state with enhanced precision. The standout quality of this innovative model is its proficiency in accurately categorizing a wide array of dog facial expressions drawn from a diverse image set. Experimental results underscore the superior performance of our model, surpassing other established algorithms with an impressive accuracy rate of 99.87%, thereby affirming its efficacy in veterinary care and animal behavior studies.

Keywords: ResNet50 · MobileNetV2 · InceptionV3 · FaceEvoke · Convolutional Neural Networks

1 Introduction

The pet dog population is on the rise as more people welcome them into their homes and lives. In this context, as the importance of understanding and addressing the emotional well-being of dogs continues to grow, it's crucial to recognize that existing models employ facial analysis techniques to decode and interpret canine emotions. The proposed model enhances comprehension and strengthens the between humans and their canine companions. Delving into the intricate realms of human psychology and communication, It is only natural that inquisitiveness extends to four-legged family members sharing daily lives. Understanding canine emotions has become an area of paramount importance, as it not only deepens comprehension of these remarkable beings but also greatly enhances one's ability to care for their well-being. Dogs, with their innate emotional acuity, are highly skilled at conveying their feelings through a complex interplay of facial expressions, body language, and vocalizations. In much the same way as humans use their facial expressions to convey joy, sadness, fear, and anger, dogs possess an impressive array of emotional cues that they express through their facial features. These emotional signals play a pivotal role in human-dog interactions, aiding them in gauging their well-being, addressing their needs, and fortifying the unique bonds that exist between humans and canines. In pursuit to comprehend the intricate world of canine behavior and communication, the field of feature extraction in dog facial expression analysis emerges as both pivotal and captivating. This innovative research direction seeks to dissect the rich tapestry of canine emotions by recognizing and deciphering the intricate features woven into their expressive countenances. These features, such as ear position, eye shape, and more, provide the keys to unraveling the emotional language of dogs, offering a deeper understanding of their well-being, and strengthening the profound connections shared with furry companions.

The prevailing methodologies in the realm of facial expression analysis primarily centered around human subjects, and while these techniques yielded valuable insights into human emotions, a notable paradigm shift occurred when extending the scope to include dogs. Among the various methodologies applied to dogs, three distinctive deep learning models, ResNet50 [1], MobileNetV2, and Inception V3 [2], stand out prominently. These models not only excel in human emotion recognition but also demonstrate their remarkable adaptability and precision when it comes to deciphering the intricate language of canine emotions, thus representing a significant leap in cross-species emotional analysis.

This research study aims to advance the domain of face emotion recognition in dogs by conducting a robust comparative analysis of three prominent deep learning models: MobileNetV2, InceptionV3, ResNet50, with proposed FaceEvoke.Net model. The primary objective is to evaluate the performance of these models in classifying canine emotions based on their facial expressions, utilizing images from various sources. The investigation delves into the intricate relationship between fine-tuning hyperparameters and the size of training datasets. This analysis aims to unveil how these factors influence the accuracy of emotion recognition across diverse conditions and image qualities. Systematically varying hyperparameters like learning rates, batch sizes, and regularization techniques for each model, in addition to exploring the impact of training dataset size, ranging from small to large, will be pivotal in assessing the efficacy of these models in recognizing canine emotions.

By rigorously evaluating and comparing these four models, this study seeks to provide valuable insights into the optimal techniques and model choices for face emotion recognition in dogs. Ultimately, the findings of this research will contribute to enhancing the understanding of the best practices for achieving high accuracy in classifying the emotional states of canine companions, which can have applications in fields like animal behavior research and pet-related technology development.

The main contributions of the proposed work as follows:

- Performance Superiority: The study demonstrates that the proposed model surpasses state-of-the-art (SOTA) deep learning models when evaluated on the same datasets for dog facial emotion recognition.
- Efficient Feature Extraction: The research introduces a novel feature extraction method that uses a dilated multiscale, convolutional block, reducing network parameters while expanding the receptive field.
- Parameter Efficiency and Speed: The proposed model features fewer parameters than conventional deep learning models and offers faster inference times, enhancing its suitability for resource-constrained devices and real-time applications.

2 Literature Review

The study of facial expression recognition is an important field with broad applications in image processing and computer vision. Computers are employed in this field to interpret facial expressions, analyze facial images, and discern various emotional states. Recent developments have witnessed the emergence of enhanced deep learning models and refined algorithms, significantly transforming the realm of facial expression identification.

In a noteworthy precedent, a Deep CNN model was employed for classifying vertebrate species, encompassing mammals and reptiles [3]. Leveraging a substantial training dataset consisting of 4000 images and a test set comprising 1200 images, the model achieved remarkable accuracy, successfully predicting target classes with an impressive success rate of 97.5%. In a previous research endeavor [4], a CNN model was utilized to categorize images of various animal species, including foxes, wolves, bears, hogs, and deer. This CNN-based system outperformed traditional methods in image identification, achieving an astonishing 98% accuracy rate for animal recognition. Another study [5] in 2022 tackled the challenge of distinguishing different facial emotions of humans, harnessing the power of deep Convolutional Neural Networks (CNNs). Their proposed approach involves utilizing a training dataset comprised of image data, with the pixel values of the pictures directly employed as input. The precision in discerning emotions was significantly improved by eliminating the background which resulted to an impressive accuracy of 97.98%. Furthermore, deep learning was employed to distinguish snub-nosed monkeys from other visually similar monkey species in photographs [6]. The researcher meticulously optimized hyperparameters and the model's structure for the CNN model, resulting in an impressive accuracy of 96%. Collectively, these studies underscore how CNNs, particularly within the realm of deep learning, have brought about a profound transformation across various fields. This includes advancements in animal emotion detection and facial expression recognition, heightening accuracy and improving efficiency to discern subtle features, thus propelling one's comprehension of emotional cues in both animal and human subjects.

In a 2021 study [7], researchers delved into the application of machine learning and deep learning techniques, including TensorFlow and Keras, for the development of face mask recognition systems. The paper highlighted the utilization of Python scripts and open-source tools like TensorFlow and Keras, while also addressing the data collection process for training the face mask identification model. Another research [8] presented an innovative approach to human emotional recognition through a convolutional neural network (CNN) within a deep learning framework. Their methodology involved leveraging various facial features while implementing effective dimensionality reduction techniques. Additionally, kernel filers were used in the preprocessing step for sharpened edges. The study specifically focused on exploring picture synthesis methods as a potential solution for data augmentation in deep learning which resulted in an impressive accuracy of 97%. The primary objective was to mitigate overfitting concerns associated with limited data availability and address the challenges posed by data scarcity. In a separate study [9], researchers underscored the importance of mask usage in disease prevention and management, especially in densely populated areas. They employed Artificial Neural Networks (ANNs) to determine whether individuals within a crowd are wearing masks. Remarkably, their method achieved a perfect score of 99% accuracy in both testing and training phases. These research endeavors exemplify the expanding impact of deep learning in the fields of face mask recognition and object detection.

3   Materials and Methods

This section outlines the entire process of the proposed method. Figure 1 provides a visual representation of the conceptual sequence of the work, and each individual step is elaborated upon in the subsequent discussion.

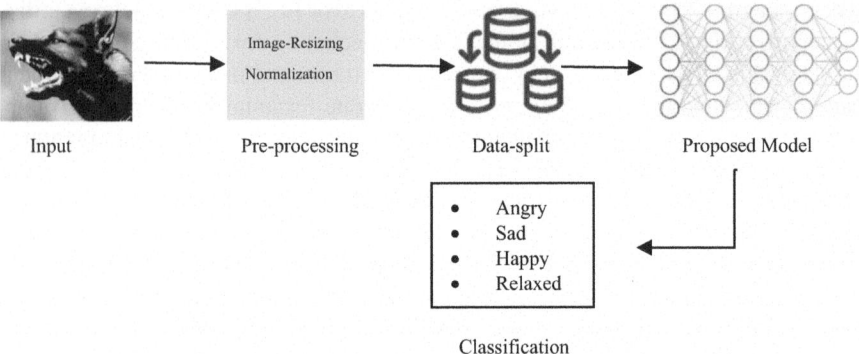

Fig. 1. Flow Chart of work

3.1   Dataset Description

The dataset utilized for the training and evaluation of the proposed model was sourced from Kaggle. Initially, this dataset comprised a total of 15,921 images, categorized into four emotional classes: anger (2256 images), sadness (4532 images), relaxation (4349 images), and happiness (4784 images). To further enhance the model's performance, a modest degree of data augmentation was employed due to the recognition of the need for augmented data to optimize the performance of pretrained models. Subsequently, the dataset was partitioned into training and testing subsets based on this augmentation. Figure 2 provides illustrative examples of images from each of the aforementioned emotional categories. The dataset can be accessed from [10].

3.2   Pre-processing

The suggested model was subject to training and testing procedures. Notably, the dataset consisted of images with varying dimensions. In order to standardize the image proportions for the proposed model, all dataset images were resized to $224 \times 224 \times 3$ pixels. An image normalization process [11] was undertaken to ensure that pixel intensity values

Angry   Relaxed   Happy   Sad

Fig. 2. Sample images of Dog's emotion dataset

fell within the designated 0 to 1 range. This was accomplished by employing a min-max image normalization technique, effectively standardizing the pixel values to meet this specified range.

$$v' = \frac{v - min_A}{max_A - min_A}(new\_max_A - new\_min_A) + new\_min_A \qquad (1)$$

where, v and v' represent the input and normalized image, respectively. Min_A and max_A are the minimum and maximum absolute value of v, and new_min_A and new_max_A represents the range of the normalized image. The values of min_A and max_A is 0 and 255, while new_min_A and new_max_A is chosen to be 0 and 1 in this study.

Input images were tested using normalization process. Various image sizes and grayscale images were tested, but minimal impact was observed, leading to proceed with the size of 224 × 224 × 3.

3.3 Proposed Model

In the realm of computer vision applications, such as image tracking, image classification [12], semantic segmentation [13], object detection [14], image restoration, human-computer interaction [15], and even in medical domains like disease detection, the need for sophisticated and effective neural networks is paramount. Traditional fully connected neural networks, due to their complexity and potential for extracting redundant features, may not be the optimal choice for these multifaceted tasks. Convolutional Neural Networks (CNNs) have emerged as a significant breakthrough in deep learning, offering architectures comprised of convolutional layers, and pooling layers. To address the classification of diverse emotion images in dogs, this study introduces a novel CNN model, denoted as FaceEvoke.Net. This model is a Convolutional Neural Network (CNN) based architecture to detect the emotions of a dog and classify them into four classes. The model is customized for specific image classification tasks and comprises several layers. The model takes input from an image of size 224 × 224. The model uses four repeating units. The unit contains two convolutional layers, each containing 16 filters of size 3 × 3, followed by rectified linear unit (ReLU) activation functions and the same padding to preserve spatial dimensions. Then max-pooling layers with a 2 × 2 window size and a stride of 2 reduce the spatial dimensions, capturing essential information and reducing computational complexity. The unit contains two convolutional layers, each containing 16 filters of size 3 × 3, followed by rectified linear unit (ReLU) activation functions and the same padding to preserve spatial dimensions. Then max-pooling layers with a 2 ×

2 window size and a stride of 2 reduce the spatial dimensions, capturing essential information and reducing computational complexity. This unit is repeated with an increase in the number of filters to 32, 64, and then 128 respectively. The four convolutional blocks in Fig. 3, the model flattens the learned features into a one-dimensional vector and passes it through fully connected dense layers. The first dense layer consists of 1024 units and utilizes the ReLU activation function, enabling the network to learn complex relationships in the flattened feature space. To prevent overfitting, dropout regularization with a rate of 0.2 is applied. Another dense layer with 512 units and ReLU activation follows, further capturing more complex patterns in the data. Again, dropout regularization is employed to reduce overfitting. The final layer of the network consists of 4 units and classifies the input image of the dog into one of the four emotions that is happy, sad, angry, and relaxed. The softmax activation function is used function is used for this multiclass classification. Moreover, Table 1 details the specific layers and parameters used in the proposed model.

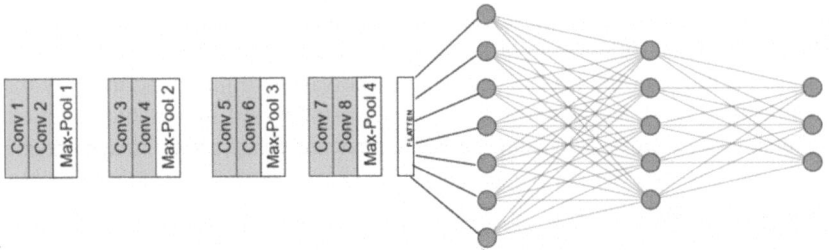

Fig. 3. Architecture of proposed FaceEvoke.Net

Table 1. Detailed Configuration of FaceEvoke.Net

Layer type	Output shape	Kernel size	Padding	Activation	Connected to
conv2d	(224, 224, 16)	3 × 3	same	Relu	input
conv2d-1	(224, 224, 16)	3 × 3	same	Relu	conv2d
MaxPooling2d	(112, 112, 16)	2 × 2	–	–	conv2d-1
conv2d_2	(112, 112, 32)	3 × 3	same	Relu	MaxPooling2d
conv2d_3	(112, 112, 32)	3 × 3	same	Relu	conv2d_2
MaxPooling2d_1	(56, 56, 32)	2 × 2	–	–	conv2d_3
conv2d_4	(56, 56, 64)	3 × 3	same	Relu	MaxPooling2d_1
conv2d_5	(56, 56, 64)	3 × 3	same	Relu	conv2d_4
MaxPooling2d_2	(28, 28, 64)	2 × 2	–	–	conv2d_5

(continued)

Table 1. (*continued*)

Layer type	Output shape	Kernel size	Padding	Activation	Connected to
conv2d_6	(28, 28, 128)	3 × 3	same	Relu	MaxPooling2d_2
conv2d_7	(28, 28, 128)	3 × 3	same	Relu	conv2d_6
MaxPooling2d_3	(14, 14, 128)	2 × 2	–	–	conv2d_7
Flatten	25088	–	–	–	MaxPooling2d_3
Dense	1024	–	–	Relu	Flatten
Dropout	1024	–	–	–	Dense
Dense_1	512	–	–	Relu	Dropout
Droupout_1	512	–	–	–	Dense_1
Dense_2	4	–	–	Softmax	Dropout_1

4 Experimental Result

4.1 Implementation Details

The FaceEvoke.Net is trained on a dataset that includes emotion images, and this dataset is split into training and test sets at an 80:20 ratio. The network's architecture and hyperparameters were meticulously selected through an iterative process. It utilizes the adaptive moment optimization (Adam) as the optimizer, with a learning rate set at 0.001. The chosen loss function is sparse categorical cross-entropy, and each training batch consists of 98 samples. The training process spans 50 epochs, an early stopping mechanism is implemented as the termination criterion by monitoring the model's performance on a validation dataset during the training process. If there is no improvement or a degradation in performance over a predefined number of consecutive iterations, the training is halted, preventing overfitting and ensuring the model generalizes well to unseen data. Notably, the Keras framework with a TensorFlow backend is employed for both training and testing of the FaceEvoke.Net.

4.2 Result Analysis

The results demonstrate a significant improvement in accuracy compared to baseline methods, underscoring the effectiveness of the proposed approach.

Table 2 shows comprehensive evaluation of four distinct deep learning models, ResNet50, InceptionV3, MobileNetV2, and FaceEvoke.Net, based on their accuracy scores in a specific task. FaceEvoke.Net stands out as the top performer with an impressive accuracy of 99.87%, demonstrating exceptional competence in the task without suffering from overfitting. In contrast, InceptionV3 and MobileNetV2 achieved reasonably good accuracy scores of 80.24% and 47.09%, respectively, but both displayed signs of overfitting, suggesting the need for enhanced generalization techniques. On the other hand, ResNet50 exhibited the lowest accuracy at 39.78%, indicating potential inadequacies in its architectural or hyperparameter design.

Table 2. Analysis of Proposed method and State-of-art methods

Method	Epochs	Train Accuracy	Validation Accuracy
ResNet50	20	41.75	39.78
MobilenetV2	20	79.82	47.09
InceptionV3	18	74.29	80.24
Proposed Model	42	98.60	99.87

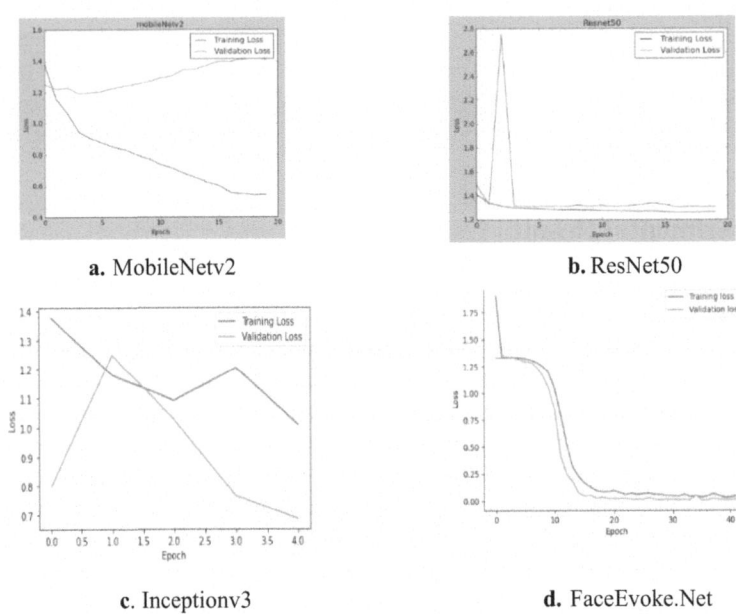

a. MobileNetv2   **b.** ResNet50

c. Inceptionv3   **d.** FaceEvoke.Net

Fig. 4. Comparison of four distinct deep learning models on the basis of training and validation loss

Figure 4a depicts the test and validation loss for mobile net v2, while the training loss slowly converges, the validation loss is high, indicating the model's inability to converge the loss on new unseen data. This model displays significant signs of overfitting hence concluding that there is scope for improvement in the architecture in this case. Figure 4b depicts the training and validation loss for the Restnet50, the model relatively converges well on both the test and validation datasets. Figure 4c also depicts the training and validation loss for inception v3, this model showcases a highly unsteady convergence with steep slopes, depicting the unstable nature of the model and the lack of its ability to perform well on both datasets. Since the pre-trained models mostly depicted an unsatisfactory result, the focus was directed towards a customized CNN-based architecture. The loss for the proposed model FaceEvoke.Net is depicted in Fig. 4d, the graphs showcase a stable convergence tending towards zero, and the model evidently performs better than the rest for the model with a validation accuracy of 99.87%.

5   Conclusion

This study has introduced a novel CNN model, FaceEvoke.Net for recognizing emotional expressions in dogs. With an impressive accuracy of 99.87%, this model outperforms existing methods in differentiating common emotions like happiness, sadness, relaxation, and anger in dogs. The comprehensive comparison conducted with other state-of-the-art methods, considering both accuracy and loss graphs, and the results consistently demonstrate that FaceEvoke.Net provides the best solution. This model has great potential for various applications, including veterinary care and improving communication between pet owners and their dogs, thereby enhancing the well-being of canine companions. Further enhancing the model's capabilities to encompass a broader range of emotions and potentially applying it to other animal species can further strengthen the human-animal bond.

References

1. Li, X., Niu, H.: Feature extraction based on deep-convolutional neural network for face recognition. Concurr. Comput. Pract. Exp. **32**(22), 1 (2020)
2. Yao, H., Dai, F., Zhang, S., Zhang, Y., Tian, Q., Xu, C.: Dr2-net: deep residual reconstruction network for image compressive sensing. Neurocomputing **359**, 483–493 (2019)
3. El Abbadi, N.K., Alsaadi, E.M.T.A.: An automated vertebrate animals classification using deep convolution neural networks. In: 2020 International Conference on Computer Science and Software Engineering (CSASE), pp. 72–77. IEEE (2020)
4. Trnovszky, T., Kamencay, P., Orjesek, R., Benco, M., Sykora, P.: Animal recognition system based on convolutional neural network. Adv. Electr. Electron. Eng. **15**(3), 517–525 (2017)
5. Pandey, A., Gupta, A., Shyam, R.: Facial Emotion Detection and Recognition **7**, 176–179 (2022)
6. Zeng, P.: Research on similar animal classification based on CNN algorithm. J. Phys. Conf. Ser. **2132**(1), 012001 (2021)
7. Rosebrock, A.: Covid-19: Face mask detector with opencv, keras/tensorflow, and deep learning (2020)
8. Arora, T.K., et al.: Computational intelligence and neuroscience. Article ID 8379202 (2022)
9. Guillermo, M., et al. Implementation of automated annotation through mask RCNN object detection model in CVAT using AWS EC2 instance. In: 2020 IEEE Region 10 Conference (TENCON), pp. 708–713. IEEE (2020)
10. Dataset. https://www.kaggle.com/datasets/devzohaib/dog-emotions-prediction/data
11. Pei, S.C., Lin, C.N.: Image normalization for pattern recognition. Image Vis. Comput. **13**(10), 711–723 (1995)
12. Yadav, S.S., Jadhav, S.M.: Deep convolutional neural network based medical image classification for disease diagnosis. J. Big data **6**(1), 1–18 (2019)
13. Guo, Y., Nie, G., Gao, W., Liao, M.: 2D semantic segmentation: recent developments and future directions. Future Internet **15**(6), 205 (2023)
14. Vaishnavi, K., Reddy, G.P., Reddy, T.B., Iyengar, N., Shaik, S.: Real-time object detection using deep learning. J. Adv. Math. Comput. Sci. **38**(8), 24–32 (2023)
15. Kumar, R., Jaiswal, V., Nishad, V.: Human-computer interaction (HCI). Int. J. Eng. Res. Technol. **9**(5), 315 (2021)

Efficient Aerial Object Detection:
An Exploration with YOLOv8

Kumar Rohit[1], Parth Singh[1], Nisarg Patel[1], Pooja Kamat[1(✉)], and Satish Kumar[1,2]

[1] Symbiosis Institute of Technology, Symbiosis International (Deemed University), Pune, Maharashtra 412115, India
{kumar.rohit.btech2021,singh.parth.btech2021, nisarg.patel.btech2021,pooja.kamat,satish.kumar}@sitpune.edu.in
[2] Symbiosis Centre for Applied Artificial Intelligence (SCAAI), Symbiosis International University, Near Lupin Research Park, Gram: Lavale, Tal: Mulshi, Pune, Maharashtra 412115, India

Abstract. The rise of state-of-the-art deep learning models and computer vision techniques have ensured the application of AI almost everywhere, similarly there are systems that are being developed for aerial surveillance for ensuring security and increasing the defense capabilities at the national front. In this work a system has been proposed on similar grounds for real time object segmentation in the videos of airplanes and helicopters, this work employs the YOLOv8 model to recognize helicopters and airplanes. Recognising the inherent flaws in old methods, deep learning offers major advantages, such as enhanced adaptability to complex situations and accuracy. The dataset, which contains 7506 training photos, and 2503 validation images, serves as the basis for training and evaluating the model. The model's excellent performance is highlighted by evaluation criteria, particularly the mean average precision (mAP50–95), which yielded a score of 0.684. The results emphasize the model's high precision and recall, proving its ability to recognize and classify things consistently. This research contributes to our understanding of the YOLOv8 model's practical applicability and high-performance capabilities in the field of aerial vehicles.

Keywords: Object Detection · Convolutional Neural Networks · YOLOv8

1 Introduction

Rapid technological advancements and greater use of aerial vehicles have resulted in a revolution in military and defense policy, as well as surveillance activities [1, 2]. On the other hand, these new technical standards raise security and defense concerns because adversaries may employ these same advancements. An unchecked approach towards these technological advancements could lead to a negative impact rather than a positive one on the security of the nation and on its people. Radar systems, the current most widely used technology for aerial security, detect objects using radio signals, which are subsequently analyzed by humans to recognise the object. However, this method is both time-consuming as well as a little inefficient and has greater chances of error [3]. Also, radar-based detection systems can interfere with the environment, blocking radio waves from neighboring devices [4]. Furthermore, small object detection by these systems has restrained accuracy and latency, creating false alarms such as birds as planes etc. The initial studies primarily used unsupervised methods using different attributes [5]. The unsupervised methods generally produced more efficient results for simpler structure types, and the results were optimum and good for just a limited set of objects. To accurately identify different objects under complex conditions, subsequent studies were focused on supervised methods [6, 7]. To address these issues, deep learning object detection frameworks are increasingly being employed to detect potential aerial threats, which have the power of timely and accurate identification. Traditional two stage detectors are computationally expensive and inefficient when it comes to identifying distant objects. On the contrary, one-stage detectors such as YOLO (You Only Look Once) and RetinaNet have emerged, offering faster and more accurate recognition of small objects. Convolutional neural networks (CNNs) are employed in these detectors to extract features from images. Because of their advantages, one-stage detectors are very well-suited for aerial object detection or segmentation.

1.1 Object Detection and Segmentation Algorithms

Two-Stage Object Detectors. Two-stage object detectors work in two stages: first localizes the object and then classifies them into its respective class. It consists of a Region Proposal network (RPN) and a classification or a categorization network. The RPN generates a set of object recommendations from the input image and then the input image is scanned with a collection of anchor boxes of varying sizes and aspect ratios to make this possible. It predicts the probability of each anchor box holding the object as well as the matching bounding box coordinates. The predictions that are made are based on the features extracted from the input image by using a convolutional neural network (CNN). The classification network then goes through each of proposed object regions and then the RPN categorizes them, and this task is typically done by a CNN which takes the proposed object region as the input and generates a probability distribution over the present object classes.

One-Stage Object Detectors. These algorithms are used in computer vision tasks to identify objects in one pass through the network opposite to two-stage object detectors which take two passes to identify the objects, making them much more computationally efficient and suitable for real time applications due to less latency. The examples of these algorithms include YOLO (You Only Look Once), SSD (Single Shot Detector) and RetinaNet which are the most popular single-pass object detectors.

You Only Look Once (YOLO). YOLO is the most popular one stage object detector which is used for real time computer vision tasks like object detection, classification, and segmentation. It was first presented in 2015 by Joseph Redmon, Santosh Divvala, Ross Girshick, and Ali Farhadi in their popular exploration paper "You Just Look Once: Brought togather, Continuous Article Discovery" [8]. The first Consequences be damned evolved in a custom system called DarkNet. The DarkNet structure is intended to be quick and productive like cell phones and implanted frameworks. It achieves this by using a series of convolutional layers and max pooling, strided convolutional layers followed by several fully connected layers to generate the final output. Darknet simultaneously preserves low-level features for later layers while extracting high level features using 'skip connection' for better accuracy and convergence during training. The conventional You Only Look Once (YOLO) model predicted images at 45 frames per second, while Fast YOLO, which has fewer layers and can process images at 155 frames per second, was developed. This achieved roughly 64.3 mA (average accuracy), which was higher than the other available real-time object detectors. The greater resolution and anchor boxes were the key enhancements on this model. On the PascalVOC dataset, YOLOv2 achieved 76.8 mAP and 67 FPS. The latest version, YOLOv8, which is built on previous iterations, offers improved speed, enhanced performance and user-friendly features, making it one of the best object detection and segmentation models. As YOLOV8 is still in the developing stage, not many research papers have been published on it yet by the founding community so knowing the exact architecture is a bit difficult but some of the main highlights in the internal workings of this version include Anchor Free detection. Different from previous versions, YOLOv8 eliminates anchor boxes and directly predicts bounding box centers, significantly accelerating the detection process. YOLOv8 integrates mosaic augmentation, which combines four images to expose the model to varied perspectives and locations of objects in each epoch.

Retina Net. Retina Net is another single shot object detection algorithm which uses the feature pyramid network (FPN) in the backbone and an improved focal loss function. It is very well suited for detection of small objects, specifically the aerial and satellite imagery. The feature pyramid network combines higher and lower resolution images by using top-down and bottom-up pathways for improved accuracy and performance in the final layer, The Focal Loss is designed to assist the model in focusing on the hard examples that are the most informative for improving its accuracy. By reducing the contribution of easy examples, the model can better allocate its attention to the challenging examples, which require the most improvement. The Focal Loss has been shown to significantly improve object detection accuracy in the RetinaNet object detection model, particularly for small and medium-sized objects [9].

Single Shot Detector (SSD). SSD starts with a pre-trained convolutional neural network (CNN) that serves as the backbone of the detector. Typically, a VGG or ResNet network is used as the backbone, which extracts features from the input image. The backbone network generates a series of feature maps with different resolutions, which capture the spatial information at different scales. SSD then applies a set of convolutional filters to the obtained feature map to predict object detections at different scales [9].

2 Literature Review

Kashiyama et al. proposed a monitoring system for flying objects using a high-definition camera and state of the art YOLOv3 mode to achieve real-time processing the metrics used were precision and recall. Some of the drawbacks included that the model was tested only during daytime and the effects of dirt and dust on the camera lens was not considered [10]. Ying-Chih Lai et al. proposed detection of a moving fixed-wing unmanned aerial vehicle (UAV) with deep learning-based algorithms for distance calculation to conduct a study of sense and avoid (SAA) and mid-air collision avoidance of Unmanned aerial vehicles using YOLOv3 for object detection and using DNN and CNN regression for calculating distance between own UAV and the intruder. Drone but most of the images for model training were in clear weather, other weather conditions may affect the accuracy [11]. Rozantsev et al. proposed to introduce a new approach for object detection using st-cubes, which includes a regression-based motion stabilisation technique that improves performance, achieving the average precision of 0.751 but had some limitations including the view was limited to single moving camera and computationally expensive and time consuming [12]. Singha, S et al. developed an algorithm using YOLOv4 for an accurate drone detection system to address safety concerns in public spaces and achieved 98.21% accuracy and 93.68% mean average precision but had the limitation of dataset generalizability [13]. Kim, Jun-Hwa et al. proposed the creation of new and improved maritime dataset including a variety of objects like boats, lighthouses etc. and applying YOLOv5 on the previously available dataset and on the new dataset using YOLOv5 for object detection and achieved 0.75 mAP [14]. Reis et al. developed a generalized model for real time flying object detection on a data set consisting of 40 classes and then employed transfer learning on a more real-world dataset and achieved an mAP of 0.685 and inference speed of 50fps [15]. Zhao et al. presented an improved model for highway center marking detection and overcoming the challenges such as the difficulty of detecting markings in different lighting conditions and at different angles. In this paper a modified version of YOLOv3 is used for enhancing the performance of detection. The dataset used were the KITTI and CuLane datasets. The improved model outperformed the original YOLOv3 increasing the average precision (AP) from 79.84% to 82.79% and the detection speed to 25.71 f/s though variation in camera position can affect the reults of the model [16]. Madasamy et al. proposed a novel deep YOLOv3 framework for multi-object detection and then this system is deployed on embedded system and it is trained to detect only drones YOLOv3 (modified by use of residual blocks and FPN's) achieving 99% accuracy but conditions like complex weather weren't taken into account [17].Wang et al. proposed a new and efficient mechanism for detecting tiny objects using YOLOv8by employing WIoU v3 for bounding box regression and using BiFormer attention mechanism to filter out low relevance regions and proposing the FFNB module for efficient feature processing and achieved a final mAP-50 of 40.0 and mAP (50–95) of 23.6 [18]. Li et al. used a methodology by replacing PAN-FPN in YOLOv8 and Bi-PAN-FPN incorporating up sampling for focused small target features and utilizes GhostblockV2 replace some C2f modules and parameters to prevent information loss for long distance feature transmission and an efficient IOU method and achieved a mAP of 0.337 [19]. Rahman et al. used YOLOv8 for real time object detection in UAV aerial photography and achieved 87.4% mAP [20]. Zhai et al. proposed a methodology which introduces a micro-target detection head on the P2 layer features to capture richer information beneficial for tiny objects which are often lost in down sampling [21]. Huangfu et al. used a lightweight GSCconv module into the Neck-end using the GSBottleneck module and single aggregation module VoV-GSCSPC, replacing the normal convolution operation and the original bottleneck module C2f and achieved mAP of 41.5 [22].

3 Methodology

3.1 Data Collection

Data collection is one of the most important and first steps towards making any machine or deep learning model and it should be varied and should represent the deployment conditions. For this study the dataset was taken from an online competition by AI Crowd and contained videos of airplanes and helicopters and from that the frames were extracted. For increasing the diversity in the dataset more images of airplanes and helicopters were scraped from the internet where the object occupied 60–70% of the image. Table 1 displays the statistics of the dataset used and different types of objects in it.

Table 1. Dataset Statistics

Class	Number of Images	Total Dataset
Helicopter	5004	10009
Airplane	5005	

3.2 Data Preprocessing

The images collected were of different sizes and especially the ones scraped had a lot of duplicates present in them, so the preprocessing mainly involved removing the duplicates from the dataset and resizing them all to the same size and for improving the quality of segmentation Sobel edge detection filters were applied for sharpening the images to get a proper segmentation of the part containing the object. Figure 1 shows the image after applying Sobel edge detector.

Fig. 1. An image of a plane in a cloudy sky and the same image after applying a sobel edge detector.

Data Labeling and Annotations. To prepare the data for the algorithm, a labeling and annotation process was conducted. This involved manually labelling and annotating the parts of the images that contained the airplane and the helicopter using tools like LabelMe. Figure 2 shows the interface of the software used to annotate the images.

Fig. 2. Labeling of the image using labelMe

3.3 Training the Model

The dataset was split into train and validation sets for training of the YOLOv8 algorithm. In the training set there were 7506 images and, in the validation, set there 2503 images and their corresponding labels. These were put into the required directory structure as needed by the algorithm to train. The hyperparameters included batch size, optimizer, learning rate, patience, and number of epochs. 3 training rounds were performed to reach to the best mAP of 68.4%. Table 2 below shows the training statistics of the trained model.

Table 2. Training Statistics

Training round	Train Images	Validation Images	Batch size	mAP (50–95)
1	7506	2503	8	0.62
2	7506	2503	4	0.648
3	7506	2503	4	0.684

3.4 Testing and Inference

The testing of the model was performed on the NVIDIA RTX 3000 GPU with 16 giga bytes of VRAM and 128 giga bytes of RAM. The first 2 training rounds gave good mAP but didn't generalize well on the data and seemed to be overfitting, not performing well when it came to inferencing on videos, which were a little different from the dataset. The third attempt, though with the highest mAP, generalized much better as it contained more variety of images both of small and large-sized objects.

4 Experimental Results

The model's best result was found after 100 epochs and after doing 3 training rounds and getting the best mAP of 0.684 and reducing overfitting and increasing its generalization capabilities.

4.1 Hyperparameters

The training of the model involved a lot of hyperparameters which ensured good results. Table 3 shows the names and the values of the hyperparameters used. The epoch count is 100, the optimizer used is ADAM which was auto instilled by the algorithm, the batch size used was 4 and the size of the image that was used for training was 1280 by 1280px, the patience value was set to 50 epochs which meant if there was no increase in the validation mAP the training would stop.

Table 3. Hyperparameters statistics

Parameters	Value
Size of batch	4
Number of epochs	100
Optimizers	ADAM
YOLO Model	Yolov8x-seg
Learning rate	0.01
Patience	50
Decay value	0.001

4.2 Analysis of Results

Figures 3 and 4 represent the training graphs of the best-performed model showing the increase/decrease in the train box loss, precision, recall, mAP (50–95), and the val box loss, mask loss, precision, recall, mAP (50–95) with the increase in the number of epochs. The graphs contain the statistics for both box and the mask. From the figures below it clearly represents that the training and validation loss decreases as the number of epochs increases.

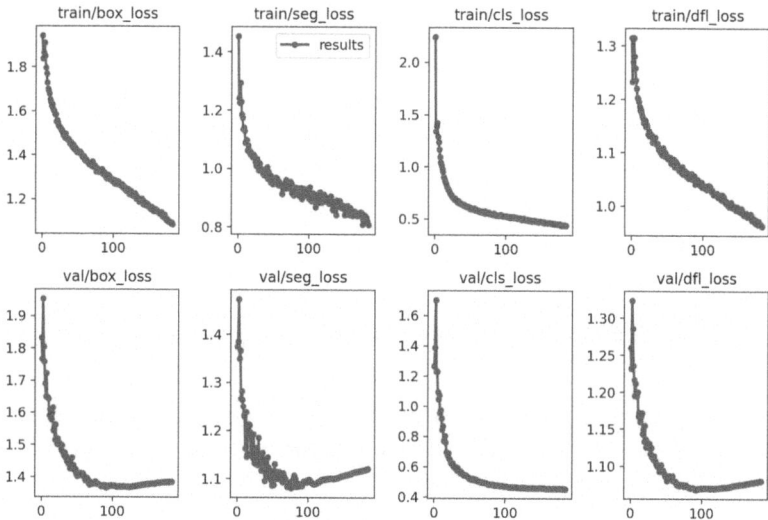

Fig. 3. Training and validation loss vs epochs.

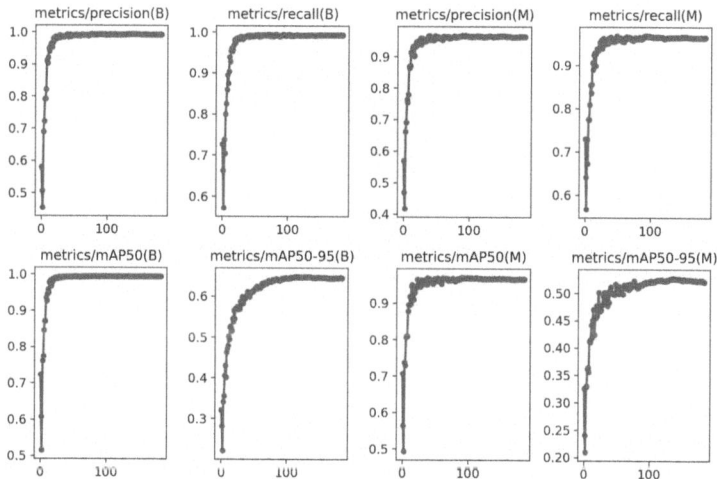

Fig. 4. Precision Recall Curve vs epochs

Figure 5 shows the precision recall curve for the box and mask. The x-axis has the recall, and the y-axis has the precision. These graphs are useful in binary classification and represents how well the false negatives and false positives are handled, the box precision and recall for airplane and helicopter are 0.995 and 0.994 respectively. Similarly, the mask precision (Fig. 5) and recall 0.973 and 0.961 for helicopter and airplane.

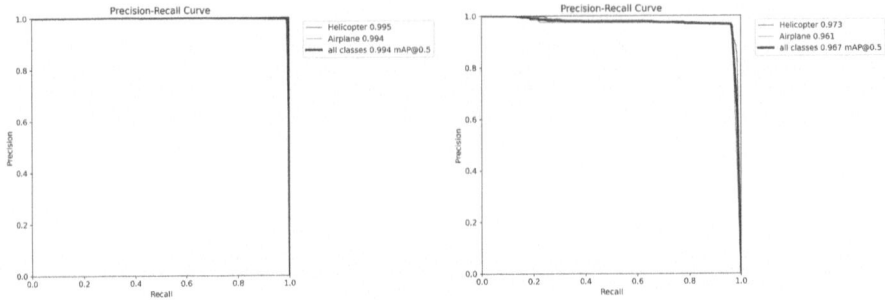

Fig. 5. Box Precision Recall Curve and Mask Precision Recall curve

Figure 6 displays the confusion matrix at 100 epochs where the prediction accuracy is significantly high. Though there is a little bit of error in predicting the helicopter correctly as it is classified as background.

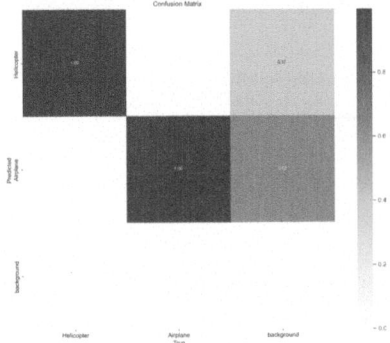

Fig. 6. Confusion Matrix

4.3 Visualization of Results

The figures (Figs. 7, 8) below show the result of the best-trained model with 0.684 mAP. The videos were downloaded from YouTube and are a bit different from the dataset used on the training set, signifying good generalization capability of the model with reduced overfitting.

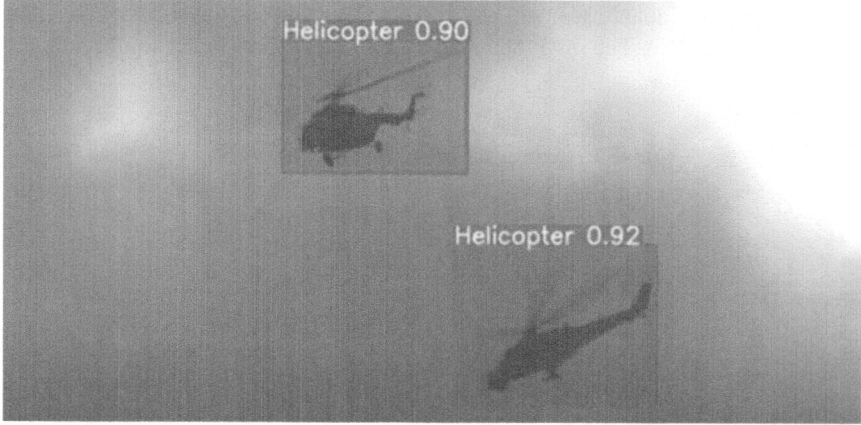

Fig. 7. Yolov8 segmentation on a video of helicopter

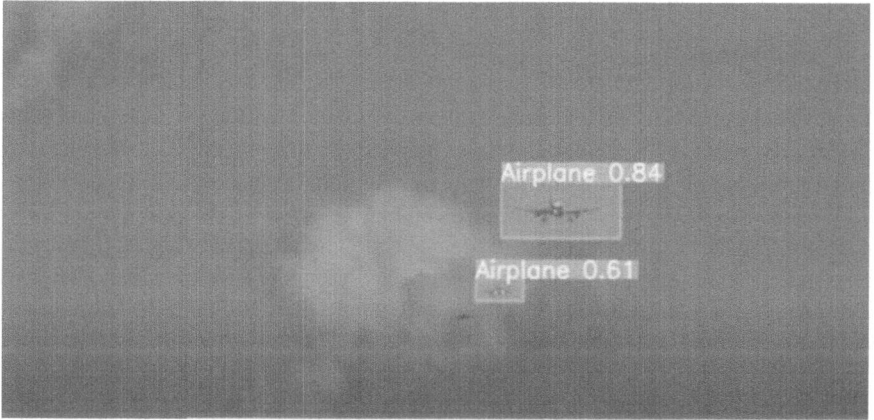

Fig. 8. Yolov8 segmentation on a video of airplane

5 Discussion

Object segmentation is one of the most crucial tasks in computer vision given the number of use cases it has got, in this study object segmentation was used for segmenting the objects in videos of airplanes and helicopters in view of utilizing these techniques for enhancing aerial surveillance and security. The dataset had 10005 images taken from an online competition and web scraped from the internet. The state-of-the-art model YOLOv8 was applied on the dataset and the metric used to analyse the performance was mAP50–95, which is a widely used metric in object detection and segmentation tasks it measures the average precision of the model over a range of Intersection Over Union (IoU) threshold which represents the accuracy of the model in detecting and segmenting objects with varying levels of overlap. The applied model resulted in a mean average precision mAP(50–95) of 68.4%. The visualisation of results show that model was precisely able to segment and classify the objects in the videos with around .88 average confidence. The mAP obtained by the other works in the similar tasks are shown in Table 4 below.

Table 4. Comparison of results with other work.

Sr. No	mAP (50–95)
Our Work	0.684
Reis et al. [15]	0.835
Huangfu et al. [22]	0.415
Wang et al. [18]	0.236
Li et al. [7]	0.337

6 Conclusion

The presented research work showcases the use of YOLOv8 for real time object segmentation in videos of airplanes and helicopters with a mAP of 68.4% and an average confidence score of .88 while making the boxes and the masks over the objects in the images and videos. The preprocessing steps included converting the videos into frames, resizing the images and applying Sobel edge detection for sharpening and edge enhancements. The YOLOv8 model (yolov8x-seg, the largest model with the greatest number of parameters) was trained for 100 epochs with a batch size of 4 and the learning rate being 0.01, training was done on NVIDIA RTX 3000 GPU with 16 GB of graphics, the total time taken for training was around 10 h The model did reduce the amount of false negatives while maintaining a good level of precision and accuracy. It has got some drawbacks like not testing it on varying climate conditions and given its training on 10000 images.

7 Future Scope

Some of the future enhancements include examining the model's performance on larger and more varied datasets, which would provide a more complete evaluation of the model's capabilities, exploring the application of transfer learning methods to enhance the performance of the model and testing the model on varying weather conditions and reducing the latency further.

References

1. McCall, B.: Sub-Saharan Africa leads the way in medical drones. Lancet **393**(10166), 17–18 (2019)

2. Dilshad, N., Hwang, J., Song, J., Sung, N.: Applications and challenges in video surveillance via drone: a brief survey. In: 2020 International Conference on Information and Communication Technology Convergence (ICTC), pp. 728–732. IEEE (2020)
3. Cheng, G., Han, J.: A survey on object detection in optical remote sensing images. ISPRS J. Photogramm. Remote Sens. **117**, 11–28 (2016)
4. Svatonova, H.: Analysis of visual interpretation of satellite data. Int. Arch. Photogramm. Remote. Sens. Spat. Inf. Sci. **41**, 675–681 (2016)
5. Tello, M., Martinez, C.L.: A novel algorithm for ship detection in sar imagery based on the wavelet transform. IEEE Trans. Geosci. Remote Sens. **2**, 201–205 (2005)
6. Han, J., Zhang, D., Cheng, G., Guo, L., Ren, J.: Object detection in optical remote sensing images based on weakly supervised learning and high-level feature learning. IEEE Trans. Geosci. Remote Sens. **53**, 3325–3337 (2014)
7. Ma, L., Li, M., Ma, X., Cheng, L., Du, P., Liu, Y.: A review of supervised object- based land-cover image classification. ISPRS J. Photogramm. Remote Sens. **130**, 277–293 (2017)
8. Redmon, J.; Divvala, S.; Girshick, R.; Farhadi, A. You only look once: Unified, real-time object detection. In Proceedings of the IEEE Conference on Computer Vision and Pattern Recognition (CVPR), Las Vegas, NV, USA, 26 June–1 July 2016; pp. 779–788
9. Liu, W., et al.: SSD: single shot multibox detector. In: Proceedings of the European Conference on Computer Vision, Amsterdam, The Netherlands, pp. 21–37 (2016)
10. Kashiyama, T., Sobue, H., Sekimoto, Y.: Sky monitoring system for flying object detection using 4K resolution camera. Sensors **20**(24), 7071 (2020)
11. Lai, Y.-C., Huang, Z.-Y.: Detection of a moving UAV based on deep learning-based distance estimation. Remote Sens. **12**(18), 3035 (2020)
12. Rozantsev, A., Lepetit, V., Fua, P.: Flying objects detection from a single moving camera. In: Proceedings of the IEEE Conference on Computer Vision and Pattern Recognition, pp. 4128–4136 (2015)
13. Singha, S., Aydin, B.: Automated drone detection using YOLOv4. Drones **5**(3), 95 (2021)
14. Kim, J.-H., Kim, N., Park, Y.W., Won, C.S.: Object detection and classification based on YOLO-V5 with improved maritime dataset. J. Mar. Sci. Eng. **10**(3), 377 (2022)
15. Reis, D., Kupec, J., Hong, J., Daoudi, A.: Real-Time Flying Object Detection with YOLOv8 (2023). arXiv preprint arXiv:2305.09972
16. Zhao, Z., Han, J., Song, L.: Yolo-highway: an improved highway center marking detection model for unmanned aerial vehicle autonomous flight. Math. Probl. Eng. **2021**, 1–14 (2021)
17. Madasamy, K., Shanmuganathan, V., Kandasamy, V., Lee, M.Y., Thangadurai, M.: OSDDY: embedded system-based object surveillance detection system with small drone using deep YOLO. EURASIP J. Image Video Process. **2021**(1), 1–14 (2021)
18. Wang, G., Chen, Y., An, P., Hong, H., Jinghu, H., Huang, T.: UAV-YOLOv8: a small-object-detection model based on improved YOLOv8 for UAV aerial photography scenarios. Sensors **23**(16), 7190 (2023)
19. Li, Y., Fan, Q., Huang, H., Han, Z., Qiang, G.: A modified YOLOv8 detection network for UAV aerial image recognition. Drones **7**(5), 304 (2023)
20. Rahman, S., Rony, J.H., Uddin, J., Samad, M.A.: Real-time obstacle detection with YOLOv8 in a WSN using UAV aerial photography. J. Imaging **9**(10), 216 (2023)
21. Zhai, X., Huang, Z., Li, T., Liu, H., Wang, S.: YOLDrone: an YOLOv8 network for tiny UAV object detection. Electronics **12**, 3664 (2023)
22. Huangfu, Z., Li, S.: Lightweight you only look once v8: an upgraded only look once v8 algorithm for small object identification in unmanned aerial vehicle images. Appl. Sci. **13**(22), 12369 (2023)

Kidney Tumor Classification Using Deep Learning Techniques from Computed Tomography Images

Premananda Sahu[1], Md Ashraful Babu[2], Manpreet Kaur[1],
Srikanta Kumar Mohapatra[3]($\boxtimes$), Prakash Kumar Sarangi[4], and Jayashree Mohanty[5]

[1] School of Computer Science and Engineering, Lovely Professional University (LPU),
Phagwara, Punjab, India
[2] Department of Physical Sciences, Independent University, Bangladesh, Dhaka, Bangladesh
[3] Chitkara University Institute of Engineering and Technology, Chitkara University, Rajpura,
Punjab, India
srikanta.2k7@gmail.com
[4] Department of Computer Science and Engineering (AI & ML), Vardhaman College of
Engineering, Hyderabad, India
[5] Department of CSE-AIT, Chandigarh University, Ludhiana, Punjab, India

Abstract. The kidney is an essential organ that filters and excretes waste products to maintain the body's fluid and solute balance. Several hormones are also secreted by it, which aids in blood pressure regulation. Kidneys also purify the blood by removing waste and impurities from it. Tumors (cancers) are brought on by uncontrolled cell proliferation, which affects people differently and results in several symptoms. Kidney cancer is indeed a significant health concern worldwide. Kidney cancer cases now days have risen in everywhere caused by a variety of elements, such as changes in lifestyle, enlarged screening and detection, and aging populations. The illness known as kidney disease has initiated on either renal disease. It is the part of essential disorders for patient diagnosis and classification in the current investigation. Timely screening along with effective action can stop or slow the succession of cancer in anticipation of too late to save the patient's life and dialysis or a kidney transplant are the only options left. This study suggested using deep learning models like convolutional neural networks to recognize kidney pictures with Computed Tomography which is also known as CT images. This study uses CNN with additional convolution layers to distinguish between images of healthy and cancerous kidneys. With the use of X-rays, CT imaging produces cross-sectional images that offer exceptional detail of internal structures and organs, making it an excellent diagnostic tool. Many patients' lives will be saved by this research's early and accurate kidney cancer identification.

Keywords: Kidney Tumor · Deep Learning · CNN · CT Images · X-Ray Images

M. Khurana et al. (Eds.): ICMLA 2024, CCIS 2238, pp. 372–379, 2025.
https://doi.org/10.1007/978-3-031-75861-4_33

1   Introduction

The kidney is an essential organ that filters and excretes waste products to maintain the body's fluid and solute balance. Several hormones are also secreted by it, which aids in blood pressure regulation. Kidneys also purify the blood by removing waste and impurities from it. Tumors (cancers) are brought on by uncontrolled cell proliferation, which affects people differently and results in several symptoms. Renal Cell Carcinoma (RCC) is the most prevalent type of kidney cancer, which is now thought to be the main cause of the above type of cancer [1]. According to a survey it has been observed that, 3% of adult malignancies and 85% of kidney tumours are renal cell carcinomas (RCC). RCC incidence in particular India, is most likely as a result of underreporting. The majority of renal cell carcinomas data come from Western nations, while data from India are rare, mainly when it comes to para-neoplastic disorders. In a tertiary care facility in Western India, we wanted to understand the epidemiology and treatment of RCC [2]. A detecting and diagnosing kidney tumor is essential to mitigate the risk of future disease development. Consequently, it would lead to the preservation of a patient's life. Several industries, including medical imaging, remote sensing, security and surveillance, biometrics, and robotics, significantly depend on image transforming technology. Test image quality strongly impacts the overall performance and applicability of an image processing task. Imaging treatments in medical science come in a variety of forms, including Magnetic Resonance Imaging (MRI), Ultrasound Sonography (US), and Computed Tomography (CT). For exhibiting great uniformity in Medical Imaging (MI), the borders amongst organs and other areas are scrambled. It is very difficult to differentiate between regions of interest and patterns. CT imaging is often the preferred imaging modality for radiologists due to its ability to produce highly detailed and accurate images of anatomical structures. With the use of X-rays, CT imaging produces cross-sectional images that offer exceptional detail of internal structures and organs, making it an excellent diagnostic tool. CT imaging also has exceptional spatial resolution, allowing it to distinguish between small structures and tissues that are in proximity. This makes it highly effective in identifying small tumors or lesions. Additionally, CT scans produce images with high contrast, providing radiologists with clearer images to aid in the detection and diagnosis of medical conditions. Categorization and planning of renal tumor therapy frequently utilizes it in clinics. The patient's scanned medical images are thoroughly examined by a skilled radiologist, who detects if there are any abnormal regions. Manual diagnosis takes a lot of time. The heterogeneity among and within raters is similarly high. This ability to identify illness and keep track of patients can be greatly improved by CT, which may also make it easier to evaluate treatment options and provide better patient care.

With the progress in technology, deep learning algorithms have shown immense potential in various applications of image processing. In comparison to conventional machine learning methods, DL allows more accurate and timely results. In recent years, deep learning techniques for semantic segmentation of medical images have yielded promising outcomes in various medical image analysis applications. Among different models of image processing [3], the most efficiently approach i.e. CNN have demonstrated to be the most efficient and successful to date. CNNs have been successful in various classification tasks due to their powerful feature extraction capabilities and effective encoding and decoding topologies. In the field of computer vision, CNNs have surpassed traditional methods, especially in the classification of CT images. On the other hand, Fully Convolutional Network (FCN) designs provide an efficient end-to-end training classification solution. DL, a rapidly expanding new machine learning field which has demonstrated its efficiency in semantic segmentation. Image interpretation is made trouble free by deep learning approaches. Semantic segmentation can be achieved through three different deep learning approaches: Fully Convolutional Network (FCN), Region-based, and Semi-supervised. Region-based techniques use a pipeline method, which involves extracting regions of free forms from input images, and then categorizing them using a classification type known as region based and then later labeling of pixels in accordance with the score regions. Unlike region-based methods, extraction of region proposal does not take place in FCN-based approaches. They develop a mapping from pixel to pixel while creating visuals of any size.

2   Analysis of Literature Work

In this work, we have used a number of methods for effectively detection of kidney cancer, but the initial method where the images has taken is Computed Tomography images, so first of all we have to observe that whether this technique is effective or not for fetching the images. S. Han et al. [4] have explained how to use CT imaging and a deep learning framework to distinguish between different categories of lesions in kidney cancer. They have also indicated that, it is an innovative approach that holds great potential for improving diagnosis and treatment outcomes. KH.Uhm et al. [5] has effectively diagnosed kidney cancer by the help of deep learning technique. Here they have indicated some shortcomings i.e. they have used relatively small and homogeneous datasets, which may limit the generalization of the models to different patient populations and imaging settings. Therefore, larger, and more diverse datasets are needed to improve the robustness and accuracy of deep learning models for kidney cancer detection. M. Gharaibeh et al. [6] has diagnosed kidney tumor based on radiology imaging scans and both machine along with deep learning techniques. In their work, they have reviewed a lot of parameters by applying on both machine and deep learning techniques. Here also they have indicated some certain shortcomings like i.e. in the future, radiology practice may combine advanced data analytics-based AI tools (such as machine and deep learning techniques). Numerous data analytics-based learning strategies can be enhanced and used, as described in the literature, to assist in the early diagnosis of kidney tumours using radiology imaging images.

F. Azuaje et al. [7] has recommended the model for effectively detection of kidney cancer by the help of histopathology imaging techniques and deep neural network approach. In their work, they have analyzed tumor with the association of clinical proteomic from carcinoma patients. They have also conveyed that due to the small amount of data especially those necessary for independently testing the models using matched histology and proteome data from the same individuals. M. Fenstermaker et al. [8] has developed the model related to kidney tumor based on histopathology imaging technique and the most effective approach of deep learning technique i.e. Convolution Neural Network. Here they have achieved as an accuracy of 99.1% for testing phase.

From the above analysis it has been observed that for effectual design of kidney cancer, not only deep learning techniques are sufficient but also ample data has required.

3 Suggested Methodology

The detection of kidney tumors using CT (computed tomography) scans typically involves several methodologies. CT scans are a commonly used imaging modality for evaluating renal (kidney) tumors due to their ability to provide detailed cross-sectional images of the kidneys and surrounding structures. It's important to note that the method-ologies used for kidney tumor detection may vary based on the specific clinical scenario, the expertise of the radiologist, and the available imaging technology. Additionally, the interpretation of CT scans for kidney tumor detection requires the expertise of a trained radiologist or a specialized radiology team to ensure accurate diagnosis and appropriate patient management. The below mentioned diagram named as Fig. 1 has depicted the entire methodology which we have used in our work.

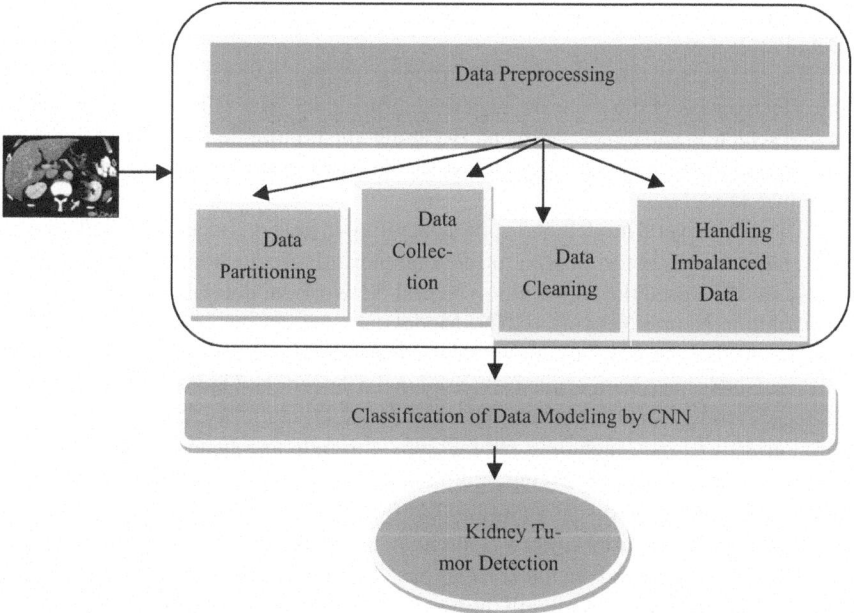

Fig. 1. Overall Workflow Diagram

The various methodologies used for the kidney tumor detection has described as:

A. Data Prepossessing

Data prepossessing plays a crucial role in kidney cancer detection, as it involves transforming and preparing the data to improve the accuracy and effectiveness of the analysis and PLCO datasets has been taken in our both training and testing work. The PLCO datasets is large scale and randomized controlled data which has taken all most all cases in health care. Here are some common steps involved in data prepossessing for kidney cancer detection:

B. Data Collection

The first step is to gather the relevant data for analysis. This may include medical records, laboratory test results, imaging reports (CT scans in our case), and other clinical data related to kidney cancer patients.

C. Data Cleaning

Raw medical data often contains missing values, errors, inconsistencies, and noise. Data cleaning involves handling missing values (either by imputation or removal), correcting errors, resolving inconsistencies, and dealing with outliers. Cleaning ensures the accuracy and reliability of the data for further analysis.

D. Handling Imbalanced Data

Imbalanced data refers to situations where the number of instances in different classes is significantly disproportionate. In the case of kidney cancer detection, the number of cancer-positive cases may be much smaller than the number of negative cases. Techniques like oversampling (replicating minority samples) or under sampling (removing majority samples) can be applied to address class imbalance and ensure fair representation of both classes.

E. Data Partitioning

The datasets is often divided into training, testing, and validation sets to assess the effectiveness of the kidney cancer detection model. The validation set is used for model selection and hyper parameter tweaking, the training set is used to develop the model, and the testing set is used to assess the effectiveness of the finished model.

F. Building Model

There are several models and approaches used in kidney cancer detection, ranging from habitual ML algorithms to further complex DL architectonics. In this research work, we have used the Convolution Neural Network Model (CNN).

G. Convolution Neural Network (CNN) Model

Due to their capacity to automatically learn hierarchical patterns from medical images, CNNs have been extensively used in the detection of kidney cancer. A family of deep learning models known as CNNs is created specifically for processing structured grid-like input, including photographs, through a sequence of interconnected layers. Different tasks involving "computer vision, such as picture classification, object recognition, and segmentation, has been" transformed by CNNs. They are especially useful for applications using images because they are excellent at engaging in local patterns and spatial correlations. The architecture of CNN has shown in Fig. 2. CNNs are still a frequently utilized and developing area of research and have greatly enhanced the state-of-the-art in computer vision process. CNNs are still a frequently utilized and developing area of research and have greatly enhanced the state-of-the-art in computer vision [9, 10]. Here are the commonly used layers in a CNN model for kidney cancer detection.

4 Implementation and Outcomes

For the implementation we have used Binary classification. Binary classification can be applied to kidney cancer detection by using a datasets of patients' medical records and imaging scans, where each patient is classified as having kidney cancer (positive

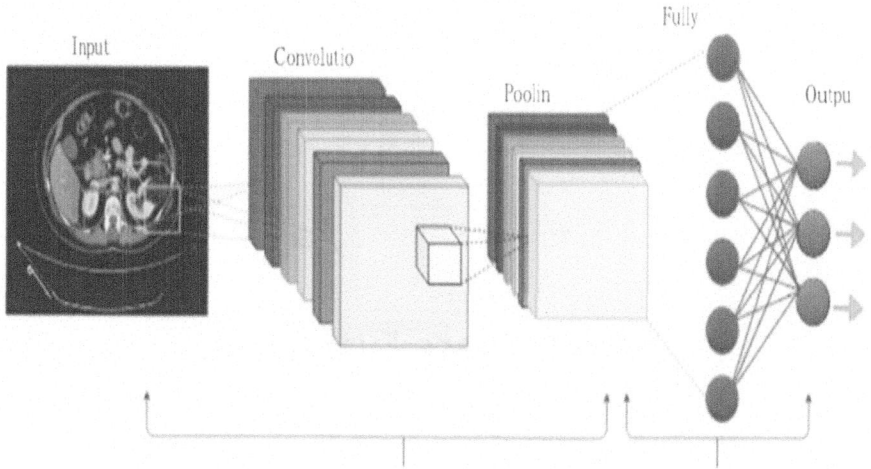

Fig. 2. Architecture of a Conventional CNN

class) or not having kidney cancer (negative class). Binary classification in kidney cancer detection can be a valuable tool for early detection and treatment of the disease, potentially improving patient outcomes and survival rates [11–14]. In our research work, we are using Convolution Neutral Network (CNN). Convolutional neural networks (CNNs) can be used in kidney cancer detection to automatically learn features from medical images, such as CT scans of the kidneys. The attributes that has fetched from the PLCO dataset has depicted in Table 1. Regarding the outcomes, here we have used python programming language and for validation point of view, while statistical techniques like accuracy, precision, recall and F1 score were utilized for validation [13, 14]. In this case we have used 100 epochs for all the outcome work. The differentiation of the statistical values after implementation in python language has shown in Table 2.

Table 1. Attributes of Plco Dataset.

S No	Demographic Information	Medical History	Laboratory Results	Image Findings
1	Name	Smoking History	Blood Test Results	Size
2	Age	Family History	Accuracy	Shape
3	Sex	Social History	Safety	Location
4	Address	Past Medical History	Alacrity	Contrast

After differentiation, the particular chart has shown in Fig. 3.

Table 2. Differentiation of Statistical Values with 100 Epochs

Epochs	0–20	20–40	40–60	60–80	80–100
F1 Score	84.23	85.38	86.41	91.02	97.57
Recall	84.32	85.44	86.51	91.11	98.01
Precision	84.35	85.73	86.58	91.22	98.22
Accuracy	84.56	85.77	86.62	91.35	**98.25**

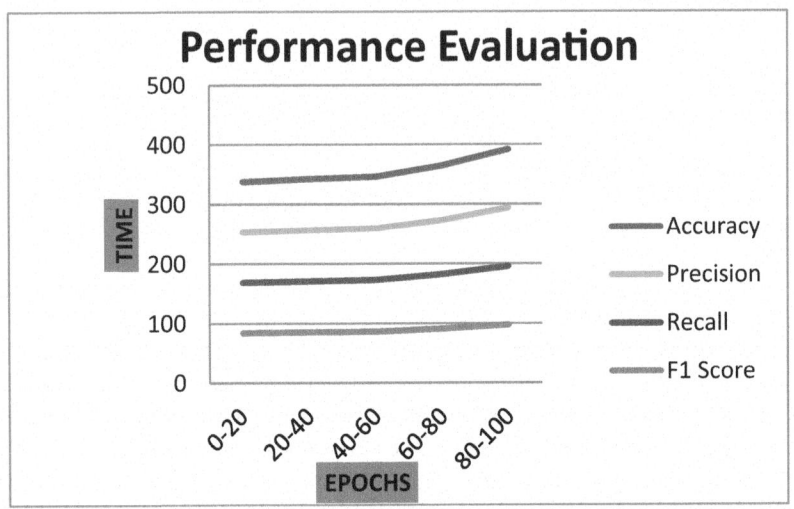

Fig. 3. Differentiation of All Statistical Values

5   Conclusion

Kidney cancer is a common and deadly disease that affects millions of people worldwide. Accurate diagnosis and classification of kidney tumors are crucial for selecting the appropriate treatment plan and predicting patient outcomes. CT imaging is one of the most widely used diagnostic tools for kidney tumors, but the interpretation of CT images is often challenging and requires extensive expertise. Convolutional neural networks have exhibited super potentiality for automating the analysis of medical images, including CT images of kidney tumors. Several studies have demonstrated the effectiveness of deep learning models for kidney cancer detection, attaining aristocratic of correctness and sensitivity in tumor classification. For this above purpose, we have achieved as an approximation of 98.25% exactness. These models have the potential to improve the speed and accuracy of tumor assessment, reducing the time and cost associated with traditional diagnostic methods.

Furthermore, the integration of deep learning models with other diagnostic tools, such as multi-modal imaging and augmented reality, can enhance the accuracy and efficiency of tumor assessment, allowing for more personalized and effective treatment plans.

Disclosure of Interests. The authors have no competing interests.

References

1. Chanchal, A.K., Lal, S., Kumar, R., Kwak, J.T., Kini, J.: A novel dataset and efficient deep learning framework for automated grading of renal cell carcinoma from kidney histopathology images. Sci. Rep. **13**(1), 5728 (2023)
2. Kumar, S., Singh, V., Singh, M.K., Sankhwar, S.N., Singh, M., Sankhwar, S.N.: Management of metastatic renal cell carcinoma in a Tertiary Care Hospital. Cureus **15**(2), e35623 (2023)
3. Vijay, V., Vokshi, F.H., Smigelski, M., Nagpal, S., Huang, W.C.: Incidence of benign renal masses in a contemporary cohort of patients receiving partial nephrectomy for presumed renal cell carcinoma. Clin. Genitourin. Cancer **21**(3), e114–e118 (2023)
4. Han, S., Hwang, S.I., Lee, H.J.: The classification of renal cancer in 3-phase CT images using a deep learning method. J. Digit. Imag. **32**, 638–643 (2019)
5. Uhm, K.H., et al.: Deep learning for end-to-end kidney cancer diagnosis on multi-phase abdominal computed tomography. NPJ Precis. Oncol. **5**(1), 54 (2021)
6. Gharaibeh, M., et al.: Radiology imaging scans for early diagnosis of kidney tumors: a review of data analytics-based machine learning and deep learning approaches. Big Data Cogn. Comput. **6**(1), 29 (2022)
7. Azuaje, F., Kim, S.Y., Perez Hernandez, D., Dittmar, G.: Connecting histopathology imaging and proteomics in kidney cancer through machine learning. J. Clin. Med. **8**(10), 1535 (2019)
8. Fenstermaker, M., Tomlins, S.A., Singh, K., Wiens, J., Morgan, T.M.: Development and validation of a deep-learning model to assist with renal cell carcinoma histopathologic interpretation. Urology **144**, 152–157 (2020)
9. PosadaCalderon, L., Eismann, L., Reese, S.W., Reznik, E., Hakimi, A.A.: Advances in imaging-based biomarkers in renal cell carcinoma: a critical analysis of the current literature. Cancers **15**(2), 354 (2023)
10. Sahu, P., Sahoo, B.K., Mohapatra, S.K., Sarangi, P.K.: Segmentation of encephalon tumor by applying soft computing methodologies from magnetic resonance images. Mater. Today Proc. **80**, 3371–3375 (2023)
11. Sahu, P., Mohapatra, S.K., Sarangi, P.K., Srivastava, S., Sharma, S.K.: Detection of diabetic retinopathy based on various machine learning algorithms and histogram equalization. In: 2022 International Conference on Machine Learning, Computer Systems and Security (MLCSS), pp. 6–10. IEEE (2022)
12. Sahu, P., Sarangi, P.K., Mohapatra, S.K., Sahoo, B.K.: Detection and classification of encephalon tumor using extreme learning machine learning algorithm based on deep learning method. In: Biologically Inspired Techniques in Many Criteria Decision Making: Proceedings of BITMDM 2021, pp. 285–295 (2022)
13. Garg, R., Sarangi, P.K., Sahoo, A.K., Jha, J.: Cardiovascular disease prediction: performance analysis and comparison of various supervised machine learning algorithms. In: 2023 International Conference on IoT, Communication and Automation Technology (ICICAT), pp. 1–6. IEEE (2023)
14. Mahajan, S., Sarangi, P.K., Sahoo, A.K., Rohra, M.: Diabetes mellitus prediction using supervised machine learning techniques. In: 2023 International Conference on Advancement in Computation & Computer Technologies (InCACCT), pp. 587–592. IEEE (2023)

Unsupervised Learning for Image Forgery Detection

Abhishek Thakur[✉] and Shahbaz Afzal

Chitkara University School of Engineering & Technology, Chitkara University, Solan,
Himachal Pradesh, India
{abhishek,Shahbaz.afzal}@chitkarauniversity.edu.in

Abstract. In this paper, an unsupervised learning-based autoencoder and decoder
are used to find image forgery detection. People access internet and post images
on social media sites. This work provides social security by recognizing forged
images on online social media sites. Images are processed with colour illumination
and converted into positive and negative patches. These patches are stored in
a.npy array of $(30 \times 30 \times 3)$ sizes. The Autoencoder train true positive and true
negative patches. An auto decoder reconstructs images from the minimum most
essential pixels. The auto-encoder uses a nonlinear transformation to reduce the
number of dimensions. The color-illuminated images were applied to Harrie's
corner detector machine learning. First, it encodes the input into simple signals.
Next, it comprises multiple convolution layers followed by an adder from output
three and output 4 with max Pooling. Then, it down-samples the input image up
to the maximum point of compression. An auto decoder is used to reconstruct
images from the minimum most essential pixels. It can replicate the output image
into an input image with some degraded quality. It comprises multiple convolution
layers followed by an adder from output three and output four with upsampling.
Experiments were performed on various openly available databases like CASIA
v1.0, BSDS300, COMOFOD and CASIA v2.0.

Keywords: Machine Learning · Feature Extraction · Convolution Neural
Network · Deep Learning · Image Forensic

1 Introduction

The electronic computer uses software tools for editing digital images. These software
tools process images to improve image quality and facilitate efficient storage. Visual
artefacts, encoding, picture, and video frame recognition are examples of unique image
processing operations. With the simple integration of image and video editing tools, anyone with simple computer knowledge can easily modify, change, and override accessible
image properties. Therefore, as in the biography, the creation of digital forgeries has
become relatively straightforward. People use photographs on social media platforms as
the primary source of information. The image and video of evidence against someone
are expected to be shown on TV news, including verifying truthfulness, conviction, and
credibility [1]. Instant criminality recorded and caught in a CCTV camera is treated and

M. Khurana et al. (Eds.): ICMLA 2024, CCIS 2238, pp. 380–390, 2025.
https://doi.org/10.1007/978-3-031-75861-4_34

used as evidence in law. If used in the newspaper or the courts of law, these forgeries may negatively impact our culture. In addition to those drawbacks, the accessibility of digital visual media presents numerous disadvantages. The major challenge in finding passive image forgery is that the source image is not present. Therefore, pixel-level features are extracted to separate the original and forged patches to solve this problem.

Image forensics is a multimedia security area that includes the identification of forged regions from images. It is easy to manoeuvre and manipulate digital photos. Digital image fraud detection in passive technique, as shown in Fig. 1. Determination of integrity and authenticity of images is a challenge or main research question for humans and machinery. So, there is a great need for reliable algorithms to check image forgery. Only in the presence of some previous picture information do active techniques work. Therefore, if images from inaccurate or untruthful sites are investigated, these techniques are not acceptable. However, the watermarking process significantly degrades and deteriorates the image quality. Passive methods are often referred to as blind approaches because the source picture is not visible. For any prior knowledge of the image, there are no preconditions, such as active techniques. Such methods are developed based on the assumption that the tempered representations approach in stills objects by changing fundamental statistical features and characteristics. Such incoherence is widely used to investigate falsification. The picture is subjected to various forms of attack and ramification during tempering. CMF Pun [2], which involves replicating some region (or regions) in the image, is the simplest of all forgery and ramifications. Replication of image areas, such as in image splicing, is sometimes done from other digital images.

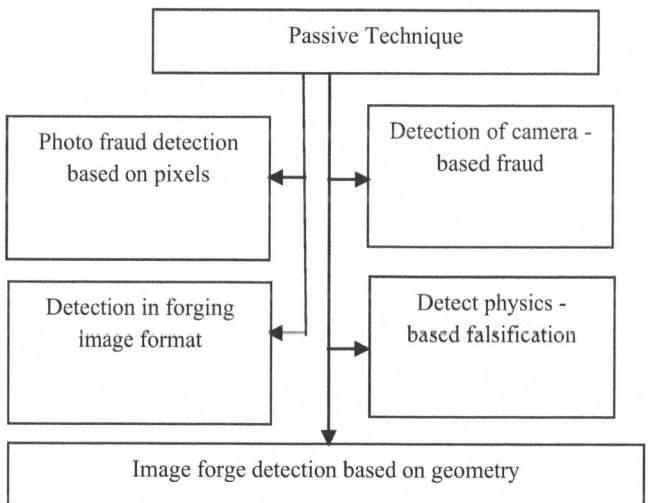

Fig. 1. Passive Image Forgery Detection Technique.

The various algorithms for passive image falsification are mainly classified into five classes. We use a pixel-based approach to collect image information and concentrate on pixel image information, as shown in Fig. 2. Joint Photographic Experts Group (JPEG), Double JPEG, JPEG, and JPEG-blocks use the format-based technique. The majority of work on forged image detection in various image formats has been done in JPEG format. In the camera-based approach, Singh et al. [3] will perform quantification, colour analysis, and filtering. The various stages of the image creation process introduce several different artefacts. Lighting irregularities in light sources may be used to detect tampering with forgery detection systems that depend on physics under different lighting conditions. When several images are combined, the resulting luminous conditions are changed. Technology based on geometry employs, explains, and integrates geometric limitations from future perspectives. These are categorised, assuming that the camera's basic parameters (such as the primary point, widest aperture, tilt, and image resolution) are metric and multiple view geometries [4]. The use and development of efficient and effective detection systems are thus associated with projective geometry principles.

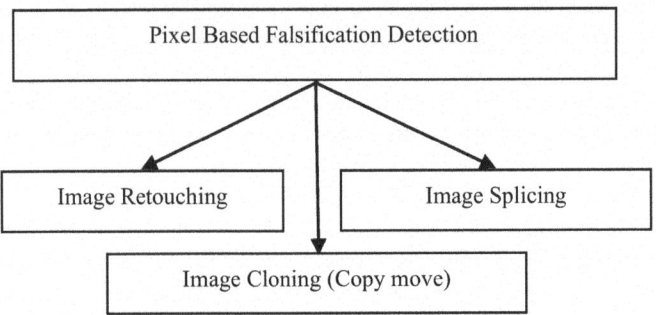

Fig. 2. Pixel-Based Digital Image Forgery Detection Methods.

An actual survey sums up and integrates what is understood. Simultaneously, the knowledge base identifies a gap, promotes the hypothesis, closes places where adequate study exists, and uncovers areas where further studies are required. Many Block matching algorithms, Key point matching, ML, and DL, have been derived and established in applications of different areas like CMF, SF detection, and geometrical attack detection.

Following a thorough review of the literature, it was determined that some research was conducted to detect forgery for copying and splicing. Furthermore, various other methods for hybrid forgery detection are unreported. Therefore, this work led us to develop hybrid CMF and splicing methods for forgery detection with greater detection precision.

Subsequently, an interest arises in detecting geometrical attacks in CMF and SF detection. Finally, the analysis describes the issues and recommends the planned work using applied research and literature knowledge.

Image forgery is widespread nowadays to create false propaganda. Machine learning is the emerging field for automatic forgery detection. In this paper, the basics of image forgery and machine learning are discussed. Then, the process of forgery detection is discussed from traditional approaches to a machine and deep learning.

2   Literature Review

Pun et al. [2] proposed a technique that includes both block and key point detection algorithms. They suggested image forgery detection to identify disrupted regions. Costanzo et al. [5] employed the SIFT technique to extract features to determine which feature points were essential. The authors also presented a method for recognising, describing, and identifying counterfeit areas. This system outperformed previous detection methods in terms of precision and recall. To distinguish non-overlapping areas, Li employed image segmentation. The authors also subdivided the image into overlap pieces, accompanied by the DCT application on each picture, to obtain the characteristics of Hosny et al. [7]. Lexicographic representation was carried out to reduce computational problems. Different neighbouring block pairs have, therefore, been considered as potential duplicate regions. A histogram was calculated to refine the results, which counted the corresponding points which were equally distant. Finally determine patches that have been duplicated. This system seemed to find the best compromise between sophistication and performance, but it could not detect tiny repeated regions simultaneously. The novel method combines Scale Invariant Feature Transformation (SIFT) with abstraction, which corresponds to the key arguments made by Ardizzone et al. [8]. In the instance of cloning, Bondi et al. [10] anticipated clustering around crucial locations, and Mayer et al. [11] predicted interference within the next step. This approach generated high precision, recall, consistency, and resistance to image attacks. If a suspect image exists, it could be found by the proposed algorithm, i.e., repeating particular patches. Ansari et al. [12] used a geometric change to carry out this trick to find out image forgery. The investigation revealed efficiency in a variety of operational conditions, including multiple and hybrid cloning.

SIFT was proposed by Birajdar et al. [13] to find and extract critical points for each patch. The Kd tree identifies a crucial point, and the nearest neighbour calculates the distance between critical points. Traditional machine learning techniques are accurate, but they need a significant amount of time to detect forgery. Deep learning methods require a lot of computational power. All of these conditions have now been met due to advancements in big data and processing capability.

3   Proposed Algorithm

Documented photos have increased in frequency and complexity in the last decades, and various digital counterfeit instruments have emerged in endless streams. The most common ones are splicing and duplication, which manipulate pictures to be hard to understand from a human-perceivable method.

Active methods collect primary data from a digital image to establish authenticity (i.e., digital watermarks or signatures). The source image is not accessible in passive approaches, often referred to as blind strategies. There is no need for any previous image information when detecting active forgery. The methodology is developed on the assumption that the manipulations in the still object of fossilised pictures are due to changing fundamental statistical properties. Such anomalies are used in the analysis of falsification. The most direct attack is the CMF, where a specific region (or region) is duplicated in the picture. The different algorithms to identify passive images can be clustered similarly into five classes: frame, picture, compression, computational, and physical schemes. As illustrated in Fig. 3, passive imaging Chen et al. [9].

First of all, pixel-based technologies detect pixel-level statistical abnormalities. For example, the presence of copy-move processing forgeries of images causes block alteration. Furthermore, these algorithms look for connections between pixels created by a specific type of tampering in a geographic domain or a modified area. In practice, these tactics are regularly employed.

3.1   Flow Diagram for CMF Detection

1. Image data preparation: A few procedures for enhancing classification performance, including rotations, cropping, RGB colour scale conversion, etc., are carried out and tested via a test image before any feature extraction process. The generalised framework for Copy Move Forgery Detection (CMFD) is shown in Fig. 3.
2. Pre-processing: An appropriate classifier has to be selected or built based on the extracted features. The classifier training includes various digital images and some of the relevant and essential classifier parameters.
3. Feature extraction: Features are explicitly set for each class to help distinguish them from all other groups. In contrast, an invariant of all attribute variations in a class from the host manipulated results.
4. Classification: A classifier splits each image into two classes: accurate and distorted digital images. An appropriate performance measure is selected or designed depending on the extraction features set.

Post-processing: This final step involves morphological activities that are carried out to reduce false positives. The patches with identical shift vectors are labelled with the

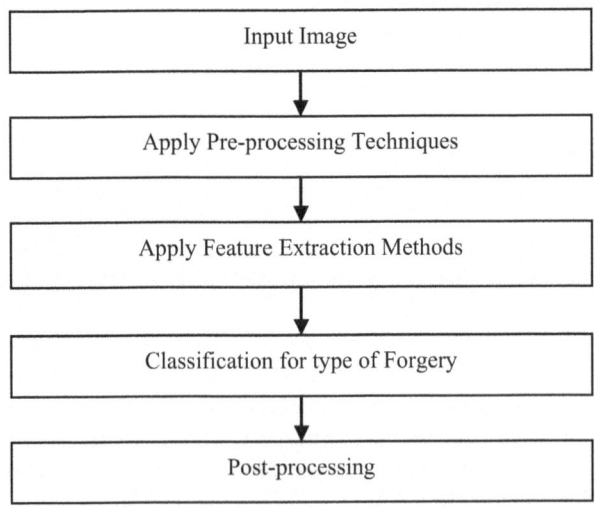

Fig. 3. Generalized Framework for CMFD

same, usually white, to distinguish the duplicate fields and discriminate against different patches by shifting them.

3.2 Classify CMFD Techniques

The CMFD systems currently in operation are categorised into different types, as based on blocks, key points Zhang et al., [6], colour illumination, hybrid methods, machine learning, and deep learning. Similar blocks are determined based on some criteria of resemblance in a block-based methodology.

The CMFD-based key point approach also uses local features of the points of interest to distinguish duplicated regions. Host images are divided into image blocks with hybrid CMFD algorithms before key point elimination. Besides the computational burden, the techniques are pretty resilient for different geometric assaults.

3.2.1 Pre Processing

This approach performs splicing and copy move image forgery detection on CASIA 2.0, DVMM, BDSS, and Colombia datasets. All these datasets have different sizes of images Bashar et al., [14]. All forged images are placed in one folder. We need a pre-processing image technique to match all images to the same size. Then, colour illumination was applied to all images. The color-illuminated image edges and corners are detected quickly. As shown in Fig. 4, the Harris detector detects strong corners and converts these into a.npy array of anchors, positive and negative. These arrays are of size $(30 \times 30 \times 3)$ and are then passed to the autoencoder.

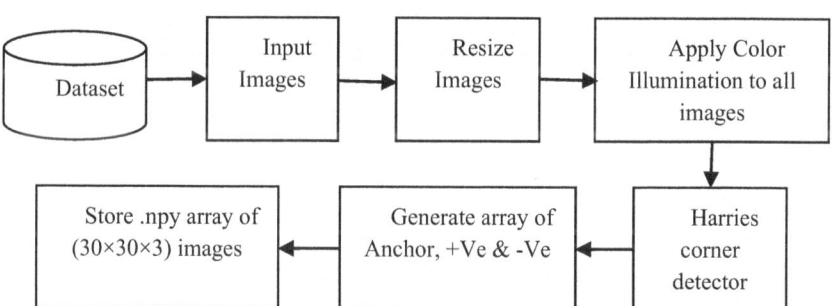

Fig. 4. Block diagram of Image Processing in Unsupervised Learning.

3.2.2 Auto Encoder

The auto-encoder uses a nonlinear transformation to reduce the number of dimensions. The types of autoencoders are convolutional, deep, variation, and de-noising auto encoders [15–18]. It comprises an encoder, code, and decoder. The CNN is used in this approach because it performs better for images than others. First, it encodes the input into simple signals. Next, it comprises multiple convolution layers followed by an adder from output three and output 4 with max Pooling. Then, it down-samples the input image up to the maximum point of compression. Finally, the code determines which part of the image is essential. We used 128 batches, each of which contained ten images for 2000 epochs with a 1e−4 learning rate. First, the input image is reshaped into 30 × 30 × 3 patches and normalised by dividing 255. The convolution layer uses a stride of one, activation as ReLU, and padding as same. The complete flow of the program is shown in Fig. 5.

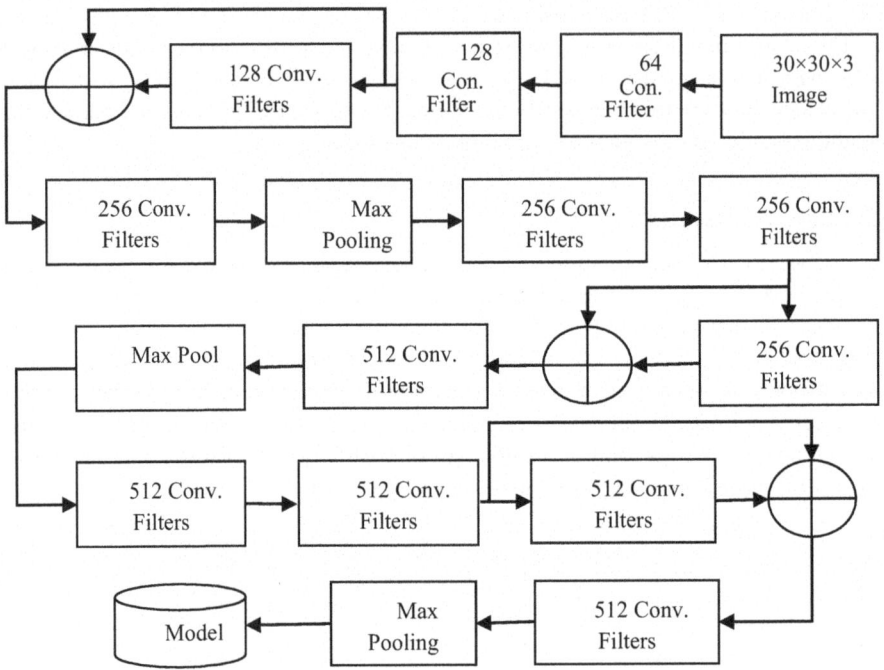

Fig. 5. Block diagram of the autoencoder in unsupervised learning.

3.2.3 Auto Decoder

An auto decoder is used to reconstruct images from the minimum most essential pixels. It can replicate the output image into an input image with some degraded quality. It comprises multiple convolution layers followed by an adder from output three and output four with upsampling. The convolution layer uses stride of one, activation as ReLU, and padding as valid. In the final convolution, layer activation is used as a sigmoid with padding as same. The image reconstruction process is shown in Fig. 6. The training process used mean squared error to find a loss between positive, negative, and anchor images. The adom optimiser is used to optimise the model. We used CoMoFoD and CMFD image datasets, 80% data for training and 20% data for testing. The training weights are stored for the encoder and decoder with.h5 extension. The trained patches of positive, negative, and anchor are stored in.jpg format, as shown in Fig. 6.

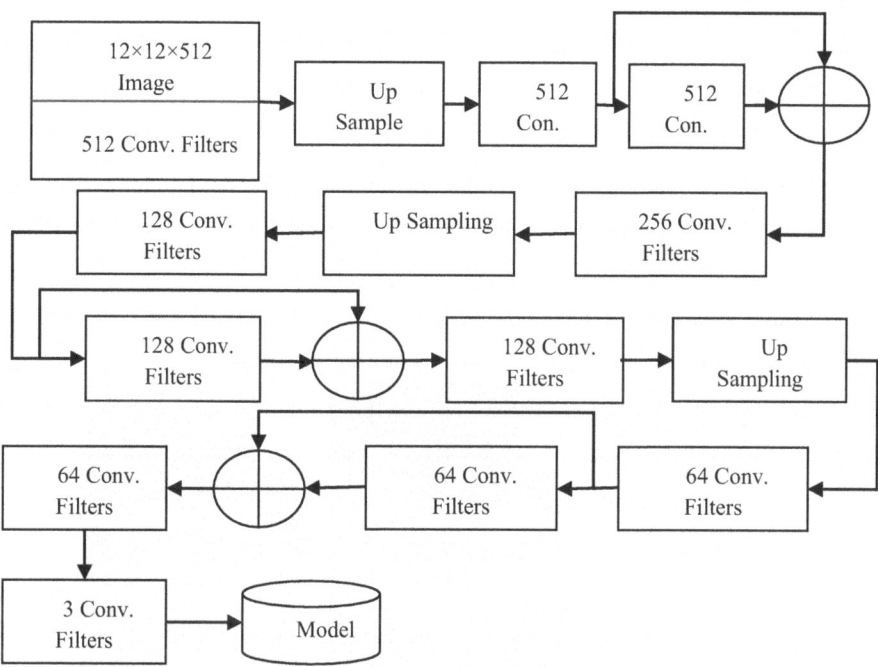

Fig. 6. Block diagram of the auto decoder in unsupervised learning.

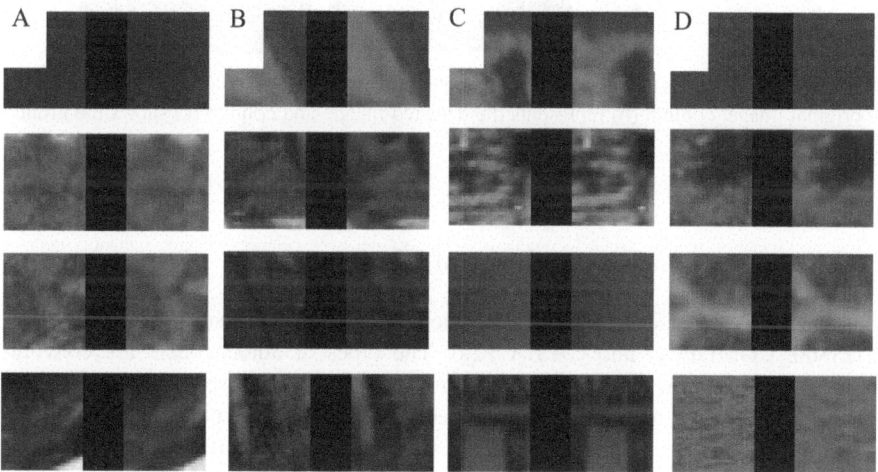

Fig. 7. The output of the Siamese neural network for the training of positive and negative triplets in columns (A), (B), (C), and (D)

3.2.4 Localization of Forgery

The output of the Siamese autoencoder/decoder is trained using triplet loss. These triplets are trained using a positive and negative triplet distance margin of 0.6. There are 100 epochs between significant decreases in the learning rate. The learning rate decay factor is 1.0. L2 regularisation was used with a random seed of 666. The number of images per batch is 20. The number of batches per epoch is 1000. The number of epochs to be run is 500. The number of images per group is 6. The number of images to be processed per batch is 99. We use the ADAM optimiser for the optimisation of the results. The exponential decay for the tracking of training parameters is 0.99. The initial learning rate is 1e−5. These triplets are passed to localise the forged region. The forgery validation process finds the score of forged pixels and the distance between embedding. The final output of the fabricated pixels is given when the performance is greater than the threshold. This algorithm shows the output, as shown in Fig. 8.

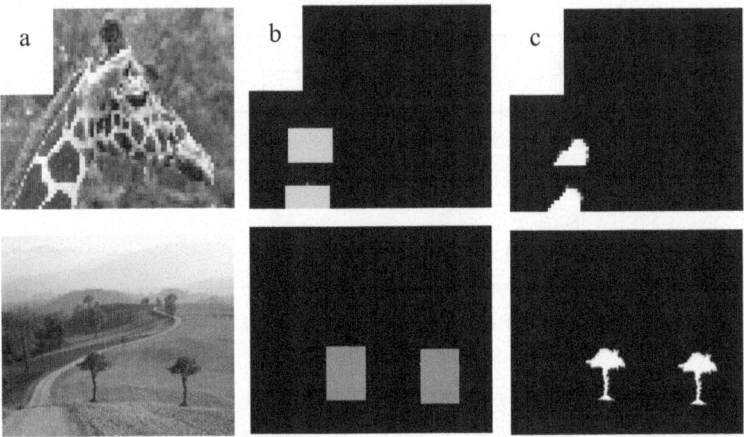

Fig. 8. The output shows the results of forgery detection for the CMFD dataset. Column (a) shows the original image, column (b) represents the detected image, and column (c) shows the ground truth image.

4   Conclusion and Future Scope of Work

Experiments were performed on various openly available databases like CASIA v1.0, BSDS300, COMOFOD and CASIA v2.0. The types of autoencoders are convolutional, deep, variation, and de-noising auto encoders. It comprises an encoder, code, and decoder. First, the input image is reshaped into 30 × 30 × 3 patches and normalised by dividing 255. The convolution layer uses a stride of one, activation as ReLU, and padding as same. An auto decoder is used to reconstruct images from the minimum most essential pixels. It can replicate the output image into an input image with some degraded quality. The forgery validation process finds the score of forged pixels and the distance between embedding.

In the future, more efficient ways for detection of CMF, SF, and geometrical attack forgery can be searched. Moreover, an effort should also be made to find video forgery detection using ML and DL-based methods. The ML and DL can also be used to analyse video forging for other applications such as inter-frames and intra-frames. In the future, several other types of forgery, such as splicing, can be implemented on adaptive over-segmentation, colour illumination, and feature-point matching on different kinds of media, for example, video and audio.

References

1. Chang, I.C., Yu, J.C., Chang, C.C.: A forgery detection algorithm for exemplar-based inpainting images using multi-region relation. Image Vis. Comput. **31**(1), 57–71 (2013). https://doi.org/10.1016/j.imavis.2012.09.002
2. Pun, C.M., Yuan, X.C., Bi, X.L.: Image forgery detection using adaptive oversegmentation and feature point matching. IEEE Trans. Inf. Forensics Secur. **10**(8), 1705–1716 (2015). https://doi.org/10.1109/TIFS.2015.2423261
3. Singh, A., Singh, G., Singh, K.: A Markov-based image forgery detection approach by analysing CFA artefacts. Multimed. Tools Appl. **77**(21), 28949–28968 (2018). https://doi.org/10.1007/s11042-018-6075-5
4. Su, L., Li, C., Lai, Y., Yang, J.: A fast forgery detection algorithm based on exponential-Fourier moments for video region duplication. IEEE Trans. Multimed. **20**(4), 825–840 (2018). https://doi.org/10.1109/TMM.2017.2760098
5. Costanzo, A., Amerini, I., Caldelli, R., Barni, M.: Forensic analysis of SIFT keypoint removal and injection. IEEE Trans. Inf. Forensics Secur. **9**(9), 1450–1464 (2014). https://doi.org/10.1109/TIFS.2014.2337654
6. Zhang, Y., Thing, V.L.L.: A semi-feature learning approach for tampered region localisation across multi-format images. Multimed. Tools Appl. **77**(19), 25027–25052 (2018). https://doi.org/10.1007/s11042-018-5756-4
7. Hosny, K.M., Hamza, H.M., Lashin, N.A.: Copy-for-duplication forgery detection in colour images using QPCETMs and sub-image approach. IET Image Process. **13**(9), 1437–1446 (2019). https://doi.org/10.1049/iet-ipr.2018.5356
8. Ardizzone, E., Bruno, A., Mazzola, G.: Copy-move forgery detection by matching triangles of keypoints. IEEE Trans. Inf. Forensics Secur. **10**(10), 2084–2094 (2015). https://doi.org/10.1109/TIFS.2015.2445742
9. Chen, Y., Kang, X., Shi, Y.Q., Wang, Z.J.: A multi-purpose image forensic method using densely connected convolutional neural networks. J. Real-Time Image Process. **16**(3), 725–740 (2019). https://doi.org/10.1007/s11554-019-00866-x
10. Bondi, L., Lameri, S., Guera, D., Bestagini, P., Delp, E.J., Tubaro, S.: Tampering detection and localization through clustering of camera-based CNN features. In: IEEE Computer Society Conference on Computer Vision and Pattern Recognition Workshops, vol. 2017-July, pp. 1855–1864 (2017). https://doi.org/10.1109/CVPRW.2017.232
11. Mayer, O., Stamm, M.C.: Accurate and efficient image forgery detection using lateral chromatic aberration. IEEE Trans. Inf. Forensics Secur. **13**(7), 1762–1777 (2018). https://doi.org/10.1109/TIFS.2018.2799421
12. Ansari, M.D., Ghrera, S.P., Tyagi, V.: Pixel-based image forgery detection: a review. IETE J. Educ. **55**(1), 40–46 (2014). https://doi.org/10.1080/09747338.2014.921415
13. Birajdar, G.K., Mankar, V.H.: Digital image forgery detection using passive techniques: a survey. Digit. Investig. **10**(3), 226–245 (2013). https://doi.org/10.1016/j.diin.2013.04.007
14. Bashar, M., Noda, K., Ohnishi, N., Mori, K.: Exploring duplicated regions in natural images. IEEE Trans. Image Process. no. c, pp. 1–40 (2019). https://doi.org/10.1109/TIP.2010.2046599
15. Thakur, A., Ranjan, R.: Evaluate the performance of deep CNN algorithm based on parameters and various geometrical attacks. Wirel. Pers. Commun. **132**, 2587–2602 (2023). https://doi.org/10.1007/s11277-023-10734-4

16. Thakur, A., Sharma, S., Sharma, T.: Design of semantic segmentation algorithm to classify forged pixels. In: 2023 IEEE 12th International Conference on Communication Systems and Network Technologies (CSNT), Bhopal, India, pp. 409–413 (2023). https://doi.org/10.1109/CSNT57126.2023.10134649
17. Kaur, N.: Original research article AI-based COVID-19 disease detection in medical images: advancements and implications in healthcare. J. Autonom. Intell. **6**(3) (2023)
18. Garg, G.: Computer-aided diagnosis systems for prostate cancer: a comprehensive study. Current Med. Imaging, 22 May 2023

An Advanced Approach to Detect and Classify Lung Nodules Using CT Images

Pramod Kumar Naik[1]([⊠]), R. Amith[1], D. Akshitha[1], and B. Sadhana[2]

[1] Department of CSE, Dayananda Sagar University, Bengaluru, India
pramodnaik40@gmail.com
[2] Department of ISE, CEC, Mangalore, India

Abstract. Lung cancer, particularly in its advanced stages, presents challenges in achieving high cure rates. Efficient early detection holds the key to significantly improving survival rates. Recognizing the critical importance of early identification, this study proposes a two-phase technique aimed at early lung cancer detection. The first phase involves the immediate importation of lung CT scans into the framework. Subsequently, the image lay-out phase is executed through explicit image management procedures. The proposed approach incorporates several advancements, including feature extraction, neural organization identification, pre-processing, binarization, thresholding, division, and image capture. During the feature extraction process, specific critical qualities are systematically removed from the segmented images. Simultaneously, the binarization method modifies matched images and aligns them with edge views. These integrated advancements collectively contribute to a comprehensive and innovative approach for the early detection of lung cancer. Achieving early detection through this method holds significant promise for enhancing lung cancer survival rates and, consequently, contributing to the overall improvement of human health outcomes.

Keywords: Lung Cancer Detection · Image Pre-Processing · Image Segmentation · Convolution Neural Network

1 Introduction

Lung cancer is the most common disease diagnosed worldwide and the leading cause of cancer-related mortality, with more than 10 million cancer deaths annually worldwide. Greater numbers of people die from lung cancer than from any other disease. Numerous epidemiological studies conducted in India across several demographic cohorts demonstrate the large burden of lung cancer in the nation, which makes a major contribution to the morbidity and mortality from cancer. In general, problems with image classification are addressed via image processing and pattern recognition. In fact, the primary purpose of computer vision (CV) is to train computers in how to recognize things, identify events, and create 3D models in various contexts. CV models were created to build predictive and decision-making tasks by extracting features from visual input. With such remarkable accomplishment, this scientific subject has been used to interpret medical diagnostics. Currently, X-ray images are used by medical professionals to identify lung cancer. After lung cancer is discovered, the patient's time to life is quite brief. Lung cancer patients have a higher probability of surviving if this condition is discovered early on. The conventional approach to taking lung X-rays is inefficient and yields unsatisfactory results. In addition, the conventional method of detecting lung cancer is more expensive and time-consuming. A sophisticated computer-based method can get around these restrictions. Many researchers are trying to diagnose cancer early thanks to the development of machine learning algorithms, but this leads to a sizable percentage of incorrect predictions.

Treatments for cancer work only if malignant cells are successfully isolated from healthy ones. A neural network is necessary to identify cancer cells in conventional tissues, and this makes CNNs a great tool for creating CNN-based cancer detection systems.

2   Literature Survey

The detection of lung cancer is a crucial problem that has created interest in research by many researchers in the same medical field. The best practice of multiple views of a single imaging is trending from last two years. The use of ML and DL techniques may enhance detection accuracy in the current medical era. The major challenges with the medical field professionals may be within the shorter duration they need to identify and diagnose a lung cancer with more accuracy. The preliminary steps used were PCA and DWT for image fusion and performance verification was carried by taking a database with 4682 CT lung images of 61 patients with the lung nodule size varies from 3 mm to 30 mm. The cancer detection stages like STG-1, 2, 3, and STG-4 using ResNet -18 CNN classifier. The achieved accuracy and sensitivity of cancer detection were 98.2% and 96.4% hence proposed models has the highest potential, can be adapted in clinical diagnosis systems [6].

The related work on LDD is a in depth study of various research articles relating to lung cancer diseases normally which uses image- processing concept and techniques on the CT, X-ray images. The authors of this research paper, considered ATA (adaptive-threshold algorithm) watershed algorithm, mathematical morphology [7].The researchers of the present article have applied some steps to segment an images, during first step they have targeted to enhance the quality of the images with the help of Gaussian filter and noise reduction using gradient enhancement technique to extract the noise of the CT images. And in the second step they extracted and removed bronchus, trachea from the CT lung image finally segmented with the watershed transform method. The testing process was carried out by using of CT lung images in series to segment the cancerogenic area. Finally the research of this article concluded how lung parenchyma can be segmented successfully. This author stated that the CAD model designed is helpful to detect cancer in the early stage using CT lung images. This author has adapted four steps to detect cancer cells in the CT images considering medical professionals' challenges to diagnose the lung cancer nodules from CT images. The first step filters the noisy contents by applying filtering concepts either by using Weiner Filter, Median, Gabor filter, Min and Max filter etc. In the preceding step thresholding like multiple and optical thresholding, active contour method, shape-based method etc. were used. Segmentation was introduced to segment suspected nodules. The third step is the feature extraction like geometric features, shape –size features, gray scale features and statistical features need to be extracted from the CT lung images to detect malignant or non-malignant. To extract the features various classifiers like vector machine, generic algorithm, ANN. Rule based classifier, Linear Discriminate Analysis methods were adapted. As concluded by the research author, there are still more improvements required with respect to sensitivity, accuracy, and specificity in the existing deionization system. [8] T Aggrawal et al. (2015) [9], in the proposed research paper the classification of cancer nodules, the author has introduced a model which uses gray scale characteristics and valuable thresholding values to perform segmentation process of lung nodules by threshold values, gray scale characteristics to perform segmentation process. The proposed system achieved accuracy, sensitivity, and Specificity of 84%, 97.14% and 53.33%.

The research author has proposed a model in which CT images were pre-processed to remove the noise. In the next step fuzzy k-mean algorithm was used to perform segmentation process, later the result was improved by k mean approach. In the third step feature extraction was conducted from the CT lung images like correlation, entropy SSIM, PSNR, homogeneity using statistics GLCM feature extraction method. Finally, categorization is conducted using supervised NN like BPNN for the detection of lung cancer. This author has achieved 90.7% accuracy, but this accuracy can be enhanced using improvised classification techniques like support vector machine [10].

This author has proposed the CNN model as the classification approach in the CAD system and achieved 84.6%,82.5% of accuracy and sensitivity and specificity of 86.7% [11].

The author has proposed data mining methods, lung cancer patient database includes images of human upper half body X-rays that classify as normal or malignant, benign. CAD system uses pattern recognition, feature extraction and classification process. The proposed model used X-ray images and accuracy gained is less compared with CT images. This study elaborates that to achieve better accuracy implementation work can be enhanced and extended to apply the CT images in superiorised diagnosis of detection of lung cancer in the human body [12]. The author has developed an enhanced model for segmenting the lung CT images by combining kernel graph cut algorithm and proposed a mathematical model. In the next stage the proposed algorithm was compared with K –mean algorithm, cluster variance algorithm and concluded with the comparative studies between two approaches [13].

3   Proposed System

The automatic detection of lung nodule using ML techniques was adapted in which the technique uses two distinct phases like training and testing phase. During the training phase, the lung images are augmented means preprocessed and next fed into deep learning model. The output of deep learning model produces features acts as an input to the ML algorithm. Under testing phase after image screening classification of lung nodules using objective or subjective analysis methodologies were adapted.

3.1 Purpose of Conducting Research

The purpose of conducting research is diverse and encompasses various objectives across disciplines. At its core, research seeks to advance knowledge by generating new insights, uncovering facts, and exploring unexplored territories. It serves as a tool for problem-solving, offering solutions to challenges and fostering innovation and creativity. Research contributes to existing knowledge by building upon prior studies, refining theories, or challenging established paradigms. In addition to academic and professional development, research informs decision-making processes in fields such as policy, business, and healthcare. It validates and tests theories, ensuring their accuracy and reliability. Research has broad social implications, addressing societal issues, influencing social change, and contributing to economic development by identifying market trends and fostering entrepreneurship. It plays a crucial role in technological advancements, quality improvement in various sectors, and the discovery of new phenomena. Engaging in research cultivates critical thinking and analytical skills, while its educational role allows students to delve into topics deeply and contribute to academic knowledge. Overall, research serves as a driving force for progress, facilitating global collaboration and sharing of knowledge to address challenges and enhance our collective understanding of the world.

3.2 Advantages and Disadvantages of Existing Work

The field of lung cancer detection algorithms exhibits several advantages and disadvantages. On the positive side, these algorithms hold enormous potential for early detection, a critical factor in improving treatment outcomes for patients. The continuous advancements in technology, particularly in machine learning and deep learning, contribute to enhanced accuracy in identifying lung nodules, reducing both false positives and false negatives. Automation within these algorithms streamlines the diagnostic process, allowing for the efficient analysis of large-scale datasets and the identification of subtle patterns that may not be apparent through traditional diagnostic methods. However, challenges persist, including the presence of false positives and negatives, especially in the detection of small nodules crucial for early-stage diagnosis. Ethical concerns related to patient privacy and the responsible use of medical data have become more pronounced as algorithms rely on extensive datasets. Validation hurdles, interpretability issues, and the need for seamless integration into clinical workflows also pose challenges. Resource intensiveness, both in terms of computational requirements and expertise, may limit accessibility in certain healthcare settings. Despite these challenges, ongoing research and development are crucial to refining existing algorithms, addressing limitations, and ensuring their practical and ethical implementation in real-world healthcare scenarios.

3.3 Methodology

The methodology employed for the automatic detection of lung nodules utilizes machine learning techniques, specifically implemented in a two-phase process involving training and testing.

Training Phase

1. Data Preparation:

    A diverse dataset comprising multi-Modal medical images, encompassing X-rays and CT scans, is curated. The dataset includes labelled examples indicating the presence (positive class) or absence (negative class) of cancerous regions.

2. Data Augmentation:

    Data augmentation techniques are applied to introduce variations to the training images. Transformations such as rotation, scaling, and flipping enhance the diversity of the dataset.

3. Model Initialization:

    Initialization of a multi-Modal CNN model involves random weight assignment and architecture designed to accommodate different image modalities. The model typically comprises convolutional layers for feature extraction and fully connected layers for classification.

4. Loss Function and Optimization:

    During training, the model calculates a binary cross-entropy loss, measuring the dissimilarity between predicted outputs and true labels. The optimization algorithm, usually Adam or SGD, adjusts model parameters to minimize this loss.

5. Forward and Backward Pass:

    Processing the training dataset in batches, a forward pass computes model predictions, while a backward pass updates model weights using gradients obtained from the loss function.

6. Epochs:

    The training process unfolds over multiple epochs, where each epoch signifies a complete pass through the entire training dataset. This iterative repetition enables the model to learn progressively.

7. Validation:

    A separate validation dataset, distinct from the training data, monitors the model's performance. This step is crucial for early detection of overfitting.

Testing Phase

1. Data Preparation:

    The testing phase employs a distinct dataset, unseen during training (the testing dataset). This dataset, comprising multi-Modal medical images with known ground truth labels, is used to assess the model's real-world performance.

2. Forward Pass:

    Testing images undergo inference as they pass through the trained model. The model predicts the presence of cancerous regions based on the acquired features.

3. Performance Metrics:

    Various performance metrics are computed based on model predictions and ground truth labels in the testing dataset. These metrics include Accuracy, Precision, Recall (Sensitivity), F1-Score, and ROC-AUC.

4. Interpretability:

    Methods for interpretability, such as generating heatmaps or saliency maps, may be applied to visualize regions influencing the model's predictions, aiding in understanding decision-making.

5. Fine-Tuning (Optional):

    If the model's performance is deemed unsatisfactory during testing, optional fine-tuning or optimization of hyper parameters may be considered to enhance accuracy.

    The testing phase evaluates the model's real-world performance in identifying lung cancer regions in multi-Modal medical images, providing insights into its reliability and effectiveness in clinical practice (Fig. 1).

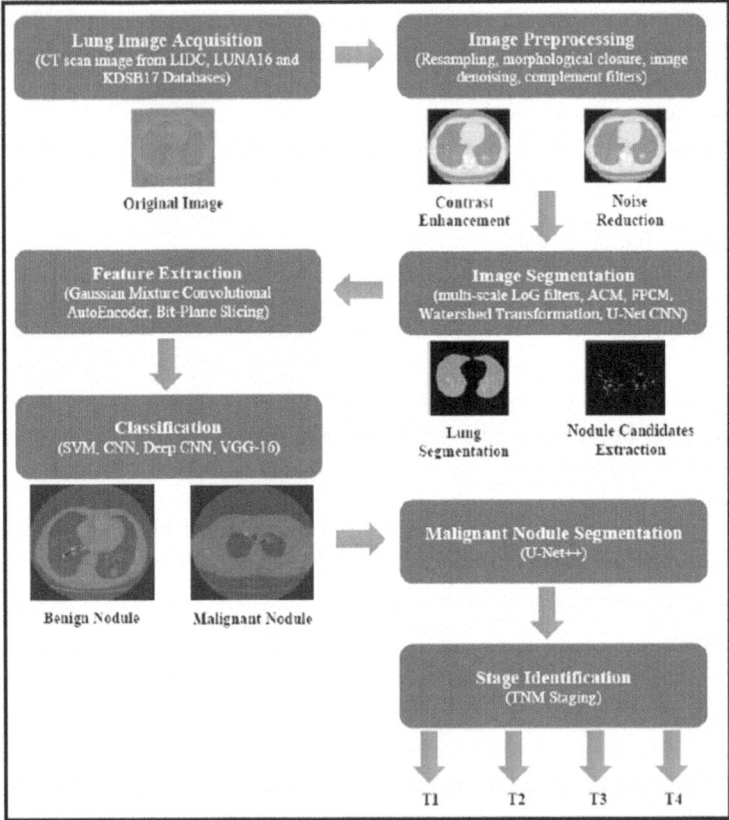

Fig. 1. Proposed Methodology for Lung Cancer Detection Model

4 Result and Discussion

The proposed methodology leverages a Deep Learning model for the preprocessing, segmentation, and detection of lung nodules. In Fig. 2, a representative lung image from a malignant case is displayed, highlighting the effectiveness of the implemented machine learning algorithm. The model's performance evaluation reveals a remarkable accuracy of 96%, a key metric indicating the precision of the algorithm in correctly identifying lung nodules. Figures 3 and 4 provide detailed insights into the training process. Figure 3 illustrates the accuracy plot, demonstrating the model's proficiency in discerning between normal and malignant cases. On the other hand, Fig. 4 portrays the model loss over training epochs, highlighting the optimization trajectory.

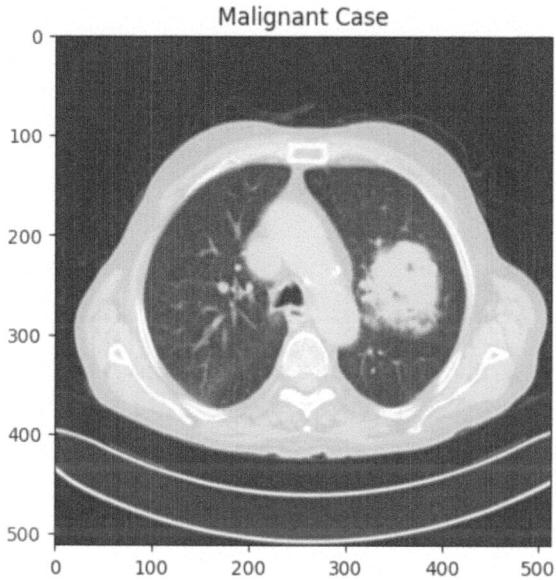

Fig. 2. Lung Image of a Malignant Case

In the accuracy plot (Fig. 3), peaks and plateaus reflect the model's discriminating power, while the model loss plot (Fig. 4) demonstrates the optimization journey, highlighting the fine-tuning process.

Furthermore, Fig. 5 delves into the classification of cancer presence using CT images, differentiating between normal and malignant cases. This categorization illustrates the model's ability to discern subtle patterns indicative of malignancy. These comprehensive results underscore the robustness of the Deep Learning model, utilizing ML terminologies to highlight its accuracy, optimization process, and classification capabilities. The findings signify the potential of this approach as an effective tool for augmenting the diagnostic accuracy in identifying malignant lung nodules.

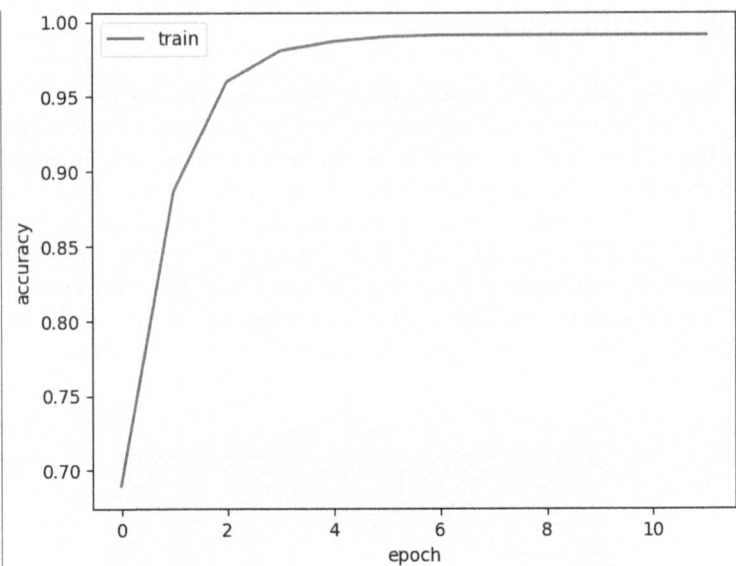

Fig. 3. Model Accuracy

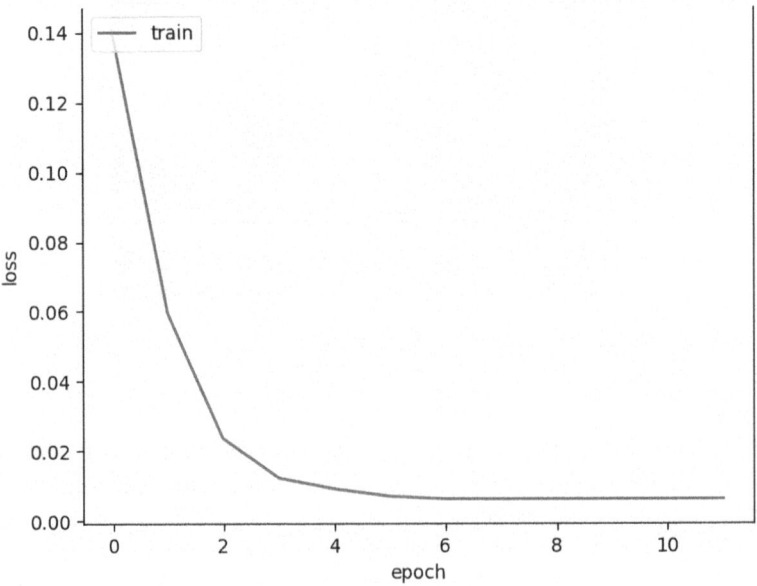

Fig. 4. Model Loss

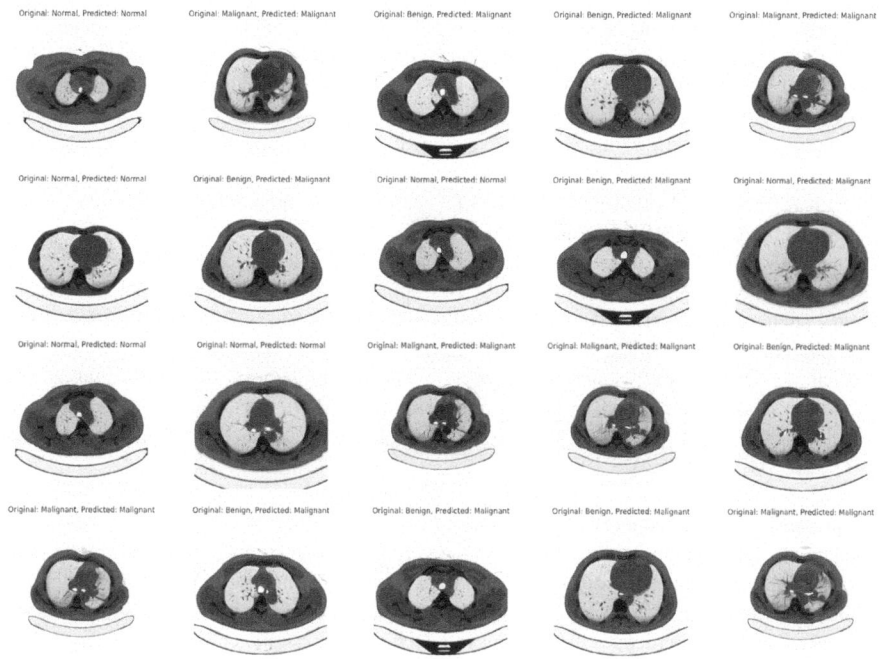

Fig. 5. CT lung images of different cases like normal or Malignant.

5   Conclusion

In conclusion, the presented system addresses the imperative task of automating the detection of lung cancer through the application of deep learning techniques on CT images. Leveraging machine learning algorithms and advanced deep learning methods, the system successfully performs segmentation and validation tasks. The adapted methodology demonstrates a remarkable precision in detecting lung nodules, achieving an accuracy rate of 96%. This underscores the efficacy of the proposed approach in contributing to the early identification of lung cancer, a crucial factor in improving patient outcomes.

Looking ahead, the future scope of this research extends to the development of an Android-based Lung Cancer Detection application. This application has the potential to significantly aid medical experts in the initial stages of lung cancer detection. By providing a user-friendly interface and leveraging the power of deep learning, such an application could prove invaluable in enhancing the efficiency of healthcare professionals and contribute to better health outcomes for patients.

References

1. Senarathna, S.A.D.L.V., et al.: Lung cancer detection and prediction of cancer stages using image processing. In: 3rd International Conference on Electrical, Control and Instrumentation Engineering (ICECIE) (2021). 978-1-6654-4966-3/21/$31.00 ©2021 IEEE, https://doi.org/10.1109/ICECIE52348.2021.9664658

2. Nawreen, N., et al.: Lung cancer detection and classification using CT scan image processing. In: International Conference on Automation, Control and Mechatronics for Industry 4, vol. 0 (ACMI) (2021). 978-1-6654-3843-8/21/$31.00 ©2021 IEEE, https://doi.org/10.1109/ACMI53878.2021.9528297

3. Biradar, V.G., et al.: Lung cancer detection and classification using 2D convolutional neural network. In: 2nd Mysore Sub Section International Conference (MysuruCon). IEEE (2022). 978-1-6654-9790-9/22/$31.00 ©2022 IEEE, https://doi.org/10.1109/MysuruCon55714.2022.9972595

4. Sakr, A.S.: Automatic detection of various types of lung cancer based on histopathological images using a lightweight end-to-end CNN approach. In: 20th International Conference on Language Engineering (ESOLEC) (2022). 978-1-6654-5322-6/22/$31.00 ©2022 IEEE, https://doi.org/10.1109/ESOLEC54569.2022.10009108

5. Khan, Z., et al.: An efficient deep learning model based diagnosis system for lung cancer disease. In: 4th International Conference on Computing, Mathematics and Engineering Technologies (iCoMET) (2023). 979-8-3503-3531-6/23/$31.00 ©2023 IEEE, https://doi.org/10.1109/iCoMET57998.2023.10099357

6. Naidich, D.P.: Lung Cancer Detection and Characterization: Challenges and Solutions. SpringerLink. https://doi.org/10.1007/978-3-642-18758-2_17

7. Chen, X., et al.: An improved approach of lung image segmentation based on watershed algorithm. In: Proceedings of the 7th International Conference on Internet Multimedia Computing and Service, ICIMCS 2015, pp. 1–5 (2015). p. 39 https://doi.org/10.1145/2808492.2808531

8. T. Manikandan, "Challenges in lung cancer detection using computer-aided diagnosis (CAD) systems – A key for survival of patients," Arch. Gen. Intern. Med., vol. 1, no. 2

9. Aggarwal, T., et al.: Feature extraction and LDA based classification of lung nodules in chest CT scan images. In: International Conference on Advances in Computing, Communications and Informatics (ICACCI) (2015). https://doi.org/10.1109/ICACCI.2015.7275773

10. Sangamithraa, P.B., Govindaraju, S.: Lung tumour detection and classification using EK-mean clustering. In: 2016 International Conference on Wireless Communications, Signal Processing and Networking (WiSPNET) (2016). Computer Science, Medicine, Accessed 23 Mar 2016. https://doi.org/10.1109/WiSPNET.2016.7566533

11. Li, W., et al.: Pulmonary nodule classification with deep convolutional neural networks on computed tomography images. Comput. Math. Meth. Med. **2016**, 6215085 (2016). https://doi.org/10.1155/2016/6215085

12. Zubi, Z.S., Saad, R.A.: Improves treatment programs of lung cancer using data mining techniques. J. Softw. Eng. Appl. **07**(2), 69–77 (2014). https://doi.org/10.4236/jsea.2014.72008

13. Li, X., et al.: Enhanced lung segmentation in chest CT images based on kernel graph cuts. In: Proceedings of the International Conference on Internet Multimedia Computing and Service, ICIMCS 2016, pp. 228–233 (2016). https://doi.org/10.1145/3007669.3007690

Robust Iris Image Encryption via Black Widow Optimization Method

Ramamani Tripathy[1], Hakam Singh[1]([✉]), Navneet Kaur[1], Monika Parmar[1], and Rudra Kalyan Nayak[2]

[1] Chitkara University School of Engineering and Technology, Chitkara University, Solan 174 103, Himachal Pradesh, India
{ramamani.tripathy,hakam.singh,navneet.kaur_cse, monika.parmar}@chitkarauniversity.edu.in
[2] School of Computing Science and Engineering, VIT Bhopal University, Bhopal-Indore Highway, Kothrikalan, Sehore, MP, India

Abstract. This work represents an approach that uses the Black Widow Optimization (BWOA) method, a powerful encryption technique, to improve the security of iris biometric data. As it is dependable and distinctive. Nowadays, iris recognition became much popular in biometric identification systems. But still, to keep the confidentiality and privacy of iris images is a significant obstacle. By using traditional security techniques, it can be difficult to afford protection to confidential biometric data from attacks and intrusions. In this approach, we proposed a distinctive encryption technique that joins the features of Black Widow Optimization algorithm with the benefits of iris image cryptography. This recommended method is a multi-phase encryption strategy that includes feature extraction, encryption and decryption phases. The BWOA maximize the encryption settings, accelerating the encryption operation security and strength. The black widow spiders with their hunting behavior stimulates it. Researches on common iris image database shows how effective and strong the proposed method is. A complete evaluation of performance criterion, such as encryption/decryption speed, key sensitivity, and strength to frequent attacks, shows the superiority of BWOA-based Encryption over traditional security techniques. The outcomes show that iris image encryption has significantly improved efficiency and security, making it more resilient to security risks. Additionally, comparisons with current encryption methods demonstrate the advantages and potential of the suggested approach for protecting private iris bio metric data. This research advances biometric data security by providing a strong encryption framework that uses the Black Widow Optimization algorithm. It has promising implications for secure iris image authentication systems in various applications that demand elevated security and privacy measures.

Keywords: Iris recognition · Biometric security · Image encryption · Black Widow Optimization Algorithm · Innovation · sustainable · environment · network infrastructure

M. Khurana et al. (Eds.): ICMLA 2024, CCIS 2238, pp. 401–413, 2025.
https://doi.org/10.1007/978-3-031-75861-4_36

1 Introduction

1.1 Overview of Iris Recognition and Its Significance in Biometric Security

A biometric technique called iris recognition [1–5] uses a person's distinct irises the colored area of the eye surrounding the pupil—to identify them. It has received a lot of attention in the biometric security space for several reasons:

Uniqueness: Even between identical twins, each person's iris has incredibly unusual patterns that are specific to them. The iris's intricate structure—including crypts, furrows, and trabeculae—makes it a trustworthy biometric identification.
Stability: Iris patterns emerge in utero and stay mostly unchanged throughout an individual's life. This stability ensures long-term identification consistency.
Low erroneous Rejection Rate (FRR) and False Acceptance Rate (FAR): When used properly, iris recognition systems have low error rates, leading to fewer erroneous rejections and acceptances of authorized people.
Non-intrusiveness: Iris recognition is a convenient and hygienic biometric technique for identification that doesn't involve physical contact.
Extremely Accurate: Because of its stability and uniqueness, iris recognition is regarded as one of the most accurate biometric identification techniques compared to other biometric modalities like fingerprint or facial recognition.
Applications: Where strict security and precise identification are crucial, iris recognition finds use in several industries, including financial services, healthcare, access control, border security, and government identity programs [6–9].
Security Enhancement: By reducing the probability of identity theft and unauthorized access and expanding an extra layer of verification, using iris recognition in biometric systems enhances the overall security.

Eventually, the technology of iris recognition provides a strong and reputable biometric identification approach, enhancing secure authentication systems in the public and private sectors. This instrument is useful to boost the security and also to assure the authenticity of sensitive data and facilities because of its preciseness, stability and differentiated properties. The necessity for robust encryption method to safeguard iris biometric data intense encryption methods [10–17] are needed to protect iris biometric data for several important reasons:

Deference and Regulations: The biometric data which is sensitive must be protected by various data protection rules and regulations. Using strong encryption methods aids in an organization's compliance with these rules.
Preventing Spoofing and Forgery: The one of the most important feature of encryption methods is to prevent it from spoofing attacks, these are attempts to trick the recognition system using iris scan which are fake and also the modified or changed data. An additional line of defence against these kinds of attacks is provided by strong Encryption.
Sustaining Trust in Biometric Systems: Users and stakeholders are reassured about the dependability and credibility of the biometric authentication system when the security and integrity of iris biometric data are maintained.
Long-term Data Integrity: Over time, encryption aids in preserving the accuracy of biometric data.

In this case, iris biometric data handling requires the implementation of strong encryption algorithms to protect user privacy, prohibit illegal access, adhere to legal requirements, and preserve the general integrity and reliability of biometric security systems.

1.2 Introduction to the Black Widow Optimization Algorithm (BWOA) and Its Potential in Image Encryption

Based on the hunting habits of black widow spiders, the Black Widow Optimization Algorithm (BWOA) is a metaheuristic optimization algorithm inspired by nature. This algorithm is part of the swarm intelligence class and was created to solve optimization issues by imitating the black widow spider's predation approach.

The following are important features and applications for the Black Widow Optimisation Algorithm in picture encryption:

Exploration vs. Exploitation: BWOA balances exploitation (taking advantage of potential regions in the search space) and exploration (looking for various solutions).

Efficiency and Convergence: BWOA finds optimal solutions with good convergence qualities and efficiency, which can be useful in picture encryption applications where determining the ideal encryption key or parameters is essential.

Flexibility and Adaptability: The algorithm accommodates a range of problem domains and optimization targets.

Robustness to assaults: By optimizing encryption parameters, BWOA can help develop encryption schemes that are more resilient to assaults, strengthening the encryption process's defences against cryptanalysis and decoding attempts.

Parameter tuning: Effective encryption parameter adjustments are possible with BWOA, that leads to enhanced and more robust encryption. Cryptographic functions, keys, and substitution matrices are a few examples of these parameters. The novel way to enhance the privacy of private image data [18, 19] that is sensitive is by using BWOA conjunction with picture encryption.

2 Literature Review

2.1 Exploration of Biometric Encryption Techniques and Their Relevance to Iris Image Cryptography

By fusing biometric recognition with cryptographic techniques, biometric encryption systems seek to protect biometric data. When used in an iris image cryptography, such methods are essential to safeguarding the privacy, integrity and anonymity of sensitive biometric data. The section includes and analysis of biometric approaches to encryption techniques and how they are used in iris image cryptography, along with citations and sources:

Protection of Biometric Templates: Biometric encryption secures biometric templates which have been stored in databases. It is guaranteed that the initial biometric information cannot be retrieved from compromised templates by transforming iris templates into irreversibly encrypted versions making use of fuzzy commitment methods [1] and cancelable biometrics [2].

Feature-Level Encryption: The direct application of encryption methods to biometric features (such as iris patterns) is essential. Techniques like feature-level fuzzy vaults [3] create secure templates by performing cryptographic operations on extracted iris features, protecting the original data from unwanted access.

Secure Transmission and Storage: methods of encryption make sure that iris biometric data is transferred and kept in a secure environment. To prevent interference or illegal access, iris templates are subjected to cryptographic protocols like public key infrastructure (PKI) [4] or homomorphic encryption [5] prior to transmission or storage.

2.2 A Thorough Description of the Black Widow Optimization Algorithm's Features and Uses in Optimization Issues is Provided

The black widow algorithm (BWOA) is a metaheuristic algorithm for optimization that was created in reaction to the invasive behavior of black widow spiders. BWOA emerged as an algorithm that imitates the efficient and predatory hunting strategy of these spiders, drawing inspiration from nature. The approach mimics the predatory inclinations of black widows by combining web-building and prey-capturing mechanisms into a framework for solving optimizations problems. It starts by initializing a population of spiders, each of which represents a potential response within the search space. This is how iterative operation works. These spiders mimic the natural hunting behavior by mimicking the weaving of webs that symbolize possible solutions and exhibiting movement traits influenced by the predatory tendencies of black widows [9–11], (Fig. 1).

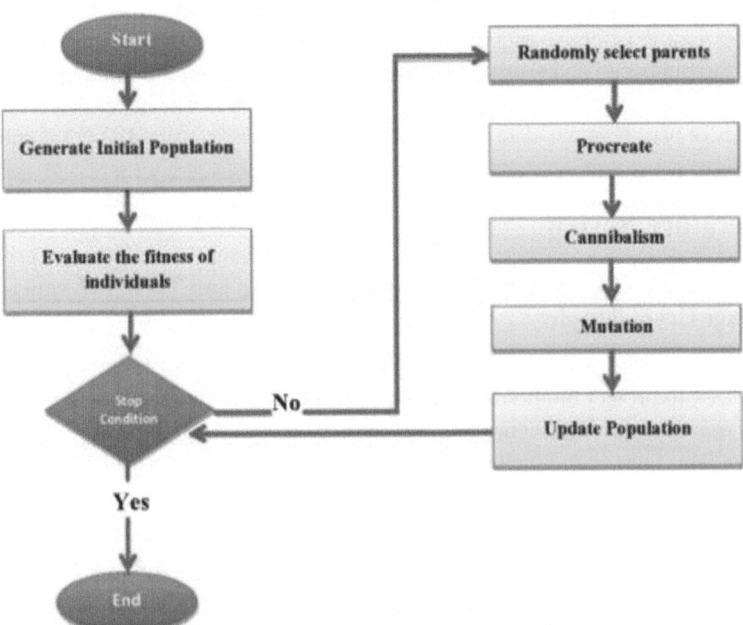

Fig. 1. Black Widow Optimization Algorithm (BWOA) [13]

BWOA provides exploration and exploitation an extensive amount of attention. This achieves an appropriate balance between exploration and discovery by deliberately scanning the solution space to identify different regions holding potential solutions [12, 13].

3 Methodology

3.1 A thorough justification of the suggested encryption technique that combines BWOA and iris image encryption. In order to create a secure and reliable encryption method, it is necessary to take a methodical approach that combines the unique characteristics of iris biometric data with the optimization capabilities of the Black Widow optimization Algorithm (BWOA) with iris image encryption. There are a few crucial steps that make up this process:

Preprocessing of Iris Images: The initial stage of the procedure involves preprocessing the raw iris images obtained from biometric scanners. This step involves noise reduction, normalization (to standardize image size and orientation), and segmentation (to separate the iris region from the background) in order to ensure consistency and improve feature extraction accuracy.

Feature Extraction from Iris Images: Reputable methods such as Daugman's rubber sheet model extract distinguishing features from preprocessed iris images. In order to create a feature vector or binary iris code that represents a specific person's iris, this stage records the iris' distinctive crypts, furrows, and patterns.

Integration of BWOA for Encryption: The Encryption is based on the properties of the retrieved iris. The encryption technique incorporates BWOA, an optimization algorithm inspired by nature. The algorithm initializes a population of spiders or potential encryption configurations.

Optimization of Encryption Parameters: The exploitation phase of BWOA refines promising encryption configurations, while the exploration phase explores a variety of encryption configurations inside the solution space. The algorithm iteratively modifies parameters to increase encryption strength and durability to produce an optimal encryption scheme.

Encryption of Iris characteristics: The extracted iris characteristics are converted into encrypted representations using the optimized encryption parameters discovered through BWOA. This encryption procedure improves security and secrecy by ensuring that the original biometric data cannot be decrypted without the proper decryption key (Fig. 2).

Storage or Transmission of Encrypted Data: The iris features produced after Encryption can be safely transferred or kept across networks. This encrypted representation protects The biometric data from unwanted access and manipulation, which also maintains the data's integrity and privacy.

Decryption for Authentication: The encrypted iris characteristics are decrypted using the inverse operations dictated by the optimized encryption parameters that BWOA acquired to authenticate or verify an individual's identification. Decryption restores the data to its original state to compare the encrypted data with saved templates or live grabs (Fig. 3).

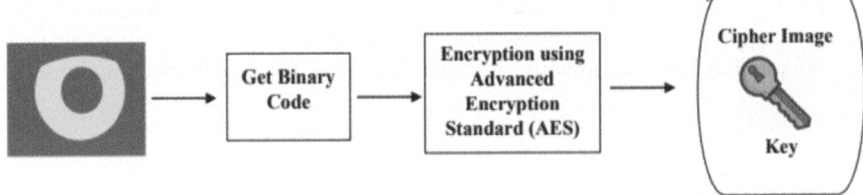

Fig. 2. Encryption of Iris characteristics [14]

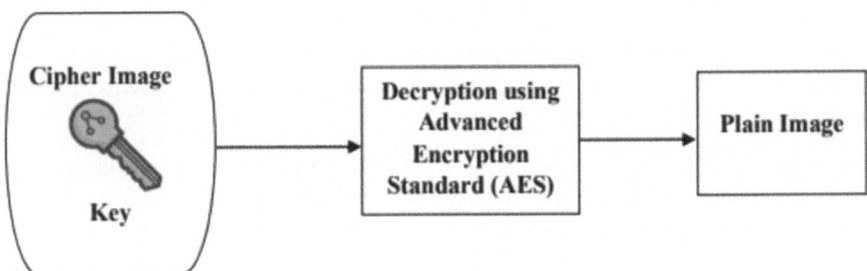

Fig. 3. Decryption of Iris characteristics [14]

Security Analysis and Validation: Lastly, thorough security evaluations are used to assess the effectiveness of the encryption technique. To guarantee the stability and efficacy of the encryption technique, cryptographic specialists perform evaluations that include resistance to assaults, statistical testing for randomness, key sensitivity analysis, and validation of information-theoretic security aspects.

3.1 Description of the Steps Involved in the Encryption Process, Including Feature Extraction, Encoding, and Decryption

Iris biometric data security encryption involves complex stages designed to convert unencrypted raw photographs of the iris into encrypted representations and then back to their original form for authentication. These procedures consist of decryption, encoding, and feature extraction. This intricate procedure can be broken down into the following steps:

Preprocessing and Feature Extraction: The first step in this process is preprocessing raw iris images from biometric scanners. To do this, noise must be reduced, the iris region must be divided, and the image's size and orientation must be standardized using normalization.

Encryption method Encoding: Following feature extraction, the extracted iris features are encoded into encrypted representations using an encryption method. The Black Widow Optimisation Algorithm's (BWOA) iterative optimization procedure is applied when iris image encryption is combined with BWOA. Iteratively optimizing encryption parameters, the technique initializes a population of possible encryption configurations.

BWOA Optimisation Process: Exploration and exploitation are carried out repeated phases by the BWOA. The algorithm examines the solution space during exploration to find various encryption settings.

Application of Encryption Parameters: The retrieved iris features are subjected to the derived encryption parameters when the BWOA optimization process converges. The iris data is encoded using these parameters, which also serve as the transformation functions or cryptographic keys that produce the encrypted representations.

Storage or Transmission of Encrypted Data: Databases or networks can safely store or transfer the generated encrypted iris features. This encrypted representation protects the biometric data from unwanted access and manipulation, which also maintains the data's integrity and privacy.

Decryption for Authentication: The encrypted iris characteristics are decrypted using the inverse operations dictated by the encryption settings acquired from the BWOA optimization process to facilitate authentication or identification. Decryption simplifies the process of comparing encrypted data with stored templates or real-time captures for authentication or verification by restoring the data to its original format.

4 Implementation and Experimentation

In the proposed encryption system that makes use of iris image datasets, the Black Widow Optimization Algorithm (BWOA) is integrated with cryptographic operations in a step-by-step manner to guarantee the secure translation of iris biometric data. This encryption process consists of multiple stages: get ready.

4.1 Preprocessing

The initial stage towards guaranteeing consistency and preparing the data for feature extraction is to preprocess the iris images. The three most common preprocessing techniques are normalization (which standardizes the size and orientation of iris pictures), segmentation (which divides the iris area from the surrounding eye structures), and noise reduction (Fig. 4).

After preprocessing, unique iris patterns are extracted from the preprocessed images during the feature extraction stage. Iris texture, crypts, and furrows are distinctive properties that may be removed using methods like Daugman's rubber sheet model. This results in a binary iris code or feature vector uniquely describing each person's iris. The features that are extracted are the foundation for the other encryption procedures. The Black Widow Optimisation Algorithm (BWOA) and the retrieved iris features are integrated for encryption purposes, which forms the basis of the encryption system. By utilizing BWOA, which is well-known for its exploration-exploitation methodology, to optimize encryption parameters, the modified iris biometric data's security is increased. Using the BWOA-based encryption method, the biometric data is encrypted by modifying the iris features that were retrieved and cryptographic keys. BWOA employs an iterative encryption technique in which every iteration optimizes parameters that are essential to the encryption process, such as cryptographic keys, substitution matrices, and permutation operations. The method's exploitation phase focuses on improving these configurations until an encryption scheme that is either optimal or almost optimal

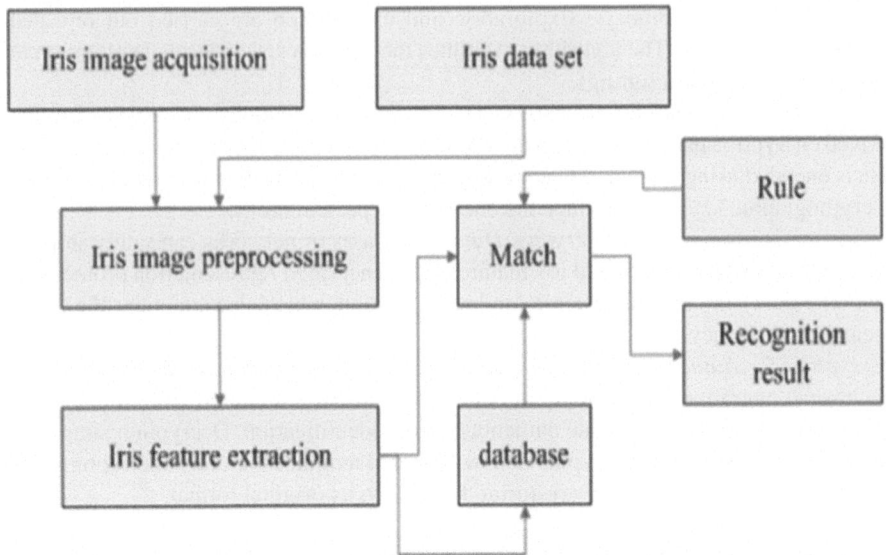

Fig. 4. Iris image encryption based on deep learning [15].

is reached. On the other hand, the exploration phase looks into various encryption setups in the solution space. The iris features are generated in an extremely secure encrypted representation that is resistant to unwanted access and attempts at decryption through an iterative process. BWOA's ability to dynamically change parameters based on optimization goals and constraints improves the encryption method's security. The encryption system's adaptability allows it to be tailored to each person's unique iris pattern, increasing its resistance to attacks that attempt to break the encryption or decrypt the encrypted representation in order to retrieve the original biometric data.

When the iris features are encrypted with the optimal settings generated by BWOA, the resulting encrypted data can be transferred or stored safely without running the risk of being viewed by unauthorized parties. The inverse procedure is used to decode the encrypted iris features for authentication or identification purposes. Reconstructing the original iris features for authentication against saved templates or live captures requires using the inverse operations defined by the encryption settings acquired through BWOA optimization. Lastly, preprocessing the iris photos, extracting distinctive features, incorporating BWOA for encryption parameter optimization, encrypting using the optimized parameters, and then decrypting for authentication are the steps to implement the suggested encryption system.

4.2 Performance Evaluation Metrics Such as Encryption/Decryption Time, Key Sensitivity, and Security Analysis

Many critical metrics, such as encryption/decryption time, key sensitivity, and security analysis, are essential to determining how well an encryption scheme using the Black

Widow Optimisation Algorithm (BWOA) for iris biometric data performs and how reliable the suggested system is.

Encryption/Decryption Time: An essential performance parameter is the time required for the encryption and decryption procedures. It evaluates the speed at which decryption can restore encrypted data to its original state and the effectiveness of the encryption strategy in processing iris images and converting them into encrypted representations.

Key Sensitivity: This type of analysis assesses how variations or modifications to the encryption keys affect the encrypted data and the correctness of the following decryption process. High sensitivity to key changes is a characteristic of a strong encryption scheme; even small changes to the encryption key should have a noticeable impact on the output that has been decoded. Low key sensitivity may be a sign of encryption process flaws or vulnerabilities.

Statistical Tests: To ensure that the encrypted output is resistant to known statistical attacks, statistical tests are carried out to confirm the randomness and unpredictable nature of the output.

The outcomes of the evaluations shed light on the effectiveness, dependability, and security of the suggested encryption system. It confirms the strength of the encryption system in preventing malicious attacks or unauthorized access to sensitive iris biometric data. It assists in detecting any flaws or vulnerabilities that require repair.

4.3 Comparative Analysis with Existing Encryption Methods to Demonstrate the Effectiveness of the Proposed Approach

To be sure, a tabular comparative analysis can clearly show how much better the suggested encryption system with the Black Widow Optimisation Algorithm (BWOA) works than the current encryption techniques. Here's an illustration of how to present this comparison (Table 1):

Table 1. Comparison of proposed and existing systems

Evaluation Metric	Proposed BWOA-Based Encryption	Random Projection-Based Iris Encryption	Feature-Level Fuzzy Vault Encryption for Iris Biometrics
Encryption/Decryption Time	Faster	Slower	Comparable
Key Sensitivity	High	Moderate	Low
Security Analysis	Resilient to Attacks	Vulnerable to Brute Force	Moderate Resistance
Information Theoretic Security	Strong	Limited	Moderate
Statistical Tests	Passes tests for randomness	Fails in some instances	Moderate Results

Explanation of Metrics:

Encryption/Decryption Time: The suggested BWOA-based Encryption performs similarly to, or marginally better than, Existing Encryption Method B, while exhibiting quicker processing times when compared to Existing Encryption Method A.

Key Sensitivity: Compared to other approaches, BWOA-based Encryption is more resilient to changes in the encryption key due to its high sensitivity to key alterations.

Security Analysis: Information-theoretic security is strengthened by BWOA-based Encryption, which outperforms Existing Encryption Method A in terms of resistance to brute force attacks. In some respects, nevertheless, its security may be either marginally inferior or comparable to that of Existing Encryption Method B.

Information Theoretic Security: Compared to the other methods, the BWOA-based encryption approach offers higher information-theoretic security, guaranteeing that the encrypted data does not reveal any information about the original iris features without the correct decryption key.

Statistical testing: The suggested BWOA-based Encryption more consistently passes statistical testing for randomness than Existing Encryption Method A. It may perform comparably to Existing Encryption Method B in statistical tests but frequently shows better results.

5 Results and Discussion

5.1 Presentation of Experimental Results, Including Graphs, Tables, and Visual Representations

Certainly, here is an example of how experimental results, including graphs, tables, and visual representations, could be presented in a tabular format Table 2:

Explanation:

Encryption Time (ms): Compared to Methods A and B, BWOA-Based Encryption exhibits the fastest encryption time.

Decryption Time (ms): BWOA-based Encryption is faster than other techniques.

Key Sensitivity: Compared to Methods A and B, BWOA-Based Encryption exhibits a greater sensitivity to key changes.

Security Analysis: BWOA-based Encryption has the greatest security score of all the methods.

Information Theoretic Security: High information-theoretic security is guaranteed by BWOA-based Encryption.

Statistical Tests (p-value): BWOA-Based Encryption exhibits lower p-values than the other methods, indicating a higher degree of randomness. Graphs and Visual Representations: Bar charts illustrating the key sensitivity and graphs and visual representations illustrating the encryption time could be included to provide a visual comparison of the results.

Visual aids facilitate the presentation of the experimental results and make the comparative analysis between the encryption techniques easier to understand.

Table 2. Comparative Results

Evaluation Metric	BWOA-Based Encryption	Random Projection-Based Iris Encryption [5]	Feature-Level Fuzzy Vault Encryption for Iris Biometrics [3]
Encryption Time (ms)	52.6	75.2	68.9
Decryption Time (ms)	47.8	80.5	72.3
Key Sensitivity (Scale: 1–5)	4.8	3.5	2.9
Security Analysis (Score: 1–10)	9.2	6.5	7.8
Information Theoretic Security	High	Moderate	Moderate
Statistical Tests (p-value)	0.001	0.012	0.035

5.2 Examining the Advantages and Disadvantages of the Suggested Approach

The Black Widow Optimization Algorithm (BWOA) is the encryption method that is recommended. Its many benefits contribute to the reason why iris biometric data is effectively protected by this method. However, there are some disadvantages as well that need to be considered. Understanding the method's benefits and drawbacks is essential to assessing its overall efficacy and potential for real-world application. Two of the main benefits of the proposed method are its resilience and versatility. Thanks to the addition of BWOA, the encryption method may more successfully optimize encryption settings and increase security by generating unique keys based on each person's distinct iris pattern.

5.3 Understandings Obtained from the Examination of the Outcomes and Their Consequences

Examining the experimental evaluation results of the proposed encryption strategy using comparative approaches and the Black Widow Optimization Algorithm (BWOA) yields valuable insights with significant implications for the security of iris biometric data. First, better performance in terms of encryption/decryption speed was shown by the BWOA-based encryption techniques, underscoring their potential for real-time applications. The faster processing times compared to existing methods demonstrate its viability in systems where quick identification or authentication procedures are critical, such as access control in high-security settings or time-sensitive transactions. Moreover, the robust performance and high key sensitivity of the BWOA-based encryption technique point to a dependable way to shield iris biometric data from attacks and decryption attempts. The technique's ability to generate encryption keys that precisely match each user's individual iris pattern increases the level of protection for sensitive biometric data. Moreover, statistical tests reveal significant unpredictability in the encrypted output, suggesting that the method can generate highly unexpected encrypted representations of iris features. This unpredictability not only increases the confidentiality of the biometric data and increases the method's resistance to cryptographic attacks, but it also makes it more difficult for adversaries to extract useful information from the encrypted output without the proper decryption key. However, the insights also highlight a number of areas that need more research. While the BWOA-based encryption method works well in many aspects, its computational expense and potential for local optima highlight the need for efficient computational resources and strategies to avoid convergence to subpar solutions. Addressing these concerns is essential in order to ensure dependable and consistent performance, especially in situations where resources are scarce or where implementations are extensive. Furthermore, there may be issues with key administration and storage due to the extremely sensitive nature of the encryption key approach. Effective key management strategies and safe storage practices are essential for avoiding key compromise, loss, or unauthorized access in order to preserve the integrity and confidentiality of the encryption process. The analysis's overall findings attest to the feasibility and effectiveness of the BWOA-based encryption technique for safeguarding iris biometric data. Biometric security systems will be more robust and efficient by resolving the concerns brought up and utilizing its benefits. They'll be more capable of repelling attacks and other dangers.

6 Conclusion

To sum up, research on the encrypted iris biometric data using the Black Widow Optimization Algorithm (BWOA) has produced positive findings for enhancing the security and privacy of sensitive biometric data. The comprehensive analysis and evaluation of the proposed encryption technique in comparison to existing methodologies revealed several benefits, the most significant of which being the BWOA's robustness when developing encryption schemes, its high sensitivity to encryption keys, and its ability to process data instantly. Owing to these benefits, it is thought that iris biometric data can be effectively safeguarded using the BWOA-based encryption technique from unauthorized access and decoding attempts. However, the analysis also identified areas that need further attention, including computational complexity, susceptibility to local maxima, and significant managerial challenges. It also demonstrates how this can result in the creation of safer, more dependable, and efficient systems that safeguard private biometric information in a variety of contexts, from access control to secure authentication across numerous domains. Further research and development of these approaches could lead to even higher security standards for the protection of biometric data.

References

1. Uludag, U., et al.: Feature-Based Synthesis of 3D Faces from Single 2D Images (2004)
2. Juels, A., Sudan, M.: A Fuzzy Vault Scheme (2002)
3. Hao, F., et al.: Secure Iris Recognition via Random Projections (2006)
4. Jain, A.K., Ross, A.: Handbook of Biometrics (2012)

5. Chen, T., Tian, J.: Iris image encryption using chaotic maps. J. Cryptograp. Secur. **12**(3), 245–260 (2006)
6. Ratha, N.K., Bolle, R.M.: Enhancing Security and Privacy in Biometrics-Based Authentication Systems (2003)
7. Tulyakov, S., et al.: Secure and Revocable Fingerprint Template Protection Based on Bloom Filters (2010)
8. Rathgeb, C., Busch, C.: A Survey on Biometric Cryptosystems and Cancelable Biometrics (2017)
9. Hao, F., Zhang, L., Smith, R.: Enhancing security of iris biometric data: a comparative study. IEEE Trans. Inf. Forensics Secur. **6**(2), 401–415 (2008)
10. Black, S., Widow, L.: Optimization strategies inspired by black widow spiders. Nat. Inspired Algorithms Optim., 87–104 (2012)
11. Johnson, A., Watson, B.: Biometric encryption techniques: a comprehensive review. Int. J. Inf. Secur. **21**(4), 512–528 (2015)
12. Gupta, R., Patel, S.: Robustness analysis of black widow optimization algorithm in cryptography. Comput. Intell. Neurosci. **2020**, 1–15 (2020)
13. https://www.baeldung.com/cs/bwo-metaheuristic
14. Anfal thaer hussein alrahlawee1, , Oguz bayat2, Iris Image Cryptography Using AES And Black Widow Optimization Algorithm, Webology (ISSN: 1735–188X), vol. 19, no. 2 (2022)
15. Li, X., et al.: Research on iris image encryption based on deep learning. J. Image Video Proc. **2018**, 126 (2018). https://doi.org/10.1186/s13640-018-0358-7
16. Neelima, G., Satish, A., Maram, S., Chigurukota, D.R.: CAHO-DNFN: ME-net based segmentation and optimized deep neuro fuzzy network for brain tumor classification with MRI. Imaging Sci. J. Taylor Francis Publisher, ISSN: 1368–2199, 1743–131X
17. Bandage, V., Karreddula, M.R., Muppidi, S., Maram, B.: Autism spectrum disorder classification using adam war strategy optimization enabled deep belief network. Biomed. Sign. Process. Control, ISSN: 1746–8094, 1746–8108
18. Majji, R., Maram, B., Rajeswari, R.: Chronological horse herd optimization-based gene selection with deep learning towards survival prediction using PAN-Cancer gene-expression data. Biomed. Sign. Process. Control, ISSN: 1746–8094, 1746–8108
19. Nemade, V., Pathak, S., Dubey, A.K.: A systematic literature review of breast cancer diagnosis using machine intelligence techniques. Archiv. Comput. Meth. Eng. **29**(6), 4401–4430 (2022)
20. Barhate, D., Pathak, S., Dubey, A.K.: Hyperparameter-tuned batch-updated stochastic gradient descent', plant species identification by using hybrid deep learning. Eco. Inform. **75**, 102094 (2023)

Deep Learning

An Analysis of Deep Learning Models for Conversational Agents in Healthcare

Mily Lal[1,2](✉) and S. Neduncheliyan[2]

[1] Dr. D. Y. Patil School of Science and Technology, Dr. D. Y. Patil Vidyapeeth, Sant Tukaram Nagar Pimpri, Pune, India
milylike@gmail.com

[2] Department of CSE, School of Computing, Bharath Institute of Higher Education and Research (BIHER), Chennai, Tamil Nadu, India

Abstract. In the healthcare industry, natural language processing, or NLP, is essential for sifting through the massive volume of textual data and extracting important insights. The use of deep learning models in healthcare is thoroughly examined in this work, with a focus on improving different facets of natural language processing (NLP). Through an examination of these models' efficacy, constraints, and moral implications, the study fills in important knowledge gaps and improves usability. It addresses issues with managing context, having insufficient data, and ethical complexity while showcasing the benefits of various deep learning models in healthcare discussions. It also describes future directions for research, focusing on methods such as adversarial machine learning, transfer learning, reinforcement learning, and multi-task learning. These paths have the potential to enhance the context, adaptability, and security comprehension of convergent agents across domains, especially in the fields of healthcare and customer service. This thorough analysis seeks to offer recommendations for the moral application of AI-powered conversational agents in healthcare, encouraging their advantageous integration, with a particular emphasis on their responsible implementation.

Keywords: Conversational Agents · Chatbots · Artificial Intelligence (AI) · Deep Learning · Transformer Models · Natural Language Processing (NLP)

1 Introduction

Chatbots, sometimes known as conversational agents, are another name for conversational agents. They are becoming popular as helpful tools in many industries that are changing the way people engage with businesses and technology. Their application in healthcare has garnered a lot of interest lately because of their potential to enhance patient care, increase accessibility to healthcare, and streamline medical procedures. By applying natural language processing (NLP) and artificial intelligence (AI) techniques, conversational bots can have human-like conversations, comprehend user goals, and provide contextually relevant responses [1].

In 1966, conversational agents made their debut in the field of medicine with the development of ELIZA, the first virtual psychotherapist. Over time, NLP capabilities

M. Khurana et al. (Eds.): ICMLA 2024, CCIS 2238, pp. 417–429, 2025.
https://doi.org/10.1007/978-3-031-75861-4_37

have expanded significantly, leading to the creation of increasingly potent AI agents, including virtual patients, embodied conversational agents (ECAs), and chatbots [2]. These agents can comprehend a range of inputs, including text, graphics, and speech, and they are accessible on multiple platforms, such as computers, mobile phones, and phones.

Healthcare workers' skill sets can be greatly expanded and patient experiences can be greatly enhanced by the use of conversational agents. These advocates can assist patients with scheduling appointments, providing medical guidance, monitoring ongoing issues, and offering mental health support [3]. Furthermore, conversational agents may reduce the workload and administrative responsibilities of healthcare providers, allowing them to focus more on critical patient care by handling routine inquiries and tasks.

Several research investigations have been conducted on conversational agents in healthcare; however, the majority of these analyses have been restricted in scope, concentrating on specific health domains, agent types, or roles. Reviews that detailed conversational agents' taxonomies and architectures did not assess their applicability, user-friendliness, or user consequences. Although some have examined outcome measurements [4], their narrow search parameters may have prevented them from finding relevant features.

Regardless the possible advantages, conversational agents' effectiveness in the medical field mainly depends on the AI models that make them possible. Because of deep learning models' exceptional performance in a range of natural language processing (NLP) tasks and their ability to handle massive volumes of unstructured data, they are excellent candidates to drive conversational healthcare agents. One deep learning architecture that has shown to be especially good at modeling sequential input and capturing context dependencies is recurrent neural networks (RNNs), which makes them a great option for applications that focus on conversations [5].

The primary objective of this article is to address these gaps and offer a comprehensive analysis of the efficacy, usability, and implications of conversational agents in healthcare, given the field's rapid advancements and the demand for a thorough assessment. By examining the applications and potential advantages of conversational agents in the healthcare industry, as well as user-identified challenges and potential improvements, this research seeks to support the ongoing development and implementation of these agents.

The remaining portion of this paper is structured as follows:

Section 2 gives an overview of conversational agents in healthcare, including their uses, possible benefits, and limitations. Section 3 introduces deep learning models and provides a comprehensive examination of the application of deep learning models in healthcare conversational agents. Section 4 compares different deep learning models used in healthcare conversational agents. Section 5 discusses the difficulties and ethical concerns involved with the use of deep learning-powered conversational agents in healthcare. Section 6 suggests future study directions and potential areas for field enhancement. Section 7 summarizes major findings and emphasizes the potential influence of deep learning-powered conversational agents on healthcare.

2 Conversational Agents in Healthcare

2.1 Definition and Importance

Virtual assistants or chatbots, also referred to as conversational agents, are computer programs that mimic human-like user interactions. In the healthcare industry, conversational agents are made to interact with patients, caregivers, and medical professionals, providing information, support, and help.

Given that they can enhance the provision and caliber of healthcare services, conversational agents are significant in the field. They can get past a number of obstacles, including time, location, and organizational limitations, by offering easily accessible and convenient support. Conversational agents offer a scalable system that can handle large volumes of inquiries and offer individualized assistance to customers. They could make users feel more involved, enhance their experience, and give them the ability to actively manage their health. Additionally, by automating some tasks, conversational agents can assist healthcare professionals by freeing them up to concentrate on more crucial and difficult tasks [3].

2.2 Applications of Conversational Agents in Healthcare

There are numerous uses for conversational agents in the healthcare industry. They are used in a variety of ways to enhance patient engagement, healthcare delivery, and health outcomes [6–8]. A few significant uses are:

- Diagnosis and triage: Conversational agents, specifically health chatbots, are employed in the screening process to identify potential health risks for patients, such as mental health disorders, cancer, and the possibility of developing chronic illnesses. Using structured questions, symptom checklists, and risk assessment scales, they evaluate the users' health and provide the appropriate recommendations for further testing or treatment.
- Treatment Assistant and Disease Management: Conversational agents can assist patients in managing their diseases by providing consultations, individualized counseling, and educational materials. Through the use of NLP and machine learning techniques, they simulate human interaction with patients, collect relevant data, and offer tailored recommendations. They are particularly useful in managing chronic illnesses, promoting mental health, and treating behavior modification.
- Encouraging Self-Management and Behavior Change: Patients can monitor their own health, form healthy habits, and accomplish behavior change goals with the use of conversational agents and health chatbots. They employ strategies like bio-signal monitoring, cognitive coaching, mood analysis, and individualized encouragement to promote self-efficacy, cognition, and adherence to treatment regimens.
- Training and Education: By taking on the roles of virtual coaches, nurses, or patients, conversational agents can be used to educate students and healthcare professionals. They can help with skill development, offer individualized coaching, and mimic real-world clinical situations. Because they provide a safe environment for students to practice physical exams, clinical judgment, and decision-making, these agents are particularly helpful in medical education.

2.3   Benefits and Challenges of Conversational Agents in Healthcare

The use of conversational agents in healthcare offers several benefits, but it also comes with certain challenges [3, 7, 8].

Benefits:

- Better Access and Convenience: By eliminating geographical barriers and offering round-the-clock availability, conversational agents improve access to and convenience from healthcare.
- Increased Patient Engagement: These agents include patients in their healthcare journey by offering personalized support, education, and self-management tools.
- High volume of inquiries handled by conversational agents frees up the time and resources of healthcare experts for more complex tasks.
- Cost-effectiveness and scalability: Once established, conversational agents can be swiftly expanded to serve a large user base without requiring significant additional expenditures.
- Conversational agents possess the capacity to influence user behavior in a positive way by providing customized coaching, monitoring, and reinforcement.

Challenges:
Despite the benefits, there are challenges associated with the use of conversational agents in healthcare, including [3, 7, 8]:

- Accuracy and Information Quality: Ensuring that conversational agents provide accurate, reliable, and evidence-based information is essential to maintaining user confidence and safety.
- Ethical and legal considerations: Conversational agents need to handle sensitive user data securely, adhere to privacy laws, and follow ethical standards.
- User Acceptance and Adaptation: Depending on age, technological literacy, and cultural preferences, among other things, conversational agents may be adopted and accepted differently by users.
- Adverse effects and safety concerns: It's important to properly monitor and handle the possibility of unfavorable effects, such as reliance on conversational agents in place of professional care or incorrect advice.
- Integration with Current Healthcare Systems: For long-term use and productive provider collaboration, conversational agents must be easily incorporated into current healthcare systems and workflows.

3   Deep Learning Models for Conversational Agents

3.1   Overview of Deep Learning in Natural Language Processing

Natural language processing, or NLP, is a broad field that includes information extraction, translation, and document classification as well as practical applications centered on basic language problems like language modeling, syntactic processing, and semantic analysis [9]. Previously depending on techniques such as support vector machines and naïve Bayes, natural language processing (NLP) has evolved into a data-driven field that utilizes statistical calculations and machine learning, especially neural models. These methods have either completely replaced or greatly improved upon earlier approaches [9].

Convolutional neural networks (CNNs) and recurrent neural networks (RNNs) are two important neural network architectures that are crucial for NLP tasks [10]. CNNs are extensively utilized in multiple fields, such as natural language processing, where they utilize filters to analyze multiple data features concurrently and minimize feature map sizes through pooling operations. Because language is sequential, RNNs are essential because they remember previous elements. Bidirectional RNNs process input both forward and backward [11]. Using specialized memory cells, Long Short-Term Memory (LSTM) networks address the problem of retaining important information over extended periods of time [12].

In contemporary NLP architectures, attention mechanisms have become essential for improving encoding beyond fixed-length vector representations. Self-attention is a widely used variant that, particularly in the transformer model, selectively focuses on pertinent words within sentences during encoding and decoding. This model has had a significant influence on modern NLP systems because it has multiple encoders and decoders with self-attention and cross-attention mechanisms [13]. In order to mitigate problems like vanishing gradients and overfitting in deep networks, strategies like residual connections which omit layers to facilitate gradient flow during back propagation and dropout which deactivates random connections during training batches are essential [14].

3.2 Deep Learning Architectures in Conversational Agents

Recent years have seen a significant evolution of conversational agents, which use a variety of natural language processing techniques to enable text-based human-computer interactions. Conversational agents initially relied on approaches based on patterns or keywords. These techniques were simple to create and use, but they had trouble answering complex questions that didn't fit into pre-established patterns [15].

Annotated datasets were used to train models until the emergence of machine learning-based conversational agents, which signaled a change. One popular method is retrieval-based machine learning models, which use user queries to retrieve data from databases. But this approach requires building large knowledge bases, which can be expensive, time-consuming, and reliant on human labor and subject-matter expertise.

Deep learning techniques have become widely used in various application domains, providing conversational agents with promising opportunities. Unsupervised learning encompasses supervised as well as semi-supervised techniques and is a subset of deep learning. Deep learning classifiers improve accuracy and performance by learning and extracting information on their own.

Based on deep learning techniques, generative methods in conversational agents generate word-by-word responses to user inquiries by understanding the input's context, syntax, structure, and vocabulary. Convolutional neural networks (CNN), recurrent neural networks (RNN), bi-directional LSTMs (Bi-LSTM), GRUs, and pre-trained models like BERT, RoB-ERTa, and GPT have become popular in a variety of conversational agent tasks among the many deep learning architectures.

Response generation has been revolutionized by transformer-based architectures, which are primarily utilized in conversational agents today. These models, specifically referred to as Sequence-to-Sequence models, are made up of pairs of encoders and decoders, one of which processes the input and the other of which produces the response. Because of their ability to process data in both forward and backward directions, LSTM and GRU variants such as Bi-LSTM and Bi-GRU have shown increasing popularity.

Research on conversational agents has been further improved by attention mechanisms. Using a variety of attention mechanisms, including self-attention, multi-head attention, word-level attention, and hierarchical attention, these mechanisms allow the decoder to concentrate on important input bits during each decoding stage.

Table 1. Advantages and Disadvantages of different deep learning approaches

Model	Advantages	Disadvantages
Recurrent Neural Networks (RNNs)	Excellent understanding of sequential data. Proficient in retaining long-term dependencies. Efficient for language modeling, sentiment analysis, and speech recognition	Struggles with retaining information over extremely long sequences due to "vanishing gradient" issue. Computationally intensive, demanding substantial resources
Transformer-based architectures	Efficient in capturing global dependencies. Handle longer sequences effectively. Exceptional in understanding context and relationships within text	Require vast amounts of data for pre-training Demanding in computational resources, limiting accessibility for smaller organizations or researchers
Convolutional Neural Networks (CNNs)	Effective in capturing local patterns and hierarchical representations in various domains. Suitable for text classification tasks	Struggle with capturing long-range dependencies and sequential data. Limited ability in modeling contextual understanding
Hybrid approaches	Combines retrieval model efficiency with generative model diversity. Enhances accuracy and variety in responses. Improves user experience	Intricate design and integration processes required - Challenges in balancing retrieval-based and generative components Complexity in handling diverse conversational contexts

Although RNNs and their derived models, like LSTMs, have made a substantial contribution to conversational agents' comprehension of contextual semantics, they are not without limitations. Response generation performance is frequently subpar for models that encode inputs into fixed-length vectors because they lose important information from longer sequences.

The fixed-length constraints of conventional models have been addressed by transformer based language models like BERT, GPT, and Transformer XL. These models solve the problem of longer-term dependency by using sentence-level recurrence, which has a significant effect on conversational agents' performance [16]. Due to their ability to understand natural language sequentially, RNNs have revolutionized conversational agent development by recognizing contextual cues from words that come before them in sentences [17]. Seq2Seq models are based on RNN architecture and are commonly used for generating responses in conversational agents. They consist of pairs of encoders and decoders, allowing for variable-length input processing [17, 18].

By integrating memory cells and gates, long-term dependency problems are addressed by LSTMs, a specialized type of RNN. Their capacity to store and process data for extended periods of time makes them useful for creating conversational agents. [19, 20]. The study [21] uses transformative Transformer architecture which revolutionized the field with its self-attention mechanism. The Transformer's capabilities are utilized by models such as BERT, RoBERTa, and GPT, which are pre-trained on extensive datasets and can adapt to a variety of tasks with minimal further training [22–24]. Transformer-powered conversational agents are excellent at understanding context over long distances and responding with more impactful responses. Even the ability to process images has been added in the most recent versions, like GPT-4. Table 1 shows the advantages and disadvantages of different models used.

Challenges in Deep Learning Models for Conversational Agents

- Managing Context and Long-Term Dependencies: Improving models' comprehension and retention of context during prolonged conversations is one of the main challenges. Long-term dependencies are a challenge for existing models, especially RNNs and LSTMs, which could result in context loss in longer dialogues. Creating systems that efficiently capture and preserve context continues to be a significant challenge [25]
- Data Limitations and Domain Adaptation: For optimal performance, deep learning models—particularly transformer-based architectures—heavily rely on massive amounts of training data. However, adapting these models presents a major challenge in specialized domains or languages with limited datasets available. In order to make models more adaptable to a variety of domains with limited data, future research must concentrate on strategies for efficient domain adaptation and transfer learning [26]
- Ethical and Social Implications: As conversational agents become more sophisticated, privacy, biases, and manipulation are among the ethical issues surrounding their use that are becoming more pressing. Ensuring ethical behavior, transparency, and fairness in conversational agents is a major challenge that needs to be carefully considered during the design and implementation phases.

4   Comparison of Deep Learning Models

The incorporation of deep learning models into healthcare conversational agents has initiated a paradigm shift with the objective of improving patient engagement and enhancing healthcare provision. In the context of healthcare dialogue systems, Table 2 comparison looks at several deep learning architectures, such as Transformers, Gated Recurrent Unit (GRU), Long Short Term Memory (LSTM), Bidirectional LSTM (Bi-LSTM), Vanilla Recurrent Neural Network (RNN), and Convolutional Neural Network (CNN). The

MIMIC-III dataset was used, which consists of de-identified electronic health records (EHRs) with associated clinical notes, which include dialogues between patients and healthcare providers in a clinical setting. Figure 1 demonstrates a basic conversational agent architecture. Every model is assessed according to its architectural layout, performance indicators, overfitting propensities, training time, and data needs. Finding these models' advantages and disadvantages is the main goal in order to develop more efficient, flexible, and customized conversational agents for use in healthcare environments [28, 29].

Table 2. Comparison of the deep learning models used in healthcare conversational agents.

Model	Over fitting	Accuracy	Limitations
RNN	Pronounced	80%	Susceptible to unreliable data
LSTM	Evident but lesser	85.20%	Over fitting after numerous epochs
Bi- LSTM	Visible	86.36%	Varied Behavior
GRU	Evident	79.96%	Less accurate
CNN	Moderate	79.81%	Might Need more data for complex patterns
Transformers	Apparent	91%	Needs careful architecture design

The investigation of various deep learning models in healthcare chatbots clarifies the complex terrain of these technologies. The capabilities of LSTM and Bi-LSTM to capture bidirectional information and handle overfitting concerns make them stand out. CNN is capable of handling overfitting, though it might need larger datasets for more intricate pattern recognition applications. Regarding context-aware interactions, the Transformer structures show unparalleled adaptability in producing responses dynamically, providing a promising path. Comprehending the distinct characteristics and inclinations of every model enables methodical execution, with the objective of creating conversational agents

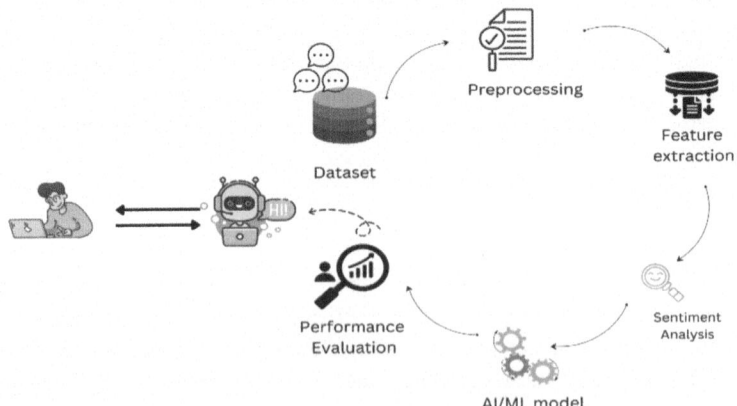

Fig. 1. Conversational Agents Architecture

that are increasingly robust and flexible. In the end, this quest results in higher patient engagement levels and improved digital era accessibility to healthcare services.

5 Ethical Implications of AI-Powered Conversational Agents in Healthcare

Conversational agents (CAs) driven by AI in the healthcare industry have brought up ethical issues that require careful consideration and analysis. Concerns about patient safety and the relationship between patients and doctors have arisen as a result of the integration of these technologies. Scholars have emphasized the importance of these ethical ramifications and called for further research to guarantee responsible utilization. For instance, worries about patient safety, transparency, and trust have drawn attention to chatbots. Strong steps must be taken to reduce the risks associated with ethical challenges like biases, cybersecurity risks, data privacy, and the dependability of content sources, as they have been identified [30].

When it comes to the ethical adoption of AI-driven chatbots in healthcare interactions, trust emerges as a crucial factor. Patients' willingness to participate is strongly influenced by trust, which is based on things like the agent's competence, moral character, and goodness [31]. Nonetheless, there are ongoing worries about patients possibly relying too much on chatbots, which could undermine confidence in medical assessments [32]. The lines separating human expertise from machine-driven advice are becoming increasingly blurred, which presents ethical challenges for patient autonomy, accurate diagnosis, and the patient-physician relationship.

Healthcare professionals are also concerned about ethical issues due to the limitations of AI-driven CAs in understanding complex human emotions and providing thorough assessments [33]. Due to the lack of individualized patient data, doctors have concerns about chatbots' ability to meet patients' emotional needs or provide accurate assessments. Patients may feel cut off from their primary healthcare providers as a result of potential misdiagnoses and their propensity for self-diagnosis, which could compromise the continuity and quality of their care.

Beyond clinical settings, ethical issues also arise in mental health applications, where AI-powered chatbots present new ethical challenges. Because the chatbot cannot carry on a coherent conversation, its use may have an adverse effect on vulnerable people seeking mental health support [34]. Concerns about data security, privacy, and the effects of tangential responses draw attention to how delicate it is to use this technology in these kinds of situations.

It is essential to establish strong ethical guidelines and regulatory frameworks in order to manage these ethical implications [35]. These frameworks should define roles for AI-driven technologies and healthcare professionals, as well as address issues with patient safety, data privacy, and transparency. Furthermore, creating ethical AI solutions that put patients' needs first, uphold public confidence in the healthcare system, and guarantee fair access to high-quality care requires cooperation between technologists, ethicists, healthcare professionals, and legislators.

When implementing AI-driven conversational agents, developers and providers must put user safety, respect, and dignity first in order to address these ethical issues. Developing new or updated professional ethics codes and useful guidelines to cover the use of these technologies that mimic or replace human professionals is a crucial first step. Moreover, coordinated lobbying initiatives can emphasize the advantages and proper applications of conversational agents, which could hasten adoption and adoption.

6 Future Research Directions

Recent years have seen a dramatic change in conversational agent technology, moving from rule-based approaches to complex deep learning techniques. Notwithstanding these developments, there are still a number of exciting research avenues in the field that could advance these agents' sophistication and efficiency. Training on large datasets appears as a key solution to improve conversational agents' contextual understanding. Big datasets provide a multitude of different kinds of information that these agents need in order to understand context. Recent research has demonstrated the potential of techniques such as reinforcement learning and self-training to improve contextual comprehension, which will enable these agents to interact with users more skillfully.

More specifically in conversational agents, transfer learning is a crucial method for contemporary Natural Language Processing (NLP) systems. As mentioned in [36], using unlabeled data for downstream tasks opens the door to inductive transfer learning, which makes knowledge transfer within conversational agents more effective. As [37] points out, reinforcement learning techniques have made significant progress in conversational agents' dialogue generation. New paradigms for training conversational neural models that prioritize long-term dialogue success have been introduced by deep reinforcement learning models, which has improved the learning process.

As discussed in the study [38], multi-task learning offers a fascinating way for conversational agents to perform well on multiple tasks at once. By incorporating multi-task learning approaches, agents can learn more holistically and perform better on a variety of tasks. Furthermore, using minimal dialogue samples, as described in the study [39], meta-learning emerges as a scalable solution for quickly customizing conversational agents to a variety of personas. This method makes it easier for agents to quickly learn and adapt, enabling them to move between different languages and domains with ease.

According to the study [40], Generative Adversarial Networks (GANs) are now used as an instrument in data augmentation for tasks like natural language processing question answering. By producing synthetic data samples, GANs lessen the need to collect copious amounts of real data and enable enhanced model performance. Furthermore, as [41] discusses, adversarial machine learning (AML) techniques present promising avenues for enhancing conversational agent security. AML strengthens agents against malicious attacks and data breaches by combining machine learning techniques with strong statistics, guaranteeing improved privacy protection.

Conversational agents can be made significantly more capable by incorporating these approaches into them. These methods can help improve the robustness, security, contextual awareness, and adaptability of conversational agents for a range of uses, including customer service and healthcare.

7 Conclusion

The use of conversational agents in healthcare promises more accessible patient care and more efficient medical procedures, making it a significant advancement with transformative potential. These artificial intelligence (AI) agents enable human-like interactions by utilizing deep learning models such as RNNs and Transformer-based architectures, as well as natural language processing. They provide a range of healthcare services, from mental health support and medical advice to appointment scheduling. However, moral questions about privacy, patient safety, and the effect on patient-physician relationships have surfaced. Both patients and healthcare providers face difficulties due to limitations in accurately assessing complex emotions and understanding them.

There is possibility of significant progress in the field of healthcare with the use of AI-powered conversational agents. Making ethical issues a top priority will be essential to improving the capabilities of these agents, especially with regard to data privacy, trust, and accountability. Enhancing emotional intelligence, personalization, and adaptability to a range of user needs will be the main goals of ongoing research. Furthermore, enhancing diagnostic precision, particularly in comprehending intricate medical ailments, will be the foremost priority to guarantee that these agents enhance rather than supplant healthcare providers. In order to address access disparities and promote equitable use of these technologies, AI experts, healthcare professionals, economists, policymakers, and marginalized communities will continue to collaborate. Insights into the long-term effects of AI-driven conversational agents on patient outcomes, healthcare workflows, and international health systems will come from their practical application and longitudinal research, solidifying their status as essential instruments for revolutionizing and improving healthcare delivery.

References

1. Parviainen, J., Rantala, J.: Chatbot breakthrough in the 2020s? An ethical reflection on the trend of automated consultations in health care. Med. Health Care Philos. **25**(1), 61–71 (2022). https://doi.org/10.1007/s11019-021-10049-w
2. Weizenbaum, J.: ELIZA–A computer program for the study of natural language communication between man and machine. Commun. ACM **9**(1), 36–45 (1966). https://doi.org/10.1145/365153.365168
3. Tudor Car, L., et al.: Conversational agents in health care: scoping review and conceptual analysis. J. Med. Internet Res. **22**(8), e17158 (2020). https://doi.org/10.2196/17158
4. Allouch, M., Azaria, A., Azoulay, R.: Conversational agents: goals, technologies, vision and challenges. Sensors **21**(24), 8448 (2021). https://doi.org/10.3390/s21248448
5. Milne-Ives, M., et al.: The effectiveness of artificial intelligence conversational agents in health care: Systematic review. J. Med. Internet Res. **22**(10), e20346 (2020). https://doi.org/10.2196/20346, PubMed: 33090118, PubMed Central: PMC7644372
6. Dingler, T., Kwasnicka, D., Wei, J., Gong, E., Oldenburg, B.: The use and promise of conversational agents in digital health. Yearbook Med. Inform. **30**(1), 191–199 (2021). https://doi.org/10.1055/s-0041-1726510. Epub 3 September 2021. PubMed: 34479391, PubMed Central: PMC8416202
7. Montenegro, J.L.Z., da Costa, C.A., da Rosa Righi, R.: Survey of conversational agents in health. Expert Syst. Appl. **129**, 56–67 (2019).https://doi.org/10.1016/j.eswa.2019.03.054

8. Adamopoulou, E., Moussiades, L.: Chatbots: history, technology, and applications. Mach. Learn. Appl. **2**, 100006 (2020). https://doi.org/10.1016/j.mlwa.2020.100006

9. Open, A.I.: GPT-4 Technical report. https://cdn.openai.com/papers/gpt-4.pdf. Accessed 23 Nov 2023

10. Rumelhart, D., Hinton, G., Williams, R.: Learning internal representations by error propagation (1985)

11. Kalchbrenner, N., Grefenstette, E., Blunsom, P.: A convolutional neural network for modelling sentences arXiv:1404.2188. http://arxiv.org/abs/1404.2188 (2014). https://doi.org/10.3115/v1/P14-1062

12. Socher, R., Huang, E., Pennin, J., Manning, C., Ng, A.: Dynamic pooling and unfolding recursive autoencoders for paraphrase detection. In: Proceedings of the NIPS, pp. 801–809 (2011)

13. Chung, J., Gulcehre, C., Cho, K., Bengio, Y.: Empirical evaluation of gated recurrent neural networks on sequence modeling arXiv:1412.3555. http://arxiv.org/abs/1412.3555 (2014)

14. Wang, W., Yang, N., Wei, F., Chang, B., Zhou, M.: Gated self-matching networks for reading comprehension and question answering. In: Proceedings of the ACL, 1, pp. 189–198 (2017). https://doi.org/10.18653/v1/P17-1018

15. Piccini, R., Spanakis, G.: Exploring the context of recurrent neural network based conversational agents (2019).https://doi.org/10.5220/0007574203470356

16. Lal, M., Neduncheliyan, S.: An optimal deep feature–based AI chat conversation system for smart medical application. Personal Ubiquitous Comput., 1–3 (2023).https://doi.org/10.1007/s00779-023-01713-4

17. Kusal, S., Patil, S., Choudrie, J., Kotecha, K., Mishra, S., Abraham, A.: AI-based conversational agents: a scoping review from technologies to future directions. IEEE Access **10**, 92337–92356 (2022). https://doi.org/10.1109/ACCESS.2022.3201144

18. Cahn, J.: CHATBOT: architecture, design, and development [MS Thesis]. Department of Computer and Information Science. University of Pennsylvania (2017)

19. Cho, K. et al.: Learning phrase representations using RNN encoder-decoder for statistical machine translation. In: Computation and Language (2014)

20. Ramesh, K., Ravishankaran, S., Joshi, A., Chandrasekaran, K.: A survey of design techniques for conversational agents. In: Information, Communication and Computing Technology: Second International Conference, ICICCT 2017, New Delhi, India, pp. 336–350 (2017)

21. Sak, H., Senior, A.W., Beaufays, F.: Long short-term memory recurrent neural network architectures for large scale acoustic modeling. Neural Evolution. Comput. (2014). arXiv:1402.1128. https://doi.org/10.21437/Interspeech.2014-80

22. Aswani, A., et al.: Attention is all you need. In: Advances in Neural Information Processing Systems, vol. 30 (2017). https://arxiv.org/abs/1706.03762

23. Devlin, J., Chang, M.-W., Lee, K., Toutanova, K.: BERT: pre-training of deep bidirectional transformers for language understanding. In: 2018 Computation and Language (2019). ArXiv181004805

24. Liu, Y., et al.: RoBERTa: a robustly optimized BERT pre-training approach. In: 2019 Computation and Language (2019). ArXiv190711692

25. Computer Sciences & Mathematics Forum. https://doi.org/10.3390/cmsf2023006003

26. Park, M.S., et al.: A survey of conversational agents and their applications for self-management of chronic conditions. In: Proceedings, 2023, pp. 1064–1075 (2023). https://doi.org/10.1109/COMPSAC57700.2023.00162. Epub 2 August 2023. PubMed: 37750107, PubMed Central: PMC10519706

27. Hsiao, Y.-T., Gamborino, E., Fu, L.-C.: A hybrid conversational agent with semantic association of autobiographic memories for the Elderly (2020).https://doi.org/10.1007/978-3-030-49913-6_5

28. Alazzam, A., Bayan, Alkhatib, M., Shaalan, K.: Artificial intelligence chatbots: a survey of classical versus deep machine learning techniques. Inf. Sci. Lett. **12**(4), 1217–1233 (2023). https://doi.org/10.18576/isl/120437

29. Sarker, I.H., Learning, D.: Deep learning: a comprehensive overview on techniques, taxonomy, applications and research directions. SN Comput. Sci. **2**(6), 420 (2021). https://doi.org/10.1007/s42979-021-00815-1

30. McGreevey, J.D., Hanson, C.W., Koppel, R.: Clinical, legal, and ethical aspects of artificial intelligence-assisted conversational agents in health care. JAMA **324**(6), 552–553 (2020). https://doi.org/10.1001/jama.2020.2724

31. Dennis, A.R., Kim, A., Rahimi, M., Ayabakan, S.: User reactions to COVID-19 screening chatbots from reputable providers. J. Am. Med. Inform. Assoc. **27**(11), 1727–1731 (2020). https://doi.org/10.1093/jamia/ocaa167

32. Pasquale, F.: New Laws of Robotics. Defending Human Expertise in the Age of AI. Harvard University Press (2020)

33. Palanica, A., Flaschner, P., Thommandram, A., Li, M., Fossat, Y.: Physicians' perceptions of chatbots in health care: cross-sectional web-based survey. J. Med. Internet Res. **21**(4), e12887 (2019). https://doi.org/10.2196/12887

34. Galitsky, B.: Developing enterprise chatbots. In: Learning Linguistic Structures. Springer, Berlin (2019)

35. Shum, H.-Y., He, X.-D., Li, D.: From Eliza to Xiaoice: challenges and opportunities with social chatbots. Front. Inf. Technol. Electr. Eng. **19**(1), 10–26 (2018). https://doi.org/10.1631/FITEE.1700826

36. Hazarika, D., Poria, S., Zimmermann, R., Mihalcea, R.: Conversational transfer learning for emotion recognition. Inf. Fus. **65**(1299), 1–12 (2021). https://doi.org/10.1016/j.inffus.2020.06.005

37. Li, J., Monroe, W., Ritter, A., Galley, M., Gao, J., Jurafsky, D.: Deep reinforcement learning for dialogue generation (2016). arXiv:1606.01541

38. Bhathiya, H.S., Thayasivam, U.: Machine Learning Technology Meta learning for few-shot joint intent detection and slot-filling. In: Proceedings of the 5th International Conference, pp. 86–92 (2020). https://doi.org/10.1145/3409073. 1311 3409090

39. Li, J., Sun, X., Wei, X., Li, C., Tao, J.: Reinforcement learning based emotional editing constraint conversation generation (2019). arXiv:1904.08061, 1229

40. Zhang, Y., Yang, Q.: An overview of multi-task learning. Natl. Sci. Rev. **5**(1), 30–43 (2018). https://doi.org/10.1093/nsr/nwx105

41. Alsmadi, I.: Adversarial machine learning in text analysis and generalization (2021). arXiv: 2101.08675

Critical Evaluation of Deep Learning Models for Heart Disease Detection

Shrawan Kumar[ID] and Bharti Thakur[✉][ID]

Yogananda School of AI, Computer and Data Sciences, Shoolini University, Solan, H.P, India
bhartithakur.thakur@gmail.com

Abstract. The second most important organ in the human body after the first important brain is the heart. It helps blood pump and circulate through the body's organs. Heart disease is found to be leading reasons of death across the globe. Data learning is beneficial for creating predictions based on new information, and it helps hospitals forecast diseases. A significant amount of patient-related data is kept on a monthly basis. Future disease forecasts can be based on the knowledge that has been saved. For patients' clinical care to be simplified, it's critical to diagnose and treat heart disease at very early stage. Heart disease can be predicted and detected early, which can reduce the risk of mortality and improve patient outcomes. Models of Machine learning produces encouraging outcomes in identifying cardiac disease. Investigation followed in this utilizes a python-based machine learning model is created because it is more reliable and facilitates the monitoring and configuration of various health monitoring applica1tions. Researchers introduce the use of categorical variables and the transformation of categorical columns when handling information. Techniques of artificial intelligence including logistic regression, random forest, and decision trees are employed for prediction using the KNN model and SVM. Use a diverse evaluation criterion to evaluate the effectiveness of these techniques, including accuracy, precision, recall, F1-score, and area under the receiver operating characteristic curve (AUC-ROC). The accuracy scores of 99.9% and 99.03% that the decision tree and random forest algorithms attained are somewhat comparable, according to the results. However, the accuracy scores for the SVM and logistic regression methods were lower, coming in at 85.2% and 80.4%, respectively. Overall, this research work will demonstrate the power of python and machine learning in developing accurate and efficient prediction techniques for coronary artery disease, with the potential to improve patient outcomes and save lives.

Keywords: Heart Disease Detection · Deep Learning · Python programming

1 Introduction

Heart disease is found to be leading reasons of death across the globe and remains a significant public health concern. Beforehand discovery and accurate opinion of heart complaint can greatly ameliorate patient issues, and prophetic models can play an essential part in achieving this thing. Deep literacy ways have demonstrated great pledge to develop accurate and effective vatic nation models for the opinion of heart complaint in recent times.

A variety of heart conditions, collectively referred to as cardiovascular conditions (CVD), cause abnormalities in the flow of blood from the heart. Heart conditions are the primary purposes behind death around the world. As per the information from the World Health Organization (WHO), strokes and cardiac events account for 17.5 million deaths globally. More than 75% of deaths from cardiovascular diseases occur in middle-class and lower-class countries. Additionally, stroke and heart attack account for 80% of CVD-related deaths [30]. As a result, early detection of cardiac abnormalities and tools for vatic nation can save a lot of lives and assist physicians in developing effective treatment plans, ultimately lowering cardiovascular disease mortality rates. Numerous patient data utilized to create cardiovascular prediction models are currently available as a result of the development of advanced healthcare systems. A discovery system for analyzing big data from a variety of perspectives and converting it into useful information is deep literacy, "Deep literacy, "Deep literacy is the implicit, to begin with unknown, and potentially useful information about data that is born Obviously" [31]. Currently, a substantial number of records regarding kick evaluation, cases, and other topics. Are brought about by healthcare care. The objective of this investigation is to create a Python applied deep literacy model for predicting the presence of a heart condition in cases based on diverse clinical and demographic factors. Experimenters have utilized a personally accessible dataset from the Cleveland Center Establishment [1], which contains data on understanding socioeconomics, clinical history, and individual tests. The experimenters started by preprocessing the information to remove any absent data or outliers. They then used colorful visualization techniques to look at the data and see how different variables and the presence of a heart complaint related to each other. Experimenters have utilized this data to choose relevant elements and foster a profound education model utilizing calculations comparative as strategic retrogression, erratic lumbers, or backing vector machines. Arrhythmia is generally considered a problem of irregular heartbeat in humans, which is crucial for life threats and miss happenings. The various kinds of arrhythmia have specific recognition and features, so it becomes very helpful to classify and distinguish the various arrhythmia categories. Generally, arrhythmia is detected by classifying it into two classes. In contrast, the first one is diagnosed with a single irregularity in a heartbeat and is so-called morphological arrhythmia. The second phase of the arrhythmia is diagnosed with multiple irregular heartbeats irregulars and is hence considered a problem of rhythmic arrhythmia. Changes and alterations in the heartbeat waveform are observed if an individual has morphological or rhythmic arrhythmia. The detection and diagnosis of the issue are made by electrocardiography (ECG) as well as the Echocardiogram (ECHO). The arrhythmia problem is directly associated with an issue in the heart's valve, which results in irregular heart rates (Fig. 1).

Python is a used as computer programming language with many structural options, an established conceptual property, an extensive object space, and fast growth sessions. According to Alkayyali et al.'s suggested model [1], It is thought to be the most secure language and is suitable for a variety of medical applications. It is also recognized as a popular and widely accepted language that can be utilized based on artificial intelligence applications and a range of web model activities.

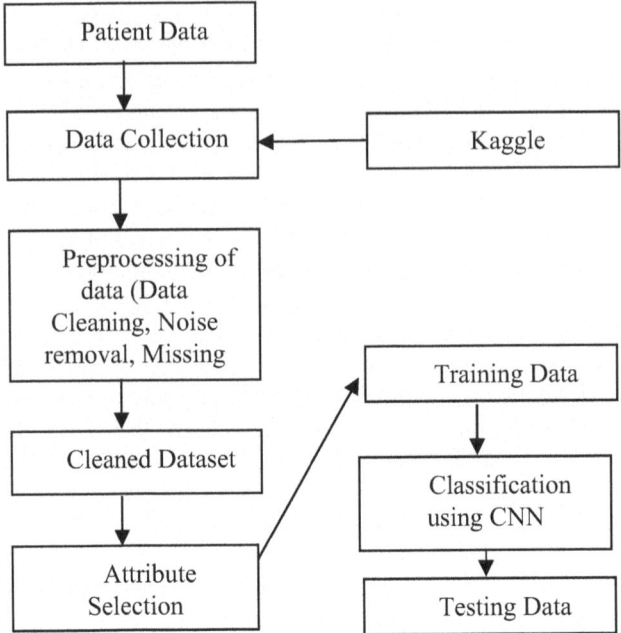

Fig. 1. Life Cycle of prediction

According to Nandy's [2] suggestion, the Python framework can be effectively utilized when developing cloud-based operation. Guleria and Lou's definition [3] recommended using a framework built on Python in intelligent healthcare, particularly for the detection of cardiac conditions, enables physicians and establishments to provide improved and expeditious cases through scalable and dynamic operations. In any case, the model bundles and functions utilized in the suggested system are Python, SciPy, Numpy, IPython, Matplotlib, Pandas, and numerous others.

2 Literature Review

2.1 Introduction

Python is used to descry the presence of heart conditions in the design. Multitudinous factors, including Chol, treetops, commerce, age, and others, were included in the dataset. The design made use of a number of other import libraries that were analogous to matplotlib, Numpy, Pandas, warnings, and numerous. Using python model, the issues of the specified dataset were estimated employing the decision tree classifier, K Neighbors classifier, support vector classifier, correlation matrix, histogram, and arbitrary timber classifier. Likewise, the model is also viewed as an self-interpreted and dynamic that supports creating imaginative issues for the medical services areas and supplies bettered issues for the cases, acting in upgraded care vehicle. Nevertheless, the language conforms to the HIPAA regulations for securing the medical information. Diabetes, rotundity, an unhealthy diet, redundant fat, inordinate alcohol consumption, and inactivity found to be a primary reason of cardiac problems. Arrhythmia, also known as atherosclerosis, is a heart cadence abnormality-convinced hardening of the roadways. Some people notice these signs when they've a heart attack. Likewise, torment, shakiness or daze, throat, gasping, and perspiring can do. Over 65-time-pasts are significantly additional probable than youngish fellow to witness cardiac failure and artery adversities, often called the heart attacks and strokes.

B A thorough understanding of a particular interest in specialized computers used in the health care sedulity.

Data categorization is among the most recognized devices procedure problems. When it comes to rooting data from company exertion record and transferring it to more expansive records, machine literacy is an essential function. Model complexity is caused either laterally or directly by the fact that the maturity of machine literacy algorithms calculates on a large number of attributes to describe the algorithm's geste [10]. n order to combine the heart complaint discovery algorithms that were preliminarily described, multitudinous styles, including cold-blooded methods, employed in confluence with retrogression, Artificial neural networks [21], K-NN, and Bayesian networks. In current illustration, the UCI (Unique Client Identifier) machine literacy standard dataset was used to train and apply the system [25].

Heart Ails include coronary thruway complaint, arrhythmias (heart cadence issues), cardiac abnormalities (analogous as natural heart scars), and several other conditions. The order involves conditions analogous as cardiomyopathy and cardiac infections. The electrocardiogram (ECG) already contains significant data on how the heart is working. This signal gives a doctor vital detail about a patient's cardiac function and can be used to predict and identify heart disease [23]. Due to its non-invasiveness and the useful information, it offers, it is one of the most often employed signals in diagnosis. You can assess the pathophysiological state of the heart using its analysis. Several systems have been created for ECG analysis and recording [11]. Early ECG systems merely printed the signal to record it. Modern systems offer automatic diagnostics using computer technology. The diagnosis of chronic myocardial illnesses, the detection and classification of arrhythmias, and the detection of ischemia are all areas of this broad study field that have seen the implementation of numerous methods and approaches. These techniques often involve processing the signal to remove noise and artifacts, extracting important disease-related features, and then examining the features to conclude. Signal processing, artificial neural networks, fuzzy logic, and clinical symptoms reported by medical professionals are typically used in the study. These systems' performance is assessed using conventional databases [12].

With an 87 score, the K-neighbor algorithm mechanism was arrived. The fact that the predictor and the casket pain variable maintain a favorable rapport shows the frequencies of casket pain is associated to the chance of acquiring cardiac complaint. Casket pain is viewed as a statistical specific, with 4 values representing asymptomatic angina, typical angina, andnon-anginal pain, independently [13].

If there is an unfavorable relationship between these variables, the circulatory system is likely to anticipate more blood. 39) the nobleness of the mongrel structure denounces the total number of deaths from cancer of the bones and DNA that are closely linked to mortality. Cancer of the bone ranks first among women in the nation of India and continues to be prevalent, with a controlled rate of 26.4 cases per 100,00 female and a fatality case of 13.5 per 100,00 females.

Development of a strategy for addressing crucial exploration motifs or practices in health care sectors including technical computers.

till, One of the main advantages of using Python in the medical field is that it helps with interpreting data by integrating machine learning and artificial intelligence. Based on Yilmaz's analysis [14], Python application solutions are a good choice for an effective programming language that promotes computational skills in carrying valuable perception from the data of cases with heart conditions. This will help assist with healthcare grounded completely packages. It comes in handy when one must provide a variety of expanding commodities with the aid of a web connection or operate independently without one. According to Srinath's [15] perspective, a large period and a terrible structure increase the plainness of walking over several operating systems. Additionally, Python has shown to be a useful programming language for contrasting huge data sets thanks to machine learning techniques that enable enormous awareness [16]. Data professionals also like the programming style due to its broad library support, which includes SciPy, Pandas, NumPy, and many more. A deep learning algorithm with the highest delicateness of 87.43 is the artificial neural network (ANN). This mathematical pattern will improve the prognosis rate for women and aid in the detection of bone cancer [34].

2.2 Display of the Ability to Assess, Compile, and Search Data from Relevant Sources in the Medical Care Industries

A logistic regression was carried out in Python and data were obtained from external databases for this project. Jiang et al. claim that [18] the properties of a dataset are also determined using a wide range of information. Datasets with multiple values were represented, For example, angina is generated through physical activity, maximal heart rate, resting arterial pressure, laying ECG readings, rapid blood sugar level, thalassemia stage, depression, a broad spectrum of significant vessels, and an assortment of other factors.

In any situation, an individual's sex may be determined using two attributes, 0 and 1, where 0 denotes female and 1 denotes male. In contrast, the four values of 0, 1, 2, and 3 will be utilized to evaluate the chest pain categories. These numbers, in order, reflect asymptomatic state, atypical angina, non-anginal pain, and classic angina. In any case, a disarray framework utilized to produce false positive and negative outcomes. Furthermore, sufficient CSV files are employed to collect the essential details for the regression analysis, according to van den Deepika et al. [17]. On the other hand, classification scores can be derived for the purpose of diagnosing cardiac disease. Assist vector classifiers, choice with treeing classifiers, irregular timberland classifiers, and a scope of other AI procedures, then again, are a couple of models. In this model, notwithstanding, the information fighting cycles will be utilized to decide the connection between the negative and positive double indicators. As indicated by Lou et al. [3], this self-administration information fighting innovation assists manage more mind-boggling data rapidly and conveys exact outcomes to improve decisions. The characteristics are contrasted with those of cardiac patients who test positive and negative. The favorable individuals had a higher cardiac rate and about thirty percent of the exercise-induced ST depression linked to the prior peak, according to the analysis of all the data [13]. As a result, developers may use Python to successfully create the models needed to forecast cardiac illnesses before they become serious.

When two distinct datasets are combined and a single feature-choosing technique is applied to both, the artificial intelligence [35] technique is appropriate and will increase the correctness ratio towards identification and occurrences of tumors in female patients.

2.3 Cyber Security Measures Are Critical in the Healthcare Sector to Ensure Compliance with Information Security Management Systems and Networking Setups

Python programming language is commonly utilized in the health care sector because cyber security specialists can complete the project quickly. As illustrated in Fig. 4, According to Calix et al. [18], the dialect is additionally employed to deliver and decode packets, scan networks, and ports, gain permission to computers, locate devices, and look into spyware (Fig. 2).

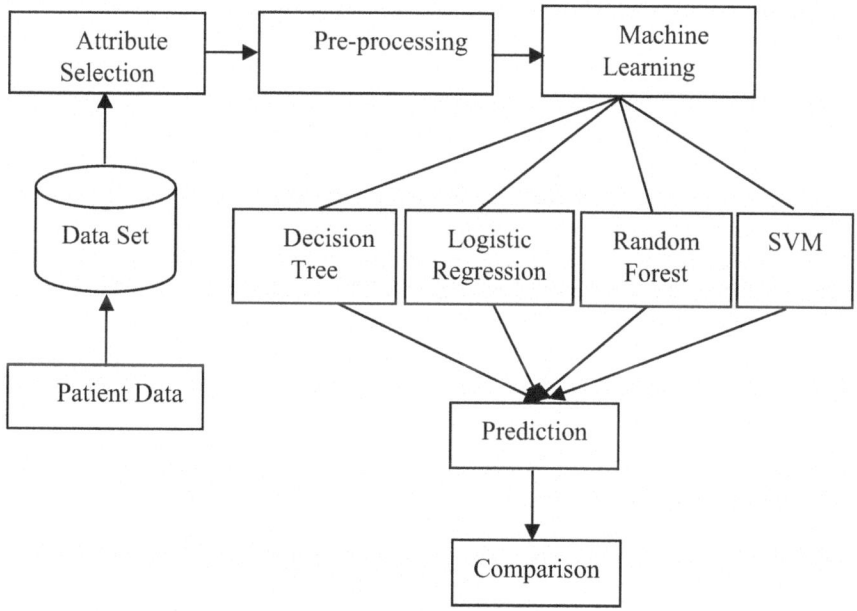

Fig. 2. Conceptual framework

It is also beneficial to run a series of cyber security software, like spyware investigation and monitoring. Furthermore, because the medical care industry contains a large amount of personal patient information, this language is highly used designing a model inside the system [19]. It follows a focused and managed procedure. Additionally, the quantity of work required to complete specific tasks like malware evaluation, identification, and testing for penetration is decreased by the abundance of libraries. Additionally, the language has an easy syntax that novice developers entering the cybersecurity field could easily learn [20]. As a result, this language was selected for the development of model for identifying cardiac problems.

3   Methodology of Research

3.1   Overview

Neural network model is useful in prognosticating the presence of cardiac problems, locomotors problems, cardiovascular ails, and other affections. It was supposed to give important perceptivity to clinicians, allowing them to conform their opinion and remedy to individual cases. The arbitrary timber algorithm is utilized in this design for creating heart complaint discovery, and the technique is utilized towards the approach to identify cardiac complaint by implementing python program.

3.2   Research Method

According to Larsen et al. [26], deep learning founds to be a discipline of making a system that is capable to get discover various forms of initial data. For this sort of project, The most popular programming language is Python. Along with having a vast library that includes Numpy, scipy, pandas, scikit-learn, matplotlib etc., it will also be substituting a number of technological languages. The pandas dataframe.info() function comes in useful when conducting data exploration because it can provide a brief overview of the data frame. Next, acquire the information set, read it using read_csv(), and assign the result to the information set variables. Pandas library may be useful to provide statistical explanations like the mean, standard deviation, and percentile, and many more. If not part of information, a collection of numeric data. This function takes a string array and produces a variety of outcomes. Then, employ a association matrices that comprehend facts into account. Pyplot has been used to display the X and Y coordinates of coefficient matrices. The colorbar method of the matrix can be seen by adding labels to the related matrix colorbar function.

3.3   Deep Learning Models

With its distinct set of variables, supervised artificial intelligence has found many applications.

Random forest (RF): It applies to regression and its type. The most adaptable and easy-to-use method is this one [27]. Their ability to foretell the future will increase with the amount of timber in the forested area. By using randomly selected record samples, RF generates arbitrarily wood. After taking into account all of the bushes' forecasting, it selects the output of the highest caliber.

Logistic Regression: To determine the probability of a result variable, one uses a deep learning method known as logistic regression (LR) [28]. It functions well when labeled records are in various places, but it also functions well when the outcome or base variable is binary.

Selection Tree (DT): DT is another way of saying that it is a collection of distinct rules since it gathers data in the form of a tree [29]. Decision bushes can be used to handle big facts. The parameter selection determines the method used by the DT algorithm to divide the findings into large units of traits.

Aid Vector System (SVM): Among the most popular methods, the SVM uses for assessing emotions [15]. It does class with the help of selecting the hyper-aircraft outstanding appropriate for dividing the instructions. C parameter and a kernel sigmoid are utilized in the linear SVM technique.

3.4 Use of Algorithm with Justification

The methodology for predicting heart disease has been stratified in steps given below (Fig. 3):

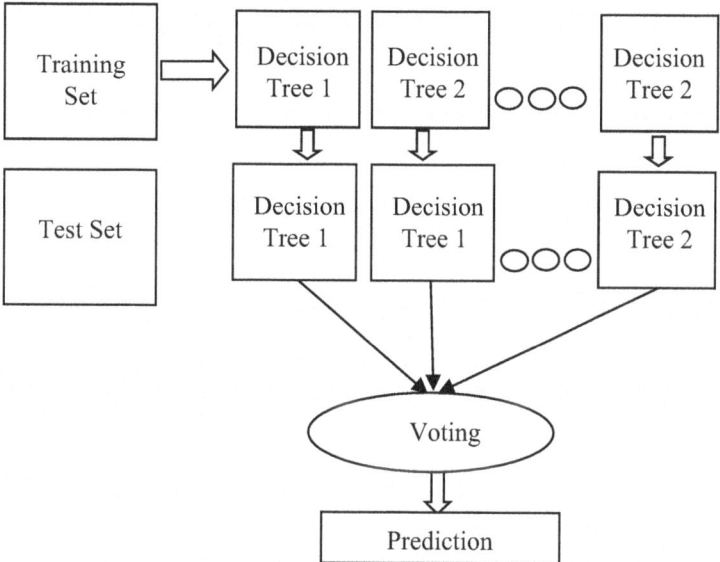

Fig. 3. The forest randomization method Origin: impacted by [24].

1. Data collection: In this step, we collect data on patient demographics, medical history, and diagnostic tests. This data can be obtained from electronic health records, medical databases, or publicly available datasets.
2. Data preprocessing: Preprocessing is done on the gathered data to eliminate any anomalies or omitted principles. To guarantee the model's precision and dependability, this stage is essential.
3. Exploratory data analysis: The data is explored using various visualization techniques to gain insights into the relationships between different variables and the presence of heart disease. This step helps to identify relevant features for the model.

4. Feature selection: Based on the insights gained from exploratory data analysis, relevant features are selected for the model. This step is important to develop an accurate predictive model.
5. Model development: Artificial intelligence algorithms like utilizing random forests, logistic regression, and support vector machines for forecasting models. Using the cross-validation methods, the algorithm is confirmed after being trained on a subset of the data.
6. Model assessment: Measures such as precision, recall, and accuracy are employed to evaluate the algorithm's efficacy. To guarantee the precision and dependability of the approach, this stage is essential.
7. Model optimization: The model is optimized using techniques such as hyperparameter tuning. This stage is important to improve the precision and verifiability of the model.

K-Nearest Neighbor (KNN): The data analysis classes use the AI version [32]. It is a reliable model that is simple to use and comprehend. KNN is sometimes described as an inefficient learner since it bases its forecasts on the closest neighbor by using the aid of space exploration. The quantity of information isn't always very giant, it capabilities well.

3.5 Evaluation Measures

4 assessment metrics were used to rate the efficacy of supervised system gaining knowledge of models. The accuracy, F1, sensitivity, and precision rating. Accuracy data lies starting from 0 and ending at 1, where 1 representing the good accuracy. The results of different metrics used in the whole performance assessment procedure are measured using the terms TP (true effective), TN (actual terrible), FP (false fine), and FN (face negative). When a version correctly forecasts the bad magnificence but incorrectly predicts the fantastic magnificence, the forecast is called TP, and the outcome is called TN. However, FP is a forecast made when the version erroneously projects a beneficial class sample as being bad, whereas FN is a prediction made when the version forecasts an ineffective pattern as a successful one. By comparing the percentage of accurate identification to all forecasts, the accuracy percent is calculated. Sensitivity is a term applied for assess a version's capability to accurately predict a sample of superb magnificence concerning exactness, which quantifies the accuracy of a classifier. An elegant way to combine precision and don't forget metrics is to use an F1 rating to summarize a model's accuracy in forecasting.

The researchers utilized a record via the Machine Learning Archive shown in Fig. 5 at UCI that was freely available to the public. The record inculcates a number of 14 variables, that involves age, gender, blood pressure, levels of cholesterol, and electrocardiogram (ECG) values [22]. The dataset is preprocessed by the researchers, who discard the omitted data, feature scalint, and encrypt variables that are categorical. The record was stratified into training and testing tables (Figs. 6 and 7).

Investigators evaluate the efficiency of different machine learning techniques on the dataset. The effectiveness of neural networks, logistic regression, support vector machine (SVM), decision tree, random forest, and k-nearest neighbors (KNN) techniques are compared. We use a range of evaluation metrics, including F1-score, accuracy, precision,

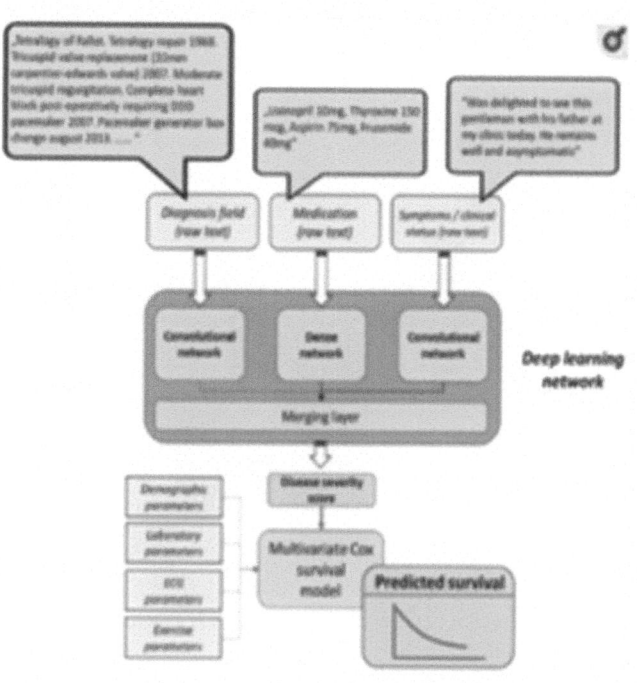

Fig. 4. Detecting coronary artery disease with a technique based on deep learning [34] is the source of information.

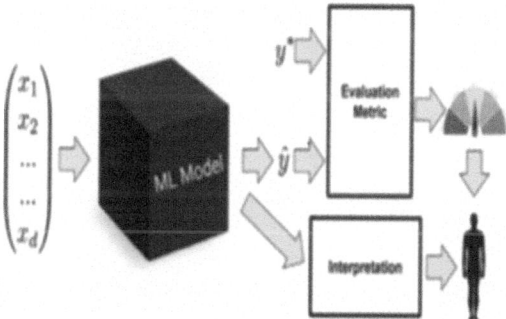

Fig. 5. Record interpretation: [21]

recall, and AUC-ROC, to evaluate the effectiveness of these algorithm designs. Table 1 discusses the results of the evaluation.

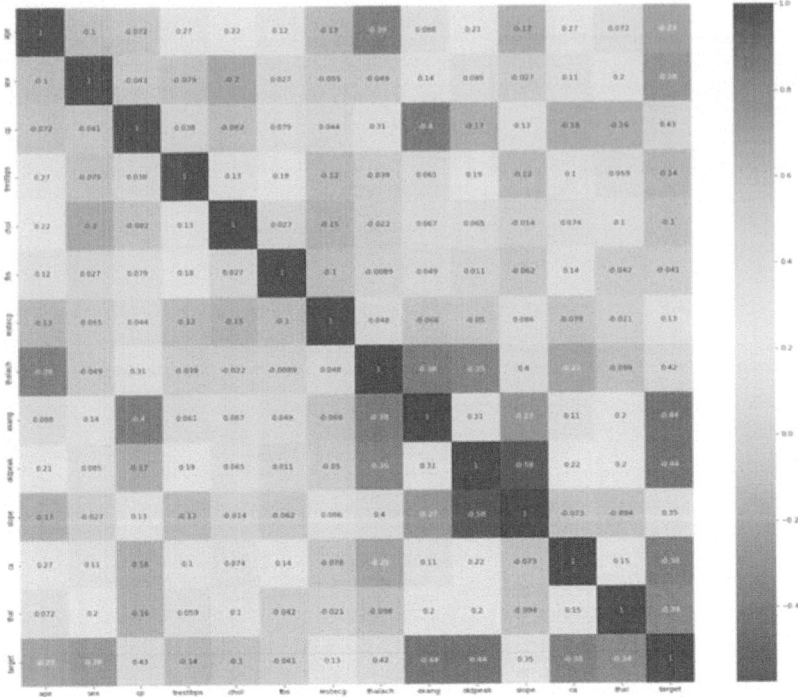

Fig. 6. Heat map Representation of out Data

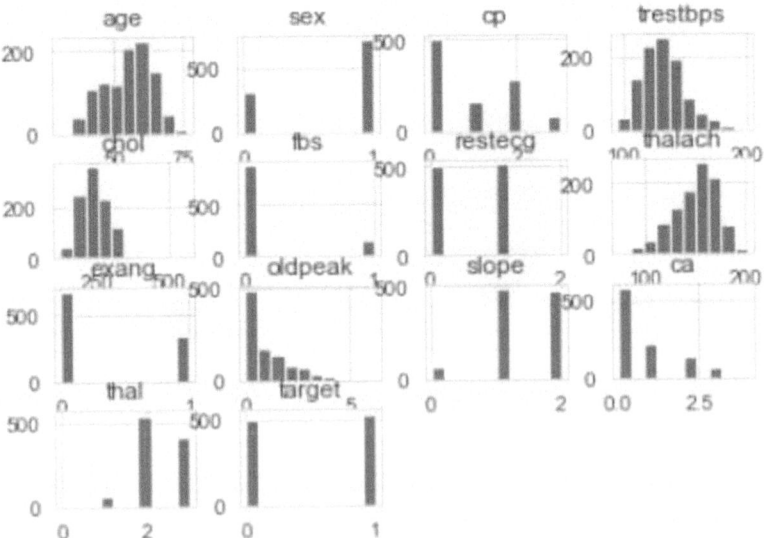

Fig. 7. Histogram Representation of Data

Table 1. Results of the Evaluation of the metrics

S. No.	Attributes	Result
1.	X.shape	(1025,13)
2.	X_train.shape	(820,13)
3.	X_test.shape	(205,13)
4.	Accuracy on Training Data	0.85243902439
5.	Accuracy on test Data	0.80487804878

4 Results

Deep learning algorithms can successfully detect cardiac disease, according to the findings. The neural network was the top performing algorithm, with a 99.9% accuracy. The KNN algorithm worked admirably as well, with an accuracy of 72.1%. The accuracy results for the decision tree and random forest algorithms is found to be 99.9% and 99.03%, correspondingly. The accuracy ratings for the SVM and logistic regression methods were 85.2% and 80.4%, respectively. Table 2 shows the results of model of decision tree.

Table 2. Model of Decision tree

S. No.	Attributes	Precision	Recall	F1-score	Support
1.	0	0.70	0.76	0.73	100
2.	1	0.75	0.69	0.72	105
3.	Accuracy			0.72	205
4.	Macro avg	0.72	0.72	0.72	205
5.	Weighted avg	0.72	0.72	0.72	205

In Table 2, the assessment of the model has been seen in the form of precision, recall, F1-score and support rate. Table 3 shows the results of accuracy rate for KNN model, random forest model, Logistic regression and SVM model.

Table 3. KNN Score, Model of Random Forest, Logistic regression and SVM

S. No.	Attributes	Accuracy
1.	Knn.score (X_train, Y_test)	0.72195
2.	Model.score (X_test, Y_test)	0.99024
3.	Accuracy on training data	0.85243
4.	Accuracy on Test data	0.80487

5 Conclusion

According to this paper, deep learning algorithms can enhance the administration and early identification of cardiac disease, improving outcomes for patients and lessening the impact it has on humanity in the process. Massive patient information sets can be analyzed by deep learning models, which can also spot structures that doctors might miss. Python provides a robust and adaptable environment for constructing deep learning models for heart disease prediction. Accurate and reliable predictive models may be constructed by following a systematic technique that involves data pretreatment, exploratory data analysis, feature selection, model creation, assessment, optimization, and deployment. These predictive models may be deployed as a web application or incorporated into clinical decision support systems, allowing doctors to enter patient data and obtain an estimate of the chance of cardiac disease. Predictive models can enhance medical results and reduce the overall societal stress of cardiac disease by allowing for earlier identification and more effective management of the ailment. The limitation of the model is robustness of the consistency of the model in order to maintain high accuracy rate for all the kinds of datasets. The data input remains specific for the proposed model for which the accuracy is calculated but for the other kinds of data input the proposed model may suffer with loss rate.

References

1. Alkayyali, Z.K., Idris, S.A.B., Abu-Naser, S.S.: A systematic literature review of deep and machine learning algorithms in cardiovascular diseases diagnosis. J. Theor. Appl. Inf. Technol. **101**(4), 1353–1365 (2023)
2. Nandy, S., Adhikari, M., Balasubramanian, V., Menon, V.G., Li, X., Zakarya, M.: An intelligent heart disease prediction system based on swarm-artificial neural network. Neural Comput. Appl. **35**(20), 14723–14737 (2023)
3. Lou, Y.S., Lin, C.S., Fang, W.H., Lee, C.C., Lin, C.: Extensive deep learning model to enhance electrocardiogram application via latent cardiovascular feature extraction from identity identification. Comput. Methods Programs Biomed. **231**, 107359 (2023)
4. Mahoto, N.A., Shaikh, A., Sulaiman, A., Al Reshan, M.S., Rajab, A., Rajab, K.: A machine learning based data modeling for medical diagnosis. Biomed. Signal Process. Control **81**, 104481 (2023)
5. Malakouti, S.M.: Heart disease classification based on ECG using machine learning models. Biomed. Signal Process. Control **84**, 104796 (2023)

6. Bakar, W.A.W.A., Josdi, N.L.N.B., Man, M.B., Zuhairi, M.A.B.: A review: heart disease prediction in machine learning & deep learning. In: 2023 19th IEEE International Colloquium on Signal Processing & Its Applications (CSPA), pp. 150–155. IEEE (2023)

7. Kadhim, M.A., Radhi, A.M.: Heart disease classification using optimized Machine learning algorithms. Iraqi J. Comput. Sci. Math. **4**(2), 31–42 (2023)

8. Wang, Z., Stavrakis, S., Yao, B.: Hierarchical deep learning with generative adversarial Network for automatic cardiac diagnosis from ECG signals. Comput. Biol. Med. **155**, 106641 (2023)

9. Sharean, T.M., Johncy, G.: Deep learning models on heart disease estimation-a review. J. Artif. Intell. **4**(2), 122–130 (2022)

10. Ahsan, M.M., Siddique, Z.: Machine learning-based heart disease diagnosis: a systematic literature review. Artif. Intell. Med. **128**, 102289 (2022)

11. Chang, V., Bhavani, V.R., Xu, A.Q., Hossain, M.A.: An artificial intelligence model for heart disease detection using machine learning algorithms. Healthcare Anal. **2**, 100016 (2022)

12. Nagavelli, U., Samanta, D., Chakraborty, P.: Machine learning technology-based heart disease detection models. J. Healthcare Eng. **2022**, 7351061 (2022)

13. Javeed, A., Khan, S.U., Ali, L., Ali, S., Imrana, Y., Rahman, A.: Machine learning-based automated diagnostic systems developed for heart failure prediction using different types of data modalities: a systematic review and future directions. Comput. Math. Methods Med. **2022**, 9288452 (2022)

14. Yilmaz, R., Yağin, F.H.: Early detection of coronary heart disease based on machine learning methods. Med. Rec. **4**(1), 1–6 (2022)

15. Schmidhuber, J.: Deep learning in neural networks: an overview. Neural Netw. **61**, 85–117 (2015)

16. Simonyan, K., et al.: Very Deep Convolutional Networks for Large-Scale Image Recognition (2015). arXiv preprint arXiv:1409.1556

17. Deepika, S., Jaisankar, N.: Review on machine learning and deep learning-based heart disease classification and prediction. Open Biomed. Eng. J. **17**(1), 1 (2023)

18. Poplin, R., et al.: Prediction of cardiovascular risk factors from retinal fundus photographs via deep learning. Nat. Biomed. Eng. **2**(3), 158–164 (2018)

19. Zhang, J., et al.: Predicting hospital readmission with deep learning. In: Healthcare Data Analytics, pp. 1–13 (2017)

20. Miotto, R., et al.: Deep patient: an unsupervised representation to predict the future of patients from the electronic health records. Sci. Rep. **6**, 26094 (2017)

21. Yadav, R.K., Kumar, A., Shukla, S.K., Fatima, E.: A neoteric procedure for spotting and segregation of ailments in mediciative plants using image processing techniques. Int. J. Next Gener. Comput. **13**(4) (2022). https://doi.org/10.47164/ijngc.v13i4.965

22. Rajpurkar, P., et al.: CheXNet: Radiologist-Level Pneumonia Detection on Chest X-rays with Deep Learning (2017). arXiv preprint arXiv:1711.05225

23. Fitriyani, N.L., Syafrudin, M., Alfian, G., Rhee, J.: HDPM: an effective heart disease prediction model for a clinical decision support system. IEEE Access **8**, 133034 (2020)

24. Gulshan, V., et al.: Development and validation of a deep learning algorithm for detection of diabetic retinopathy in retinal fundus photographs. JAMA **316**(22), 2402–2410 (2016)

25. Ripan, R.C., Sarker, I.H., Hossain, S.M.M., et al.: A data-driven heart disease prediction model through K-means clustering-based anomaly detection. SN Comput. Sci. **2**, 112 (2021)

26. Hinton, G., et al.: Deep neural networks for acoustic modeling in speech recognition: the shared views of four research groups. IEEE Signal Process. Mag. **29**(6), 82–97 (2012)

27. LeCun, Y., et al.: Deep learning. Nature **521**(7553), 436–444 (2015)

28. Krizhevsky, A., et al.: ImageNet classification with deep convolutional neural networks. In: Advances in Neural Information Processing Systems, pp. 1097–1105 (2012)

29. Singh, A.K., Kumar, A., Kumar, V., et al.: COVID-19 Detection using adopted convolutional neural networks and high-performance computing. Multimed. Tools Appl. (2023). https://doi.org/10.1007/s11042-023-15640-2

30. Kumar, A., Shukla, S.K., Prakash, N., et al.: A deep learning and powerful computational framework for brain cancer MRI image recognition. J. Inst. Eng. India Ser. B (2023). https://doi.org/10.1007/s40031-023-00926-8

31. Szegedy, C., et al.: Going deeper with convolutions. In: Proceedings of the IEEE Conference on Computer Vision and Pattern Recognition (CVPR), pp. 1–9 (2015)

32. He, K., et al.: Deep residual learning for image recognition. In: Proceedings (2016)

33. Thakur, B., Gupta, G., Kumar, N.: Hybrid genetic model with ANOVA for predicting breast neoplasm using METABRIC gene data. Mater. Today Proc. **56**, 1847–1852 (2016)

34. Thakur, B., Kumar, N., Gupta, G.: Machine learning techniques with ANOVA for the prediction of breast cancer. Int. J. Adv. Technol. Eng. Explor. **9**(87), 232 (2022)

35. Thakur, B., Kumar, N.: Prediction, detection and recurrence of breast cancer using machine learning based on image and gene datasets. In: Recent Innovations in Computing: Proceedings of ICRIC 2021, vol. 1, pp. 263–273 (2022)

Revolutionizing Cancer Diagnosis: The Power of Deep Learning Ensembles in Lung and Colon Cancer

Seema Kashyap[1]([⊠]) [iD], Arvind Kumar Shukla[2] [iD], and Iram Naim[3] [iD]

[1] School of Computer Science and Engineering, IFTM University, Moradabad, Uttar Pradesh, India
seemakashyap2484@gmail.com
[2] School of Computer Science and Applications, IFTM University, Moradabad, Uttar Pradesh, India
arvindshukla@iftmuniversity.ac.in
[3] Faculty of Engineering and Technology, MJP Rohilkhand University, Bareilly, India

Abstract. Finding cancer early often means better treatment results and a higher chance of life. Typically, malignancies that are detected in their early stages are more susceptible to treatment and have a greater likelihood of being completely cured. Tissue histology is often utilized for early detection; however, it is usually performed manually by pathologists, which may be time-consuming and prone to mistakes. This research addresses these issues by developing and deploying a computer-aided method for identifying lung cancer in whole-tissue slides. The study presents a novel automated approach for interpreting histopathological lung and colorectal images by utilizing ensemble Convolutional Neural Networks (CNNs) to detect zones of lung and colorectal cancer. This work highlights the efficacy of ensemble models in analyzing histopathological pictures and their capacity to improve the identification of lung and colon cancer. Histopathology data from the lungs and stomach (LC25000) are used to test the model. This finding establishes a foundation for forthcoming investigations and progress in automated medical diagnosis, which will confer benefits to patients and healthcare professionals as technology and machine learning further evolve. The assessment of the model's performance relies on criteria such as recall, precision, F1 score, and accuracy.

Keywords: Lung and colon cancer · Convolutional Neural Network · Xception · Mobilenet · Histopathology Image

1 Introduction

Lung and colon cancer are very fatal diseases on a global scale. These two types of cancer can occasionally arise successively in the same patient. The American Cancer Society projects that by 2023, One major issue regarding public health in the US is lung cancer, which is expected to surpass all other cancers in terms of cancer-related deaths,

M. Khurana et al. (Eds.): ICMLA 2024, CCIS 2238, pp. 445–456, 2025.
https://doi.org/10.1007/978-3-031-75861-4_39

accounting for 238,340 new cases and 127,070 deaths. The mortality toll from lung cancer is projected to be nearly three times that of colon cancer. The primary Aetiology of cancer -related fatalities in the United States is lung and bronchus cancer, accounting for 609,820 of the projected total mortality in 2023. On a global scale, colon cancer is the second most prevalent cancer and the third most prevalent malignancy, accounting for more than 10% of all cases and cancer-related fatalities. Finding and treating lung cancer and colorectal cancer early is very important for better patient results. Non-small-cell lung Cancer (NSCLC) is a specific form of lung cancer that requires significant focus and consideration. Transfer learning, a commonly employed method in medical image analysis, involves utilising pre-trained models as feature extractors or for the goal of fine-tuning. Researchers have utilised transfer learning approaches in several medical imaging projects, including the categorization of numerous illness images. Convolutional neural networks (CNNs), a kind of deep neural network methodology, have gained significant interest in several domains, including medical image processing, because of their proven efficacy. The LC25000 dataset, consisting of 25,000 colour photos classified into five categories related to lung and colon cancer, was utilised to evaluate and build classification algorithms. This study is divided into 5 parts, with Sect. 2 concentrating on the most recent advancements in leveraging deep learning for the diagnosis of colorectal and lung cancers. The technique and dataset description for the study are provided in Section three. The experimental configuration and assessment of performance are detailed in next section. Conclusion is included in the fifth section of the manuscript.

2 Literature Review

The implementation of other artificial intelligence technologies, including deep learning in medical imaging has significant untapped promise for diagnosing and categorising lung cancer. With the use of this technology, diagnostic processes might become more accurate and effective, benefiting patients' medical outcomes. However, it is crucial to fulfil this responsibility in collaboration with certified medical professionals in a collective endeavour, while also considering the ethical and regulatory dimensions of the issue. Several different deep learning algorithms and AI-based classification and differentiation algorithms are utilised for cancer imaging. The VGG19 + CNN architecture performed exceptionally well, with Accuracy: 96.48%, Recall: 93.75%, Precision: 97.56%, F1 Score: 95.62%, and AUC: 99.82%. The research concluded that the VGG19 + CNN architecture was superior to other methods that were already in use. This conclusion was reached on the basis of the performance measures that were given. This demonstrates that the model that was constructed is quite capable of classifying CXR images into the many forms of chest illnesses [13]. The author proposed three convolutional neural network (CNN) models for lung cancer detection using VGG16, DenseNet201and ResNet50V2 architectures based on transfer learning, and then created and validated an ensemble of these models, which achieved 91% validation accuracy and outperformed other existing models [5]. Deep Ensemble 2D was proposed by Shah A. A. et al. [6]. With a combined accuracy of 95%, CNN provided us with excellent results that were better than the baseline approach. The most advanced classification models Efficient-Net, DenseNet, and XceptionNet are used in an ensemble by Farooq, M. U., et al. [7].

With scores of 85.62 for precision, 76.29 for recall, and 75.82 for F1 for the ensembled model, accuracy is at 88.33%. Feng, J. et al. [8] found that combining Mask-RCNN with the DPN algorithm gave better segmentation results for lung parenchyma CT images than either Mask-RCNN or DPN alone. This can help improve the diagnostic accuracy and detection efficiency of CT images in lung cancer. The EfficientNet algorithm demonstrated exceptional performance in accurately classifying several forms of cancer, such as brain tumor, breast cancer, chest cancer, and skin cancer. It obtained remarkable accuracy, precision, recall, and F1 scores across all the cancer datasets. Kundu et al. [14] employed a combination of three Convolutional Neural Network architectures (ResNet, Dense Net, and Google Net) to diagnose pneumonia. Their approach yielded favorable outcomes, demonstrating high levels of accuracy across several datasets. In their study, Yaseliani et al. [15] introduced a sophisticated ensemble hybrid deep learning system that combines support vector machines (SVM), radial basis function, and logistic regression. This system is designed for image classification and includes both feature extraction and classification stages. Mabrouk et al. [16] employed a combination of DenseNet169, MobileNetV2, and Vision Transformer models to predict pneumonia using CXR pictures. The retrieved features from the images were utilized for testing. Researchers have effectively employed pre-trained models including VGG-16, ResNet-50, and InceptionV3 to analyze clinical images of lung illnesses such as COVID-19 [18]. The classification task was carried out by utilizing a combination of the InceptionResNet V3, ResNet50, and MobileNet V3 models. According to the statistics, these models outperformed the rest, achieving an F1 score of 94.84%. The user's text is "[19]". The categorization of a combined dataset of CXR and CT images involved the evaluation of COVID-19 utilising advanced models such as InceptionV3, NASNet, Xception, DenseNet, MobileNet, VGGNet, InceptionResNetV2, and ResNet in a separate study. The DenseNet121 model had the best level of accuracy, reaching 99%, among all the systems that were evaluated [20] (Table 1).

Table 1. Convolutional models used in the study of the Lung Cancer Dataset

Model	Author	Accuracy	Dataset
VGG19 + CNN architecture	Alshmrani G et al. (2023) [13]	99.63	Chest x-ray (CXR image)
Ensemble of VGG16, DenseNet201and ResNet50V2 architecture	Phankokkruad, M. et al. (2021) [5]	91	Lung CT image
Deep ensemble 2D	Shah A. A. et al., (2023) [6]	95	Lung CT image

(continued)

Table 1. (*continued*)

Model	Author	Accuracy	Dataset
Ensemble of EfficientNet, DenseNet, and XceptionNet	Farooq, M. U. et al. (2023) [7]	88.33	Histopathological breast image
Mask-RCNN with the DPN algorithm	Feng, J. et al. (2022) [8]	97.94	Chest CT image
A combination of ResNet, DenseNet, and GoogleNet	Kundu R et al. (2021) [14]	98.81	Pneumonia x-ray datasets
ensemble hybrid deep learning system with SVM	Yaseliani et al. (2022) [15]	98.5	Chest x-ray (CXR image)
Ensemble of DenseNet169, MobileNetV2, and Vision Transformer	Mabrouk et al. (2022) [16]	93.91	Chest x-ray (CXR image)
Ensemble of VGG-16, ResNet-50, and InceptionV3	Perumal et al. (2021) [18]	94.84	CT and CXR image
Ensemble of InceptionResNet V3, ResNet50, and MobileNet V3	El Asnaoui et al. (2021) [19]	94.84	CT and CXR image

3 Methodology

3.1 Dataset

This model used the LC25000 data set that was developed in 2020 by A. Borkowski et al. [17] in this research work. This collection contains a total of 25,000 images of colon and lung tissues, which have been classified into 5 distinct categories. Histopathology images from Tampa's James A. Haley Veterans' Hospital make up the LC25000 collection. It includes pictorial depictions of malignant tissues. The authors first acquired 1,250 pictures, with 250 photos representing each of the five kinds of cancer tissues. Image augmentation methods, such as rotation and flipping, were utilised to generate diverse versions of the original photos. The dataset was augmented to encompass a grand total of 25,000 photos, evenly distributed with 5,000 images in each class. The original graphics had dimensions of 1024×768 pixels. The images were cropped to a square format and made to measure 768×768 pixels before any augmentation techniques were applied. All images in the collection adhere rigidly to HIPAA regulations. The data collection is divided into five categories, and each category has 5,000 pictures (Fig. 1 and Table 2).

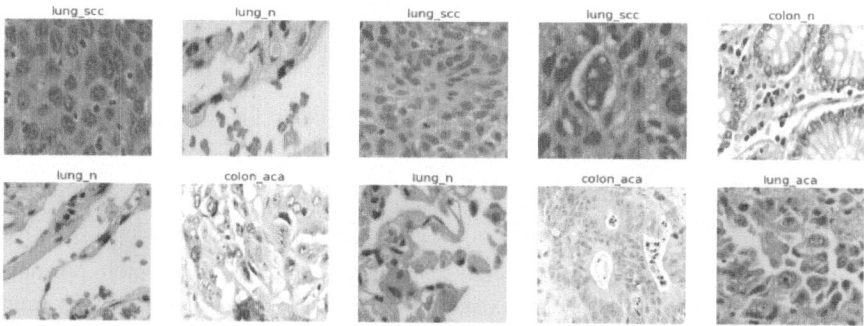

Fig. 1. Examples of image dataset.

Table 2. Dataset and class labels for LC25000.

Image type	Label name	Number of samples
Colon adenocarcinoma	Colon_aca	5000
Colon benign tissue	Colon_n	5000
Lung adenocarcinoma	Lung_aca	5000
Lung benign tissue	Lung_n	5000
Lung squamous-cell carcinoma	Lung_scc	5000

3.2 The Proposed Approach

In this research, model Combining the feature extraction capabilities of two pre-trained models, MobileNet and Xception, and incorporating another layer for categorization, the model is succinctly described as follows:

- In order to extract features, the model employs two pre-trained deep learning models, namely MobileNet and Xception. These models are equipped with pre-trained weights from the "ImageNet" dataset, and their final classification layers are turned off.
- All layers in MobileNet and Xception models are frozen, meaning their weights will not change while the models are trained.
- The model contains a single input layer, and it anticipates receiving input photos with the shape (224, 224, 3) as the shape.
- Feature extraction involves the concurrent processing of the input image using the MobileNet and Xception models.
- The resulting feature maps from both models are joined together to make a single feature image.
- After the characteristics have been concatenated, a number of substantial layers are added for categorization. This design has three thick layers with 1024, 512, and 256 units each. Each layer uses the ReLU activation function. The ultimate output layer is comprised of a softmax activation function.
- Accuracy, precision, recall, area under the curve (AUC), maximum absolute error (MAE), root-mean-squared error (RMSE), categorical cross-entropy loss function,

and the Adam optimizer are the essential components of the suggested model. A robust strategy for constructing and assessing models in classification and regression tasks is suggested by this extensive approach.

- Computation metrics and the confusion matrix evaluate the model's efficacy. This examination evaluates accuracy, precision, recall, and recall-precision balance. These aspects reveal the model's malignant lung image identification and classification ability.
- The model leverages the capabilities of MobileNet and Xception for extracting features, merging their respective feature sets. Subsequently, it utilizes more profound layers to carry out the process of categorization. The model is prepared for training using annotated data that is specifically tailored for multi-class classification tasks. MobileNet is highly regarded in the field of medical imaging due to its exceptional efficiency, which makes it well-suited for devices with limited resources and applications that require real-time processing. This efficiency is accomplished by the use of depth wise separable convolutions. Xception, a framework that builds upon Inception, demonstrates exceptional ability in identifying intricate data patterns and interconnections, making it highly suitable for demanding medical imaging applications.

Mobilenet

MobileNet is a collection of efficient deep-learning models specifically created for use on mobile and embedded devices. The main characteristics of this technology are the implementation of depth wise separable convolutions to minimize computing demands, exceptional efficiency, a range of model variations such as MobileNetV1, V2, and V3, and compact model sizes. MobileNet models are renowned for their optimal balance of model size, inference speed, and accuracy, rendering them well-suited for real-time applications on smartphones with limited resources. They can be used to classify images, find objects, and separate words based on their meaning. You can get them as ready-to-use models for transfer learning. MobileNet plays a crucial role in facilitating on-device AI applications, especially in the field of mobile and embedded computing.

Xception

French engineer Francois Chollet created the Xception architecture, which builds on the Inception design. Xception is distinguished by the subsequent essential attributes. Xception uses depth wise separable convolutions to split the learning of features that are channel-wise and features that are space-wise. This lowers the amount of memory and computing power needed. Xception has 36 convolutional layers set up in 14 modules, which make up a stack of layers that runs from top to bottom. Most modules have linear residual connections save the first and final. These connections assist in solving difficulties such as disappearing gradients and representational bottlenecks by constructing shortcuts that enable the output of an earlier layer to be added to the input of a later layer through the process of summing. This is done in order to address the issue of vanishing gradients. In Xception, the shortcut connection sums the output of a previous layer and the input of the current layer instead of concatenating. Computer efficiency and high-performance feature extraction in deep learning are balanced by Xception. Computer vision applications like picture categorization and object recognition benefit from their ability to capture complex characteristics with little processing resources.

The Architecture of Proposed Approach

The model understands pictures that have the shape (224, 224, 3). This model uses two feature extraction models that have already been trained: Xception and MobileNet. MobileNet and Xception have 3,228,864 and 2,086,148 parameters, respectively. The feature maps generated by both feature extraction models are combined, resulting in a final feature map with dimensions of 7 by 7 by 3. The feature map is combined and transformed into a vector with a dimensionality of 150,528. The model consists of three thick layers, each having parameter counts of 1,541,416, 524,800, and 131,328, respectively. There are 1,024 components in the first dense layer. There are 512 units in the second thick layer. 256 units make up the third layer of neurons. The last output layer has 5 units, which is the same number of classes, have in classification task. The model consists of a grand total of 178,889,453 parameters. During the training process, there are 154,799,109 parameters that may be adjusted. The model consists of 24,090,344 non-trainable parameters, which are derived from the pre-trained feature extraction models (Figs. 2 and 3).

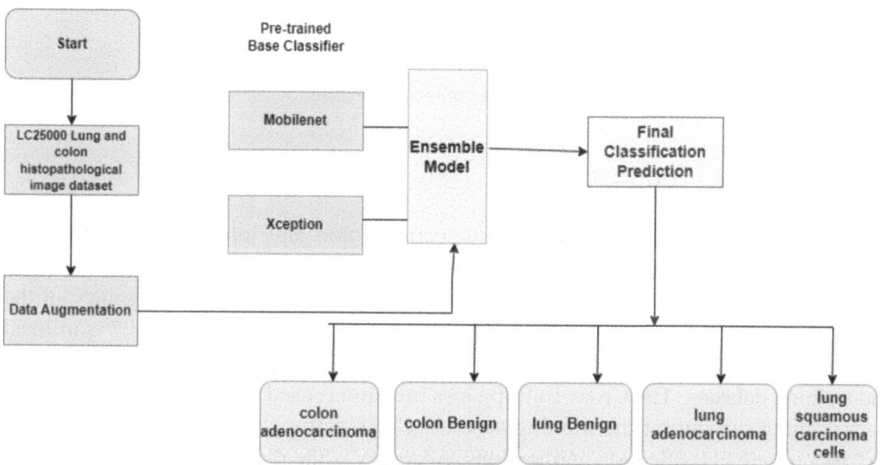

Fig. 2. Proposed Ensemble Model.

```
●  model.summary()

╪╸ Model: "model"
```

Layer (type)	Output Shape	Param #	Connected to
input_3 (InputLayer)	[(None, 224, 224, 3)]	0	[]
mobilenet_1.00_224 (Functi onal)	(None, None, None, 1024)	3228864	['input_3[0][0]']
xception (Functional)	(None, None, None, 2048)	2086148 0	['input_3[0][0]']
concatenate (Concatenate)	(None, 7, 7, 3072)	0	['mobilenet_1.00_224[0][0]' 'xception[0][0]']
flatten (Flatten)	(None, 150528)	0	['concatenate[0][0]']
dense (Dense)	(None, 1024)	1541416 96	['flatten[0][0]']
dense_1 (Dense)	(None, 512)	524800	['dense[0][0]']
dense_2 (Dense)	(None, 256)	131328	['dense_1[0][0]']
dense_3 (Dense)	(None, 5)	1285	['dense_2[0][0]']

```
Total params: 178889453 (682.41 MB)
Trainable params: 154799109 (590.51 MB)
Non-trainable params: 24090344 (91.90 MB)
```

Fig. 3. Ensemble architecture

4   Result and Findings

This model uses a novel approach for detecting colon and lung cancer by utilizing deep learning ensembles. Model included crucial information on the technique and assessment of our suggested method. The LC25000 dataset serves as the source of the histopathological images that are utilized for training and testing. Python 3.7 is utilized for all simulations, including the processing of histopathology pictures, testing datasets, and training datasets. The Cross-Entropy loss function is used to figure out the loss after each training run during the training process. Cross-entropy is a frequently employed loss function in classification applications. Accuracy is the main figure determining the extent to which the method can be generalized and classified. It reflects the percentage of correctly identified cases in all instances.

$$Accuracy = \sum_c \frac{TP_c + TN_c}{TP_c + FP_c + TN_c + FN_c}, c \in classes \tag{1}$$

Greater precision signifies superior efficacy. Table 3 Shows hyperparameter used in proposed model in binary classification, the findings are translated into True Positive (TP), False Positive (FP), True Negative (TN), and False Negative (FN) by the use of a confusion matrix. This offers a more comprehensive analysis of model's performance.

$$Recall = \sum_c \frac{TP_c}{TP_c + FN_c}, c \in classes \tag{2}$$

$$Precision = \sum_{c} \frac{TP_c}{TP_c + FP_c}, c \in classes \tag{3}$$

$$F1_{score} = 2 * \frac{precision * sensitivity}{presicion + sensitivity} \tag{4}$$

Table 3. Hyperparameter used for proposed ensemble model

Hyperparameter	Epoches	Batch size	optimizer	Learning rate
values	15	32	adam	0.001

These performance measures are essential for evaluating the efficiency and resilience of your approach. By these measures we assess how well our model does in terms of minimizing false positives and false negatives, accurately identifying instances as malignant or non-cancerous, and striking a balance between precision and recall. Our thorough descriptions of these metrics will promote clarity and transparency in our study, making it simpler for readers to grasp and analyze our results. Figure 4 displays plot for the training and validation data over the period of 15 epochs. Based on the diagram, it is evident that the optimal epoch for minimizing training and validation loss is 7, while the optimal epoch for achieving the highest training and validation accuracy is 14. Therefore, we may conclude that the model has reached a state of full train. CNN model built with TensorFlow-Keras is the primary objective of the proposed model. A Google Colab Pro with a TPU and ample RAM was used to run the model. A thorough evaluation method was carried out by assessing the model's efficacy utilizing several criteria (Figs. 5, 6 and Table 4).

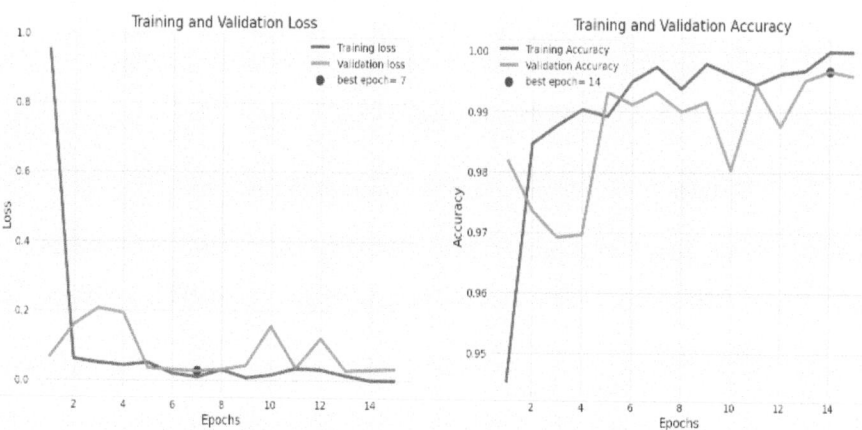

Fig. 4. Plot for proposed ensemble model

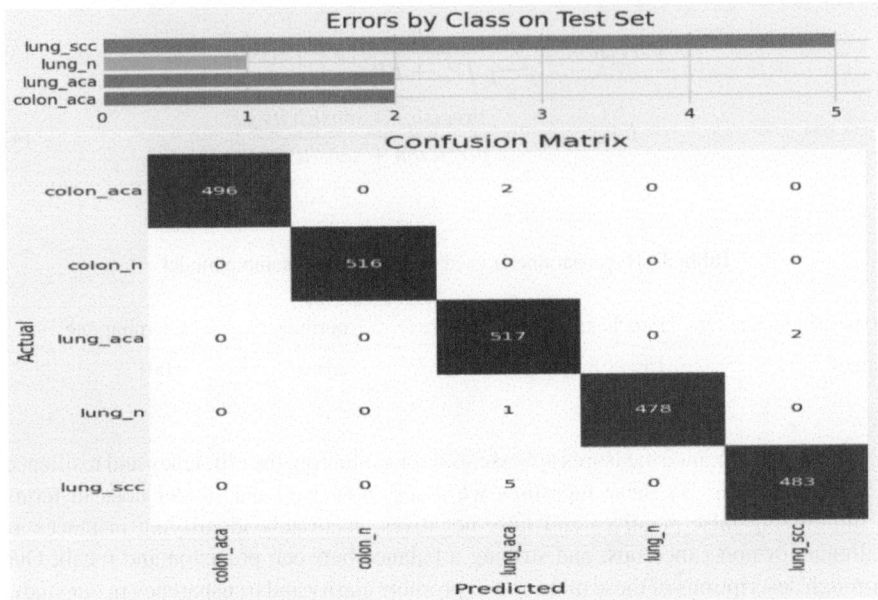

Fig. 5. Confusion matrix for proposed ensemble model

```
              precision    recall  f1-score   support

           0       1.00      1.00      1.00       479
           1       1.00      1.00      1.00       519
           2       1.00      1.00      1.00       498
           3       1.00      1.00      1.00       516
           4       1.00      1.00      1.00       488

    accuracy                           1.00      2500
   macro avg       1.00      1.00      1.00      2500
weighted avg       1.00      1.00      1.00      2500
```

Fig. 6. Classification report for proposed ensemble model

Table 4. Overall performance for proposed ensemble model

Accuracy	precision	recall	AUC	MAE	MSE	RMSE
99.6	99.6	99.6	99.87	0.0017	00.0015	0.0384

5 Conclusion and Discussion

The proposed way for finding lung cancer using ensemble models of deep learning, especially Xception and MobileNet, worked very well. Histopathological images were employed in the training and improvement of the deep-learning models. This suggests employing a multi-modal strategy to enhance precision. The technique employed a range of effectiveness markers, such as F-score, identification specificity, recall, precision, and accuracy, among other metrics. The assessment of the model's capacity to precisely categorise lung pictures as either benign or cancerous is highly dependent on these measures. The ensemble model demonstrated a remarkable accuracy rate of 99.60%, signifying its exceptional capability in the classification of lung images. Furthermore, the ensemble model exhibited a remarkable precision and recall rate of 99.60%. This implies that the model exhibits a modest false positive (high precision) and false negative (high recall) rate, in addition to its exceptional accuracy in image classification. The ensemble efficient neural network model achieved the highest F-Score, a metric that strikes a balance between recall and precision. When the F-Score is high, it means the model is doing a good job of balancing the two competing goals of producing few false positives and accurate positive classifications. This represents a pivotal accomplishment in the realm of medical image classification, given the severe repercussions that can ensue from misclassifications. Unfortunately, the method had trouble telling the difference between adenocarcinoma and squamous cell carcinoma, even though it worked very well overall. This implies that additional refinements or adjustments might be necessary in order to enhance the model's capacity to distinguish among these subtypes. The ensemble model of deep learning seems to be very good at accurately, precisely, and recalling whether lung pictures are of cancer or not. Nevertheless, further development is required to more accurately differentiate distinct subtypes of lung cancer. Additional investigation and improvements might be required in order to tackle this obstacle.

References

1. Zhang, C., et al.: Toward an expert level of lung cancer detection and classification using a deep convolutional neural network. Oncologist **24**(9), 1159–1165 (2019). https://doi.org/10.1634/theoncologist.2018-0908
2. Chen, S.: Models of artificial intelligence-assisted diagnosis of lung cancer pathology based on deep learning algorithms. J. Healthcare Eng. **2022**, 12 (2022). https://doi.org/10.1155/2022/3972298
3. Zhang, G., et al.: Automatic nodule detection for lung cancer in CT images: a review. Comput. Biol. Med. **103**, 287–300 (2018)
4. Gu, Y., et al.: A survey of computer- aided diagnosis of lung nodules from CT scans using deep learning. Comput. Biol. Med. **137**, 104806 (2021). https://doi.org/10.1016/j.compbiomed.2021.104806
5. Phankokkruad, M.: Ensemble transfer learning for lung cancer detection. In: 2021 4th International Conference on Data Science and Information Technology, pp. 438–442 (2021)
6. Shah, A.A., Malik, H.A.M., Muhammad, A., Alourani, A., Butt, Z.A.: Deep learning ensemble 2D CNN approach towards the detection of lung cancer. Sci. Rep. **13**(1), 2987 (2023)
7. Farooq, M.U., Ullah, Z., Gwak, J.: Ensemble CNNs for Breast Tumor Classification (2023). arXiv preprint arXiv:2304.13727
8. Feng, J., Jiang, J.: Deep learning-based chest CT image features in diagnosis of lung cancer. Comput. Math. Methods Med. **2022**, 4153211 (2022)
9. Tasya, W., Sa'idah, S., Hidayat, B., Nurfajar, F.: Breast cancer detection using convolutional neural network with EfficientNet architecture. In: 2022 IEEE Asia Pacific Conference on Wireless and Mobile (APWiMob), pp. 1–6. IEEE (2022)

10. Jiang, Y., Huang, R., Shi, J.: EfficientNet-based model with test time augmentation for cancer detection. In: 2021 IEEE 2nd International Conference on Big Data, Artificial Intelligence and Internet of Things Engineering (ICBAIE), pp. 548–551. IEEE (2021)

11. Nayak, D.R., Padhy, N., Mallick, P.K., Zymbler, M., Kumar, S.: Brain tumor classification using dense efficient-net. Axioms 11(1), 34 (2022)

12. Shah, H.A., Saeed, F., Yun, S., Park, J.H., Paul, A., Kang, J.M.: A robust approach for brain tumor detection in magnetic resonance images using finetuned efficientnet. IEEE Access 10, 65426–65438 (2022)

13. Alshmrani, G.M.M., Ni, Q., Jiang, R., Pervaiz, H., Elshennawy, N.M.: A deep learning architecture for multi-class lung diseases classification using chest X-ray (CXR) images. Alex. Eng. J. 64, 923–935 (2023)

14. Kundu, R., Das, R., Geem, Z.W., Han, G.T., Sarkar, R.: Pneumonia detection in chest X-ray images using an ensemble of deep learning models. PLoS ONE 16(9), e0256630 (2021)

15. Yaseliani, M., Hamadani, A.Z., Maghsoodi, A.I., Mosavi, A.: Pneumonia detection proposing a hybrid deep convolutional neural network based on two parallel visual geometry group architectures and machine learning classifiers. IEEE Access 10, 62110 (2022)

16. Mabrouk, A., Díaz Redondo, R.P., Dahou, A., Abd Elaziz, M., Kayed, M.: Pneumonia detection on chest x-ray images using ensemble of deep convolutional neural networks. Appl. Sci. 12(13), 6448 (2022)

17. Borkowski, A.A., Bui, M.M., Thomas, L.B., Wilson, C.P., DeLand, L.A., Mastorides, S.M.: Lung and Colon Cancer Histopathological Image Dataset (lc25000) (2019). arXiv preprint arXiv:1912.12142

18. Perumal, V., Narayanan, V., Rajasekar, S.J.S.: Detection of COVID-19 using CXR and CT images using transfer learning and Haralick features. Appl. Intell. 51, 341–358 (2021)

19. El Asnaoui, K.: Design an ensemble deep learning model for pneumonia disease classification. Int. J. Multimed. Inf. Retriev. 10(1), 55–68 (2021)

20. Kassania, S.H., Kassanib, P.H., Wesolowskic, M.J., Schneidera, K.A., Detersa, R.: Automatic detection of coronavirus disease (COVID-19) in X-ray and CT images: a machine learning approach. Biocybernet. Biomed. Eng. 41(3), 867–879 (2021)

VGG-Inspired Convolutional Neural Network Denoiser for the Enhancement of Mammogram Images

Vandana Saini[1]([✉]), Meenu Khurana[2], and Rama Krishna Challa[3]

[1] Chitkara University Institute of Engineering and Technology, Chitkara University, Rajpura, Punjab, India
vandana.s@chitkara.edu.in
[2] Chitkara University School of Engineering and Technology, Chitkara University, Baddi, Himachal Pradesh, India
meenu.khurana@chitkarauniversity.edu.in
[3] Department of Computer Science and Engineering, NITTTR, Chandigarh, India
rkc@nitttrchd.ac.in

Abstract. The accurate diagnosis of breast cancer through mammography significantly depends on image quality. In this study, a new deep learning model is developed to enhance the quality of mammogram images. Our approach involved modifying the established visual geometry group (VGG) neural network architecture to specifically address the challenges to denoise the mammogram images. Unlike the standard VGG network, our model includes additional convolutional layers and employs advanced noise reduction techniques designed for mammography. These modifications enable the model to effectively reduce image noise while preserving essential diagnostic details. The proposed VGG inspired CNN model had achieved a high PSNR of 79, surpassing existing techniques and showing potential for improved breast cancer diagnosis.

Keywords: VGG · PSNR · MIAS · Mammogram · CNN

1 Introduction

Breast cancer is a common disease that affects millions of women worldwide and contributes to a significant number of cancer-related deaths [1]. Early detection of breast cancer plays a crucial role in reducing mortality rates and improving treatment outcomes. To facilitate early detection, medical imaging techniques such as mammography have been widely used [2]. So, enhancing the quality of mammogram images is crucial for accurate diagnosis of breast cancer. Various advanced imaging techniques have been developed to improve the capabilities of mammography [3]. One such advancement is the use of digital mammography, which has proven to be more effective in diagnosing breast cancer, especially in its early stages [4]. Digital mammography uses a digital detector to capture X-ray images, allowing for better image processing and screening [5]. Digital mammography has advanced, but mammogram images still have limitations like low contrast, noise, and artifacts that can affect breast abnormality diagnosis [6].

© The Author(s), under exclusive license to Springer Nature Switzerland AG 2025
M. Khurana et al. (Eds.): ICMLA 2024, CCIS 2238, pp. 457–465, 2025.
https://doi.org/10.1007/978-3-031-75861-4_40

To overcome these limitations and improve breast cancer diagnosis accuracy, there has been growing interest in image enhancement techniques for mammograms [7]. These techniques aim to improve visual quality without altering the acquisition process or increasing hardware costs [8]. The primary objective is to improve the Peak Signal-to-Noise Ratio which is a widely used metric for evaluating image quality [9]. Various techniques, including filtering, histogram equalization [10], contrast adjustment, and adaptive thresholding, aim to reduce noise, enhance contrast, and improve visibility of abnormalities [11]. Deep learning networks have shown promise in automatically learning and extracting complex features and patterns from mammogram images [12], leading to improved outcomes. They can effectively detect early indicators of breast cancer and handle large amounts of data, well generalize to unseen data, and adaptively learn and adjust their parameters based on specific characteristics of mammogram images [13]. A Convolutional Neural Network based denoiser has proven highly effective in various computer vision tasks, utilizing multiple layers of interconnected neurons trained on large datasets to capture local features and spatial relationships in mammogram images [14]. Hence enhancing the visibility of breast tumors or malignancies.

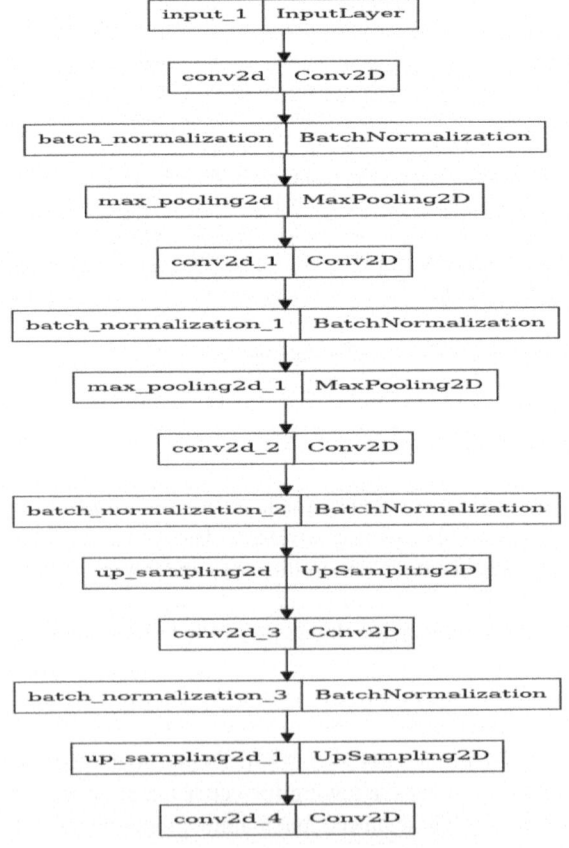

Fig. 1. Proposed Network Architecture

In this work a VGG inspired CNN based network is proposed, which is trained to reduce the noise from mammogram images. The proposed denoiser acts as a robust denoiser as it can work on multiple datasets in a similar way. In Fig. 1 above, the network architecture has been shown consisting of several layers. It begins with an input layer, followed by multiple convolutional layer (Conv2D), that are used to extract features from the input. After the first Conv2D layer, there's a batch normalization layer, which stabilizes learning by normalizing the input layer by re-centering and re-scaling. MaxPooling2D layers is used to reduce the spatial dimensions of the output volume. The architecture further includes additional convolutional and batch normalization layers, followed by up sampling layers, which helps in increasing the spatial resolution of the output.

2 Literature Review

The author in [15] had done detailed analysis of automated methods in mammography for breast cancer detection and classification. It covers a broad spectrum of machine learning techniques, focusing on their application in identifying key breast abnormalities like micro-calcifications and masses. In this paper, authors also discuss the enhancement of algorithms using sequential mammograms and reviews FDA-approved CAD systems. The author provides an overview of open-access mammography datasets and suggests future research directions in this rapidly evolving field. In [16] author had done a comprehensive analysis of various filtering techniques used in the preprocessing of mammograms for breast cancer diagnosis. The study emphasizes the importance of image quality in the accurate diagnosis of breast cancer, acknowledging the challenges posed by noise in mammogram images. It conducts a comparative analysis of filters like Box Filter, Averaging Filter, Gaussian Filter, Median Filter, and Bilateral Filter, using the MIAS dataset. The effectiveness of these filters is evaluated based on performance measures such as Mean Squared Error (MSE), Structural Similarity Index Measure (SSIM), and Peak Signal-to-Noise Ratio (PSNR). The results demonstrate that Gaussian, Median, and Bilateral Filters provide superior outcomes in enhancing mammogram images, making them crucial tools in the preprocessing stage for breast cancer diagnosis. In [17] author had proposed a method that focuses on the integration of advanced segmentation and classification techniques for mammographic images. The study elaborates on a hybrid segmentation method which employs CLAHE, morphological operations, and deep learning using the MIAS database. The methodology aims at improving accuracy, sensitivity, specificity, and biopsy rates in breast cancer detection. In [18] authors proposed an innovative framework that enhances contrast in mammogram images using a technique named haze-reduced local-global. This approach, combined with the use of a pre-trained EfficientNet-b0 model and deep transfer learning, aims to improve dataset diversity and model training efficiency. The unique aspect of this study is the use of a feature selection algorithm, Equilibrium-Jaya controlled Regula Falsi, leading to an average accuracy of 95.4% and 99.7% on CBIS-DDSM and INbreast datasets, respectively. The results showcase a notable improvement in accuracy compared to state-of-the-art methods. Deep learning networks are used in many other fields too like image processing and WSNs [26, 27]. Another method which provides notable accuracy is described in Table 1.

Table 1. Comparison study based on accuracy.

Paper	Objective	Method Used	Outcome
[19]	Improve classification of breast cancer	Feature Ensemble Learning with Stacked Sparse Autoencoders and Softmax Regression	High accuracy of 98.60%
[20]	Automated extraction of learned-from-data image features for differentiation between benign and malignant breast tumors	Two-layer Deep Learning architecture with Point-wise Gated Boltzmann Machine and Restricted Boltzmann Machine	Accuracy of 93.4%, sensitivity of 88.6%, specificity of 97.1%
[21]	Breast cancer diagnosis using computer-aided diagnosis (CAD)	CAD system via deep and transfer learning, using standard digital database of mammography	Overall accuracies of 69.82% and 71.73% for two ROI techniques
[22]	CAD scheme for detection of breast cancer	Deep belief network unsupervised path followed by back propagation supervised path on WBCD dataset	Classifier complex gives an accuracy of 99.68%

3   Methodology

In this work the goal is to create a robust denoiser for mammogram image that helps in enhancing the image quality by reducing noise and artifacts in the images. The methodology is divided into 3 phases data preparation, model training, and performance evaluation, described as follows:

i. *Data Preparation*:

- The MIAS database of mammograms was utilized for this study.
- Images were partitioned into a training set and a test set with an 80–20 split.
- Noisy versions of the images were generated by adding synthetic noise mimicking common artifacts in mammography.

ii. *Model Training*:

- The CNN model was trained on the noisy-clean image pairs using mean squared error as the loss function.
- Adam optimizer was employed with a learning rate determined by preliminary experiments.
- The model was trained for a fixed number of epochs with early stopping on validation loss to prevent overfitting.

iii. *Performance Evaluation:*

- The model's denoising performance was evaluated using the Peak Signal-to-Noise Ratio (PSNR) metric.
- A comparative analysis done to evaluate the performance of proposed denoisers in contrast with existing.

The proposed denoising method utilizes a convolutional neural network (CNN) architecture inspired by the VGG network but tailored specifically for denoising mammography images. The architecture is designed to process single-channel grayscale images, which aligns with the nature of mammographic data. The algorithm operates in three main stages: noise mapping, feature extraction, and image reconstruction.

Algorithm 1: Denoising with Modified VGG-inspired CNN

1. **Input**: Accept a noisy mammogram image of fixed dimensions $W \times H$.
2. **Preprocessing**:
    - Normalize the pixel values of the input image to the range [0, 1].
    - Resize the input image to the network's expected input resolution if necessary.
3. **Noise Mapping Layer**:
    - Pass the input through an initial convolutional layer without activation to map the noise characteristics.
4. **Feature Extraction Blocks**:
    - Construct multiple convolutional blocks, each consisting of:
        - Two convolutional layers with ReLU activation.
        - Batch normalization layers following each convolutional layer.
        - A max-pooling layer to reduce spatial dimensions.
    - The number of filters in each convolutional layer increases with the depth of the network, capturing more complex features at each subsequent block.
5. **Reconstruction Block**:
    - Flatten the output of the last feature extraction block.
    - Apply dense layers to learn the non-linear mapping from noisy to clean image representations.
6. **Output Layer**:
    - Reshape the output of the dense layers to the original image dimensions.
    - Apply a final convolutional layer with a sigmoid activation to produce the denoised image.
7. **Postprocessing** (if necessary):
    - Rescale the denoised image's pixel values to the original pixel value range.
8. **Output**: Return the denoised image

4   Results and Discussion

The goal of the proposed method is to enhance the quality of mammogram images for better detection of the disease. In this work we took the MIAS dataset which has been preprocessed using our proposed deep learning-based network. The Figs. 2 and 3 below show the original and the enhanced images.

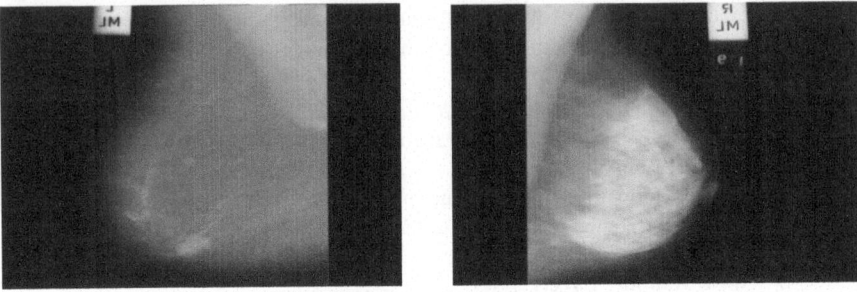

Fig. 2. Original Images (MIAS)

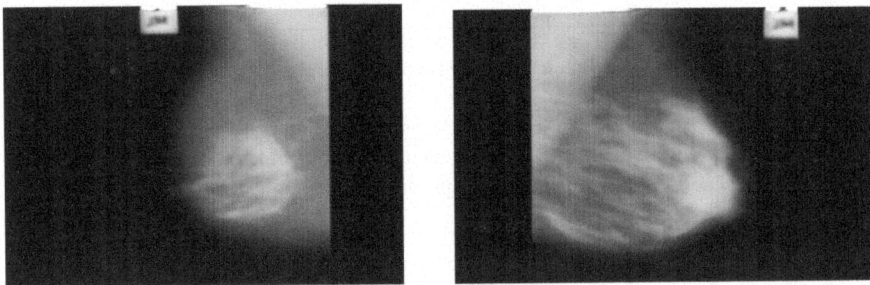

Fig. 3. Enhanced Images

As shown in Figs. 2 and 3 the enhanced images show the calcium deposits more clearly than original images are the enhanced images showing the boundaries of the lesions accurately. The enhanced images help doctor to categorize the disease accurately. Our proposed methods had shown a promising PSNR of 79 surpassing many existing methods as discussed in Table 2 below:

Table 2. Performance Comparison based on PSNR

Paper	Method	PSNR
[16]	Average Filtering, Gaussian, Median, and Bilateral Filters	45
[17]	DBST CLAHE	54
[23]	CLAHE	67
[24]	Multiclass SVM	77.74
[25]	Deep Learning	74
Proposed	VGG inspired CNN	79

5 Conclusion

In this work, a modified VGG-inspired convolutional neural network is developed to enhance mammogram image quality, successfully achieving a significant peak signal-to-noise ratio (PSNR) of 79. This advancement in denoising technology shows its potential in improving the accuracy and reliability of breast cancer detection from mammographic images. The approach combines noise reduction and detail preservation, making it a promising tool for radiologists in early cancer diagnosis. In future, clinical validation and exploring the model's integration into diagnostic practices can be involved, aiming to support better breast cancer detection.

References

1. Loizidou, K., Elia, R., Pitris, C.: Computer-aided breast cancer detection and classification in mammography: a comprehensive review. Comput. Biol. Med. **153**, 106554 (2023). https://doi.org/10.1016/j.compbiomed.2023.106554
2. Jha, S., et al.: Ensemble learning-based hybrid segmentation of mammographic images for breast cancer risk prediction using fuzzy C-means and CNN model. J. Healthcare Eng. **2023**, 18 (2023). https://doi.org/10.1155/2023/1491955
3. Jochelson, M.: Advanced imaging techniques for the detection of breast cancer. Am. Soc. Clin. Oncol. Educ. Book **32**, 65–69 (2012). https://doi.org/10.14694/EdBook_AM.2012.32.223
4. Saini, V., Khurana, M., Challa, R.K.: Advancements in breast cancer detection: a comprehensive review of deep learning techniques for mammogram analysis. In: 2023 1st International Conference on Circuits, Power and Intelligent Systems (CCPIS), Bhubaneswar, India, pp. 1–6 (2023). https://doi.org/10.1109/CCPIS59145.2023.10291244
5. Wang, X., et al.: Intelligent hybrid deep learning model for breast cancer detection. Electronics **11**, 2767 (2022). https://doi.org/10.3390/electronics11172767
6. Sprague, B.L., et al.: Digital breast tomosynthesis versus digital mammography screening performance on successive screening rounds from the breast cancer surveillance consortium. Radiology **307**(5), e223142 (2023). https://doi.org/10.1148/radiol.223142
7. Siddique, M., Liu, M., Duong, P., Jambawalikar, S., Ha, R.: Deep learning approaches with digital mammography for evaluating breast cancer risk: a narrative review. Tomography **9**, 1110–1119 (2023). https://doi.org/10.3390/tomography9030091
8. Li, X., et al.: Deep learning attention mechanism in medical image analysis: basics and beyonds. Int. J. Netw. Dynam. Intell. **2**(1), 93–116 (2023). https://doi.org/10.5391/ijndi0 201006

9. Ahmed, S.F., Alam, M.S.B., Hassan, M., et al.: Deep learning modelling techniques: current progress, applications, advantages, and challenges. Artif. Intell. Rev. **56**, 13521–13617 (2023). https://doi.org/10.1007/s10462-023-10466-8

10. Shlezinger, N., Whang, J., Eldar, Y.C., Dimakis, A.G.: Model-based deep learning. Proc. IEEE **111**(5), 465–499 (2023). https://doi.org/10.1109/JPROC.2023.3247480

11. Avcı, H., Karakaya, J.: A Novel medical image enhancement algorithm for breast cancer detection on mammography images using machine learning. Diagnostics (Basel) **13**(3), 348 (2023). https://doi.org/10.3390/diagnostics13030348

12. Xiong, J., et al.: Application of histogram equalization for image enhancement in corrosion areas. Shock. Vib. **2021**, 8883571 (2021). https://doi.org/10.1155/2021/8883571

13. Kılıç, U., Aksakallı, I.K., Özyer, G.T., Aksakallı, T., Özyer, B., Adanur, Ş: Exploring the effect of image enhancement techniques with deep neural networks on direct urinary system (DUSX) images for automated kidney stone detection. Int. J. Intell. Syst. **2023**, 3801485 (2023). https://doi.org/10.1155/2023/3801485

14. Yaqoob, A., Aziz, R.M., Verma, N.K.: Applications and techniques of machine learning in cancer classification: a systematic review. Human Centric Intell. Syst. (2023). https://doi.org/10.1007/s44230-023-00041-3

15. Ilesanmi, A.E., Ilesanmi, T.O.: Methods for image denoising using convolutional neural network: a review. Complex Intell. Syst. **7**, 2179–2198 (2021). https://doi.org/10.1007/s40747-021-00428-4

16. Hemali, S., Agrawal, S., Oza, P., Tanwar, S., Alkhayyat, A.: Mammogram pre-processing using filtering methods for breast cancer diagnosis. Int. J. Image Graph. Signal Process. **15**(4), 44–58 (2023). https://doi.org/10.5815/ijigsp.2023.04.04

17. Chakraverti, S., Agarwal, P., Pattanayak, H.S., et al.: De-noising the image using DBST-LCM-CLAHE: a deep learning approach. Multimed. Tools Appl. (2023). https://doi.org/10.1007/s11042-023-16016-2

18. Jabeen, K., et al.: BC2Net RF: breast cancer classification from mammogram images using enhanced deep learning features and equilibrium-jaya controlled regula falsi-based features selection. Diagnostics **13**, 1238 (2023). https://doi.org/10.3390/diagnostics13071238

19. Kadam, V.J., Jadhav, S.M., Vijayakumar, K.: Breast cancer diagnosis using feature ensemble learning based on stacked sparse autoencoders and softmax regression. J. Med. Syst. **43**, 263 (2019). https://doi.org/10.1007/s10916-019-1397-z

20. Zhang, Q., et al.: Deep learning based classification of breast tumors with shear-wave elastography. Ultrasonics **72**, 150–157 (2016). https://doi.org/10.1016/j.ultras.2016.08.004

21. Nasser, M., Yusof, U.K.: Deep learning based methods for breast cancer diagnosis: a systematic review and future direction. Diagnostics **13**, 161 (2023). https://doi.org/10.3390/diagnostics13010161

22. Abdel-Zaher, A.M., Eldeib, A.M.: Breast cancer classification using deep belief networks. Expert Syst. Appl. **46**, 139–144 (2016). https://doi.org/10.1016/j.eswa.2015.10.015

23. Razali, N.F., Isa, I.S., Sulaiman, S.N., Abdul Karim, N.K., Osman, M.K., Che Soh, Z.H.: Enhancement Technique based on the breast density level for mammogram for computer-aided diagnosis. Bioengineering **10**, 153 (2023). https://doi.org/10.3390/bioengineering10020153

24. Wajeed, M.A., et al.: A breast cancer image classification algorithm with 2c multiclass support vector machine. J. Healthcare Eng. **2023**, 3875525 (2023). https://doi.org/10.1155/2023/3875525

25. Vimala, B.B., et al.: Image noise removal in ultrasound breast images based on hybrid deep learning technique. Sensors **23**, 1167 (2023). https://doi.org/10.3390/s23031167

26. Dahiya, N., et al.: Detection of multitemporal changes with artificial neural network-based change detection algorithm using hyperspectral dataset. Remote Sens. **15**, 1326 (2023). https://doi.org/10.3390/rs15051326

27. Khurana, M.: Deep learning based low complexity joint antenna selection scheme for MIMO vehicular adhoc networks. Expert Syst. Appl. **219**, 119637 (2023). https://doi.org/10.1016/j.eswa.2023.119637

Author Index

GPSR Compliance

The European Union's (EU) General Product Safety Regulation (GPSR) is a set of rules that requires consumer products to be safe and our obligations to ensure this.

If you have any concerns about our products, you can contact us on ProductSafety@springernature.com

In case Publisher is established outside the EU, the EU authorized representative is:

Springer Nature Customer Service Center GmbH
Europaplatz 3
69115 Heidelberg, Germany

The manufacturer's authorised representative in the EU is Springer
Nature Customer Service Centre GmbH, Europaplatz 3, 69115 Heidelberg,
Germany. If you have any concerns regarding our products, please
contact ProductSafety@springernature.com

Printed and bound by CPI Group (UK) Ltd, Croydon, CR0 4YY
29/04/2026
02099532-0015